Virtopsy Update

Virtopsy Update: The Next Level of Forensic Imaging is a comprehensive reference providing the latest developments, applications, and emerging research in forensic imaging in order to advance the field. It offers a state-of-the-art reference designed for forensic, criminal justice, and medico-legal professionals worldwide.

The book addresses the critical need for updated forensic imaging knowledge, as capabilities have dramatically improved in the last 10 to 15 years and will continue to impact and contribute to forensic science practices—both in the U.S. and globally. The content includes the work and research conducted by the internationally recognized members of the Virtopsy team from the Forensic Institute at the University of Zürich. The book examines current imaging methods for human body analysis and crime scene documentation while also exploring reconstruction techniques that synthesize physical and behavioral evidence. In addition, it evaluates how modern forensic imaging methods can potentially reconstruct crime scene events, providing professionals with advanced tools for hypothesis testing and investigation.

In this, techniques presented cover all current forms of imaging including sonography, angiography, CT, MR, MRI, and MRS for human body analysis. Coverage looks at strategies to apply to cases, crime reconstruction applications, and diagnostic techniques for diseases, trauma, and other diagnoses. The contributing author team comprises experts who are both internationally known and have an excellent reputation broadly within the field—having previously produced *The Virtopsy Approach* (CRC Press, 2009) and *Brogdon's Forensic Radiology, Second Edition* (CRC Press, 2010). Their combined expertise ensures that this book stands as a definitive, forward-looking resource for practitioners, educators, and researchers seeking to navigate the next era of forensic imaging.

Virtopsy Update

The Next Level of Forensic Imaging

Edited by
Michael Thali

CRC Press
Taylor & Francis Group
Boca Raton London New York

CRC Press is an imprint of the
Taylor & Francis Group, an **informa** business

Cover image designed by Silvana Pasquier and used with permission

First edition published 2026
by CRC Press
2385 NW Executive Center Drive, Suite 320, Boca Raton FL 33431

and by CRC Press
4 Park Square, Milton Park, Abingdon, Oxon, OX14 4RN

CRC Press is an imprint of Taylor & Francis Group, LLC

ISBN: 978-1-032-53858-7 (hbk)
ISBN: 978-1-032-53861-7 (pbk)
ISBN: 978-1-003-41399-8 (ebk)

DOI: 10.4324/9781003413998

Typeset in Berling
by Apex CoVantage, LLC

CONTENTS

Chapter 12 Visualization

Lars Ebert and Wolf Schweitzer

Chapter 13 Machine Learning Techniques

Akos Dobay and Raffael Golomingi

PART 2
Current Developments and Future Perspectives

Chapter 14 The Virtobot System

Lars Ebert and Till Sieberth

Chapter 15 Forensic Holodeck

Till Sieberth and Lars Ebert

Chapter 16 Augmented Reality

Raffael Golomingi

Chapter 17 The VirtoScan Project

Sören Kottner and Dominic Gascho

Chapter 18 Imaging-Ring Forensics—A Mobile CT Scanner with Integrated Cameras for Simultaneous 3D Surface Documentation

Dominic Gascho

PART 4
Disease Diagnosis

Chapter 30 Cardiovascular System

Garyfalia Ampanozi, Carlo Tappero, and Wolf Schweitzer

PART 5
Other Diagnoses

Chapter 31 Normal Postmortem Changes

Garyfalia Ampanozi, Carlo Tappero, and Wolf Schweitzer

Chapter 32 Stay Cool!—Freezing and Thawing in Forensic Medicine

Dominic Gascho and Niklaus Zoelch

PART 6
Radiologic Identification

Chapter 33 Identification of Human Remains

Wolf Schweitzer, Rilana Baumeister, and Jakob Heimer

PART 7
Visualization and Documentation for Reporting

Chapter 34 What Software to Use in Forensic Imaging

Lars Ebert

Chapter 35 Reading and Documentation

Wolf Schweitzer, Garyfalia Ampanozi, and Carlo Tappero

PART 8
Strategies and Cases

Chapter 36 The Forensic Evaluation of Evidence in Legal Settings

Dilan Seckiner

Chapter 37 Postmortem Imaging Workflows

Lars Ebert and Wolf Schweitzer

Chapter 38 PMCT for Triage

Wolf Schweitzer, Vasiliki Chatzaraki, and Garyfalia Ampanozi

ABOUT THE EDITOR

Michael J. Thali, MD, Executive MBA HSG, is a Swiss forensic physician, professor, and internationally recognised pioneer in forensic imaging, virtual autopsy, and technology-enhanced forensic medicine. For more than 30 years, he has contributed to redefining medico-legal investigations through the integration of imaging, engineering, digitalisation, and minimally invasive procedures into forensic practice. Internationally, he is widely recognised as one of the initiators and scientific driving forces behind the Virtopsy® approach, which fundamentally transformed how forensic examinations can be performed and interpreted in the twenty-first century.

Professor Thali has served as Chair and Full Professor of "Forensic Medicine and Relation to Jurisprudence" at the Institute of Forensic Medicine, University of Zürich (IRM-UZH), Switzerland, since 2011. Under his leadership as managing director 2011–2023, the institute expanded interdisciplinary collaboration between medicine, law, engineering, computer science, radiology, public health, and nursing sciences, providing scientific and strategic leadership for approximately 180 employees.

He studied human medicine at the University of Bern, receiving his medical degree in 1995, and subsequently trained in forensic medicine (1995–1999) at the Institute of Forensic Medicine, University of Bern. Recognising early the future importance of imaging technologies for medico-legal diagnostics, he completed training in diagnostic radiology (1999–2000) at the University Hospital Bern. In 2000, he obtained board certification in forensic medicine from the Swiss Society of Forensic Medicine.

His scientific and international development was shaped through specialised fellowships in ballistics at Armasuisse, Thun, Switzerland (2001) and at the internationally renowned Armed Forces Institute of Pathology (AFIP), Office of the Armed Medical Examiners, Washington, D.C., USA (2001–2002). Returning to Switzerland, he worked as a forensic pathologist and project leader of the Virtopsy® initiative at the University of Bern (2002–2005), where he and his interdisciplinary collaborators pursued what was initially considered an unconventional and provocative vision: combining three-dimensional optical surface scanning, postmortem computed tomography (PMCT), magnetic resonance imaging (MRI), computer-assisted reconstruction, and digital visualisation to complement and, in selected cases, partially replace conventional autopsy procedures.

This work culminated in his habilitation in forensic medicine in 2003, laying important academic foundations for a new discipline at the interface of forensic medicine, 3D imaging radiology, engineering, and computer sciences. In 2006, he was appointed Full Professor of Forensic Medicine at the University of Bern, serving first as Co-Director and later as Director of the Institute of Forensic Medicine (2008–2011). During this period, he further expanded scientific activities in forensic imaging, postmortem angiography, biomechanics, forensic ballistics, and workflow innovation.

Professor Thali recognised early that the future of forensic medicine would extend far beyond the traditional autopsy room. By combining forensic medicine with imaging, radiology, engineering, biomechanics, digital technologies, and artificial intelligence, he contributed to establishing one of the world's first comprehensive forensic imaging ecosystems, integrating postmortem CT, MRI, 3D surface documentation, postmortem angiography, image-guided biopsy, toxicological workflows, virtual reconstructions, and AI-supported diagnostics into forensic casework. What initially faced scepticism evolved into a globally recognised scientific field now implemented by forensic institutions worldwide.

In 2011, he co-founded the International Society of Forensic Radiology and Imaging (ISFRI) in Zürich, creating the leading international scientific society dedicated to forensic imaging and radiology. Through ISFRI, he contributed to building a global scientific community connecting forensic physicians, radiologists, anthropologists, engineers, and imaging scientists to advance education, scientific standards, and interdisciplinary collaboration in forensic imaging.

Further strengthening this emerging scientific field, Professor Thali became Founding Editor-in-Chief of the *Journal of Forensic Radiology and Imaging* in 2012, later renamed *Forensic Imaging*, thereby establishing one of the first international academic platforms dedicated exclusively to forensic imaging sciences.

Beyond research and clinical leadership, Professor Thali has been deeply committed to education and interdisciplinary postgraduate training. He co-founded the CAS MedLaw programme at the University of Zürich (2012) and subsequently founded several innovative postgraduate programmes, including the CAS Forensic Imaging & Virtopsy (2013), the CAS Forensic Nursing (2014), the CAS Medicolegal Death Scene Investigation (2016), and additional educational initiatives bridging medicine, law, science, healthcare, and technology.

His scientific achievements have been recognised through several prestigious distinctions. In 2001, he received the Science Prize of the German Society of Legal Medicine for his pioneering. In 2009, Professor Thali and members of his interdisciplinary research team were awarded the internationally renowned Ig Nobel Peace Prize for biomechanical forensic research investigating whether full or empty beer bottles are more likely to fracture and whether their impact thresholds are sufficient to produce skull fractures—research that combined humour, scientific rigour, and forensic relevance, attracting worldwide media attention. In 2015, he and the Virtopsy® team received the prestigious Swiss ICT Award for technological innovation and digital transformation in forensic medicine.

Professor Thali is editor and author of several influential books and reference works in forensic medicine, forensic imaging, and forensic ballistics, including seminal publications that helped establish Virtopsy® as a globally recognised concept. He has authored and co-authored hundreds of scientific publications, contributed numerous book chapters, and delivered more than 300 lectures and scientific presentations worldwide.

Despite his strong emphasis on technological innovation, Professor Thali remains convinced that the future of forensic medicine will not be defined by machines replacing human expertise, but rather by an empowered forensic physician—combining medical judgement, imaging literacy, technological understanding, and interdisciplinary knowledge to better serve medicine, justice, and society.

CONTRIBUTORS

All authors in this publication conduct research as part of the Forensic Medicine & Imaging, Virtopsy Research Group at the University of Zürich, Switzerland:

Garyfalia Ampanozi
Rilana Baumeister
Vasiliki Chatzaraki
Akos Dobay
Lars Ebert
Sabine Franckenberg
Martin Frenz
Dominic Gascho
Raffael Golomingi
Jakob Heimer
Rolf Kaufmann
Sören Kottner
Victor Mergen
Wolf Schweitzer
Dilan Seckiner
Till Sieberth
Carlo Tappero
Patrik Wili
Niklaus Zoelch

FOREWORD

Forensic imaging was not something I planned to build a career around. Like many of us in the field, I found my way to it in part due to the work of the Virtopsy Team.

It was in the early 2000s, in an autopsy suite of the Clark County Coroner's and Medical Examiner's Office in Las Vegas, Nevada, that I was first introduced to the idea that a field which had gone fairly unchanged since 1761 with Morgagni's *On the Seats and Causes of Diseases Investigated by Anatomy*, forensic science was ripe to be transformed by technology. One of the forensic pathologists paused during an autopsy, turned to me and stated, "You know, what we do hasn't really changed in centuries. We can do better, we *need* to do better. I think medical imaging, that's the future. There's a group in Switzerland who are changing everything." He pointed me in the direction of their old plain film X-ray machine to learn and handed me an article, published by a team in Bern, Switzerland, headed by Prof. Dr. med. Richard Dirnhofer and Prof. Dr. med. Michael Thali, which described a method that could scan the body and reconstruct it digitally, called a "Virtopsy". That moment shifted my entire career trajectory.

I didn't know how prescient those words would be or how profoundly that article would shape my life's work. In those early days, medical imaging in forensic science was still a novelty, often constrained by access, cost, or skepticism. Yet the idea of conducting postmortem investigations with CT and/or MRI, of documenting traumas in three dimensions with cutting-edge advanced visualization methods and of using imaging not just as a supplement but sometimes even as a substitute for autopsy was revolutionary.

Some years later in my medical training, the first edition of *The Virtopsy Approach: 3D Optical and Radiological Scanning and Reconstruction in Forensic Medicine* (2009) was published. As a student, I wrote our medical school's library purchasing team fervently explaining how seminal the text was and how it was critical that we get a copy so we, students, could learn their methods. That text became a one-of-a-kind manual to a new field that would transform forensic pathology around the world. It was there I learned how imaging could serve not as a replacement for traditional autopsy methods but as a powerful tool to augment and elevate them. The book helped bridge the gap between the autopsy suite and the imaging suite.

In 2011 after having moved to Zürich to expand the Virtopsy Team, they co-founded the International Society of Forensic Radiology and Imaging, which established an international society for this emerging field, one where I found a home among a global network of professionals pursuing similar goals. Together as a society, we've worked to legitimize and expand the role of medical imaging in forensics not only as a diagnostic tool but as a scientific methodology, one capable of enhancing precision, reproducibility, transparency, and ultimately justice. The transformation has been remarkable.

Today, the integration of medical imaging into forensic workflows is no longer an outlier; on the international level, it's an expectation. Institutions now routinely perform postmortem CT examinations, and the resulting three-dimensional reconstructions are increasingly utilized in judicial settings, where they serve as effective demonstrative evidence to support forensic interpretation and enhance clarity in court proceedings. This is also no longer limited to the medical image data but also includes technologies like 3D modeling, photogrammetry, postmortem angiography, and additive manufacturing, which enhance everything from pathology to anthropology to law enforcement investigations. Even artificial intelligence has begun to enter the space, offering automatic/semi-automatic pattern recognition and segmentation tools that were unimaginable when I first read that Swiss article over two decades ago. And yet, perhaps the most exciting part is that as a field, we're still just getting started.

Forensic imaging continues to evolve, and this book is evidence of that ongoing growth. The field-renowned authors and contributors who have assembled this volume are not merely documenting the present state-of-the-art methods; they are also actively shaping the future. Their work highlights the interdisciplinary nature of forensic radiology and imaging, drawing from medicine, engineering, digital technologies, and law. It is at this intersection where data meets narrative and where visualization meets justice that our field finds its strength.

This book, like the one before it, serves as both a reference and a roadmap. It captures the maturity of the field while offering insight into where we still need to go. For students, this text will be a foundation. For professionals, it will be a touchstone. For those of us who have walked this journey from the margins to the mainstream, it is a reminder of what is possible when innovation meets persistence.

The spark that began with Röntgen's rays, reignited by the Virtopsy Team in Bern, then Zürich, and that began my lifelong passion in a coroner's office in Las Vegas, now burns brightly across disciplines and continents. I am honored to contribute this foreword to a volume that continues that legacy.

—Dr. Summer J. Decker
President, International Society of Forensic Radiology and Imaging
Director, USC Center for Innovation in Medical Visualization
Grace Whisler Endowed Professor of Medicine
Professor of Radiology, Pathology and Surgery
Keck School of Medicine, University of Southern California
Professor of Biomedical Engineering, USC Viterbi School of Engineering
Professor of Cinematic Arts, USC School of Cinematic Arts

FOREWORD

Virtual Autopsy—At the Intersection of Science, Faith, Religion, and Bioethics

The rapid expansion of biotechnology—driven by digital imaging, advanced computing, and artificial intelligence—continues to reshape the practice of medicine in profound ways. Among these developments, *virtual autopsy* (or "Virtopsy") has emerged as a transformative tool in postmortem investigation. The implementation of virtual autopsies as an ancillary postmortem investigation, and even an alternative to the traditional autopsy, has the potential not only to influence medical jurisprudence but also the transcultural and interfaith values that govern society's relationship with the dead. By preserving bodily integrity and promoting dignity and respect, virtual autopsy occupies a unique position at the intersection of science, ethics, and spirituality.

When considering the introduction of any new technology, it is often more instructive to ask *where we want to be* rather than where a technology *can* take us. For the purposes of this discussion, the envisioned future is one in which virtual autopsy is not merely an ancillary investigative modality but an ethically grounded and scientifically robust alternative for selected medicolegal cases—particularly non-homicidal deaths. Such a future requires thoughtful ethical analysis. Scientific advancement does not occur in isolation; it inevitably produces moral and cultural challenges shaped by religious convictions, resource constraints, questions of justice, and competing interpretations of fundamental biological concepts, such as the beginning and end of human life (Rachels & Rachels, 2019; Beauchamp & Childress, 2019).

A responsible assessment of any new biotechnology therefore requires a dual process of due diligence: a scientific evaluation (often guided by research ethics) and a rigorous inquiry into its broader bioethical implications. While the scientific discussion extends beyond the focus of this chapter, the ethical, cultural, and religious dimensions of virtual autopsy require careful elaboration.

Previous Crossroads in Medicine

Virtual autopsy is not the first innovation to force medicine to confront ethical tension. History provides several instructive parallels (Jonsen, 2000, 1990).

The advent of the contraceptive pill in 1960 was viewed by the Roman Catholic Church as whether it was morally equivalent to physical contraception. At the same time, and from another perspective, broader society also raised concerns regarding the revolution this may have on social and sexual life.

During that same period, Belding Scribner invented a shunt which changed the fate of patients with chronic renal failure. The Scribner shunt allowed repetitive access to the vascular system of patients, enabling chronic long-term dialyses. Yet it also generated an acute ethical dilemma: dialysis machines were scarce. To ration this limited resource, the Seattle Kidney Centre created an Admissions and Policy Committee to determine access based on criteria partly relating to "social worth". This system starkly violated the principle of distributive justice, demonstrating how technological advances may inadvertently amplify inequity.

Similarly, the first heart transplant in 1967 required a reconceptualization of death. Prior to this, death was tied to the cessation of cardiac and respiratory function. This major step in transplanting a human heart still viable and suitable to sustain life needed a new set of criteria for the moment of death. In August 1968, the Harvard Definition of Irreversible Coma paved the way not only for a more scientific definition of the moment of death but also to allow these transplants to be done within an objective bioethical framework (Ad Hoc Committee of the Harvard Medical School, 1968).

These historical examples illustrate that innovation often forces medicine to re-examine deeply held assumptions and confront the moral terrain that new technologies inevitably expose.

Ethics in General

Ethics is a branch of philosophy that is a reflection on concepts and ideas about what confers value or disvalue on human action. It can be defined as a matter of knowing what the right thing is to do, while morality is a matter of doing the right thing (Moodley, 2023). Ethics is not perfect. Ethical deliberation rarely offers perfect answers; instead, it often requires identifying the least harmful option, balancing competing duties, or locating the midpoint between conflicting alternatives.

Although law and ethics frequently intersect, they remain distinct normative frameworks. In fact, something may be legal but unethical and vice versa. For instance, discriminatory laws of a country towards a specific group of individuals often fail the ethical benchmark. Nevertheless, there is a relationship between these two pointers in our everyday life, and it was best formulized by Earl Warren, Chief justice of the United States in 1962 (The New York Times, 1962): "In civilised life, law floats in a sea of ethics. Each was indispensable to civilization. Without law, the least scrupulous might prevail; without ethics, law could not persist".

Although there are older writings on moral coding systems, like the Code of Hammurabi (2500 BC), a timeline of Western philosophy usually starts around the time of Aristotle (b 384 BC) with his moral theory based on virtue ethics. Numerous other moral theories have been formulated, and some receive more prominence and are usually discussed in bioethics textbooks (Beauchamp & Childress, 2019; Moodley, 2023).

It is important to remember that although we often focus on the philosophy practiced in the Western world, there is also an important school of Eastern philosophy which finds expression in the practicing of faith or religion, like Shia and Sunni in Islam. The same applies to the moral theories of the Indian sub-continent and China, like Confucianism (Jonsen, 2000).

Tradition, Faith, and Religion

Human spirituality is shaped by a dynamic interplay between tradition, faith, and religion. Traditions are inherited cultural practices, faith reflects a deep personal conviction, religion is an organized system of belief accompanied by moral and ritual structures. Together, these elements shape an individual's worldview and influence attitudes toward death, the body, and postmortem procedures.

These concepts are often closely interrelated, as a religion cannot exist without faith, and it is merely an expression and embodiment of that faith. In this spiritual evolution the question can sometimes indeed be asked, when does tradition becomes religion.

Religion and faith together with an individual's personal experience create a three-dimensional space in which they give expression to their spiritual realization. For instance, someone may believe in Jesus Christ and belong to the Christian faith. Their religion may be based on the doctrine of the Protestant church (even though there may be a number of different churches based on the same doctrine). Their spiritual position as a human being, as a function of their personal experience of faith and religion, places them somewhere in this space.

Religion and faith guide us in our daily living. It helps us to differentiate between good and evil, and this has been expanded into a theory known as the divine command theory (Rachels & Rachels, 2019). Plato raised an interesting question, asked by Socrates in the *Euthyphro*, "is conduct right because god commands it, or do the gods command it because it is right?"

It is therefore no surprise that the views of different faiths and religion regarding death and the dead cover a vast spectrum (Huff, 2006; Taylor & Box, 1999). A literature study of this topic is difficult, as there are so many different interpretations among individuals of the same faith and religion.

Across many faiths, the question of autopsy becomes a point of tension between medical necessity and spiritual values. For example, Christianity generally adopts a dualistic view of the person, distinguishing the soul from the body. This can make autopsy more acceptable, provided the body is handled respectfully.

Islam and Judaism, both rooted in monistic understandings of the human person, emphasize the sacredness of the body and often require prompt burial. There is no specific refence to autopsies in the Islamic faith, but autopsies are not encouraged (Mulhim *et al.*, 2023). It appears as the statement of the Prophet Muhammed, "to break the bone of a dead person is like breaking the bone of a living person", is central to this position. The Sunni and Shia are the largest denominations. This is also to some extent reflected in the legal systems of Islamic countries, where some follow only Shariah law, like Saudi Arabia, while others draw some principles from the civilian judiciary system (Bamousa *et al.*, 2016). Autopsies are permitted under the principles of necessity *(darura)* and welfare, the former allowing the performance of medico-legal autopsies.

Other faiths, like Judaism, Buddhism, and Hinduism, all allow autopsies for legal purposes, albeit with some preconditions. In Judaism it is preferred that the dissection be limited only to the relevant body parts and that care be taken to return as much as possible of the organs and tissue back to the body (Dorff, 2006). Although differences exist with the distinct practices amongst the Theravada, Mahayana, Soto Zen, and Tibetan Buddhism, a general requirement seems that following death, the body is to be highly respected so that the spirit can concentrate on achieving enlightenment. This may require the body to remain undisturbed for three days. Hinduism believes that an autopsy may be disturbing to the soul, which could re-enter the body and not continue life, or even become evil (Taylor & Box, 1999).

The culture of Indigenous people can be superimposed on an existing faith. African traditional religion is often practiced alongside Christian and other faiths, creating a unique expression of worshipping, as well as adding a traditional element to events like funerals in emphasizing the continuity between the living and ancestors incorporating the principle of Ubuntu. The same applies to the Aboriginal people of Australia (Bolliger *et al.*, 2009).

Virtual autopsy offers certain benefits when viewed through a universal lens of faith and religion. First, it preserves body integrity while at the same time allowing investigation into the cause of death. Second, it enables the quick release of the body for burial or cremation. This more rapid processing of the dead, together with the knowledge that the body is treated according to the prescriptions of the faith and religion, minimizes the emotional trauma on the family of the deceased. In addition, it creates a permanent digital record of the body; in fact, a "second body". This "second body" is permanently available for review and digital autopsy, allowing a second autopsy to be performed whenever needed if necessary. Finally, in our diverse society, it offers dignity and respect to the dead as well as the living. It creates interfaith and transcultural acceptance, binding us all together in our experience of life and death.

Moral Theories

As has been stated, the foundations of bioethics are heavily embedded in philosophy, and moral theories are based on philosophical reflection (Beauchamp & Childress, 2019; Van Niekerk, 2023). Five of these deserve some reference when formulating a bioethical framework for the practice of virtual autopsies.

Utilitarianism is the most important consequentialism theory, highlighting the importance of the consequences of one's actions. It is all about doing the greater good or lesser evil to the greatest number. In many countries with high crime rates, like South Africa, the forensic services are often under pressure. It is therefore important to enable the forensic practitioner to focus more on homicide cases. Virtual autopsies provide a mechanism through which the workload can be addressed by processing the non-homicide workload. This will serve the greater community, as more time and expertise can be committed to solving murder cases in a system not flooded by unnecessary traditional autopsies. It will be an advantage to the judicial system and, through its effect on dealing with crime, to the greater number of society members.

Immanuel Kant formulated another important moral theory. Kantian deontology formulates a set of rules to live by. He formulated two categorical imperatives, both of which are in support of virtual autopsies, depending on certain prerequisites. The first categorical imperative, the so-called universal law formulation, states "act only on that maxim through which you can at the same time will that it should become a universal law". The second imperative, the humanity formulation, states, "act in such a way that you always treat humanity, whether in your own person or in the person of any other, never simply as a means, but always at the same time as an end". Both set of rules are relevant in our discussion. First, we must not implement new technology if we cannot apply it universally and in line with certain universally applied criteria. Second, new technology must be used to achieve a desired outcome, in other words, an end in itself, and not only treating the body as a means for technical or scientific experimentation. It must assist with determining the cause of death and as such support justice.

Virtue ethics is not only one of the oldest moral theories but also a simplistic moral basis to which any lay person can relate.[1] A virtue is a character trait manifested in habitual action, which is good for a person to have. Aristotle stated that virtues are those characteristics that will ensure that those endowed with them will have the best life overall, associated with *eudaimonia* (flourishing). So, to be a virtuous person is to always do good things and not to practice vices; in fact, a virtue is the midpoint between two vices, like generosity being somewhere between stinginess and extravagance. The list of virtues applicable to virtual autopsies is to a large extent identical to those that we expect from healthcare workers in general. Benevolence, fairness, prudence, compassion, honesty, justice, moderation, and diligence are just some.

One can (hopefully) argue that most humans are inherently good. This gives rise to the concept of the common morality, which is a set of universal norms shared by all persons committed to morality (Beauchamp & Childress, 2019). It consists of two different components. The one comprises a set of standards of actions or rules of obligation and includes rules like do not steal, tell the truth, and do not kill. In short, a very simple principle to live by: do not do to others what you do not want others to do to you. The second one consists of good moral character traits or virtues, things like honesty, integrity, and trustworthiness.

Since the latter part of the 1970s the *Principles of Biomedical Ethics* (Beauchamp & Childress, 2019) has become an important textbook and guide to bioethics and is also sometimes referred to as a moral theory in its own right, "principlism". It draws on the idea of a common morality. With this publication the authors have made moral deliberation a bit easier in healthcare. Four principles exist, each of them equal in standing, but with the possibility to be in opposition and conflict with one another—respect for autonomy, beneficence, non-maleficence, and justice. For the resolution of possible conflict amongst these four principles, the principles must be weighed until some equilibrium is found and the best solution to the moral dilemma identified. Sometimes we willingly know that we may be harming the patient with our chemotherapy treatment; however, the intention is to do good. There must never be an intention to do harm in medicine.

Each of these four principles is relevant in a discussion on virtual autopsies and its moral and ethical justification.

The principle of respect for autonomy is supported by the respect and dignity which a virtual autopsy offers, as it reduces the disfigurement of the body and also limits unnecessary delay in the burial of the dead. If a virtual autopsy is performed by request of the family, in other words, when it is not performed as part of a medico-legal investigation, the prescribed process to obtain consent needs to be followed. This must be informed consent and needs to include the limitations of the examination. Confidentiality and privacy of the data generated by a virtual autopsy also extend to that "second body", just as in a living individual.

Virtual autopsy offers a balance between the sometimes opposing principles of beneficence and non-maleficence. It follows a principle of proportionality as adequate information is sought through the least invasive procedure, as well as within an accepted timeframe.

The primary purpose of postmortem investigations is to serve justice. However, the principle of justice also requires equal access and an equitable and fair distribution of technology, resources, and funding. In developing countries, which are often over-burdened by high crime and homicide rates, this potential obstacle needs to be properly managed to support service delivery in a fair and just manner to all the citizens.

Finally, it is important to remember that the moral theory of communitarianism plays an important role in some communities as part of the expression of cultural relativism. For instance, in the African culture it is expressed through the philosophy of Ubuntu, i.e. "people are people through other people" (Letseka & Ganya, 2023). This is also relevant in death, as the ancestors are just as important a part of the community. Virtual autopsies allow the family to treat the deceased with the necessary dignity and respect while still fulfilling the needs of justice, extending the concept of Ubuntu into the afterlife, as "I am, because we are—even in death".

Conclusion

Forensic pathology services must remain attentive to the diverse interfaith and transcultural differences when there is a need to perform an autopsy for medico-legal purposes. It is important to proceed with the necessary dignity and respect not only to the dead but also to support the bereaved family as far as possible during a very traumatic and emotionally difficult time. Virtual autopsies offer an alternative to a traditional autopsy. As with any scientific tool, it needs to be sensitive and specific enough not only to "tell the truth, and nothing but the truth" but also to stand up against any robust scrutiny by legal practitioners to discredit its findings in a court of law.

As technology continues to advance, the integration of virtual autopsy into forensic practice will require thoughtful balancing of science, ethics, and spirituality. Ultimately, it offers a way to honor the deceased, support the living, and promote justice in a manner consistent with human dignity.

—By Izak Loftus

Note

1 It is not mere coincidence that "virtuous" and "virtual" sound so similar (Hollandbeck, 2020). These words are etymologically related through the Latin word *vir*, meaning "man". *Virtus* is essentially a Roman analogue to chivalry, and during the Middle Ages the French and English culture and language used "virtue" to describe any quality or trait thought to be excellent. In Shakespeare's time it became shorthand for a woman's chastity.

The development of the word "virtual" followed its path through Medieval Latin, where *virtus* referred to the effectiveness of a warrior to control a battle, from which the adjective *virtualis* was derived, meaning "effective with respect to inherent physical virtues or capabilities". English speakers started to use the word "virtual" in the mid-15th century as "being something in effect, though not in actuality". With the advent of information technology, "virtual" is the perfect word to describe something that does not physically exist, whether an autopsy, meeting, or just a computer game.

References

Ad Hoc Committee of the Harvard Medical School (1968) A definition of irreversible coma. *JAMA*, 205(6), p337–340.

Bamousa, M.S., *et al.* (2016) The Islamic approach to modern forensics and legal medicine issues. *Am J Forensic Med Pathol*, 37(2), p127–131.

Beauchamp, T. and Childress, J. (2019) *Principles of Biomedical Ethics*. 8th edn. Oxford: Oxford University Press.

Bolliger, S.A., Thali, M.J. and Segal, G.P. (2009) Religion. *In:* Thali, M., Dirnhofer, R. and Vock, P. eds. *The Virtopsy Approach*. Boca Raton: CRC Press. p41–47.

Dorff, E.N. (2006) Judaism and Ethical Issues in End of Life Care. *In*: Eisen, M.E. and Laderman, G. eds. *Science, Religion and Society: An Encyclopedia of History, Culture, and Controversity.* Armonk: M.E. Sharpe, Inc. p712–719.
Hollandbeck, A. (2020) In a word: Virtue in the virtual. *The Saturday Evening Post.* Available from https://www.saturdayeveningpost.com/2020/04/in-a-word-virtue-in-the-virtual/ [Accessed 28 November 2025].
Huff, T.E. (2006) Dissection and Autopsy. *In*: Eisen, M.E. and Laderman, G. eds. *Science, Religion and Society: An Encyclopedia of History, Culture, and Controversity.* Armonk: M.E. Sharpe, Inc. p746–753.
Jonsen, A.R. (1990) *The New Medicine and the Old Ethics.* New York and Cambridge: Harvard University Press.
Jonsen, A.R. (2000) *A Short History of Medical Ethics.* Oxford: Oxford University Press.
Letseka, M.M. and Ganya, W. (2023) African Philosophy and Medical Ethics. *In:* Moodley, K. ed. *Bioethics, Medical Law and Human Rights.* 3rd edn. Pretoria: Van Schaik Publishers, p33–44.
Moodley, K. (2023) *Bioethics, Medical Law and Human Rights.* 3rd edn. Pretoria: Van Schaik Publishers.
Mulhim, N.K., Almulhim, N.F., Alabssi, H.A., *et al.* (2023) The new era of forensic autopsies: A brief overview of virtual autopsy and its practice in the Middle East. *JOHS,* 3(12), p624630.
The New York Times. (1962) Warren favors profession to give advice on ethics. *The New York Times,* 12 November, p1.
Rachels, J. and Rachels, S. (2019) *The Elements of Moral Philosophy.* 9th edn. New York: McGraw-Hill.
Taylor, A. and Box, M. (1999) Multicultural Palliative Care Guidelines. *In: Palliative Care Council of South Australia.* Eastwood, South Australia: Palliative Care South Australia.
Van Niekerk, A. (2023) Ethics Theories and the Principlist Approach in Bioethics. *In:* Moodley, K. ed. *Bioethics, Medical Law and Human Rights.* 3rd edn. Pretoria: Van Schaik Publishers, p17–32.

ACKNOWLEDGMENTS

Our heartfelt thanks go to Mark Listewnik, all collaborating partners, and everyone whose open-mindedness and out-of-the-box thinking made progress possible

A Special Note and Thank You

Following the relocation of the Virtopsy® group from the University of Bern to the University of Zürich in 2011, a remarkable new chapter in forensic imaging began under the directorship of Prof. Dr. med. Michael Thali. Building upon *The Virtopsy Approach*, the years from 2011 to 2023 became the high bloom of Virtopsy in Zürich, characterized by scientific innovation, international visibility, and exceptional interdisciplinary teamwork.

The success of this period was shaped by a dedicated Virtopsy Research Group, whose members and co-authors included Garyfalia Ampanozi, Rilana Baumeister, Vasiliki Chatzaraki, Akos Dobay, Lars Ebert, Sabine Franckenberg, Martin Frenz, Dominic Gascho, Raffael Golomingi, Jakob Heimer, Rolf Kaufmann, Sören Kottner, Victor Mergen, Wolf Schweitzer, Dilan Seckiner, Till Sieberth, Carlo Tappero, Patrik Wili, and Niklaus Zoelch.

Together, this highly interdisciplinary team of forensic experts, radiologists, engineers, scientists, and clinicians substantially advanced postmortem imaging, Virtobot technologies, minimally invasive workflows, and later AI-assisted forensic approaches. During this golden era, Zürich also became the birthplace of the International Society of Forensic Radiology and Imaging (ISFRI), establishing a global platform for collaboration and scientific exchange.

In this sense, *Virtopsy Update: The Next Level of Forensic Imaging* represents both the continuation of *The Virtopsy Approach* and the further development of the second edition of *Brogdon's Forensic Radiology*. More importantly, the book stands as a tribute to the remarkable team spirit, scientific excellence, and shared vision that transformed forensic imaging from a pioneering concept into an internationally established discipline at the University of Zürich.

CHAPTER 1

VIRTOPSY UPDATE—A HISTORICAL AND FUTURE PERSPECTIVE

Michael Thali

DOI: 10.4324/9781003413998-1

1.1 INNOVATION, PERFORMANCE, AND SOCIETAL IMPACT—THE PILLARS OF SUCCESSFUL RESEARCH

Groundbreaking research rests on three fundamental pillars: innovation, performance, and societal relevance. These elements achieve their full potential when pursued through interdisciplinary collaboration, nurtured within a strong team environment, and guided by a shared and clearly articulated vision. Crucially, such a vision must extend beyond conceptual ambition and be translated into tangible actions and sustainable implementation.

Ultimately, scientific research must generate real-world impact. In the context of forensic medicine, this means enhancing everyday forensic practice as a service to law enforcement agencies, prosecutors, and the judiciary—thereby contributing directly to the functioning of the justice system and to society at large.

This book documents the innovations and developments of the Virtopsy group from 2011 onwards at the University of Zürich. It continues a unique scientific movement that originated at the University of Bern and has since evolved into an internationally recognized model for technology-driven forensic research.

1.2 THE ORIGIN OF VIRTOPSY—FROM BARRACK TO MOVEMENT

The origins of the Virtopsy initiative can be traced back to the late 1990s and the visionary leadership of Richard Dirnhofer, who at that time served as chair and director of forensic medicine at the University of Bern. What began in a simple construction barrack—an unassuming and improvised setting—marked the modest starting point of what would later evolve into a profound scientific and methodological transformation in forensic medicine.

Far from established laboratories and conventional infrastructures, this early environment fostered unconventional "out-of-the-box thinking" and laid the conceptual foundations for a movement that would fundamentally redefine forensic investigation through imaging-based technologies (Figure 1.1).

From the outset, virtopsy was characterized by genuine interdisciplinary collaboration: Forensic pathologists worked alongside 3D surface scanning specialists, radiologists, CT and MRI experts, autopsy technicians (like Urs Koenigsdorfer), and technical staff, complemented by operational police expertise, including specialists such as 3D visualization pioneer Marcel Braun. Together, these diverse professionals formed a highly dynamic team—an approach that was already prominently documented in one of the earliest Virtopsy posters presented in the late 1990s.

Figure 1.1 The birthplace of the Virtopsy project at the University of Bern was an old chalet-style barrack.

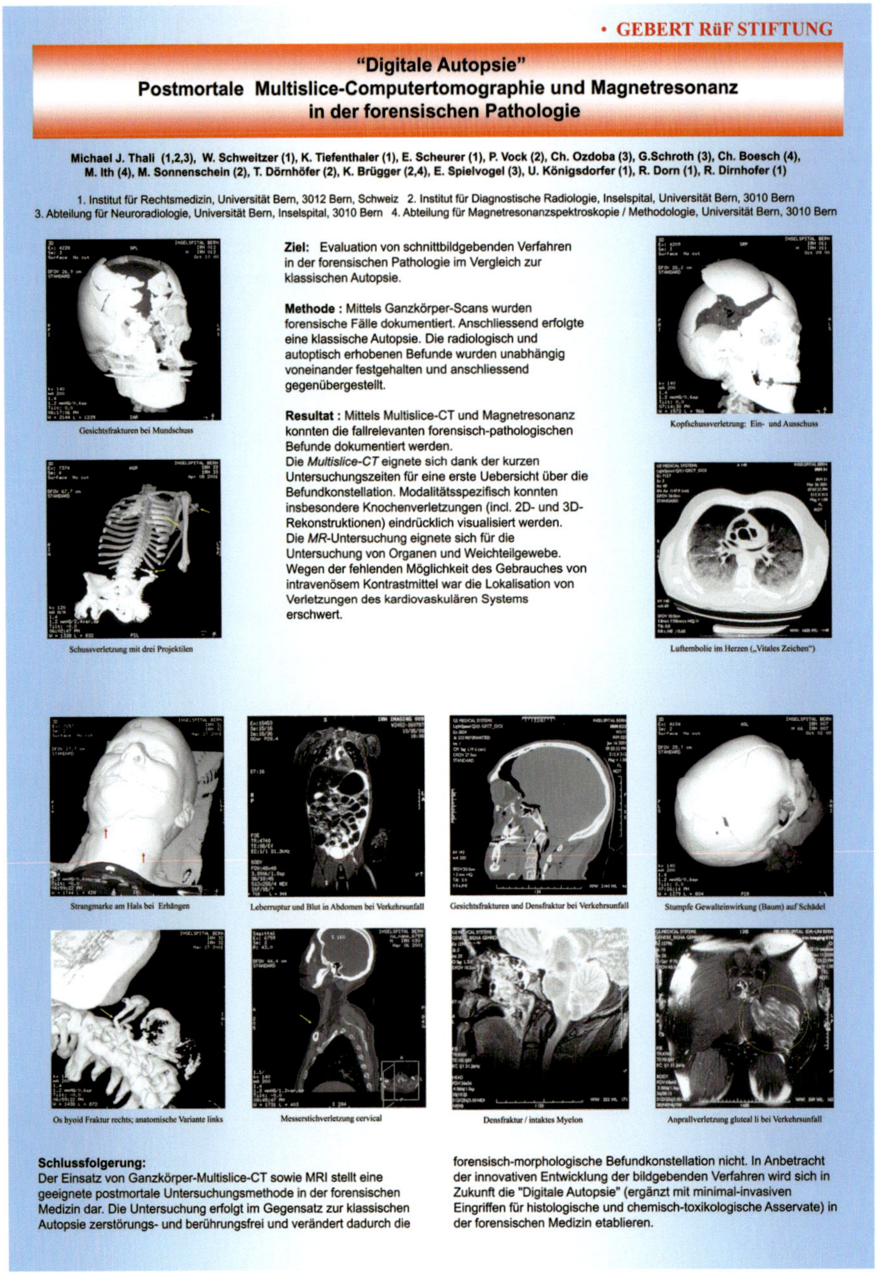

Figure 1.2 One of the earliest Virtopsy scientific posters, prominently showcasing the interdisciplinary professional teamwork that has defined the project since its inception.

1.3 PIONEERING WORK IN FORENSIC IMAGING

A landmark moment was marked by two seminal publications:

- Thali MJ, Braun M, et al. *Virtual Autopsy: A New Forensic Imaging Approach Combining 3D Surface Scanning, Postmortem Multislice CT, and MRI.* Proceedings of SPIE Medical Imaging, Washington, DC; 2002.
- Thali MJ, et al. *Virtopsy—A New Imaging Horizon in Forensic Pathology: Virtual Autopsy by Postmortem Multislice CT and MRI.* Journal of Forensic Sciences. 2003;48(2):386–403.

Together, these studies introduced for the first time the systematic integration of 3D surface scanning, postmortem multislice CT, and MRI for forensic applications. They established the scientific and conceptual foundation of image-based virtual autopsy (Virtopsy) and defined a new paradigm in forensic pathology by demonstrating how complementary imaging modalities could be combined into a coherent, reproducible investigative approach.

Subsequent milestones included:

- **The Virtopsy Approach**
  Thali MJ, Dirnhofer R, Vock P. *The Virtopsy Approach: 3D Optical and Radiological Scanning and Reconstruction in Forensic Medicine.* Boca Raton, FL: CRC Press/Taylor & Francis Group; 2009.
  This monograph has become a standard reference for digital forensic imaging, describing validated workflows for 3D documentation, radiological reconstruction, and interdisciplinary forensic analysis.
- **Brogdon's Forensic Radiology**
  Thali MJ, Viner MD, Brogdon BG (eds.). *Brogdon's Forensic Radiology.* 2nd ed. Boca Raton, FL: CRC Press/Taylor & Francis Group; 2010.
  The revised edition systematically merges traditional forensic radiology with state-of-the-art postmortem CT, MRI, and Virtopsy techniques, establishing imaging as an integral component of modern forensic medicine.

During the Bern era of Virtopsy research, key foundational technologies in modern forensic imaging were conceived, developed, and systematically implemented. This period saw the establishment of three-dimensional surface scanning, high-precision photogrammetric surface documentation, and routine integration of postmortem computed tomography (PMCT) and magnetic resonance imaging (PMMRI) into forensic diagnostics. Together, these modalities enabled non-invasive, reproducible, and metrically reliable documentation of both external and internal findings, forming the methodological backbone of the virtual autopsy concept.

Building on these core technologies, subsequent methodological innovations significantly expanded the diagnostic scope of forensic imaging. Notable advances included postmortem angiography for detailed vascular visualization,

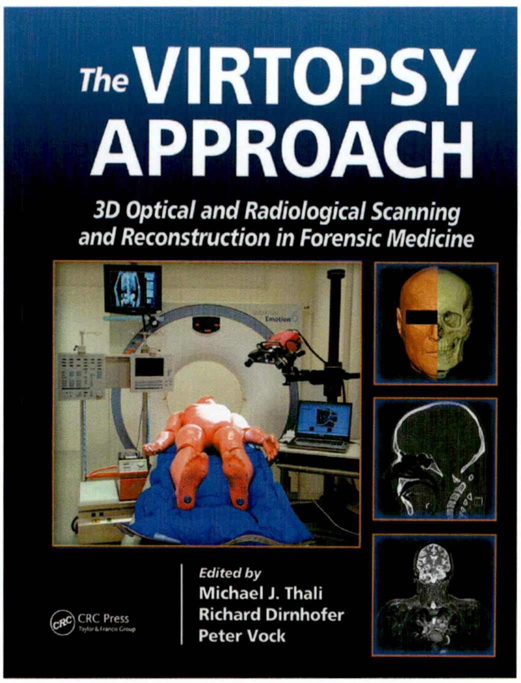

Figure 1.3a Pioneering reference textbooks in forensic imaging authored by the Virtopsy Group at the University of Bern.

From: *Virtopsy Approach 3D Optical & Radiological Scanning, Edition 1*, edited by Michael Thali, Richard Dirnhofer and Peter Vock, Copyright 2024 by CRC Press. Reproduced by permission of Taylor & Francis Group.

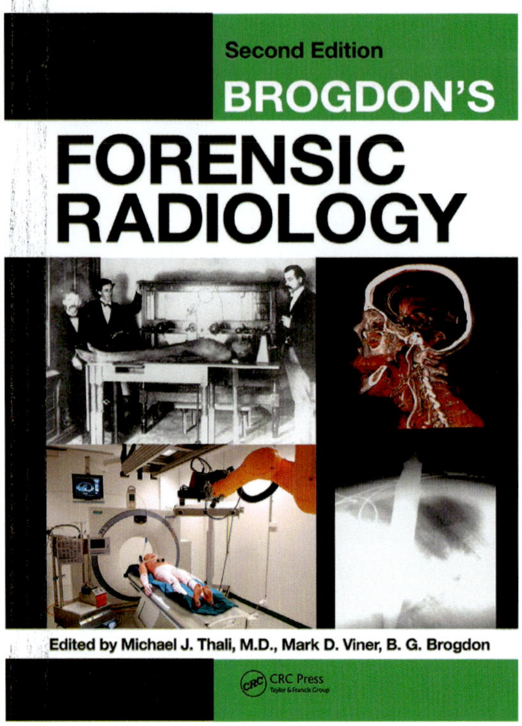

Figure 1.3b Pioneering reference textbooks in forensic imaging authored by the Virtopsy Group at the University of Bern.

From: *Brogdon's Forensic Radiology, Edition 2,* edited by Michael Thali, Mark D. Viner and B.G. Brogdon, Copyright 2024 by CRC Press. Reproduced by permission of Taylor & Francis Group.

minimally invasive biopsy techniques allowing targeted tissue sampling, and ventilation techniques applied to the deceased, facilitating realistic simulation of physiological conditions. These developments substantially enhanced the evidentiary value of forensic investigations and contributed to the international acceptance of imaging-based approaches in medicolegal practice. The scientific output arising from this phase was disseminated through numerous peer-reviewed publications, reference textbooks, and methodological guidelines, which collectively shaped forensic imaging as a distinct and internationally recognized subdiscipline.

The complete body of work produced by the Virtopsy Research Group, including contributions from all former and current members is comprehensively documented and publicly accessible at www.virtopsy.com. In recognition of these groundbreaking advancements in forensic science, the Virtopsy team was awarded the Swiss ICT Award in 2017. This accolade highlighted their pioneering technological contributions not only to forensic imaging but also to the broader healthcare sector, acknowledging the impact and innovation their work brought to both fields.

1.4 THE MOVE TO ZÜRICH—OPENING NEW HORIZONS

In 2011, Prof. Dr. med. Michael Thali and nearly the entire Virtopsy team relocated from University Bern to University Zürich. This strategic move was made possible by a generous endowment from the Emma Louise Kessler Foundation, named after a wealthy ophthalmologist renowned for her legendary enthusiasm for the lectures of the first chair of the Zürich Institute of Forensic Medicine, Prof. Heinrich Zangger.

The endowment provided the financial foundation required to continue and further develop Virtopsy technologies and to sustain the team at the highest international level. With strong strategic and infrastructural support from the University of Zürich, Zürich rapidly emerged as a global hub for forensic imaging—building firmly on the scientific and methodological foundations previously established at the University of Bern.

1.5 THE BIRTH OF A SCIENTIFIC DISCIPLINE AND THE SOCIETY/JOURNAL IN 2011

On November 27, 2011, the International Society of Forensic Radiology and Imaging (ISFRI) was formally established. Its founding was announced in a single, seminal email circulated by Michael Thali ahead of the annual meeting of the Radiological Society of North America (RSNA):

> We are pleased to announce the formation of the International Society of Forensic Radiology and Imaging (ISFRI) as well as the release of the *Journal of Forensic Radiology and Imaging.*

The inaugural ISFRI meeting took place in May 2012 in Zürich, marking the first formal gathering of the society. The founding board comprised Gary Hatch, Thomas Ruder, Gil Brogdon, Mark Viner, Robert Breitbeck, and Michael Thali. This event represented a pivotal milestone in the institutionalization of forensic radiology and imaging as an independent scientific discipline.

The International Society of Forensic Radiology and Imaging plays a pivotal role in advancing forensic radiology and imaging by:

- Developing international guidelines to ensure reproducible, standardized, and legally robust applications in forensic practice
- Organizing education and capacity building, including training courses, scientific conferences, and structured online learning formats
- Facilitating interdisciplinary exchange among radiology, forensic medicine, technology, and legal professionals to promote integrated and practice-relevant innovation

ISFRI | International Society of Forensic Radiology and Imaging

Figure 1.4 Founding board members of the International Society of Forensic Radiology and Imaging: Robert Breitbeck, Mark Viner, Gary Huch, Gil Brogdon, Thomas Ruder, and Michael Thali (from left to right).

In 2012, the official journal *Forensic Radiology and Imaging*—later renamed *Forensic Imaging* (Elsevier)—was launched. Under the leadership of its founding editor-in-chief, Michael Thali, together with Dr. Summer Decker—who became president of the International Society of Forensic Radiology and Imaging in 2025 and authored the foreword to this book—the journal rapidly achieved international scientific recognition and was successfully indexed with an impact factor.

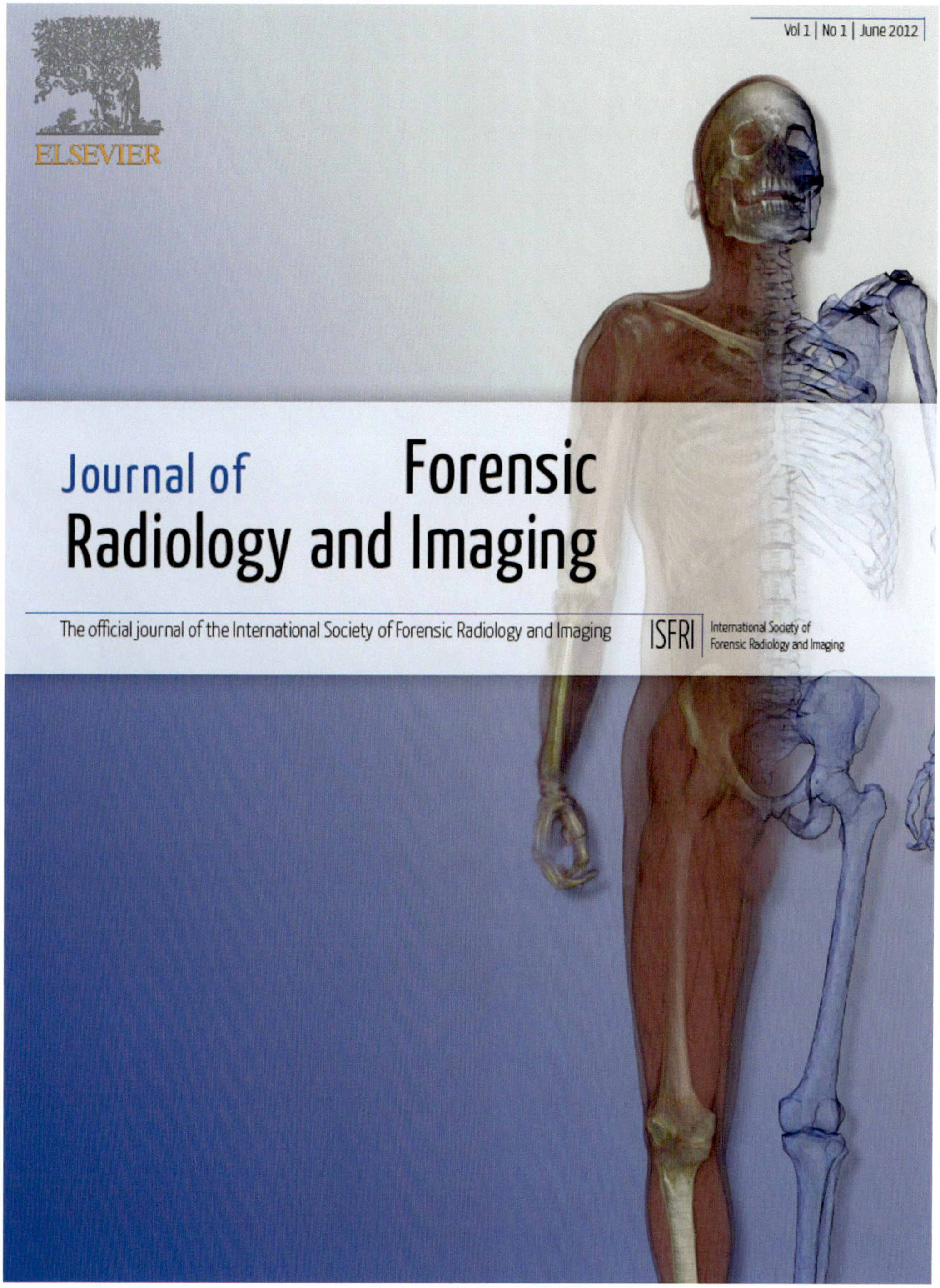

Figure 1.5a Cover of *Journal of Forensic Radiology and Imaging*, ISSN: 2212-4780, Volumes 1–19 (2013–2019), Copyright Elsevier 2026. (https://www.sciencedirect.com/journal/journal-of-forensic-radiology-and-imaging).

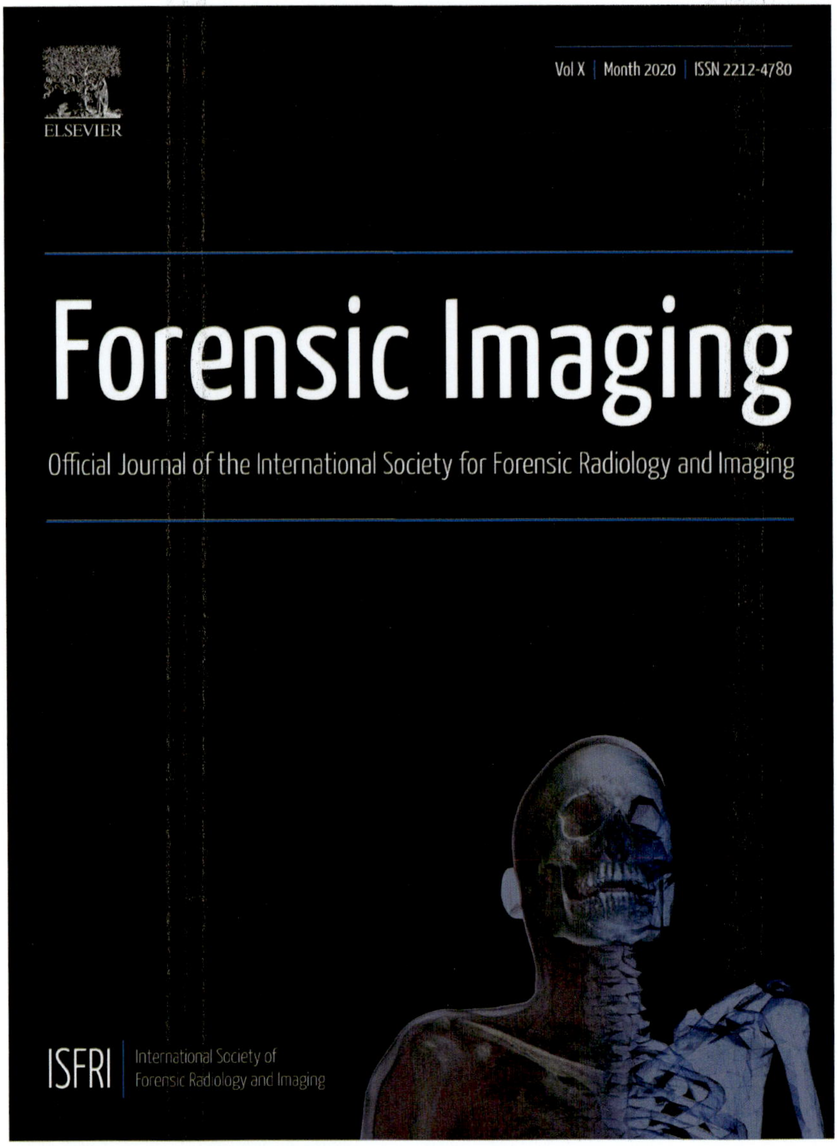

Figure 1.5b Cover of *Forensic Imaging*, online ISSN: 2666-2256; print ISSN: 2666-2264, Volumes 20–44 (2020–2026), Copyright Elsevier 2026. (https://www.sciencedirect.com/journal/forensic-imaging). Official publication of the International Society of Radiology and Imaging and International Association of Forensic Radiographers.

Figure 1.5c Summer Decker and Michael Thali pictured following lectures at the annual meeting of the Radiological Society of North America; ceremony reception held at the Field Museum, Chicago, 2022.

1.6 THE CAS VIRTOPSY COURSE—EDUCATION AND GLOBAL OUTREACH

In Zürich, the Virtopsy courses—originally developed in Bern—were systematically expanded into a comprehensive academic education program. In 2017, the CAS Virtopsy (Certificate of Advanced Studies) was formally introduced, establishing a structured postgraduate curriculum dedicated to forensic imaging. The program consolidated teaching, research, and international knowledge transfer, thereby strengthening Zürich's role as a global hub for advanced education in Virtopsy and forensic imaging.

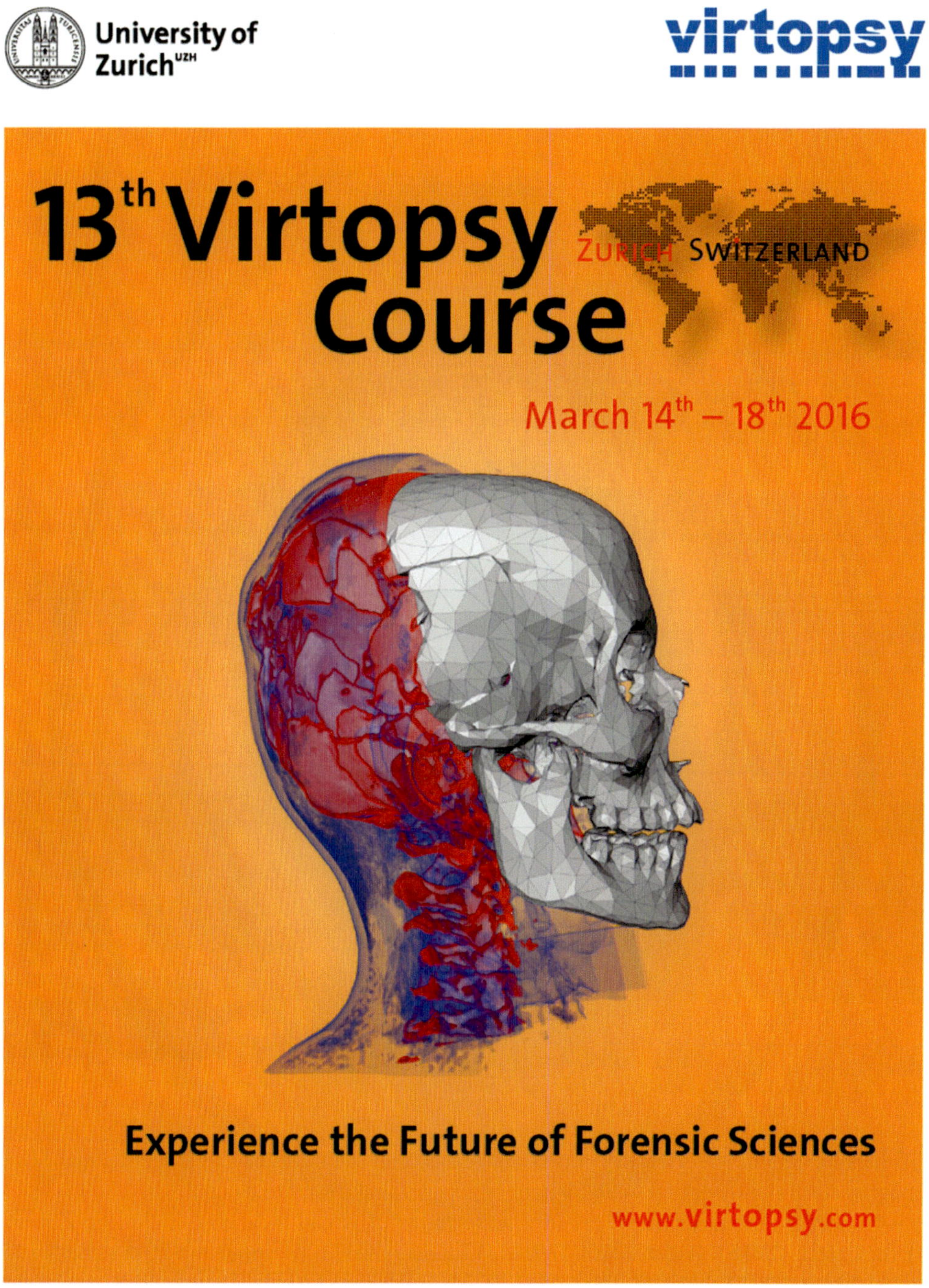

Figure 1.6ab Promotional posters for the Virtopsy Course 2016 (the annually conducted course followed a team-defined theme color; in 2016, this was burnt orange). The figures include the first official promotional poster for the CAS in Forensic Imaging and Virtopsy at the University of Zürich.

Figure 1.6ab (Continued)

More than 250 participants of the Virtopsy courses have become members of the International Society of Forensic Radiology and Imaging since the inception of the Virtopsy educational program.

ISFRI has organized international conferences in a wide range of global locations, including Zürich (2012, 2013), Marseille (2014), Leicester (2015), Amsterdam (2016), Odense (2017), Melbourne (2018), Berlin (2019), Krakow (2021, online; 2024), Tokyo (2022, hybrid), Toulouse (2023), and Fukui (2025). Future ISFRI meetings are scheduled to take place in Sofia in 2026 and Los Angeles in 2027.

1.7 INTEGRATION INTO PRACTICE AND GLOBAL ADOPTION

Beyond research, the overarching objective was always systematic implementation in daily practice. Forensic imaging rapidly achieved broad acceptance among prosecutors, law enforcement agencies, the court system, and the judiciary, becoming an integral component of evidentiary workflows.

From 2011 onward, every deceased individual in Zürich underwent standardized postmortem triage computed tomography. When indicated, this was complemented by high-resolution three-dimensional body surface documentation, enabling comprehensive, noninvasive forensic assessment.

In close collaboration with police forensic chief Thomas Fluri, the 3D Center Zürich (3DZZ) was established in 2012. This unique interdisciplinary platform integrated forensic medicine, crime scene investigation, and operational police work, creating a shared technological and methodological infrastructure. This sustained collaboration ultimately culminated, in 2022, in the foundation of the FormalTech Research Center, with a strategic focus on forensic machine learning, data-driven analysis, and advanced investigative technologies.

From 2014 onward, at the University of Zürich, within the Center for Forensic Medicine and Imaging, up to 350 forensic cases per year were resolved solely on the basis of postmortem imaging without performing a classical autopsy. At the time, this represented an international breakthrough in forensic medicine. In approximately 400 additional cases per year, conventional autopsies were carried out in combination with postmortem imaging. Consequently, in Zürich alone, more than one third of the total annual caseload was resolved through CT-based triage imaging alone, clearly demonstrating that forensic imaging had evolved from an innovative concept into routine standard practice.

Based on this principle, the QuickTox method was introduced in Zürich in 2017: QuickTox provides rapid, on-site toxicological screening, enabling the qualitative detection of more than 1000 substances as well as the quantitative determination of blood alcohol concentration. Conceptually, this approach was built upon the idea of *overnight toxicology*, first described by Gerostamoulos, Dimitri; Beyer, Jens; Woods, J. L.; George, N.; and Drummer, Olaf H. in "Overnight Toxicology—Fast Targeted Screening of Drugs and Poisons in Postmortem Blood", published in *Pathology*, 2010, Volume 42, Supplement S28 (https://doi.org/10.1097/01268031-201042001-00059). The aim of this concept was to enable rapid, targeted toxicological screening shortly after death in order to support early forensic decision-making.

Fully integrated into the Virtopsy CT triage workflow, QuickTox allows early differentiation between natural, toxicological, and traumatic causes of death and resulted in a substantial increase in efficiency and diagnostic precision in daily forensic practice.

The Zürich model combining postmortem CT triage with rapid toxicological QuickTox analysis was subsequently adopted by international forensic institutes, including those in Melbourne and Aarhus, the latter coinciding with the appointment of Michael Thali as Skou Honorary Professor in 2018. The scientific validation and further development of this integrated approach have recently been confirmed through the concept of enhanced autopsy triage (EA-Triage). In a comprehensive study on drug-related deaths, Kisbye, L. W.; Rickert, A.; Hasselstrøm, J. B.; et al. demonstrated the effectiveness of this model in "Enhanced Autopsy Triage (EA-Triage) in Drug-Related Deaths: Integrating Quick Toxicological Analysis and Postmortem Computed Tomography", published in *Forensic Science, Medicine and Pathology*, 2025, Volume 21, pages 229–238 (https://doi.org/10.1007/s12024-024-00819-2).

The 2024 international institute-based survey by Khmara et al., entitled "Virtopsy Concept Around the World: Institute-Based Survey of Worldwide Forensic Postmortem Imaging" and published in *Forensic Imaging*, confirms that postmortem computed tomography is now globally established in forensic medicine (Khmara et al., 2024). Based on responses from 100 forensic institutions across 29 countries, the study demonstrates that PMCT is the most widely implemented postmortem imaging modality worldwide and is routinely integrated into forensic workflows. These findings underscore the broad international adoption of the Virtopsy concept, with PMCT having transitioned from an experimental technique to a standard tool in contemporary forensic practice.

Virtopsy has definitely evolved into a non-invasive, evidence-based forensic methodology that is applied in numerous countries. While implementation is adapted to local legal, cultural, and infrastructural contexts, its effectiveness and reliability remain consistent across settings.

1.7.1 Virtopsy Update—The Forensic Imaging Zürich Approach

Originally conceived in 2018 under the working title "FIZA" (Forensic Imaging Zürich Approach), this book is now published as *Virtopsy Update*. It presents the latest summary of the Virtopsy Approach—a pioneering, practice-based reference grounded in more than two decades of sustained innovation and interdisciplinary collaboration in forensic imaging.

Further developed in Zürich with support from the Emma Louise Kessler Grant, and rooted in academic and research work conducted in Bern and Zürich, the book bridges the gap between cutting-edge imaging research technologies and their practical application in everyday forensic medicine.

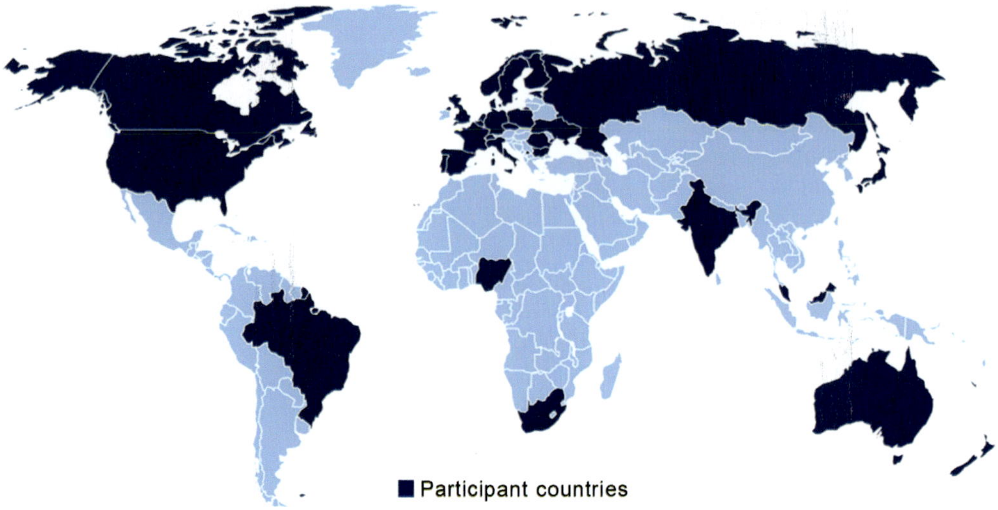

Figure 1.7 Overview of the study by Khmara et al. and global distribution of countries in which forensic imaging had established relevance by 2024. Reprinted under the Creative Commons Attribution 4.0 International Deed, CC BY 4.0: Deed—Attribution 4.0 International—Creative Commons.

The volume is structured into three parts:

- **New Core Technologies in Forensics**
  Covering key technological advances such as 3D surface scanning, augmented and virtual reality (AR/VR), Virtobot and Virtoscan systems, spectral CT, photon-counting CT, 7-Tesla MRI, MR spectroscopy (MRS), image-guided biopsy, angiography, and first AI-based diagnostic approaches.
- **Workflows, Strategies, and Future Directions**
  Addressing integrative concepts including triage imaging, digital documentation, virtual courtrooms, 3D printing, and interfaces between forensic medicine and law-tech solutions.
- **Forensic Diagnostics and Identification**
  Focusing on radiological trauma analysis, postmortem disease detection, imaging changes related to freezing and thawing processes, and radiological methods of human identification.

This book is more than a conventional textbook. It represents a future-oriented roadmap for forensic imaging, authored by members of the Virtopsy Team and associated research affiliates during the most scientifically productive and academically active phase of the Virtopsy program.

1.8 THE "THALI VIEW"—LOOKING AHEAD INTO THE FUTURE OF FORENSIC IMAGING

In Zürich, many of our early Virtopsy visions became reality—just as Peter Drucker said: *The best way to predict the future is to create it.*

Over the past seven decades, forensic imaging has undergone a transformation that is unprecedented in the history of forensic medicine. A consultation of the PubMed database reveals that more than 10,000 scientific papers addressing forensic imaging have been published since 1954. Until the year 2000, fewer than 50 such papers appeared annually. By contrast, in 2025, the publication rate exceeds 1000 papers per year. This quantitative shift reflects a deeper qualitative change: forensic imaging has evolved from a niche methodological innovation into an independent forensic subdiscipline with its own scientific standards, methodologies, and legal relevance.

A substantial proportion of this development has been shaped by the work of the Virtopsy research group, whose peer-reviewed publications are documented at www.virtopsy.com. The Zürich years demonstrated that systematic interdisciplinary collaboration—combining forensic pathology, radiology, 3D visualization, engineering, and informatics—can fundamentally redefine forensic investigation.

1.8.1 Scientific Consolidation and Strategic Orientation

Key milestones in the consolidation of forensic imaging as a distinct discipline are reflected in international consensus documents. The International Society of Forensic Radiology and Imaging white paper "Research in Forensic Radiology and Imaging" (2017) systematically identified critical scientific, educational, and infrastructural challenges facing the field (*Journal of Forensic Radiology and Imaging*, 2017; 10: 1–8). Central to this document was the call for harmonized research agendas, validated methodological protocols, and standardized reporting structures to ensure scientific reproducibility as well as legal robustness.

Complementary guidelines developed by ISFRI working groups further elaborate upon these principles and are publicly accessible via the society's official platform, www.isfri.org.

This strategic orientation was subsequently operationalized through *Rutty's Rules*, published in 2023 as baseline guidance for safe and standardized postmortem computed tomography reporting (Rutty et al., *Forensic Imaging*, Vol. 34, September 2023, Article 200558). These rules define minimum requirements for image acquisition, interpretation, documentation, and quality assurance in postmortem CT practice, thereby explicitly linking technical imaging standards with forensic accountability and courtroom applicability.

Collectively, these publications mark a decisive transition from exploratory innovation toward the establishment of a coherent normative framework in forensic imaging.

1.8.2 The "Thali View": Milestones for the Next Phase

First presented at an international SCFMS 2025 meeting in Riyadh, the "Thali View" outlines the necessary milestones for the next evolutionary phase of forensic imaging:

- **Certified training programs** to ensure global professionalization and comparability of expertise
- **Systematic integration of forensic imaging** into undergraduate and postgraduate medical-forensic curricula
- **Standardized protocols and reporting structures** that are legally sound and jurisdiction independent
- **Three-dimensional multimodal imaging** linking morphology with functional and physiological data
- **Scoring systems** to enhance clarity, transparency, and interobserver reproducibility of forensic findings
- **3D printing as well as AR/VR visualization** for forensic reconstruction, education, and courtroom communication
- **Big data analytics and machine learning** for robust pattern recognition across large case cohorts
- **AI-based diagnostic support systems** for workflow optimization, quality control, and error reduction

These milestones reflect a paradigm shift: forensic imaging is no longer defined solely by visualization but by integration, interpretation, and intelligent synthesis of multimodal data.

1.8.3 From Virtopsy to "Virtoxy" to "AI-opsy"

Forensic imaging is no longer a promising innovation; it is a mature, evidence-based, and legally validated discipline poised to shape the future of forensic science. Classical Virtopsy—non-invasive or minimally invasive imaging-based investigation-now represents a foundational layer rather than the endpoint of development.

The next evolutionary step is the Virtopsy triage integration of rapid toxicological QuickTox screening, termed "Virtoxy", allowing intoxications to be detected within hours and interpreted in direct conjunction with imaging findings.

Building upon this integration, the emerging concept of "AI-opsy" unifies Virtopsy (imaging) and Virtoxy (toxicology) within AI-assisted analytical environments. In the AI-opsy framework, artificial intelligence systems pre-analyze surface documentation, internal imaging data, and toxicological results, synthesizing them into structured (preliminary) reports.

These reports are not (yet) substitutes for expert judgment but serve as decision-support tools, enabling forensic experts to focus on interpretation, contextualization, and legal assessment. Such systems promise efficient, scalable, and cost-conscious workflows even under conditions of high case volume, while simultaneously improving consistency and quality assurance.

1.8.4 Outlook

The overarching goal remains unchanged: to further develop forensic imaging as an evidence-based, scientifically rigorous, and legally robust discipline.

The articulated and proposed "Thali View" is not a speculative vision of a distant future; rather, it constitutes a strategic roadmap grounded in decades of empirical research, sustained technological advancement, and interdisciplinary "out-of-the-box thinking" collaboration. From this perspective, the future of forensic imaging is not merely anticipated—it is actively being created.

PART 1
TECHNICAL CHAPTERS

CHAPTER 2

FORENSIC SURFACE DOCUMENTATION

Till Sieberth and Lars Ebert

DOI: 10.4324/9781003413998-3

2.1 INTRODUCTION

In forensics, documentation is of superordinate importance. Visual documentation takes on an important role and can be divided in two main expert fields, the first being documentation on a person, which usually is covered by forensic medicine, and the second being documentation of scene and objects, which is often performed by technical personal.

In particular, surface documentation addresses the procedure to record the exterior, visible surface of a body, object or scene. It aims to capture and preserve information regarding shape, structure, color and texture of an object as well as its relationship to surrounding objects or scenes. The goal is to capture the current condition, which can be done either in a two-dimensional (2D) or three-dimensional (3D) representation.

In forensic medicine surface documentation is often associated with injuries or the absence of those on the surface of the body. In this context, findings usually address the integrity of the skin, discolorations and foreign substances on or in the skin and objects attached to the body (DiMaio and DiMaio, 2001). However, depending on the circumstances of the case, findings may also address the condition of the skin that could be relevant to a legal investigation, such as scars, dermatological conditions or body modifications.

In the technical forensics, the documentation of surfaces focuses on documenting the as-is state of a scene or an object. However, in differences to body documentation, where it is easily detectable if something is normal or out of the ordinary, this can be more indiscernible at a scene or on an object. Missing or deliberately placed objects in a scene might have significant impacts on subsequent interpretation of an incident and can be much harder to detect. Therefore, appropriate and comprehensive documentation is important and should be carried out to allow the legal investigation to objectively discuss the situation.

Common methods for documentation are drawings and sketches supplemented with vocal descriptions and written text, photographs and three-dimensional scans.

2.2 SURFACE DOCUMENTATION GENERAL REQUIREMENTS

The aim of surface documentation is an objective recording of the as-is state of a person, object or scene which allows for interpretation and analysis of the content.

In order to perform an external documentation of a person that can be used in line with a legal investigation, certain documentation procedures and protocols need to be followed. The acquired dataset should allow for an overview with unambiguous identification of and assignment to a person. It is recommended to document the entire body, from front and back, to record also the absence of findings. With this overview it is also possible that subsequent close-up views of regions of interest (e.g. injuries) can be exactly located on the body (Grassberger and Schmid, 2009). Apart from that, it is necessary that the captured data include a scale to allow for size estimations.

For scene documentation similar conditions apply; overview documentation should allow for located subsequent detailed documentations. Furthermore, the documentation should follow a procedure and logical walk path through the scene. For example, adjacent rooms should be clearly connected from both rooms, which enables subsequent orientation by people who have not been to the scene before. The inclusion of a scale, especially in vicinity of traces, is also recommended.

Overall, it is important that documentation be carried out in a technically correct manner and exhibit perfect image and data quality.

2.3 2D DOCUMENTATION

Documentation based on photographs is commonly referred to as a 2D photo documentation. As cameras are frequently available and often implemented in consumer devices, a first documentation can often be easily and swiftly performed and ensure a first documentation of a scene, object or person. However, it is still recommended to perform forensic photographs following forensic standards to ensure the usability of the images in subsequent investigation and court proceedings (Evans et al., 2014).

2.3.1 Camera Settings and Acquisition Technique

To acquire technically correct photos, understanding of the camera and the appropriate settings can ensure the best achievable image quality. Even though digital cameras feature automatic modes allowing automatic acquisition of photos without manual adjustment, this might not always result in useful photographs in line with forensic requirements. Therefore, it is necessary to understand certain camera parameters and how to adjust them correctly.

Aperture is the opening within the lens and can be adjusted in steps. The further the aperture is closed, the less light reaches the image sensor and the darker the image. However, a more closed aperture allows a larger depth of field in the acquired image. For forensic images which should be focused over a large depth of field, the perfect setting would be a closed aperture.

Exposure time is how long the shutter is open. A shorter time results in less light reaching the image sensor and therefore a darker image while it also prevents motion blur in the image. For forensic images taken by hand this should be chosen to be short enough to prevent motion blur.

ISO-film speed is a value describing the sensitivity of the image sensor. The less sensitive, the more light is required on the sensor to actually be detected, which causes darker images. However, if the sensor is too sensitive, noise can increasingly appear, resulting in a grainy and often less clear image. The ISO should be chosen to be rather insensitive to prevent grain in the image; however, this strongly depends on the processing chip and hardware of the camera.

While these three parameters present the main adjustment parameters for image acquisition, they are often accompanied by other parameters such as white balance or quality of the image format. To support the operator, other functions such as metering or focusing options are implemented (Redsicker, 2001).

2.3.1.1 Forensic Medicine

Apart from camera parameters, additional requirements must be fulfilled by the documentation procedure. In terms of body documentation, it is important to include overview and detail photographs which allow localization and characterization of the type of injury. Furthermore, a scale bar, such as an ABFO No. 2 standard reference scale, should be included to allow size estimations (Ferrucci et al., 2016). However, to allow this, the scale bar needs to be in the depth of the injury, and the photograph should be taken orthogonally to the injury (Evans et al., 2014). It is also recommended to add photographs taken from different angles. Especially in clinical cases it is recommended to ask the victim to take on the body posture that they were in at the moment the injury was caused, with the image being taken from the attacking direction.

2.3.1.2 Technical Forensics

For technical forensic photographs the additional requirements are similar to the forensic medicine requirements. Photographs should include scale bars that allow one to estimate the dimension of the photographed object or scene. Furthermore, detail photographs should be accompanied by overview photos. Orthogonal photos are especially of importance when specific dimensions are to be shown; however, especially with larger objects or scenes, this is often not possible. Another special aspect here is also that images of scenes should contain a certain amount of overlap and multiple angles of view to ensure a good understanding of the scene (Redsicker, 2001).

2.3.2 Advantages and Disadvantages

Photographs are nowadays arguably the most common and available method to document and preserve information in this context, especially with the increasing availability of digital cameras such as smartphone cameras. Using these, photo documentation can be carried out almost anywhere and at any time. Therefore, it also presents one of the most important documents in subsequent case investigation, incident reconstruction and legal procedure.

Nonetheless, 2D imaging also has certain drawbacks. The lenses can change the degree of the viewing angle and can introduce aberrations such as pincushion or barrel distortion, chromatic aberration, spherical aberration, astigmatism and vignetting (Luhmann et al., 2020). Distortions especially have an influence on the geometric correctness of a 2D image, as straight lines in 3D space are projected as curved lines in the 2D image (Figure 2.1).

A photograph is a 2D projection of a three-dimensional space. Due to this projection the depth information from the original 3D scene is lost on the 2D photograph. This is why one single photograph is not sufficient to represent the spatial relationships of a 3D object or scene. The missing depth information can lead to subsequent misinterpretations. For example, a surface in a photo might appear flat and even, whereas in reality (in 3D space) the surface is actually curved.

This could be visually compensated with images from other view angles showing the curvature; however, as soon as measurements are taken based solely on the information provided by the 2D photograph, this can significantly vary from the true value (Michienzi et al., 2018).

Figure 2.1 The image taken with a wide angle lens causes distortions that make the straight pillar appear bent, as well as the ventilation pipe. The blue line shows a straight line in comparison.

2.4 3D DOCUMENTATION

3D documentation can overcome the problems and limitations of 2D images and allows digital reconstruction of the spatial relationships of an object or scene. The 3D model retains the depth information and allows interactive reconstruction with the data as well as visualization of a 3D model from varying viewpoints or in real 3D.

Common methods for 3D documentation in forensics can be either based on transmissive or reflective, non-contact methods (Figure 2.2). The transmissive methods are commonly known in medicine, such as computer tomography and magnetic resonance imaging. Reflection base methods have recently been introduced in a variety of medical fields, such as whole-body scanners for dermatological use or pattern light scanners for dental scans, as well as other applications in plastic surgery, surgery or other fields (O'Toole et al., 2019; Paar and Smolle, 2003; Son and Lee, 2020). These methods are part of the science of photogrammetry (from Greek: *phos* or *phot*, meaning light; *gramma*, meaning drawing something; and *metrein*, meaning measure). Photogrammetry is an umbrella term for "the science and technology of extracting reliable three-dimensional geometric and thematic information, often over time, of objects and scenes from image and range data" (Heipke and Halounova, 2022) to "reconstruct the position, orientation, shape and size of objects" (Kraus, 2007). Methods include laser scanning, structured light scanning, structure from motion and photogrammetry, which are currently the most relevant techniques in forensics.

2.4.1 Brief History

Police forces and forensic institutes early on already used a variety of methods, mostly tachymetry, to survey traffic accident sites or document crime scenes. The introduction of photogrammetry in forensics cannot be pinpointed exactly, with the first scientific publications mentioning photogrammetric approaches in 1993 (Chandler, 1993). Photogrammetric methods such as laser scanning were gradually implemented and have already featured in several photogrammetric expert books since then (Luhmann et al., 2020; Vosselman and Maas, 2010).

3D scanning technologies are widely accepted for crime scene documentation and traffic collisions in several countries, with guidelines and workflows available. The introduction of photogrammetry in forensic medicine goes back to a case commonly known as "Mord im Kehrsatz," 1986 to 1996. The highly discussed case ultimately caused the resignation of the director of the Institute of Forensic Medicine Bern, who was replaced by Professor Richard Dirnhofer. His task was to implement procedures to objectively document evidence and improve existing methods. This lead to first contact between forensic medicine and the Scientific Service of the City Police Zürich, represented by Walter Brüschweiler and Marcel Braun. Together they published their first work about photogrammetric injury documentation using cameras in 1995, with a subsequent publication about weapon matching in 1997 (Brüschweiler et al., 1997, 1995).

In the following years under the instruction of Dirnhofer, the Virtopsy approach was developed by Michael Thali, leading to a closer collaboration between the Scientific Service of the City Police Zürich and the Institute of Forensic Medicine Bern. As part of the Virtopsy approach, documentation of the deceased aimed to provide all-around documentation not only using medical imaging technologies such as computer tomography but also pattern light scanning for more accurate 3D scans. Automation was one of the main requirements, as the manual approach was cumbersome, required significant manual work and knowledge and was slow. Automation was realized with the Virtobot that allowed attaching different tools and automatic movement across the body (Breitbeck et al., 2013; Ebert et al., 2014).

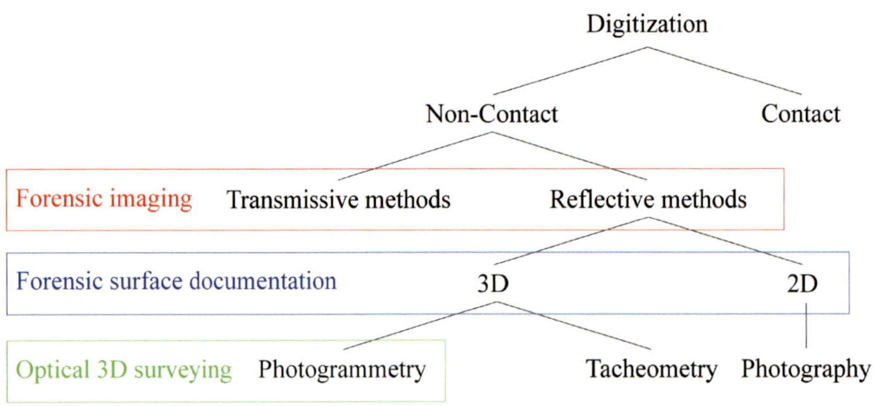

Figure 2.2 Simplified context visualization of forensic imaging, forensic surface documentation and optical 3D surveying.

Besides the use in postmortem documentation, multi-camera devices are employed in clinical applications (Leipner et al., 2016). Especially in clinical examinations, the documentation of a living, moving person requires a speedy procedure to allow whole-body 3D surface documentation without exposure to radiation.

Nowadays there is a great variety of optical measuring and imaging techniques to capture and digitize the spatial information of three-dimensional objects and scenes. However, not all of those techniques are equally useful in the context of external body documentations. At the Zürich Institute of Forensic Medicine, two main techniques are used to document the surface of human bodies, photogrammetry and structured light scanning.

2.4.2 Indications for 3D Scan

3D scanning procedures are not performed in every case but only on cases with indications.

2.4.2.1 Forensic Medicine

In forensic medicine there are several indications to perform a 3D scan on either a living or deceased person. Commonly all injuries with geometrical relevance should be documented. This includes pattern injuries, such as shoe imprints or baton patterns on the skin, or a weapon's face that was pressed against the skin while being fired. With the use of a 3D scan these can be compared to injury-causing objects. It is important that even injuries on fairly flat skin surfaces should be documented, as neither the injury-causing objects nor the area of the injury might have been flat at the time of incident.

Another indication with geometric relevance are shot and stab injures. While 3D surface documentation cannot help establish how large a blade was or which projectile was used, it can help to establish the trajectory and angle of attack. CT scans only reliably visualize the destruction of bones, while soft tissue penetrated by a projectile or blade often does not show the trajectory of the injury exactly. However, a bone injury in combination with a surface documented injury can be used to establish the trajectory inside the body, which can then also be extrapolated outside the body and used to answer forensic questions about the incident.

In addition to pattern injuries and assault injuries, traffic-related incidents should also be considered for 3D surface documentation. Here, small and barely structured injuries like a simple line or arc can often be of relevance. These small findings can often be matched with certain parts of a vehicle and allow for reconstruction of the collision.

2.4.2.2 Technical Forensics

For documentation outside of the body, there are several indications. Here also the geometric relevance is important but can have significantly different implications. Similar to the documentation of pattern injuries, technical forensics should document injury-causing objects that might have caused the relevant injury. To limit the number of objects that are scanned, this should often be limited to a certain type of weapon, like a shoe, belt or baton that might be relevant. Here it should be done according to the forensic medical examination.

In other cases like traffic incidents or homicides, 3D documentation of a scene or location should always be performed. Even though the final outcome is obvious, it is possible that in-depth forensic questions need to be answered, such as after a traffic incident if the vehicle driver was at all able to see the victim. These questions might not help to identify a suspect but rather decide on the severity of the verdict.

2.4.3 Photogrammetry

Photogrammetry allows one to digitally document a scene in 3D with the help of multiple overlapping images. In order to create a useful set of images it is essential to understand the basics of photography and follow image acquisition rules.

2.4.4 Camera Settings and Acquisition Technique

The basic camera setting should follow the previously described rules for 2D image acquisition. For an appropriate 3D documentation the ISO, aperture and shutter should be set to acquire the best possible image. In addition to the standard settings it is also necessary to consider the lens, focus and focusing mode and camera mode. The camera mode is especially important, as it is necessary to acquire images in manual mode to prevent the camera form changing any of the set parameters.

With the camera in manual mode the exposure can be set, and one should consider that it is not possible to adjust the image exposure during acquisition. Therefore, the exposure should be chosen such that both dark areas and bright areas of the object are visible and then the camera set up to be able to document either both or the more relevant area. The use of on camera flash- or speedlight or permanent lighting that is on a fixed baseline with the camera is not possible, as the created shadows negatively influence the subsequent processes and homogeneity of the final 3D model texture. This also applies for any other changes in the scene. Moving objects during the scanning procedure or changing the room lighting will cause errors during the subsequent processing.

Besides the camera settings, it is necessary to place a scale bar in the scene, which is also photographed during the image acquisition. This scale bar should not be moved during image acquisition.

During the image acquisition it is also important to ensure ~80% overlap in the images. This will allow for subsequent 3D calculation of the overlapping area. At the same time, the images should fill the complete image space, eliminating empty space in the images. It is not necessary to have the object or injury visible as a whole in each image, as this is compensated for by the overlap (Figure 2.3).

Before starting the image acquisition and considering the overlap and image filling, the operator should consider their scanning approach.

2.4.4.1 Forensic Medicine

For forensic medicine the scanning approach can be easily structured. The operator should acquire multiple images in a line sequence and then several lines consecutively, ultimately creating a raster over the injury that is supposed to be documented. The operator should make sure that they can reach all positions of this raster by positioning the injured person adequately. The scanning of living subjects also requires that the person not be moving during acquisition; therefore they should be asked to sit or lie comfortably and rest their relevant body parts, or, in a hospital bed, they should be supported by a nurse.

Furthermore, while scanning persons, either living or deceased, the relevant areas should be cleaned up as necessary and debris or hair removed before acquisition to allow unobstructed view of the relevant finding.

Another important aspect to consider is to allow the photogrammetric scan to be combined with underlying medical imaging. Especially in clinical cases, where a scan is reduced to a certain area on the body of a person, the scanned area should be chosen to be large enough to contain topological landmarks to allow the combination of surface and medical scans.

2.4.4.2 Technical Forensics

Photogrammetry for technical forensics can be used in a variety of scenarios but is especially relevant to quickly document a scene, even though it might be of less accuracy. This includes documenting the site of discovery of a corpse before

Figure 2.3 Whole-body photogrammetry with a multi-camera device.

a forensic pathologist starts the examination. Even coarse information can help plenty in subsequent reconstruction of a crime. Another reason for photogrammetry is often the absence of other, more accurate scanning methods.

For scene documentation the operator should follow similar steps to medical forensic photogrammetry. Before starting a scan a procedure should be thought of and followed to prevent forgetting areas. However, especially at a scene, the operator needs to consider occlusions and should ensure that these occluded areas are also documented in multiple overlapping areas. The larger the scene, the more scale bars should be placed, and they should be of sufficient length. This ensures correct scaling of the scene and also allows one to check whether the scaling was performed correctly.

With larger sites it can also be necessary or sensible to perform multiple scans with different focus. It is recommended to perform one overview scan, which might be less precise but provides accurate positions of objects and the larger scene. This can then be replaced by an independent image series documenting detailed information, such as a corpse, a vehicle position or relevant objects at the scene.

2.4.5 Advantages and Disadvantages

Photogrammetry has several advantages and disadvantages. The major advantage is that the method does not need any special equipment besides a camera, which is usually standard equipment of any forensics expert. Furthermore, the procedure is easy to learn and apply, making this method very reliable even for unexperienced users. Small deviations from the optimal setup can often be compensated for and have only a small impact on the final 3D result. Also, this method is mandatory for all other scanning techniques to apply accurate texture information to a 3D model. However, this is often implemented in the other scanning devices themselves, ensuring proper image acquisition.

The disadvantage is accuracy, which rarely and only under perfect conditions can match the result of pattern light or laser scanning approaches. However, with the very flexible scanning range, from small objects to large mountain ranges, everything can be 3D documented with photogrammetry. However, for the scanning of larger areas or longer distances, it might be necessary to use aerial platforms to perform the scanning. Furthermore, while the acquisition procedure is rather fast, extensive postprocessing is required to generate the 3D model.

2.4.6 Structured Light Scanning

Structured light scanning has similarities to photogrammetry. However, instead of just passively photographing an object from two or more positions, this method uses a projector and camera in conjunction. The projector projects a pattern on the object, which is recorded by the camera. The offset between camera and projector, called the baseline, is previously accurately established by calibration of the equipment. If the pattern is projected on a 3D object, the pattern appears distorted to the camera based on the shape of the 3D object. This can be used to digitize the 3D shape of an object. It is also possible to include multiple cameras with other baselines to increase the scanning volume, accuracy and precision of the subsequent 3D model.

This scanning approach largely varies in scan volume, projected pattern and projector-camera configuration. Handheld structured light scanners with white light can be used to scan small patches of an object, and by moving the scanner the whole object is documented. Other scanners allow for large scan volumes of several cubic meters to document larger objects in one scan and, due to the stable scanner position, have a very high precision and accuracy. Larger scanners often use distinctive projection colors like blue light, which subsequently requires other documentation procedures to apply true color to the object. Furthermore, the pattern can vary, from simple stripe patterns over sinus curves to coded or even randomized patterns, which are flashed at the object.

2.4.6.1 Forensic Medicine

In forensic medicine pattern light scanners can be used to document objects and corpses. Due to the flashing light during the scan procedure and the requirement of the object to remain still during the procedure, this method is not feasible for the documentation of living persons. In Zürich a structured light scanner is used in conjunction with a standard camera to document corpses with high accuracy and precision. With the pattern light scanner it is even possible to document small swellings of the skin or needle punctures if necessary. Therefore, the scanner is moved in multiple positions across the body to ensure all-around documentation of the body (Figure 2.4). The same procedure is also possible with a handheld scanner.

2.4.6.2 Technical Forensics

For technical forensic applications highly accurate surface scanners have a variety of applications. There they can be used to document mirror surfaces to subsequently allow reconstruction even of reflections in vehicle mirrors. Also very small objects such as knots can be scanned best with a structured light scanner. Areas that are barely accessible or badly illuminated such as a car underfloor can be documented with a handheld structured light scanner, as photogrammetry using a camera might not be possible due to visual obstruction and inaccessible camera positions.

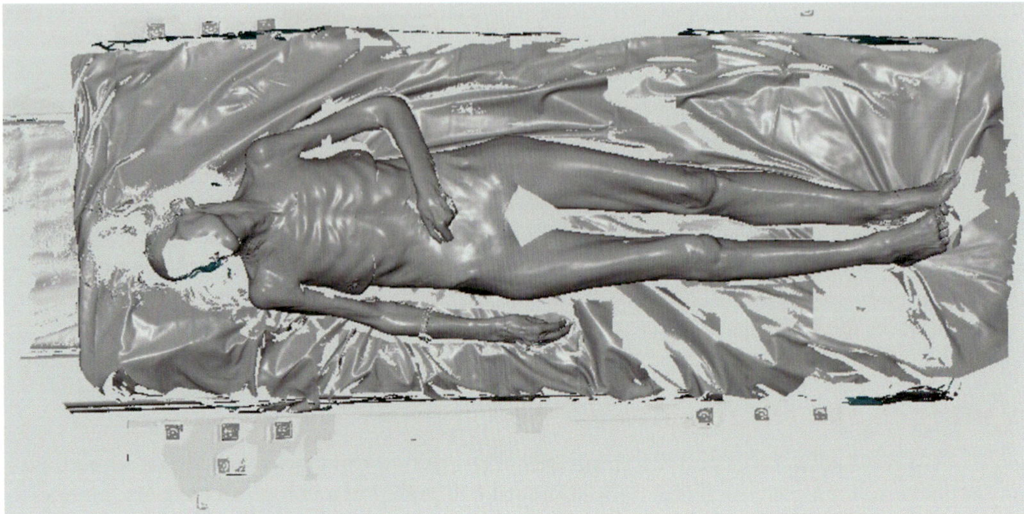

Figure 2.4 A model from a pattern light scan without coloration (anonymized).

2.4.6.3 Advantages and Disadvantages

Structured light scanners already allow the visualization of a 3D model during data acquisition, which allows the operator to check if everything important was digitized. Furthermore, a structured light scanner can allow for different scan volumes and more accurate and resolved 3D models of the scanned object. The noise in the models is significantly lower, allowing for smoother surfaces that can even be used to simulate reflections accurately. The postprocessing of the data is often limited to automatic procedures with only minor operator input.

However, to perform a structured light scan, experienced operators are required, as the handling of the equipment, the setup and the procedure are often very technical and can be difficult for inexperienced users. Especially handheld scanners, which have to be moved across the surface, require some training before adequate scans can be acquired. This can put whole 3D scans in jeopardy. Furthermore, specialized technical equipment is required, and depending on the range of scan volumes, it might be necessary to acquire multiple scanners, which allow for different scan volumes. While this can be partially compensated for by interchangeable lens systems of the projector and adjacent cameras, there is still a limitation to the volume. Also, the scanning period might be longer than with handheld photogrammetry, but postprocessing is usually shorter.

2.4.7 Laser Scanning

Another method for 3D surface documentation is laser scanning, which is based on LiDAR (backronym: light detection and ranging). As with structured light scanning, a special device is necessary to perform laser scanning. However, in contrast to the previous methods, laser scanners sample the surrounding environment with a laser beam and measure the direction in which the laser beam was sent and the distance to the object. This way a so-called point cloud is created straight from the measurements without further processing. These points are colorless, so subsequent coloring of the 3D model requires additional photographs. These are often already incorporated in the scanner so that no additional operator action is required. As this scanning procedure scans its surroundings, it is rather useful to document locations and less useful to document objects.

2.4.7.1 Forensic Medicine

In forensic medicine laser scanners are not used, as the main objective there is to document objects, which can be done more easily with the other presented methods like structured light scanning and photogrammetry.

2.4.7.2 Technical Forensics

For technical forensic applications laser scanners are of major importance. With their large scanning volume, from several millimeters in front of the scanner to several hundred meters or even kilometers, they present a viable method to document indoor and outdoor scenes. Depending on the location, there are different approaches that help to document a scene completely. In confined indoor areas such as flats it is recommended to use at least one scanning position per room, and then at the border of each room, such as doorways between rooms. This ensures enough overlap between subsequent scans so that whole 3D models can be created. For outdoor areas, a first reconnaissance of the area should be conducted to establish suitable scanning positions. However, the precise scanning procedure depends on the scanner type and manufacturer, location and scene.

2.4.7.3 Advantages and Disadvantages

The advantages of a laser scanner are their large scanning volume with comparably high resolution and precision. The reduction of postprocessing time by delivering 3D data straight out of the instrument is also an advantage. In comparison to photogrammetry and structured light scanning, a laser scanner scans its surroundings and therefore has a different field of application.

As scanning requires special devices, an experienced trained operator is necessary who should have hands-on experience before being deployed on scene. Laser scanning is not suitable to document single objects, only scenes. The scanning procedure also takes more time due to instrument setup; however, this is compensated for with shorter postprocessing.

2.5 SUMMARY

Forensic surface scanning can be performed with a variety of methods; however, these can also supplement each other to create a 3D model. In conjunction with medical imaging technologies such as PMCT, performed in the same body posture, it is possible to create whole 3D models of the surface and interior of corpses (Figure 2.5).

Photogrammetry with handheld cameras is the most-recommended method, as this is an easy first way to perform a 3D scan without additional equipment costs. The advantages of a 3D model over a 2D photograph go from better understanding and more advanced visualization to more precise measurements of injuries and the ability to perform 3D reconstructions of an incident.

However, there are some limitations to 3D surface documentations. As the presented surface scanning technologies are optical methods, certain surfaces, materials and substances are difficult or even impossible to capture. For example, reflective surfaces such as polished materials, metals and glass, as well as substances such as liquids, are usually difficult to capture.

Even though 3D surface documentation is a useful extension of forensic documentation, it is important to know that this comes with some costs. Generally, 3D surface documentation is more time consuming and requires additional equipment, which can be expensive. In addition to that, the use of 3D surface documentation might require trained personnel for the data acquisition, data processing and data analysis.

However, 3D documentation methods have recently increased in popularity, and while they still often require expert knowledge, technical advances are gradually making them accessible to non-experts.

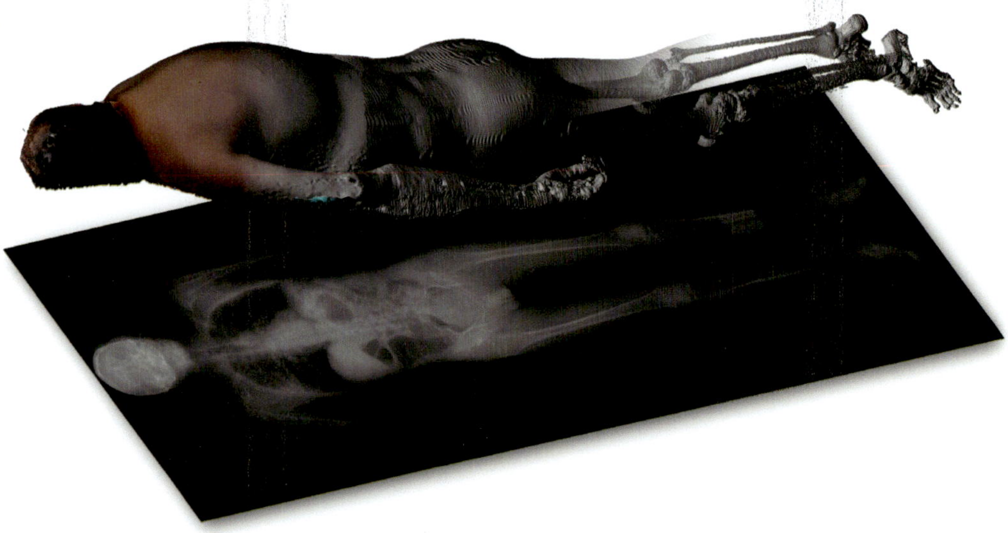

Figure 2.5 An example of whole-body documentation combining photogrammetry, pattern light scanning, CT surface and bone scan with a tomogram underneath.

References

Breitbeck, R., Ptacek, W., Ebert, L., Furst, M., Kronreif, G., Thali, M.J., 2013. Virtobot: A robot system for optical 3D scanning in forensic medicine 84–91. https://doi.org/10.15221/13.084

Brüschweiler, W., Braun, M., Fuchser, H.J., 1995. Photogrammetrische Vermessung morphologischer Befundung, in: Kompendium 4. Tagung Deutsche Gesellschaft Rechtsmedizin, Bern.

Brüschweiler, W., Braun, M., Fuchser, H.J., Dirnhofer, R., 1997. Photogrammetrische Auswertung von Haut- und Weichteilwunden sowie Knochenverletzungen zur Bestimmung des Tatwerkzeuges—grundlegende Aspekte. Rechtsmedizin 7, 76–83. https://doi.org/10.1007/bf03042360

Chandler, J.H., 1993. Height estimation using video security imagery. The Photogrammetric Record 14, 459–468. https://doi.org/10.1111/j.1477-9730.1993.tb00275.x

DiMaio, V.J., DiMaio, D., 2001. Forensic Pathology, 2nd ed. CRC Press, Practical Aspects of Criminal and Forensic Investigations, Boca Raton.

Ebert, L.C., Ptacek, W., Breitbeck, R., Fürst, M., Kronreif, G., Martinez, R.M., Thali, M., Flach, P.M., 2014. Virtobot 2.0: The future of automated surface documentation and CT-guided needle placement in forensic medicine. Forensic Science, Medicine, and Pathology 10, 179–186. https://doi.org/10.1007/s12024-013-9520-9

Evans, S., Baylis, S., Carabott, R., Jones, M., Kelson, Z., Marsh, N., Payne-James, J., Ramadani, J., Vanezis, P., Kemp, A., 2014. Guidelines for photography of cutaneous marks and injuries: A multi-professional perspective. Journal of Visual Communication in Medicine 37, 3–12. https://doi.org/10.3109/17453054.2014.911152

Ferrucci, M., Doiron, T.D., Thompson, R.M., Jones II, J.P., Freeman, A.J., Neiman, J.A., 2016. Dimensional review of scales for forensic photography. Journal of Forensic Sciences 61, 509–519. https://doi.org/10.1111/1556-4029.12976

Grassberger, M., Schmid, H., 2009. Todesermittlung—Befundaufnahme und Spurensicherung: Ein praktischer Leitfaden für Polizei, Juristen und Ärzte. Springer Vienna, New York. p388.

Heipke, C., Halounova, L., 2022. ISPRS Statutes [WWW Document]. URL https://www.isprs.org/documents/statutes.aspx (accessed 3.21.23).

Kraus, K., 2007. Photogrammetry. Walter de Gruyter GmBH & Co. KG, Berlin, Germany.

Leipner, A., Baumeister, R., Thali, M.J., Braun, M., Dobler, E., Ebert, L.C., 2016. Multi-camera system for 3D forensic documentation. Forensic Science International 261, 123–128. https://doi.org/10.1016/j.forsciint.2016.02.003

Luhmann, T., Robson, S., Kyle, S., Boehm, J., 2020. Close-Range Photogrammetry and 3D Imaging. Walter de Gruyter GmbH, Berlin/Boston.

Michienzi, R., Meier, S., Ebert, L.C., Martinez, R.M., Sieberth, T., 2018. Comparison of forensic photo-documentation to a photogrammetric solution using the multi-camera system "Botscan". Forensic Science International 288, 46–52.

O'Toole, S., Osnes, C., Bartlett, D., Keeling, A., 2019. Investigation into the accuracy and measurement methods of sequential 3D dental scan alignment. Elsevier Enhanced Reader. Dental Materials 35, 495–500. https://doi.org/10.1016/j.dental.2019.01.012

Paar, G., Smolle, J., 2003. Photogrammetric body scanning for dermatology. Presented at the 6th ISPRS Conference on Optical 3-D Measurement Techniques, Zürich, pp. 2–6.

Redsicker, D.R., 2001. The Practical Methodology of Forensic Photography, 2nd ed. CRC Press, Boca Raton.

Son, K., Lee, K., 2020. Effect of tooth types on the accuracy of dental 3D scanners: An in vitro study. Materials 13, 1744. https://doi.org/10.3390/ma13071744

Vosselman, G., Maas, H., 2010. Airborne and Terrestrial Laser Sscanning. Whittles Publishing, London.

CHAPTER 3

SPECTRAL IMAGING—BASICS AND FORENSIC APPLICATIONS

Dominic Gascho, Martin Frenz, and Sören Kottner

DOI: 10.4324/9781003413998-4

3.1 INTRODUCTION

Three basic components influence how humans perceive hue and color intensity. First, the frequency spectrum of the illuminating light (the source) or, in other words, the wavelength range of the light. Second, the optical properties of the material (the object), i.e., how the illuminating light is reflected from the surface of the object, how it penetrates the object, and how it is re-emitted from the object. Third, the spectral sensitivity of the photoreceptors of the human eye (the detector), which convert the information transmitted by reflected and re-emitted light into signals that are then transported via stimulus transmission to the brain, where the visual cortex finally converts the signals into an image with different hues and color intensities. The photoreceptors in the retina of the human eye consist of rods and cones, which absorb light and convert it into an electrical signal thanks to the visual pigments they contain, i.e. proteins such as opsin and rhodopsin. Rods are responsible for vision in dim light conditions (scotopic vision), having a low spatial acuity. The blue, green, and red cones, which are active in bright light (photopic vision), have high spatial acuity and are responsible for color vision. However, while sunlight covers a wide range of wavelengths from ultraviolet (UV) to infrared (IR), the photoreceptors of the human eye are only sensitive to a smaller portion of this spectrum, known as visible light, which ranges from about 380 nm to 770 nm. Conventional artificial light sources usually do not extend far beyond the wavelength range of visible light and also show deviations in the spectrum of visible light compared to sunlight. Quantitative measurements such as the CRI (color rendering index) or CIE R_a (*Commission internationale de l'éclairage Referenzindex allgemein*) provide information on the match of color rendering between artificial and natural light (sunlight or daylight: $R_a = 100$, maximum value). Therefore, in well-equipped examination and autopsy rooms, artificial light sources with high R_a values and thus a certain color accuracy are to be used for standardized documentation of findings. Since the visual interpretation of findings is an essential task in forensic medicine, such visual color accuracy is important for the correct assessment of findings. Optical cameras are typically used to document the findings; i.e. the camera's photosensors replace the photoreceptors of the human eye. Therefore, these photosensors in combination with filters are designed to mimic the sensitivity of the photoreceptors of the human eye, so that the image produced by the camera accurately reflects what the human eye sees in terms of hues and color intensities. In addition to the correct interpretation, however, the recognition of findings is also an essential task in forensic medicine. For this purpose, the contrast between a finding and its surroundings can be changed and increased. On the one hand, artificial light sources with narrow wavelength ranges can be used to change hues and color intensities, allowing stronger contrasts to be perceived depending on the wavelength range of the light source. On the other hand, filters can be used, e.g. in the form of special glasses that only allow a certain narrow wavelength range to pass through, so that the eye can only perceive light in this wavelength range. Or a combination of both is used. Since cameras are used for documentation, they offer the possibility to be sensitive over a broader frequency spectrum of light than the human eye. By using an alternative light source and an appropriate filter in front of the camera, findings can be highlighted and even latent findings can be made visible. This type of imaging, in which spectroscopic information is acquired in addition to the standard photographic documentation of findings, is referred to as spectral imaging.

In this chapter, the basics of spectral imaging are described, focusing on the physical description of light and its interaction with a medium. A detailed knowledge of light propagation in tissue is essential for being able to correctly interpret spectral images and draw the right forensic conclusions from them. Furthermore, the physical process of fluorescence is described, as in fluorescence photography this process is used to make latent findings visible. In addition to fluorescence photography, full-spectrum photography is described. Finally, it will be explained and illustrated how spectral imaging can be used for detecting forensically relevant findings, biofluids, and material residues.

3.2 WHAT IS LIGHT IN THE PHYSICAL SENSE?

In physics, light is defined as electromagnetic radiation. The range of electromagnetic radiation that can be detected by the human eye is called visible light, which corresponds to a frequency band of $3.9 \cdot 10^{14}$ to $7.9 \cdot 10^{14}$ Hz. However, these boundaries are not clearly defined and can vary individually. The adjacent higher range of frequencies is ultraviolet (UV) light, with a frequency band of $7.9 \cdot 10^{14}$ to $3 \cdot 10^{16}$ Hz, and the adjacent lower range of frequencies is infra-red (IR) light, with a frequency band of $3 \cdot 10^{11}$ to $3.9 \cdot 10^{14}$ Hz. The frequency (υ) is proportional to the speed of light and inversely proportional to the wavelength:

$$\upsilon = \frac{c}{\lambda}$$

where c is the speed of light in vacuum and λ the wavelength.

Thus, electromagnetic waves have shorter wavelengths (10 nm to 380 nm) in the UV range and longer wavelengths (770 nm to 1 mm) in the IR range than in the visible range. Light sources are commonly specified by their peak value and wavelength range (Figure 3.1).

So far we have considered light as a wave, but a special property of light is that it behaves in two ways, as a wave and as a particle (wave–particle duality). When light is considered as a particle, the term "photon" is used. The energy (E) of a photon is proportional to its frequency and inversely proportional to its wavelength:

$$E = h \cdot \upsilon = h \cdot \frac{c}{\lambda}$$

where h is Planck's constant (h = 6.626 10–34 J/Hz) and c the speed of light in vacuum (c = 299 792 458 m/s).

The speed at which light travels through a medium (c_m), however, is in general slower than the speed of light in vacuum. The ratio between c and c_m is determined by the refractive index (n) of the medium:

$$n = \frac{c}{c_m}$$

The refractive index of air is almost 1 (n_{air} = 1.000293), and therefore the speed of light through air (c_{air} = 299 704 645 m/s) is almost equivalent to the speed of light through vacuum. While the speed and the propagation direction of a photon change when refracted into a medium, its energy does not change and therefore neither does its frequency. Consequently, its wavelength also must change when traveling through a medium.

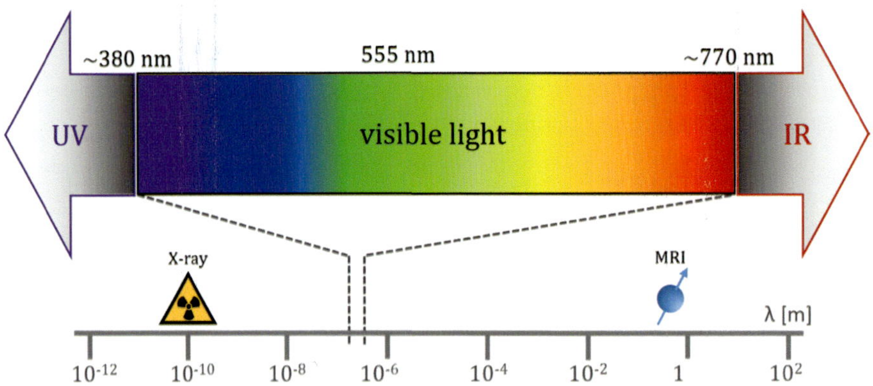

Figure 3.1 The part of electromagnetic radiation in wavelengths. The wavelength range seen by the human eye is represented by the color scale and is referred to as visible light. The exact range seen by the human eye is not defined due to individual differences, but at photopic vision it is approximately between 380 nm (violet) and 770 nm (red) with a maximum at 555 nm (green). According to the wavelengths of light represented in this spectrum, ultraviolet (UV) light is adjacent to that end of the visible light range with shorter wavelengths, and infrared (IR) light is adjacent to that end with longer wavelengths. Note that the wavelengths of X-rays are in the range of 10^{-10} nm or 1 Å, and magnetic resonance imaging, on the other hand, uses radio frequencies in the wavelength range of cm (in tissue) to m (in air) for excitation of the nuclear spins.

3.3 WHAT HAPPENS WHEN LIGHT HITS THE SURFACE OF A MEDIUM?

When light traveling through air strikes the surface of a transparent medium with a higher refractive index, two processes take place, reflection and refraction, which change the direction and radiant flux of the incident light but not its energy and frequency.

How the incident light is reflected at the surface depends on the properties of the material. For instance, if the surface is wet, then specular reflection occurs, which can be described as a mirror-like reflection where the incident angle of the light equals the reflection angle as governed by Snell's law (Figure 3.2). In contrast, Lambert diffusion describes the diffusion of a perfect scatterer leading to a homogenously distributed reflection of the light in all directions independent of the incident angle. The intensity of the reflected light is then proportional to the reflectance of the surface times the cosine of the incident angle. However, in reality, diffuse reflection is mostly observed, which occurs when the surface is rough and with a roughness slightly larger than the wavelength. In this case, more light is still reflected along the reflection angle determined by Snell's law, but fractions of the reflected light are also distributed in other directions.

Assuming that there is no absorption, the fraction of the light that is not reflected at the surface is refracted into the medium, and Snell's law can be used to derive the refraction angle (θ_t):.

$$\theta_t = \arcsin\left(\sin(\theta_i)\cdot\frac{n_1}{n_2}\right)$$

Snell's law only provides information about the angles, not about the amount of light that is reflected or refracted.

To calculate how much light is reflected from the surface and how much is refracted into the medium, the Fresnel equations must be solved. For this, light must be considered as waves taking polarization into account. Sunlight and common artificial light sources emit unpolarized light waves; i.e. their electric field vectors oscillate in all planes perpendicular to the direction of propagation. When the electric field vectors are confined to a single plane, the light is linear polarized. For example, when unpolarized light waves strike a wet surface, it is reflected from that mirror-like surface as polarized light. By using so-called polarizing filters, the reflected polarized light can be blocked and thus reduce the glare caused by a wet surface (Figure 3.3).

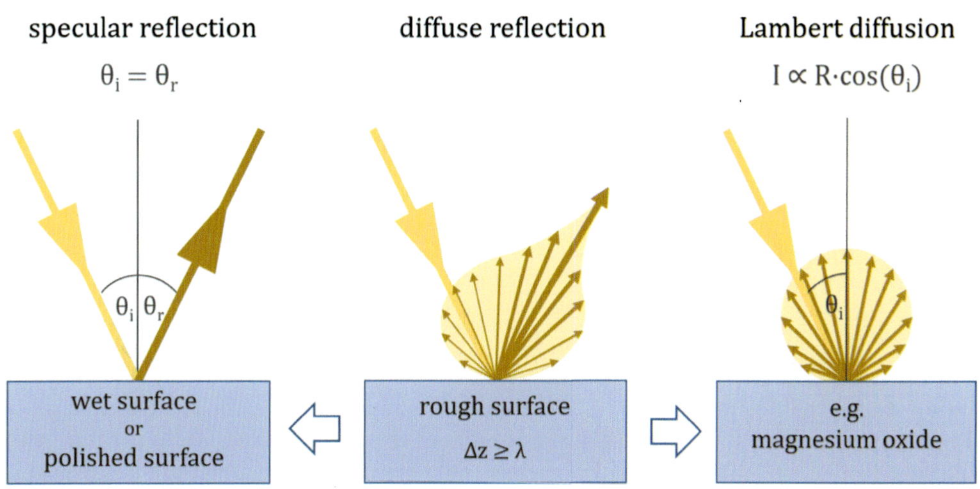

Figure 3.2 Specular reflection can be observed on wet or polished surfaces; i.e. the light is reflected at the same angle (θ_r) as the incident angle (θ_i). In contrast, Lambert diffusion reflects light in all directions, independent of the incident angle. The light intensity (I) is described by the cosine law, where the reflectance (R) contains the material and surface properties. In reality, diffuse reflection is usually observed, which occurs when the surface roughness (Δz) is greater than or equal to the wavelength (λ) of the light.

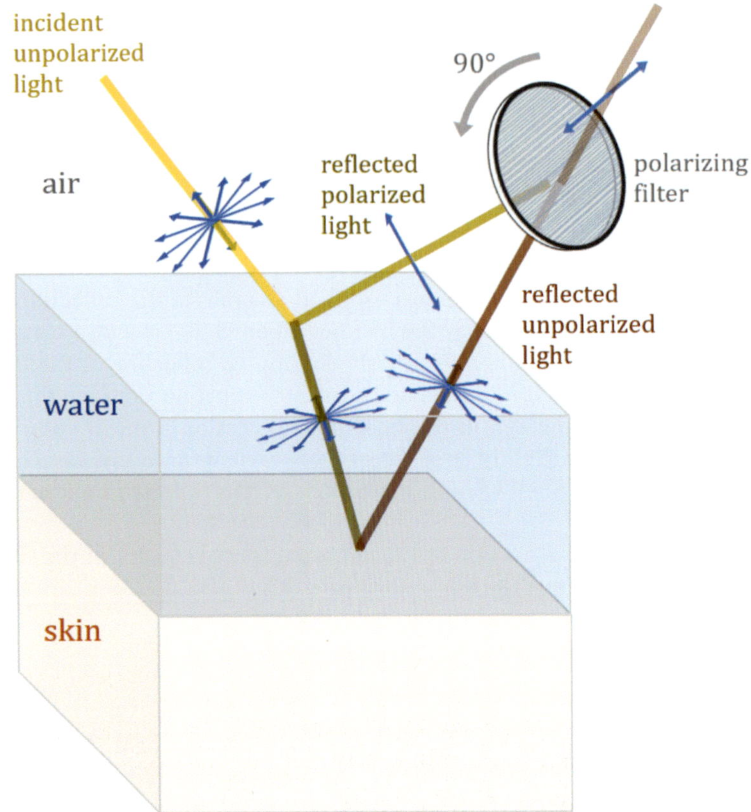

Figure 3.3 Schematic representation of unpolarized light striking a wet surface and being reflected from the surface as polarized light. By aligning a linear polarization filter perpendicular to the polarization of the reflected light, it can be blocked. Light that is refracted off the wet surface and then eventually reflected off a underlaying skin surface remains unpolarized, so some of this unpolarized light can pass through the polarizing filter and can be detected by a camera.

For the Fresnel equations, the two orthogonal polarization states termed s-polarization and p-polarization are important. The polarization is defined as s-polarization when the electrical field is perpendicular to the plane of incidence and is defined as p-polarization when the electric field is parallel to the plane of incidence. By applying the Fresnel equations, the reflection amplitude coefficient (r) and the transmission amplitude coefficient (t) can be calculated for both polarizations:

$$r_s = \frac{n_1 \cdot \cos\theta_i - n_2 \cdot \cos\theta_t}{n_1 \cdot \cos\theta_i + n_2 \cdot \cos\theta_t}$$

$$r_p = \frac{n_2 \cdot \cos\theta_i - n_1 \cdot \cos\theta_t}{n_2 \cdot \cos\theta_i + n_1 \cdot \cos\theta_t}$$

$$t_s = \frac{2 \cdot n_1 \cdot \cos\theta}{n_1 \cdot \cos\theta_i + n_2 \cdot \cos\theta_t}$$

$$t_p = \frac{2 \cdot n_1 \cdot \cos\theta_i}{n_2 \cdot \cos\theta_i + n_1 \cdot \cos\theta_t}$$

where the subscripts s and p denotes s-polarization and p-polarization, respectively.

The reflection amplitude coefficient can also be described as the ratio between the electric field complex amplitude of the reflected wave and the electric field complex amplitude of the incident wave, while the transmission amplitude coefficient can be described as the ratio between the electric field complex amplitude of the transmitted wave and the electric field complex amplitude of the incident wave.

The reflectance (R) describes the fraction of the incident light intensity that is reflected from the surface and can be derived from the Fresnel equations:

$$R_s = |r_s|^2 = \left| \frac{n_1 \cdot \cos\theta_i - n_2 \cdot \cos\theta_t}{n_1 \cdot \cos\theta_i + n_2 \cdot \cos\theta_t} \right|^2$$

$$R_p = |r_p|^2 = \left|\frac{n_1 \cdot \cos\theta_t - n_2 \cdot \cos\theta_i}{n_1 \cdot \cos\theta_t + n_2 \cdot \cos\theta_i}\right|^2$$

If the light is refracted from dielectric material, i.e. an electrically weak or non-conductive material, with a low index of refraction into dielectric material with a higher index of refraction, there is a specific angle, where all the p-polarized light is refracted into the medium, and only s-polarized light is reflected (Figure 3.4). This specific angle is termed Brewster angle (θ_B):

$$\theta_B = \arctan\left(\frac{n_2}{n_1}\right)$$

In case of unpolarized light with an equal amount of s- and p-polarization, the effective reflectance (R_{eff}) is given by:

$$R_{eff} = \frac{1}{2} \cdot (R_s + R_p)$$

Governed by the law of conservation of energy and assuming that there is no absorption at the surface, the transmittance (T), which describes the fraction of the incident light intensity that is refracted into the medium, can be derived by the reflectance (R):

$$T = 1 - R$$

Since the transmittance is given by the ratio between the refracted light intensity and the initial light intensity, and since the reflectance is given by the ratio between the reflected light intensity and the initial light intensity, the light intensity that is refracted into the medium (I_t) can be calculated from radiometer measurements of the initial light intensity (I_i) and the reflected light intensity (I_r):

$$I_t = \left(1 - \frac{I_r}{I_i}\right) \cdot I_i$$

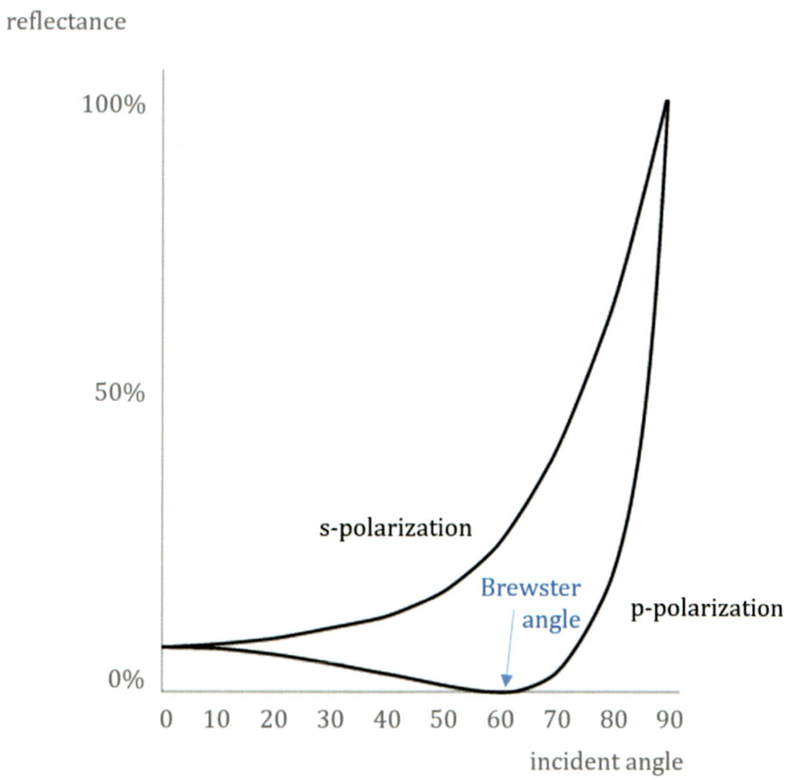

Figure 3.4 The Brewster angle describes the incident angle at which no p-polarized light is reflected but the complete part of the p-polarized light is refracted into the medium. Consequently, the reflected light is polarized, since only s-polarized light is reflected.

3.4 WHAT HAPPENS WHEN LIGHT PROPAGATES INTO TISSUE?

Light that penetrates biological tissue is essentially subject to two processes. On the one hand is absorption by endogenous chromophores. Chromophores are molecules that absorb certain wavelengths of visible light and thus give a material its color. On the other hand is scattering by the microstructures of the tissue. Both processes together lead to an attenuation of light inside the tissue.

In order to describe the process of light absorption in tissue, the light is considered a photon so that wave aspects such as polarization and interference can be ignored. In the following it is assumed that (1) there is no scattering in the isotropic quasi-homogenous medium, (2) the light is monochromatic, (3) the photon paths are parallel, and (4) the light does not change the optical properties of the medium. Under these assumptions, the intensity of the refracted light after having travelled a certain distance (d) through the medium I(d) can be expressed by the absorption coefficient (μ_a) according to the Beer–Lambert law:

$$I(d) = I_0 \cdot e^{-\mu_a \cdot d}$$

where I_t is the initial light intensity determined by the Fresnel equations that is refracted into the medium.

Therefore, the fraction of the initial photons that is left after a traveled distance of d' < d is described by a probability distribution that follows an exponential decay, meaning that with an absorption coefficient of e.g. 1 cm^{-1}, after 1 cm only ~37 % of the initial irradiance is left. Consequently, the absorption coefficient determines the absorption length, which is defined as the distance after which the irradiance is reduced to 1/e (~37 %) of its initial value. The optical penetration depth (δ_a) or the mean free path for absorption is given by μ_a^{-1}.

Looking at the absorption coefficient and its reciprocal, the optical penetration depth, as a function of the wavelength (Figure 3.5) [1], it is obvious that the penetration depth varies depending on the wavelength and the chromophores in tissue, such as water, blood, or collagen. Water, for instance, is a strong absorber in the UV and IR range. The characteristic peaks of the absorption coefficient of water in the IR range are caused by vibrational states of water.

In the range between approximately 650 nm to 1200 nm, light penetrates deepest into tissue (Figure 3.6); hence, this range is termed optical or diagnostic window.

While absorption of light strongly varies as a function of the wavelength (see Figure 3.5), scattering of light in turn decreases monotonically with increasing wavelength. Scattering of light depends on the microstructure of tissue. A good example for the dependency of scattering on the microstructure of the tissue is the outer layer of the eyeball (tunica externa bulbi), consisting of the cornea and the sclera. While the cornea is clearly transparent for the visible light, the sclera is highly scattering and therefore appears white. Although both the cornea and the sclera have a similar content of collagen, the collagen fibrils have a high degree of order and a uniformly narrow diameter in the cornea, while in the sclera the order of the collagen fibrils is irregular and their diameter is larger [2].

Basically, we can distinguish between two types of scattering; elastic scattering (the total kinetic energy is conserved) and inelastic scattering (the total kinetic energy is not conserved). In the following, two forms of elastic scattering are described, Rayleigh scattering, which is responsible for the blue color of the sky, and Mie scattering, which is responsible for the white color of the clouds. To describe these two processes of light scattering in tissue, it is assumed that a collection of discrete scattering particles is randomly distributed in our isotropic quasi-homogenously absorbing medium.

Rayleigh scattering can be described as isotropic scattering that occurs when the particle size is much smaller than the wavelength (λ). The scattering intensity in Rayleigh scattering ($I_{Rayleigh}$) is strongly inversely proportional to the wavelength.

$$I_{Rayleigh} \approx \frac{1}{\lambda^4}$$

In contrast, Mie scattering can be described as mainly forward scattering that occurs when the particle size is larger than the wavelength. There is also some backward scattering when the particle size is just slightly larger than the wavelength. When the particle size is much larger than the wavelength, the scattering angle (θ_s) becomes smaller. The scattering intensity of Mie scattering (I_{Mie}) is inversely proportional to approximately half the wavelength.

$$I_{Mie} \approx \frac{1}{\lambda^{0.5}}$$

In tissue, where the size and shape of particles is not uniform, the spectra of scattering as a function of the wavelength of Rayleigh scattering and Mie scattering overlap. For isotropic Rayleigh scattering, the probability of scattering at an angle θ_s

is derived from the solid angle of an entire sphere. Characterising a single scattering event, the probability density function for scattering a photon in direction $\vec{s}'$ that is traveling in direction $\vec{s}$ is given by:

$$p(\cos\theta_s)_{Rayleigh} = \frac{1}{4\pi}$$

For anisotropic Mie scattering, an anisotropic factor (g), which is defined as the average of scattering angles from a number of scattering events, must be considered (forward scattering: $0 < g \leq 1$; isotropic scattering: $g = 0$; backward scattering: $-1 \leq g < 0$). The probability of scattering at an angle θ_s can be described by the Henyey–Greenstein-phase function:

$$p(\cos\theta_s)_{Mie} = \frac{1}{4\pi}\cdot\frac{1-g^2}{(1+g^2-2g\cdot\cos\theta_s)^{\frac{3}{2}}}$$

In tissue, Mie scattering is mainly forward scattering, and the factor g is usually between 0.7 and ~1, corresponding to a scattering angle θ_s between 8° and 45°. The weighted scattering coefficient (μ_s'), which is actually measured in practice, results from the scattering angle (μ_s) and the factor g as follows:

$$\mu_s' = \mu_s\cdot(1-g)$$

After approximately $(1\text{-}g)^{-1}$ scattering events, the information on the initial orientation of a photon is lost, and we call this "random walk".

The total attenuation coefficient (μ_t) is given by the sum of the absorption coefficient (μ_a) and the scattering coefficient (μ_s), and the total optical penetration depth (δ_t) or mean free path (ℓ), which defines the distance between two interaction processes, or the distance a photon propagates on a straight line before it encounters an absorbing or scattering particle, is then given by μ_t^{-1}.

The albedo (a) provides information about the relation between scattering and absorption (only absorption: a = 0; evenly scattering and absorption: a = 0.5; only scattering: a = 1) and thus determines the importance of scattering versus absorption in a medium. The albedo (a) is given by:

$$a = \frac{\mu_s}{\mu_a + \mu_s}$$

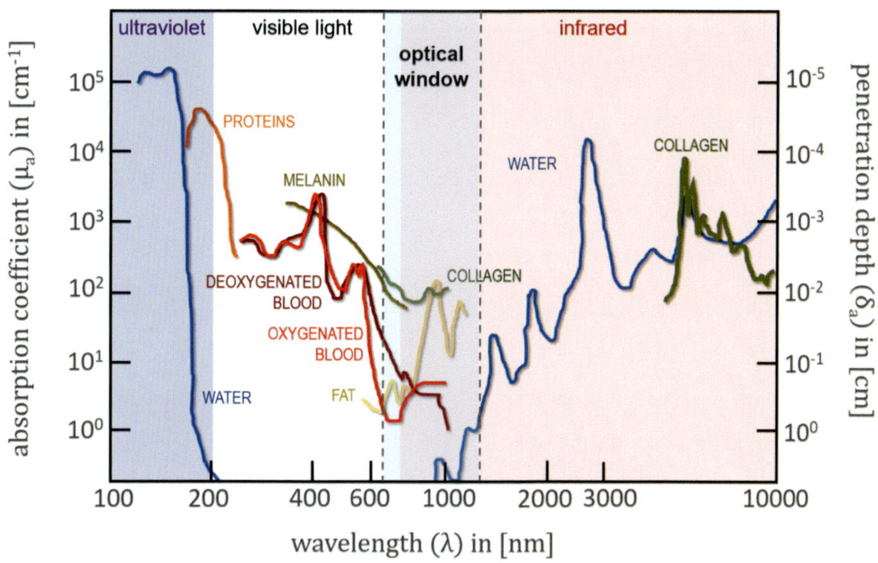

Figure 3.5 The absorption coefficient (μ_a) and the reciprocal optical penetration depth (δ_a) of different chromophores in tissue as a function of the wavelength (λ). The absorption spectra and the diagram were derived from a figure by Scholkmann et al. [1]. The range between 650 nm and 1200 nm is termed diagnostic window, as in this range light penetrates deepest into tissue.

405 nm (±10 nm) 532 nm (±10 nm) 650 nm (±10 nm)

Figure 3.6 Only light with a wavelength within the optical window penetrates deeper into the skin, in this case the fingertip, but it also quickly becomes diffuse (see image at 650 nm).

In biologic tissue, scattering generally dominates over absorption, and light exhibits strongly anisotropic forward scattering. The radiative transfer equation (RTE) is often used to analytically describe the propagation of light in tissue:

$$\frac{\partial \cdot L(\vec{r},\vec{s},t)}{c \cdot \partial t} = -\vec{\Omega} \cdot \nabla \cdot L(\vec{r},\vec{s},t) - \mu_a \cdot L(\vec{r},\vec{s},t) - \mu_s \cdot L(\vec{r},\vec{s},t) + \mu_s \int_{4\pi} p(\vec{s} \cdot \vec{s}') \cdot L(\vec{r},\vec{s}',t) d\omega + Q(\vec{r},\vec{s},t)$$

where c is the speed of light, $L(\vec{r}, \vec{s},t)$ is the energy radiance as a function of the position $\vec{r}$, the angular direction $\vec{s}$, and the time t. The scattering phase function $p(\vec{s} \cdot \vec{s}')$ describes the probability that a photon with the direction $\vec{s}'$ will be scattered in the direction $\vec{s}$ during a scattering event. The RTE shows that the rate of change of the radiance in direction $\vec{s}$ (taking into account the zenith angle θ and the azimuth angle ϕ) in the volume equals the loss of radiance due to photons that leave the volume, which is related to Gauss's theorem (first term), the loss of radiance due to absorption (second term), the loss of radiance due to scattering in another direction (third term), the gain of radiance due to scattering from any direction into the direction of interest (fourth term), and the gain of radiance due to a source term inside the volume (fifth term), such as fluorescence, which is described in the following.

But before that it should be mentioned that the analytical way for calculating light propagation in tissue often is too complex; therefore, stochastic modeling, i.e. Monte Carlo simulations, is applied to simulate light distribution in tissue.

3.5 WHAT IS FLUORESCENCE?

On the skin surface there may be fluid residues or foreign substances and material residues which are not visible to the naked eye, i.e. are latent. Some of these fluids, substances, and materials can be made visible by exciting them with a light source in the shorter wavelength range, e.g. UV or blue light, causing them to glow in the visible light range. This process is called fluorescence.

Fluorescence takes place when a substance after absorbing electromagnetic radiation subsequently emits light at a lower energy level. Fluorescence is typically a three-stage process: (1) The absorption of a photon from an external light source creates an excited electronic singlet state S_n. (2) Energy from this S_n state can dissipate, resulting in a lower energetic singlet state S_1 from which fluorescence emission originates. (3) The excited state S_1 returns to the ground state S_0 while emitting a fluorescence photon. Due to the energy dissipation within the S_n states, the energy of the fluorescence photon is lower, i.e. of longer wavelength than the energy of the excitation photon (Figure 3.7). Alternatively, with a much lower probability, intersystem crossing can yield a triplet state T_1' which again dissipates energy relaxing to T_1. Since the transition to the ground state S_0 involves a "forbidden transition" making a spin flip necessary, it happens at a significantly slower time scale. This process is called phosphorescence.

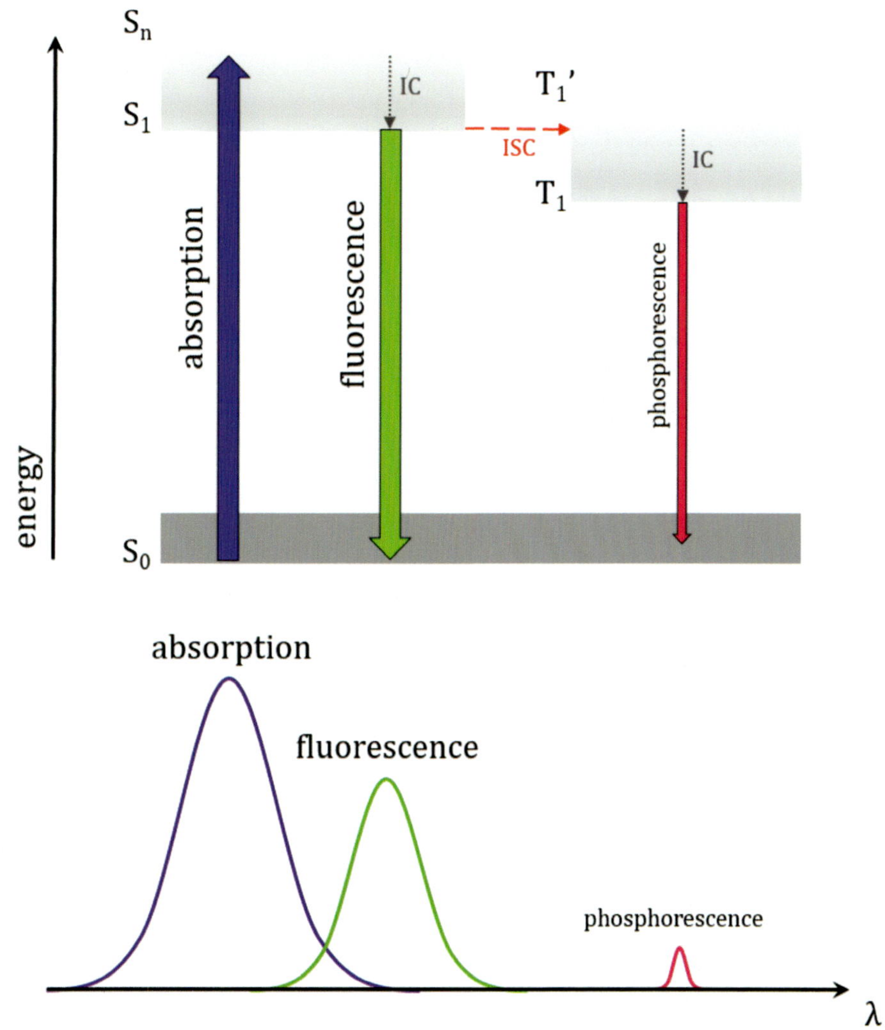

Figure 3.7 The Jablonski diagram (upper graph) illustrates the excitation of a molecule, atom, or nanostructure from the ground (S_0) to a higher energy state (S_1 to S_n) due to the absorption of a photon. By internal conversion (IC), the molecule, atom, or nanostructure falls into the lower excited state (S_1) from which it relaxes into the ground state by emitting a photon. This process is called fluorescence. With a much lower probability, the excited molecule, atom, or nanostructure turns into a triplet state (T_1 to T_n) by so-called inter-system crossing (ISC), and from the lower triplet state (T_1) it relaxes to the ground state by emitting a photon. The fluorescence spectrum can overlap with the absorption spectrum, while, in contrast, the phosphorescence spectrum can be clearly distinguished from the absorption spectrum but has a much lower intensity (lower graph).

The photon emitted during fluorescence or phosphorescence has a lower energy and thus a longer wavelength than the initially absorbed photon; therefore, the light emission of some substances, which were excited by UV radiation, can be seen in the visible spectrum. Fluorescence and phosphorescence are inelastic scattering of light, and both are forms of photoluminescence. The emitted photons have different wavelengths, and fluorescence and phosphorescence differ on the basis of the duration of the spontaneous emission of the light after the external excitating light source has been switched off. In the case of fluorescence, spontaneous emission is basically instantaneous and thus detectable only as long as the light source is switched on. In contrast, in the case of phosphorescence, the substance is able to store absorbed light energy in states T_n for up to a few milliseconds to seconds, so that spontaneous emission of light is still present even after the excitation light source is switched off.

3.6 WHAT ARE HYPERSPECTRAL, MULTISPECTRAL, AND FLUORESCENCE IMAGING?

Spectral imaging can be divided into hyperspectral imaging and multispectral imaging. Hyperspectral imaging describes the acquisition and processing of an image at many different wavelengths. Although hyperspectral imaging as an emerging imaging modality has become increasingly popular in the military, industrial, scientific, and medical fields over the past decade, it is rarely applied in forensics to investigate deceased persons. This may be due to the fact that industrial cameras are often used for this purpose, which are often not easy to handle and therefore less suitable for field work in a crime scene investigation or the autopsy room. Since hyperspectral cameras become more and more consumer friendly in shape and comfort, it can be assumed that hyperspectral imaging will also find its way into postmortem imaging of the deceased in the future. One example can be found by Edelman and Aalders [3], who used a hyperspectral camera for photogrammetric documentation of a mock crime scene. Far more common than hyperspectral imaging in forensic medicine is multispectral imaging. While hyperspectral imaging splits the image into dozens or hundreds of colors, multispectral imaging evaluates an image in only three or four colors (e.g., red, green, blue, and near-infrared [NIR]). Multispectral thus means that images are acquired within several specially selected narrow wavelength ranges. In forensic body surface documentation, either cameras equipped with special multispectral sensors or modified DSLR cameras with exchangeable filters in front of the camera lens are used. Silicium-based image sensors of ordinary digital cameras are sensitive in a wavelength range of about 350 to 1100 nm, which, however, is reduced by an IR cut-off filter placed in front of the sensor such that the spectral range of the camera corresponds to the spectral sensitivity of a human eye. To use such cameras for multispectral images, this filter must be removed. By removing this IR cut-off filter and replacing it with a special protecting filter glass, the sensitivity range of the camera sensor can be used for spectral imaging in the near IR (NIR) or more precisely IR-A range (approximately 750 to 1400 nm) and in the UV or more precisely UV-A range (approximately 315 to 400 nm).

Such modified cameras are referred to as a full-spectrum camera, and thus additionally allow so-called UV and NIR reflectance photography or in one word full-spectrum photography. To take ordinary pictures with such full-spectrum cameras, an IR cut-off filter is usually placed in front of the lens. These filters are also called hot mirror filters, and such a filter (IR-Neutralisationsfilter NG, Optic Makario GmbH) was used in combination with the installed ceiling light (Lumilux HO 49W/840 cool white, Osram) with a CIE R_a value of 80–89 for the corresponding images presented in this chapter, which are referred to as conventional images. Whenever a different artificial light source was used, the corresponding wavelength of the intensity maximum and full width at half maximum (FWHM) of the intensity peak is given. All images were taken with a modified DSLR camera (EOS 200D, Canon Inc.).

In UV and NIR reflectance photography, two key parameters can be varied, the wavelength range of the illuminating light (the source) and the wavelength range of the camera (the detector). Filters can be used to restrict the wavelength range of the illuminating light and/or the camera. When illuminating the skin with a light source in the UV range, superficial skin structures can be displayed due to the low penetration depth of the light in this wavelength range. In contrast, illumination with a light source in the NIR range allows deeper penetration into the skin (remember the optical window described in the previous chapter), which highlights blood vessels and hemorrhages in the skin layers due to the high absorption of blood compared to the surrounding tissue (Figure 3.8).

The same effect can be achieved when restricting the spectral range of the camera and using a broadband light source instead. Artificial ceiling light can already be sufficient for IR reflectance photography, while its UV content is usually too low. Filters are not only used to take images in the UV and NIR range, but the wavelength range of visible light can also be divided into narrower ranges with so-called color filters. With the help of suitable software, postprocessing can further enhance the effect of the individual filters used. However, in such cases, the postprocessing must also be clearly documented and should always be presented together with the basic image (Figure 3.9).

Color filters are also used for fluorescence imaging (Figure 3.10). Fluorescence imaging refers to the process of capturing the glow or visible light emitted by certain substances when they are irradiated with UV or blue light.

Color filters in the longer wavelength range that block the excitation light but allow fluorescence to pass make fluorescent substances visible (Figure 3.11). Knowing the specific absorption and emission spectra of different body fluids such as semen, serum, urine, or saliva and their change in time, fluorescence spectroscopy is an appropriate method for body fluid identification in forensics [4].

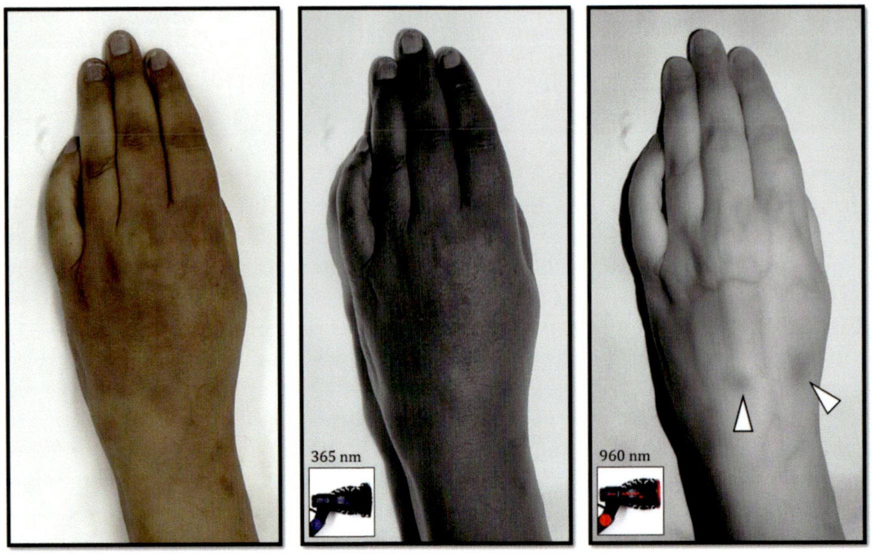

Figure 3.8 The conventional image on the left shows the hand of a deceased person in visible light range with the ceiling light (Lumilux HO 49W/840 cool white, Osram) switched on, i.e. as we would also perceive it with our eyes. The middle image was taken using a UV light source (Fluoreszilla BI, Dedolight, peak: 365 nm, FWHM: 15.4 nm) and the ceiling light switched off. The right image was taken using an IR light source (iREDZILLA BI, Dedolight, peak: 960 nm, FWHM: 71.6 nm) and the ceiling light switched off. The right image illustrates the higher penetration depth of IR light into the skin compared to the low penetration depth of UV light in the left image. Furthermore, the right image provides a good contrast between the tissue and the strong absorber blood, which makes blood vessels and hematomas (arrowheads) more visible compared to standard images in the visible light range.

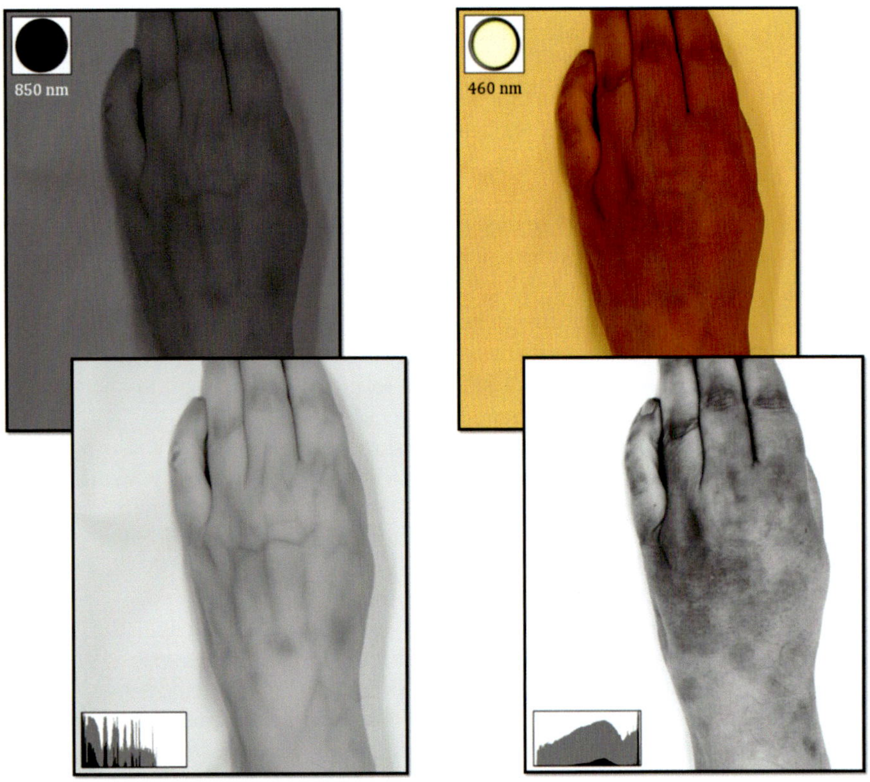

Figure 3.9 For the left images, a filter with a transmission peak at 850 nm was used, and for the right images, a filter with a transmission peak at 460 nm was used in combination with the ceiling light (Lumilux HO 49W/840 cool white, Osram). The lower images have been optimized in terms of brightness and contrast by adjusting the slope of the histogram-based transfer function using dedicated software (Fiji/ImageJ). While the 850 nm filter allows blood vessels and hematomas to be visualized in good contrast to the tissue, the 460 nm filter (yellow filter) can be used to highlight superficial skin lesions.

Figure 3.10 Three different color filters for the lens of a digital camera. The yellow filter on the left has its transmission peak at 460 nm, the orange filter in the middle at 550 nm, and the red filter on the right at 580 nm. When irradiated with a light source with shorter wavelengths, fluorescence emission can appear in the range of visible light and be highlighted with these filters.

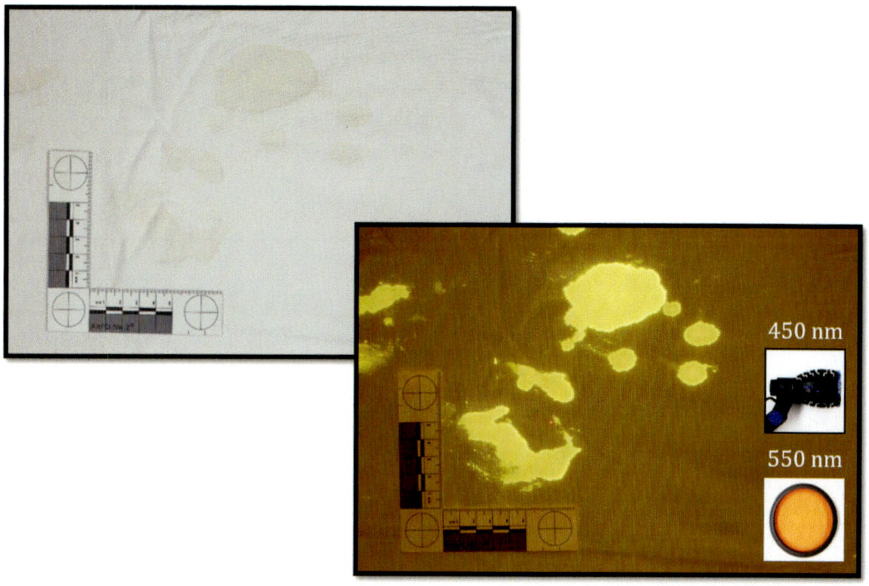

Figure 3.11 The images show sperm stains on white textile. On the conventional image, the sperm stains are barely visible. In contrast, in the fluorescent image taken with a blue light source (LEDZILLA, Dedolight BI, Dedolight, peak: 450 nm, FWHM: 22.8 nm) and a filter with a transmission peak at 550 nm, the fluorescent sperm is clearly visible.

3.7 WHAT CAN SPECTRAL IMAGING BE USED FOR IN FORENSIC MEDICINE?

In forensic medicine, the use of NIR imaging and fluorescence imaging for the detection of latent findings or traces on deceased persons or textiles is mainly described [5–17]. The authors reported the usefulness of these spectral imaging techniques for detection and documentation of latent bruises in children and adults and have specifically noted the use of spectral imaging for visualization and analysis of bite wounds and bruises associated with strangulation (Figure 3.12).

Depending on the material and dye, some textiles that are dark under visual light appear bright under IR light, so residues that are also dark under visible light, such as blood stains, become visible under IR light because blood is a strong IR light absorber (Figure 3.13).

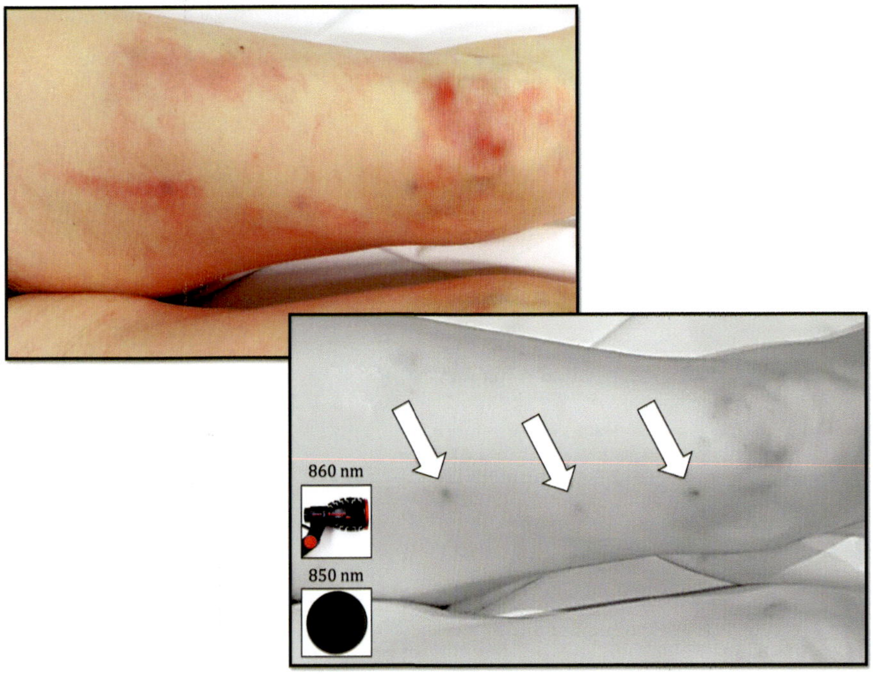

Figure 3.12 Large areas of reddish skin discoloration are visible on the conventional image with the ceiling light (Lumilux HO 49W/840 cool white, Osram). However, the NIR image shows only small dark regions (arrows) corresponding to small hematomas because of the high contrast between blood and tissue when irradiated with the IR light source (iREDZILLA BI, Dedolight, peak: 860 nm, FWHM: 71.6 nm) and using a filter with a transmission peak at 850 nm.

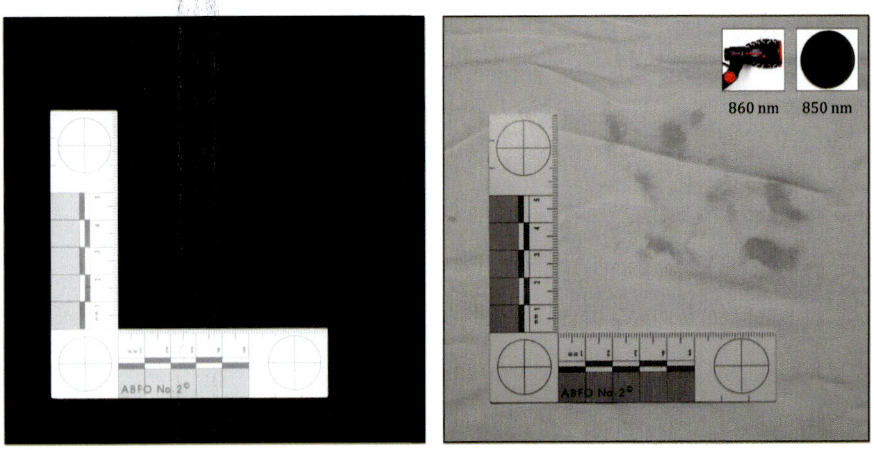

Figure 3.13 The bloodstains on the conventional image of a black T-shirt taken with the ceiling light (Lumilux HO 49W/840 cool white, Osram) are latent. In contrast, the NIR image allows clear detection of the blood stains due to the high contrast between dark blood stains and the bright textile when irradiated with the IR light source (iREDZILLA BI, Dedolight, peak: 860 nm, FWHM: 71.6 nm) and using the filter with a transmission peak at 850 nm.

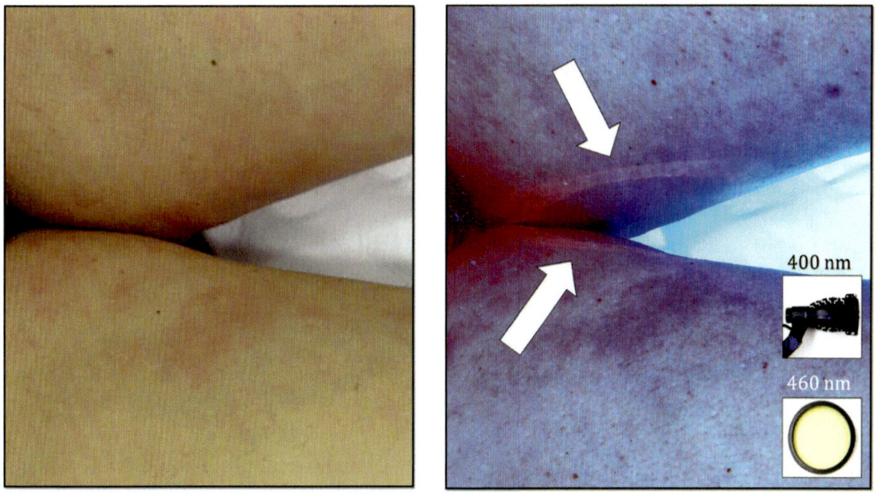

Figure 3.14 In contrast to the conventional image with the ceiling light (Lumilux HO 49W/840 cool white, Osram) on the left, the fluorescence image on the right, which was taken while irradiated with the UV light source (Fluoreszilla BI, Dedolight, peak: 365 nm, FWHM: 15.4 nm) and using a filter with a transmission peak at 460 nm, shows traces of a presumptive biofluid (white arrows).

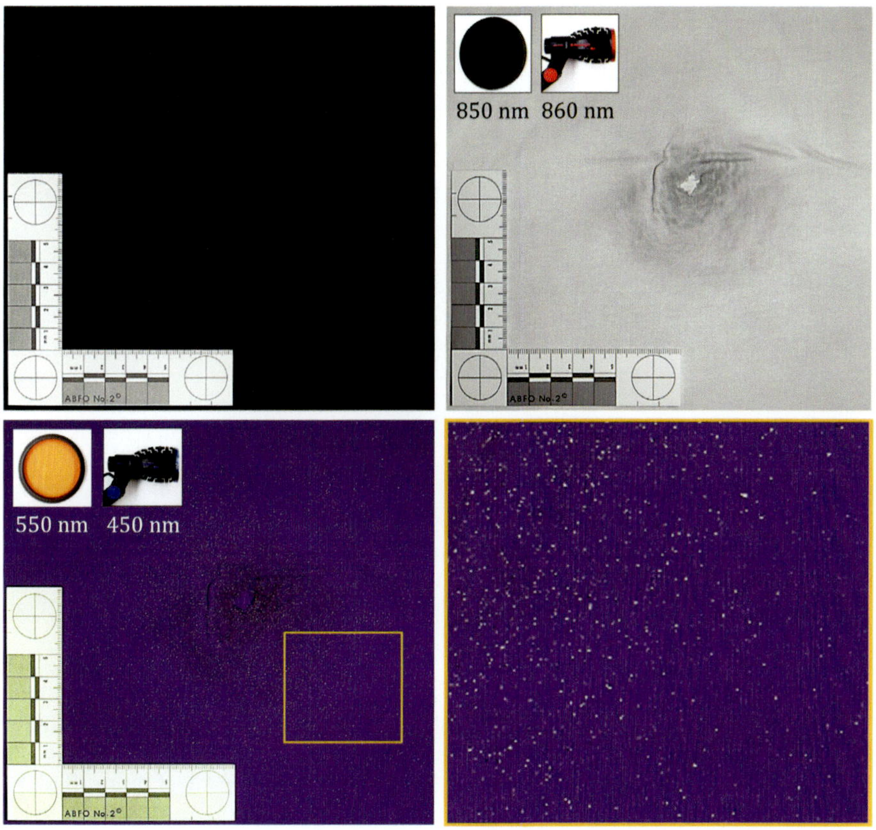

Figure 3.15 The images show a black T-shirt with a perforation defect caused by a 9 mm Quick Defense (QD_1) bullet fired with a Sig Sauer P226 from a muzzle to target distance of 2 cm. The upper left image shows the conventional image which was taken with the ceiling light (Lumilux HO 49W/840 cool white, Osram). While the defect is visible in the T-shirt, gunshot residues are not. The upper right image shows the NIR image when irradiated with the IR light source (iREDZILLA BI, Dedolight, peak: 860 nm, FWHM: 71.6 nm) and using the filter with a transmission peak at 850 nm. In this image, the black fabric of the T-shirt is shown in light gray and the soot is shown in dark, which creates a good contrast for detecting and documenting the distribution of the soot according to the shooting distance. The lower right image shows the same T-shirt under irradiation with a blue light source (LEDZILLA, Dedolight BI, Dedolight, peak: 450 nm, FWHM: 22.8 nm) and the filter with a transmission peak at 550 nm. On closer inspection in the zoomed-in image section shown in the lower right image, fine gunpowder particles are clearly visible through their fluorescence, which are not visible in the conventional image and in the NIR image.

Fluorescence imaging has been described for the detection and visualization of biofluid residues on the skin surface or on textiles [15,16,18–23]. UV- or blue-light–induced fluorescence imaging is used to highlight traces of biofluid residues (Figure 3.14). Samples can then be taken from the appropriate locations on the body surface for more detailed analyses to determine the fluid.

Other articles describe the use of spectral imaging to visualize tattoos on decomposed tissue for the identification of the decedent [12,24,25]. Discoloration and darkening of the skin in the course of decomposition causes tattoos to be barely recognizable with the naked eye. NIR images, in turn, provide an improved contrast between the tattoo ink, which strongly absorbs IR light, and the decomposed tissue, where the light is diffuse, making a tattoo clearly visible. In so-called cover-up tattoos, where colored tattoo ink is used to a cover tattoo with black ink, NIR images can also help to better identify the covered tattoo [26,27].

Furthermore, the use of spectral imaging for the detection of gunshot residues on textiles was demonstrated [28]. In addition to NIR images, which are often used to detect traces of gunshot residue, blue-light–induced fluorescence photography allows one to highlight fluorescent particles of gunshot residues when using a 550 nm filter (Figure 3.15).

In order to implement spectral imaging in routine practice, automated acquisition of images and a standardized approach with regard to the filters and light source combinations to be used are recommended. By using multi-camera systems, the entire body surface can be quickly documented multispectrally and latent traces can be detected. Furthermore, photogrammetric data allows one to calculate three-dimensional whole-body models [15,16]. The chapter on the VirtoScan project describes this approach of multispectral three-dimensional acquisition of the entire body.

References

[1] F. Scholkmann, S. Kleiser, A.J. Metz, R. Zimmermann, J. Mata Pavia, U. Wolf, M. Wolf, A review on continuous wave functional near-infrared spectroscopy and imaging instrumentation and methodology, NeuroImage. 85 (2014) 6–27. https://doi.org/10.1016/j.neuroimage.2013.05.004.

[2] K.M. Meek, N.J. Fullwood, Corneal and scleral collagens: A microscopist's perspective, Micron. 32 (2001) 261–272. https://doi.org/10.1016/S0968-4328(00)00041-X.

[3] G.J. Edelman, M.C. Aalders, Photogrammetry using visible, infrared, hyperspectral and thermal imaging of crime scenes, Forensic Science International. 292 (2018) 181–189. https://doi.org/10.1016/j.forsciint.2018.09.025.

[4] N. Achetib, K. Falkena, M. Swayambhu, M.C.G. Aalders, A. van Dam, Specific fluorescent signatures for body fluid identification using fluorescence spectroscopy, Scientific Reports. 13 (2023) 3195. https://doi.org/10.1038/s41598-023-30241-7.

[5] E. Vogeley, M.C. Pierce, G. Bertocci, Experience with wood lamp illumination and digital photography in the documentation of bruises on human skin, Archives of Pediatrics & Adolescent Medicine. 156 (2002) 265–268.

[6] F.D. Wright, G.S. Golden, The use of full spectrum digital photography for evidence collection and preservation in cases involving forensic odontology, Forensic Science International. 201 (2010) 59–67.

[7] P. Rowan, M. Hill, G.A. Gresham, E. Goodall, T. Moore, The use of infrared aided photography in identification of sites of bruises after evidence of the bruise is absent to the naked eye, Journal of Forensic and Legal Medicine. 17 (2010) 293–297.

[8] G.S. Golden, Standards and practices for bite mark photography, The Journal of Forensic Odonto-Stomatology. 29 (2011) 29.

[9] D.S. Holbrook, M.C. Jackson, Use of an alternative light source to assess strangulation victims, Journal of Forensic Nursing. 9 (2013) 140–145. https://doi.org/10.1097/JFN.0b013e31829beb1e.

[10] R.M. Limmen, M. Ceelen, U.J. Reijnders, S. Joris Stomp, K.C. de Keijzer, K. Das, Enhancing the visibility of injuries with narrow-banded beams of light within the visible light spectrum, Journal of Forensic Sciences. 58 (2013) 518–522.

[11] J. Glauche, B. Ondruschka, V. Wenzel, J. Dreßler, C. Hädrich, Examination of invisible injuries, Rechtsmedizin. 25 (2015) 543–547.

[12] T. Rost, N. Kalberer, E. Scheurer, A user-friendly technical set-up for infrared photography of forensic findings, Forensic Science International. 278 (2017) 148–155.

[13] S. Mimasaka, T. Oshima, M. Ohtani, Visualization of old bruises in children: Use of violet light to record long-term bruises, Forensic Science International. 282 (2018) 74–78.

[14] L. Trefan, C. Harris, S. Evans, D. Nuttall, S. Maguire, A.M. Kemp, A comparison of four different imaging modalities–Conventional, cross polarized, infra-red and ultra-violet in the assessment of childhood bruising, Journal of Forensic and Legal Medicine. 59 (2018) 30–35.

[15] S. Kottner, M.M. Schulz, F. Berger, M. Thali, D. Gascho, Beyond the visible spectrum—applying 3D multispectral full-body imaging to the VirtoScan system, Forensic Science, Medicine and Pathology. 17 (2021) 565–576. https://doi.org/10.1007/s12024-021-00420-x.

[16] S. Kottner, M.M. Schulz, F. Berger, M.J. Thali, D. Gascho, Multispectral 3D whole-body imaging of dressed and undressed bodies in combination with post-mortem X-Ray computed tomography, in: 2021. https://doi.org/10.15221/21.24.

[17] A. Dubey, R. Rupani, V. Sharma, R.K. Singh, S. Kumari, A.K. Verma, Reflected near-infrared photography: Digging deeper into postmortem examination, Journal of Forensic and Legal Medicine. 90 (2022) 102397. https://doi.org/10.1016/j.jflm.2022.102397.

[18] T. Gabby, M.A. Winkleby, W.T. Boyce, D.L. Fisher, A. Lancaster, G.F. Sensabaugh, Sexual abuse of children: The detection of semen on skin, American Journal of Diseases of Children. 146 (1992) 700–703.
[19] K.A. Santucci, D.G. Nelson, K.K. McQuillen, S.J. Duffy, J.G. Linakis, Wood's lamp utility in the identification of semen, Pediatrics. 104 (1999) 1342–1344.
[20] D.G. Nelson, K.A. Santucci, An alternate light source to detect semen, Academic Emergency Medicine. 9 (2002) 1045–1048.
[21] C. Carter-Snell, K. Soltys, Forensic ultraviolet lights in clinical practice: Evidence for the evidence, The Canadian Journal of Police & Security Services. 3 (2005) 79–85.
[22] J. Wawryk, M. Odell, Fluorescent identification of biological and other stains on skin by the use of alternative light sources, Journal of Clinical Forensic Medicine. 12 (2005) 296–301.
[23] C.A. Lincoln, P.M. McBride, G.R. Turbett, C.D. Garbin, E.J. MacDonald, The use of an alternative light source to detect semen in clinical forensic medical practice, Journal of Clinical Forensic Medicine. 13 (2006) 215–218.
[24] M.D. Cain, D. Roper, D.S. Atherton, Use of infrared photography to visualize a tattoo for identification in advanced decomposition, Academic Forensic Pathology. 6 (2016) 338–342.
[25] M. Cullip, V.-C. Tran, C.G. Ball, Tattoo visualization using cross-polarized lighting and infrared photography, Forensic Science, Medicine and Pathology. 17 (2021) 350–353.
[26] D. Bryson, J. Wright, K. Barker, The identification of tattoo designs under cover-up tattoos using digital infrared photography, Journal of Visual Communication in Medicine. 36 (2013) 104–110. https://doi.org/10.3109/17453054.2013.851654.
[27] F. Holz, C.G. Birngruber, F. Ramsthaler, M.A. Verhoff, Beneath cover-up tattoos: Possibilities and limitations of various photographic techniques, International Journal of Legal Medicine. 134 (2020) 697–701. https://doi.org/10.1007/s00414-019-02007-2.
[28] V. Barrera, B. Fliss, S. Panzer, S.A. Bolliger, Gunshot residue on dark materials: A comparison between infrared photography and the use of an alternative light source, International Journal of Legal Medicine. 133 (2019) 1115–1120.

CHAPTER 4

COMPUTED TOMOGRAPHY—BASICS AND FORENSIC POSTMORTEM APPLICATIONS

Dominic Gascho and Patrik Wili

DOI: 10.4324/9781003413998-5

4.1 INTRODUCTION

In 1985, Wilhelm Conrad Röntgen discovered "a new kind of ray" [1], known as "Röntgenstrahlung" or X-rays. Immediately thereafter, the first X-ray image was presented as evidence in court, as the newly discovered imaging technique made it possible to locate a projectile in a person's lower leg after attempts to locate it with a probe failed [2]. About 70 years later, the first medical X-ray computed tomography (CT) scanner was developed based on the work of Allan McLeod Cormack et al. [3,4] and eventually based on the work of Godfrey Newbold Hounsfield et al. [5–7], so that the first medical CT scanners made their way into clinical medicine starting in the late 1970s. As early as the early 1980s, the first postmortem CT examinations were performed, such as examinations of cerebral gunshot wounds or drowning deaths [8,9]. However, postmortem CT examinations for forensic purposes were performed only sporadically in the 1980s and 1990s, until the presentation of the concept idea called Virtopsy at the beginning of the 2000s [10,11]. Over the past 20 years, the use of postmortem CT has increased worldwide, CT and its diagnostic value has become increasingly important in forensic medicine. The strengths of CT lie primarily in the three-dimensional imaging of the bony skeleton and bone fractures, as well as the localization of radiopaque foreign bodies, such as projectiles. In addition, this imaging technique is suitable for the visualization of various findings in the lungs, as well as blood accumulations in body cavities, calcifications of the coronary vessels, or major cerebral hemorrhages and major injuries of the soft tissues. The speed of data acquisition over larger regions of the body up to whole-body scans and the reconstruction possibilities for three-dimensional visualizations of forensically relevant findings especially contribute to the popularity of postmortem CT in forensic medicine. In addition, the submillimeter-thick image slices can be saved and evaluated again later at any time, for example, with the inclusion of another expert opinion. Regarding three-dimensional visualization possibilities and data storage options, the reader is referred to the corresponding chapters in this book.

The present chapter briefly covers the physical and technical principles of X-ray imaging, with emphasis on CT, starting with the generation of X-rays in an X-ray tube, their interactions with tissue, and the detection of X-rays after transmission. Furthermore, it describes CT the encoding and decoding processes of image data obtained from rotating X-ray tubes as used in CT and the characteristics of the reconstructed CT image. The final part of this chapter focuses on postmortem CT scanning in practice and highlights the relevant reconstruction parameters for high-resolution postmortem CT.

4.2 HOW DO WE GENERATE X-RAYS?

X-rays are generated within a vacuum tube containing a cathode, typically a tungsten filament, and an anode, typically a frustoconical disc made of tungsten, at the opposite side. Heating the cathode to about 2200°C results in thermionic emission, i.e. the liberation of electrons from the hot tungsten filament. Now, when a voltage is applied between the cathode (negatively charged) and the anode (positively charged), the liberated electrons are accelerated towards the anode and their potential energy is converted into kinetic energy. The potential energy (E_{pot}) in Joule (J) is given by the applied tube voltage (U_{tube}), which is usually given in kilovolts (kV), and the charge of the electron (Q), which is given in coulombs (C):

$$E_{pot} = U_{tube} \cdot Q$$

The charge of an electron is $1.602 \cdot 10^{-19}$ C. Thus, at a tube voltage of 100 kV, the potential energy for the electron is $1.602 \cdot 10^{-14}$ J. Since this amount of energy is very small, energy in X-ray imaging is usually expressed in units of kilo-electron-volts (keV), where e indicates the charge of the electron ($e \equiv Q$), and thus the potential energy of an electron at a tube voltage of 100 kV is 100 keV.

The conversion into kinetic energy results in $E_{kin} = E_{pot}$, and thus:

$$\frac{1}{2} \cdot m \cdot v = Q \cdot U_{tube}$$

where m is the mass of an electron ($9.1 \cdot 10^{-31}$ kg) and v is its velocity. By transforming the equation, the velocity of the electron can be determined:

$$v = \sqrt{\frac{2 \cdot Q \cdot U_{tube}}{m}}$$

Consequently, with a tube voltage of 100 kV, the electron velocity is about $1.9 \cdot 10^{8}$ m/s (according to Newton). For comparison, the speed of light is $3 \cdot 10^{8}$ m/s.

A negatively charged focusing cup surrounds the cathode to produce a tight beam of accelerated electrons. When the accelerated electrons interact with the atoms of the anode material, X-rays are generated. Finally, the X-rays are emitted through a small glass window at the bottom side of the X-ray tube (Figure 4.1).

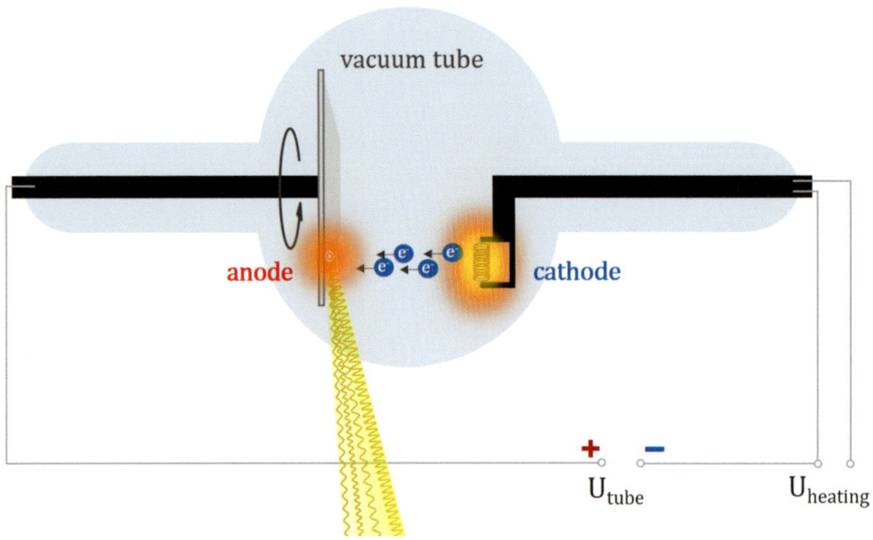

Figure 4.1 Schematic illustration of an x-ray tube. When the filament at the cathode reaches a sufficiently high temperature and a voltage (U_{tube}) is applied between the cathode (negatively charged) and anode (positively charged), electrons are accelerated from the cathode towards the anode. At the point where electrons interacting with the atoms of the anode, X-rays are generated. The X-ray photons generated in the anode material have different intensities (illustrated by the different amplitudes of the waves) and different energies (illustrated by the different wavelengths). Accordingly, common X-ray tubes generate a so-called polychromatic X-ray beam.

When a fast electron is slowed and deflected from its original trajectory as it passes close to a positively charged nucleus of the anode target material, it loses kinetic energy, and this kinetic energy is converted into an X-ray photon. In physics, an X-ray photon is defined as an elementary particle (quantum) of electromagnetic radiation with properties of both waves and particles. It is (theoretically) massless and in vacuum it moves at the speed of light. The wavelength of X-rays is in the range of angstrom i.e. 10^{-10} m (approximate range: 10^{-12} m to 10^{-9} m). The difference between the energy of the electron before and after the interaction with the anode material corresponds to the X-ray photon energy. When the total kinetic energy of an incident electron is converted into an X-ray photon, the X-ray photon reaches the maximum energy. Consequently, at a tube voltage of, for example, 100 kV, at which the kinetic energy of the electrons is 100 keV, the maximum energy of the photons is also 100 keV. Thus, the maximum X-ray energy depends on the applied tube voltage.

Rather than converting the entire kinetic energy of an electron to an X-ray photon, however, it is much more likely that an electron will undergo only a partial loss of energy, and thus a wide spectrum of X-ray energies is produced, which is referred to as Bremsstrahlung. The photon flux, the number of photons per area per time ($cm^{-2}s^{-1}$), increases almost linearly towards lower energies, but photons with very low energies are already absorbed by the anode material or deliberately absorbed with additional aluminum filters and thus do not contribute to the applied X-ray spectrum (Figure 4.2).

In addition to the Bremsstrahlung, the so-called characteristic radiation is generated, in which an electron collides with a tightly bound K-shell electron of the anode material. The bound electron is then ejected and replaced by an electron from an outer L- or M-shell. The difference in binding energies between the inner and outer atomic shells is finally emitted as discrete energy characteristic of the atomic number of the anode material. In contrast to Bremsstrahlung, characteristic radiation is not dependent on the anode voltage. As the name indicates, this type of radiation is characteristic for the anode material.

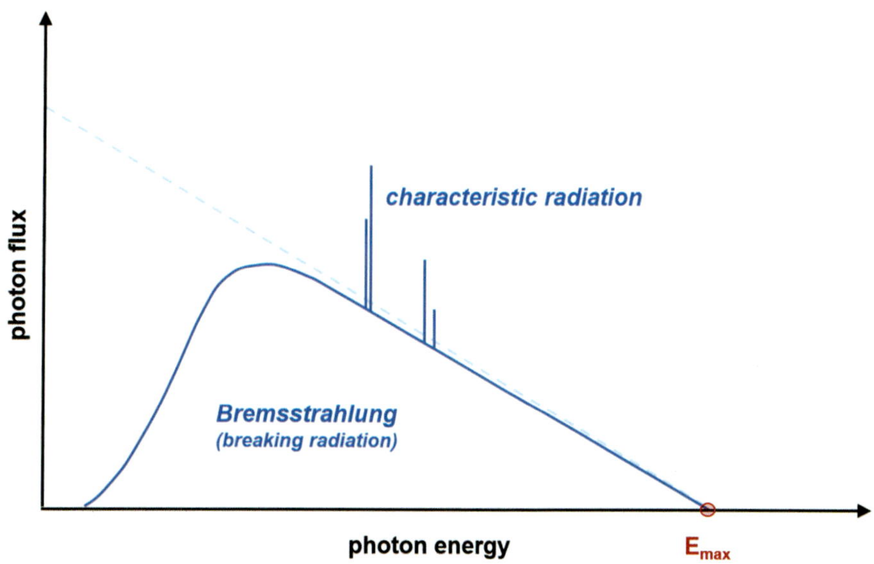

Figure 4.2 The photon flux plotted as a function of photon energy illustrates the X-ray spectrum (Bremsstrahlung) of a tungsten target with its characteristic peaks (characteristic radiation). The photon flux increases almost linearly with the decrease in photon energy. However, photons with very low energies are already absorbed by the anode material itself or deliberately absorbed with additional aluminum filters. The maximum photon energy (E_{max}) depends on the applied anode voltage. For example, with an applied anode voltage of 120 kV, a maximum energy of 120 keV is achieved.

4.3 WHAT DO THE SETTINGS OF THE TUBE VOLTAGE AND TUBE CURRENT AFFECT?

The electrical potential difference applied between the cathode and the anode of the X-ray tube, which is also called the "tube potential", is controlled by the setting of the tube voltage in kilovolt (kV) or in peak kilovoltage (kVp), which actually goes back to the early and basic X-ray devices where the applied voltage varied cyclically. Since most modern X-ray generators apply a constant potential to the X-ray tube, the kVp value and the steady-state kV value are identical. The tube voltage affects the acceleration of the electrons to the anode and thus predominantly affects the energy spectrum of the generated X-ray beam. The tube current in milliamperes (mA) affects the number of electrons released from the cathode and thus predominantly affects the intensity of the X-ray beam [12]. However, it should be accounted for that the power supplied to the X-ray tube only ~1% generates X-rays, while the remaining ~99% results in heating. Therefore, the anode must be cooled, which is achieved by a liquid surrounding and additionally by rotating the anode disk at high speed to reduce the localized heat. Usually postmortem CT scans require more frequent cooling pauses between CT scans compared with clinical CT scans as an increased tube current is used [13].

4.4 WHAT HAPPENS WHEN X-RAY PHOTONS INTERACT WITH BIOLOGICAL TISSUE?

When the generated X-ray photons pass through the body, they are absorbed (photoelectric effect) and scattered (Compton scattering), which is defined as attenuation. Rayleigh scattering (at very low photon energies) and pair production (at very high photon energies) are not described in detail here.

The photoelectric effect describes the mechanism by which an incident photon of sufficient energy overcomes the binding energy of a tightly bound electron, which is then ejected in form of a photoelectron that causes the ionizing radiation. The energy of the emitted electron is equal to the energy difference between the photon and the binding energy of the electron. The hole in the atom shell is then filled by an electron from a higher energy shell, resulting in the emission of a characteristic energy equal to the difference in the binding energies of the two atomic shells, which is absorbed by the tissue and therefore cannot be detected by the imaging system. The probability of photoelectric interactions (P_{PE}) occurring depends strongly on the effective atomic number (Z) of the tissue and the photon energy (E):

$$P_{PE} = \rho \cdot \frac{Z^3}{E^3}$$

where ρ is the tissue density.

Therefore, on the one hand, the photoelectric effect, i.e., absorption, is more likely to occur in tissues with high atomic number than in tissues with low atomic number, and, on the other hand, the photoelectric effect decreases strongly with increasing energy. Consequently, the photoelectric effect is the main cause of contrast between different types of tissues in conventional X-ray imaging, and, therefore, tissue contrast is basically higher at the lower X-ray energies used in the radiological diagnostics.

Compton scattering refers to a photon that loses a fraction of its incident energy, which is then transferred to a loosely bound electron. This recoil electron will be emitted while the X-ray photon diverges from its original trajectory and has a longer wavelength. Therefore, Compton scatter reduces image contrast and contributes to a loss of spatial resolution in the images. The probability that the Compton effect occurs (P_{CE}) is given by:

$$P_{CE} = \rho \cdot \frac{1}{E}$$

and depends moderately on the energy of the incident X-ray photon. With increasing X-ray energy, the Compton effect becomes more dominant compared to the photoelectric effect.

Both mechanisms together, thus absorption and scattering, are referred to as attenuation. The mass attenuation coefficient (μ') in cm^2/g is a measure of the attenuation caused by these two mechanisms. In practice, however, it is more common to use the linear attenuation coefficient (μ) in cm^{-1}, which is given by μ' times the density (ρ) in g/cm^3, since μ is a measure of how much attenuation can be expected per unit length or thickness of a material. The attenuation of an X-ray beam trough a homogenous material is governed by the Beer–Lambert law:

$$I = I_0 \cdot e^{-\mu \cdot d}$$

where the intensity (I) is equal to multiplying the input intensity (I_0) by an exponential function with μ times the absorber thickness (d) in the negative exponent.

For a homogenous object (monochromatic radiation), μ can be directly determined if the absorber thickness is known [14]:

$$\mu = \frac{1}{d} \cdot \ln \frac{I_0}{I}$$

Plotting μ (calculated from μ' and ρ values published by Hubbell and Seltzer [15]) as a function of energy (E) demonstrates the strong attenuation of bone compared to soft tissues and bodily fluids (Figure 4.3). Hence, X-ray imaging provides a high contrast between bone and soft tissues or bodily fluids but a low contrast between different types of soft tissues or bodily fluids. Since μ of bone decreases strongly towards higher energies, the contrast between bone and soft tissue becomes lower towards higher energies.

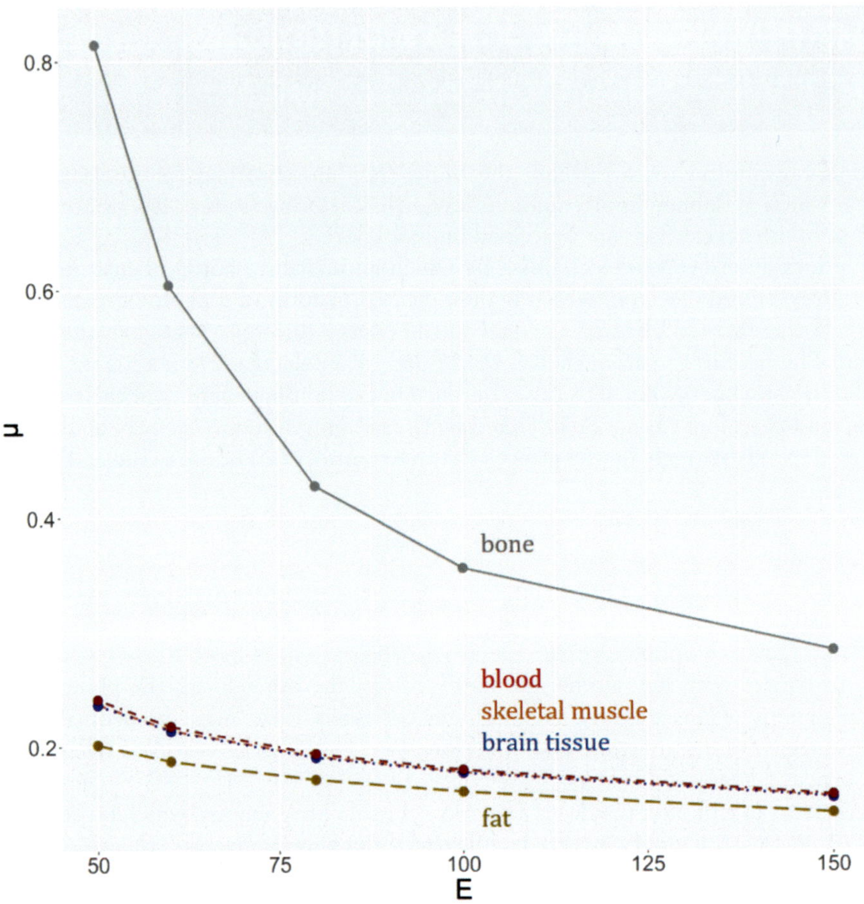

Figure 4.3 The linear attenuation coefficient (μ) in cm^{-1} as a function of energy (E) in keV for bone (grey, solid line), blood (dark red, dot-dashed line), skeletal muscle (red, dashed line), brain tissue (blue, dotted line), and adipose fat (brown, long-dashed line). The μ were calculated from μ′ and ρ values published by Hubbell and Seltzer [15].

4.5 HOW DO WE GET THE ATTENUATION INFORMATION OF THE DETECTION OF X-RAY PHOTONS?

X-ray photons that pass through the body and reach the detector system are detected. Before they reach the detector, they have to pass through an anti-scatter grid consisting of several strips coated with lead. The anti-scatter grid in front of the detector reduces the amount of Compton scattering. Finally, the remaining X-ray photons hitting the detector material are converted into a voltage, and an analog-to-digital converter is used to acquire the signal information for image reconstruction. Digital detectors typically use scintillator crystals and photodiodes to convert photons first to visible light and then to voltage, where the light intensity and ultimately the digital signal is proportional to the energy of the absorbed X-ray photon. These scintillator-based detector systems are called energy-integrating detectors, while modern photon-counting detectors use a semiconductor material in which positive and negative charges are generated directly when an X-ray photon is absorbed [16]. For a more detailed description of these two detector systems, the reader is referred to the corresponding chapter on photon-counting CT in this book.

4.6 WHAT IS COMPUTED TOMOGRAPHY?

A classic X-ray image, which is called a radiograph, is a two-dimensional projection image. Hence, the contrast information on the radiograph is based on the summation of the different intensities along the X-ray beam. Such a projection exhibits greatly reduced soft tissue contrast when bone is superimposed and does not provide information on the X-ray attenuation of individual soft tissue structures along the X-ray beam. However, a CT image, which is based on projections from multiple angles around the object (Figure 4.4), provides information on X-ray attenuation of individual soft tissue structures, even if they are surrounded by bone. But how can this be achieved?

The basic principle of CT is the measurement of the spatial distribution of a physical quantity, in this case the linear attenuation coefficient μ, using a series of projections at different angles, followed by the computation of superposition-free images from the obtained signals, where the process of signal encoding and decoding consists of complex mathematical calculations.

Applying Beer–Lambert's law on a polychromatic X-ray beam penetrating an inhomogeneous object (polychromatic radiation), as is the case for X-ray imaging in medicine, gives [14]:

$$I = \int_0^{E_{max}} I_0(E) \cdot e^{-\int_0^d \mu(E) ds} \, dE$$

where the intensity (I) is equal to the integral with respect to the energy (since the radiation spectrum generated by bremsstrahlung is continuous) of the input intensity (I_0) as a function of the energy (E) times the tissue property as an exponential function with the integral across the absorber thickness (d) of μ as a function of the energy (as μ is strongly dependent on the energy) along the way of the X-ray beam through the object (ds) in the negative exponent.

The mathematical task of CT is to determine μ as a function of the spatial positions x, y, and z to compute the CT image, which is achieved by recording a sufficient number of projections over an annular range of at least 180°. The

Figure 4.4 The basic principle of CT is a series of one-dimensional X-ray projections at different angles by rotating an X-ray tube around the object and measuring the spatial distribution of the linear attenuation coefficient.

principle of CT image reconstruction is explained in the following mathematical equation assuming two-dimensional parallel beam geometry. The basic encoding principle of these projections is based on the Radon transform ($\mathcal{R}$), which is the mathematical description for the transformation of the object function $\mu(x,y)$ (cross-sectional plane) into its sinogram $P_\varphi(r)$, which represents the measured signal during CT acquisition [17,18]:

$$\mathcal{R}\{\mu(x,y)\} = P_\varphi(r) = \int_{-\infty}^{\infty}\int_{-\infty}^{\infty} \mu(x,y)\cdot\delta(x\cdot\cos\varphi + y\cdot\sin\varphi - r)\,dxdy$$

where $P_\varphi(r)$ contains projection data (sinogram), φ is the given angle with respect to the coordinate axes x and y, and r is the distance to the origin along the detector plane given by the angle φ. Hence r is defined by $x\cdot\cos\varphi + y\cdot\sin\varphi$. As shown in the equation, the CT measurements correspond to an integration of the object function $\mu(x,y)$ sampled along the directions of the X-ray beam expressed by the Dirac's delta function (δ) [17]. The object function $\mu(x,y)$ is the desired information to represent the CT image. Therefore, finding the inverse Radon transform ($\mathcal{R}^{-1}$) is the decoding principle.

One way to reconstruct the data is the so-called Fourier reconstruction. In this reconstruction method, the Fourier domain is filled with individual 1D Fourier transforms of each projection plane, where the angle φ remains the same but the projection plane r changes into a projection plane through the origin (u) due to the Fourier slice theorem.

The Fourier transform of the projection data with respect to r is given by [17]:

$$\mathcal{F}\{P_\varphi(u)\} = \int_{-\infty}^{\infty} P_\varphi(r)\cdot e^{-2\pi\cdot i\cdot u\cdot r}dr = \int_{-\infty}^{\infty}\int_{-\infty}^{\infty} \mu(x,y)\cdot e^{-2\pi\cdot i\cdot u(x\cdot\cos\varphi + y\cdot\sin\varphi)}dxdy$$

The 2D Fourier transform of the object function, on the other hand, is given by [17]:

$$\mathcal{F}\{\mu(u_x,u_y)\} = \int_{-\infty}^{\infty}\int_{-\infty}^{\infty} \mu(x,y)\cdot e^{-2\pi\cdot i(u_x\cdot x + u_y\cdot y)}dxdy$$

Comparing these two equation reveals an identity, which is known as the Fourier slice theorem and states that the Fourier transform of the projection data with respect to r is equal to a 2D Fourier transform of the object function expressed in polar coordinates, with $(u_x, u_y) = (u\cdot\cos\varphi + u\cdot\sin\varphi)$ [17]:

$$\mathcal{F}\{P_\varphi(u)\} = \mathcal{F}\{\mu(u\cdot\cos\varphi, u\cdot\sin\varphi)\}$$

Once the Fourier domain is filled by the individual 1D Fourier transforms of each projection, the data defined in polar coordinates is transformed to Cartesian coordinates, followed by an inverse 2D Fourier transform to obtain the image in the image domain (Figure 4.5).

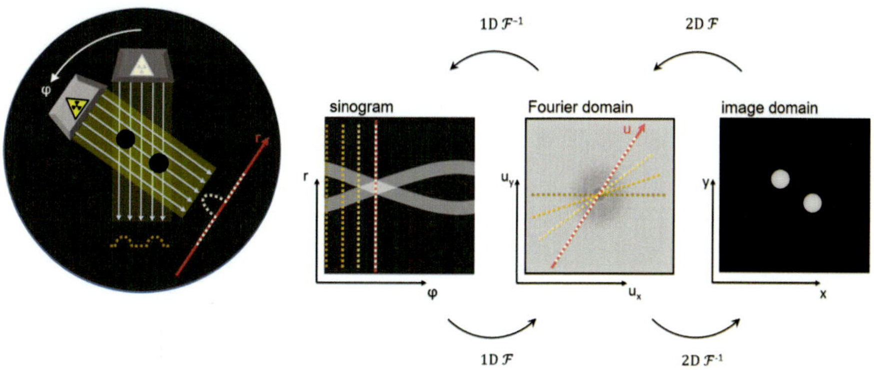

Figure 4.5 The Radon transform, the Fourier slice theorem, and the Fourier transform reconstruction. After transmission, the X-ray photons captured by the detector generate the attenuation profile of the object, i.e. the attenuation coefficient μ as a function of x and y is recorded along the detector plane in direction r at an angle φ with respect to a Cartesian coordinate system. Thus, the object function $\mu(x,y)$ is Radon transformed into the projection data $P_\varphi(r)$. When the function is plotted, the sinogram is obtained. By applying the Fourier transform on the projection data $P_\varphi(r)$, the domain changes and the projection along r is a projection u through the center in the Fourier domain but at the same angle φ, which is known as the Fourier slice theorem. The Radon transform thus allows the connection between the Cartesian coordinate system and the polar coordinate system since it gives us the projection as a function of φ. After filling the Fourier domain, μ is obtained as a function of x and y by applying the inverse 2D Fourier transform.

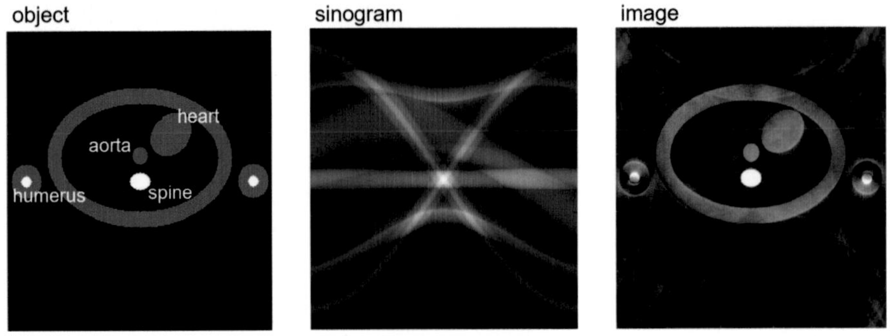

Figure 4.6 Simulation of the reconstruction problems with the Fourier transform method. For this purpose, an object was created which represents the cross-section through the thorax at the level of the heart. A sinogram was calculated by the Radon transformation, which corresponds to the data acquisition in CT scanning. The Fourier transform method was used to calculate the image from the sinogram data. Image artifacts are clearly visible in the reconstructed image.

Besides the fact that 2D inverse Fourier transform involves high computational costs, this Fourier transform method is associated with significant problems in realizing an efficient computer implementation [17]. Discretization problems combined with the conversion of polar coordinates to Cartesian coordinates, can lead to large errors in the spatial domain, resulting in image artifacts (Figure 4.6).

Another way to solve this problem is the so-called filtered-back projection (FBP) algorithm, which is given by the following equation [17,18]:

$$\mu(x,y) = \int_{0}^{\pi}\int_{-\infty}^{\infty} \mathcal{F}\{P_{\varphi}(u)\} \cdot e^{2\pi \cdot i \cdot u \cdot r} \cdot |u| \, du \, d\varphi$$

where the inner integral is a 1D inverse Fourier transform of the Fourier transform of the projection $P_{\varphi}(u)$ multiplied by a filter $|u|$, r gives the direction of a projection at angle φ, and the outer integral is the integral of all projections over the annular range of projections from 0° to 180° $(0 - \pi)$.

In general, each projection taken at angle Æ is 1D Fourier transformed and multiplied by filter $|u|$. The 1D inverse Fourier transforms of these filtered projections are then successively smeared out along r at the specified angle φ and summed up over the annular range of projections $(0 - \pi)$, which then represents the reconstructed image. Without pre-filtering with a ramp filter $|u|$, which corrects for oversampling in the middle of the Fourier domain, the image would be blurred (Figure 4.7). Filtering can be performed either by a simple multiplication in the Fourier domain, while 1D Fourier forward and back transformations are computed with fast Fourier transform algorithms or can be performed directly in the image domain by convolving the filter kernel with the projection [17,19]:

$$\mu(x,y) = \int_{0}^{\pi} P_{\varphi}(r) * k(r) \, d\varphi$$

where k(r) is the ramp filter in the spatial domain, which is given by [17,18]:

$$k(r) = \mathcal{F}^{-1}\{|u|\} = \int_{-\infty}^{\infty} |u| \cdot e^{2\pi \cdot i \cdot u \cdot r} du = \frac{-1}{2\pi^2 \cdot r^2}$$

Since the sampled data is discrete and the imaging system is band-limited, the ramp filter is regulated in the Fourier domain by multiplying a rectangular filter. By modulating the ramp filter, e.g., by rolling off in the range of higher frequencies, a CT image ultimately has higher or lower contrast or higher or lower noise. Thus, there are many different reconstruction filters or kernels, and ultimately the kernel selected for the filtered back projection determines the point spread function of the CT image.

Another image reconstruction method, used in particular to reduce image noise in low-dose CT scans or metal artifacts, is based on iterative algorithms [20]. This reconstruction method starts from an image assumption and compares the assumed values with the values measured in real time, making constant adjustments until the adjusted and measured values match. Iterative reconstructions (Figure 4.8), such as sonogram-affirmed iterative reconstruction (SAFIRE) [21], which require high computing power, are increasingly available today according to Moore's law [22].

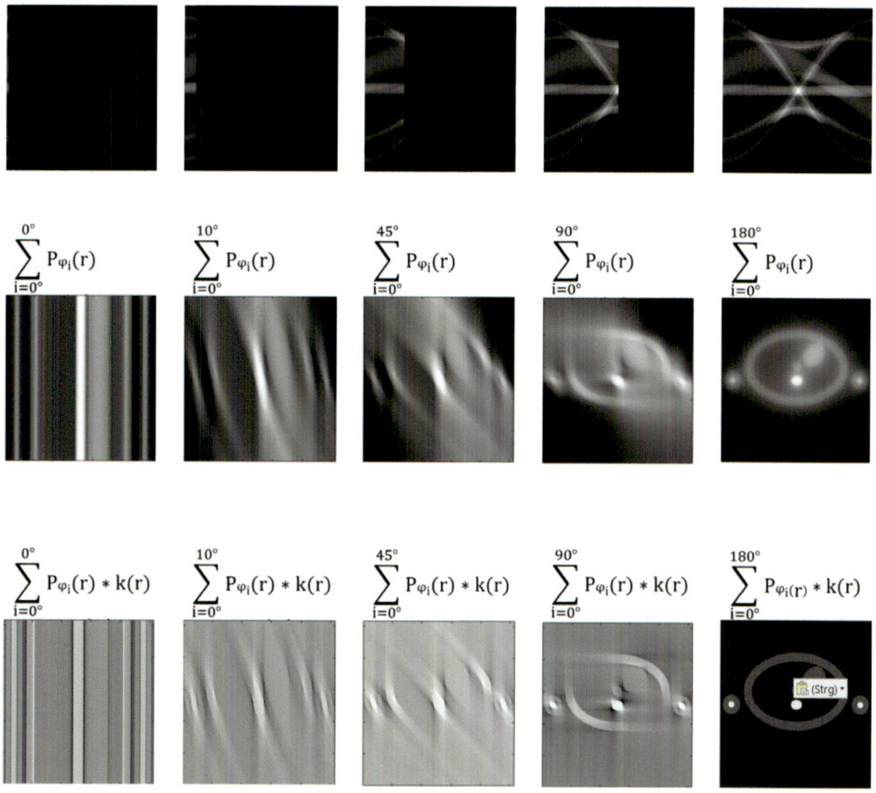

Figure 4.7 Simulation of the reconstruction of a transverse cross-section through the body at the level of the heart from acquired projection data (upper row) using back projection (middle row) and filtered back projection (bottom row). In back projection, each individual projection taken at angle φ is smeared over the image plane corresponding to the angle at which it was taken. The individual smeared projections are summed up. Without the application of an additional filter $k(r)$, a blurred image will result. However, if each projection is filtered, a sharp image is obtained.

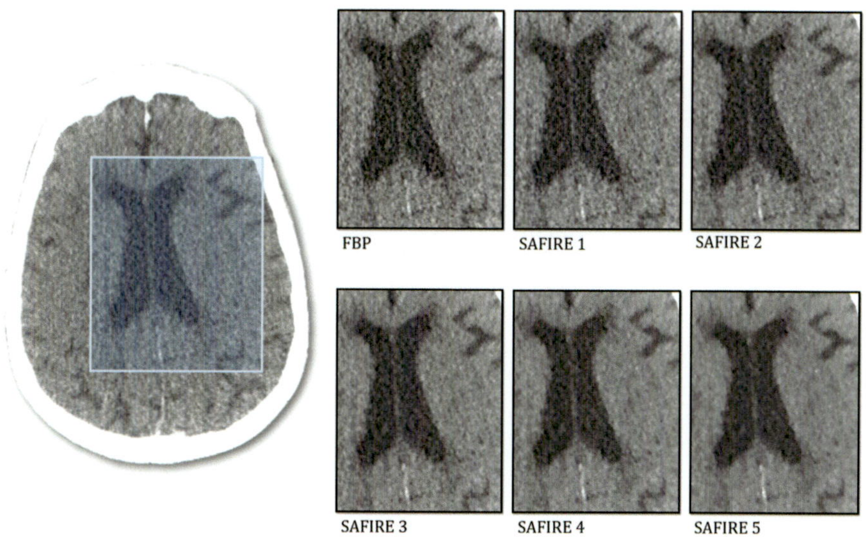

Figure 4.8 Reconstructions of a cross-sectional slice at the level of the lateral ventricles using filtered back projection (FBP) and sonogram-affirmed iterative reconstruction (SAFIRE) with strengths 1 to 5. The SAFIRE algorithm works in the raw data domain and was developed mainly for noise reduction [22]. In the images it is clearly visible how the image noise decreases with an increase of the SAFIRE strength.

This mathematical excursion describes the basic principles of image reconstruction in computed tomography. For further mathematical explanations concerning fan beam and conebeam CT image reconstructions, the reader is referred to the further explanations by Kalender [17] and Suetens [18].

In summary, by rotating the X-ray tubes around the object and then reconstructing images based on the individual projections, the linear attenuation coefficients of the individual tissue types can be reproduced and, in contrast to the projection images in planar X-ray, the soft tissue around the entire bone can also be imaged.

4.7 WHAT DOES THE HOUNSFIELD UNIT MEAN?

Although the linear attenuation coefficients (μ) are measured and computed as a function of the spatial distribution μ(x,y,z), this physical quantity is not represented in the final CT image because it depends strongly on the spectral density, resulting in high variability between CT scanners [14]. Therefore, in CT imaging, the so-called CT numbers serve as attenuation values, which are relative to the attenuation coefficient of water (W). The CT number for an arbitrary tissue (T) is given by [14]:

$$\text{CT number} = \frac{(\mu_T - \mu_W)}{\mu_W} \cdot 1000$$

The unit of CT numbers is called the Hounsfield unit (HU) to honor the work of Godfrey Newbold Hounsfield. Therefore, the CT number of water is 0 HU and the CT number of air is about −1000 HU, and both the CT number of water and that of air are considered to be independent of the applied tube voltage and thus energy [14]. Accordingly, lung tissue has high negative CT numbers, fat has low negative CT numbers, soft tissue has low positive CT numbers, and bone has high positive CT numbers.

In postmortem imaging, it must be taken into account that the HU values change as decomposition progresses [23]. Furthermore, body temperature, which has a wide range from frozen to over 30 °C with a very short postmortem interval, is an additional factor that must be considered in quantitative CT examinations. In general, HU values decrease with increasing temperature, and for bodily fluids and blood, a variation of up to 10 HU between 4 and 40°C was observed in a phantom study [23]. For further information on the influence of temperature on the measured HU values, see also the corresponding chapter on freezing and thawing in this book.

At an X-ray energy of about 80 keV in CT imaging, μ of soft tissue, which consists mainly of hydrogen, is dominated by the Compton effect, and the photoelectric effect contributes only a few percent. Therefore, μ is directly proportional to the density (ρ) of soft tissue and the proportionality factor is determined by simple calibration by the manufacturer:

$$\mu \propto \rho \cdot \frac{Z}{A} \cdot \frac{1}{E}$$

where Z is the atomic number and A is the mass number, and their ratio is equal to 1 for hydrogen, and E is the energy.

In contrast, μ of bone, where the hydrogen content is much lower compared to soft tissue, is affected by the Compton effect and the photoelectric effect, and beam hardening occurs. Therefore, a separate calibration with a corresponding calibration phantom is required to determine the bone mineral density (BMD). This technique of BMD determination is called quantitative CT (qCT) and forms the basis for biomechanical analyses. In addition, beam hardening must be taken into account since polychromatic X-ray radiation is used in medical CT. For this reason, dual-energy CT is also shown for a phantomless calculation of BMD with a reduced beam hardening effect [24]. However, after a longer postmortem interval, air or gas occurs in the bones, making phantom-based BDM assessment less suitable and multi-energy–based bone mineral assessment seems more suitable. In this method, the CT numbers at different energies are expressed by fractional volumes of the material components in the bone relative to a total volume (the voxel volume) and their energy-dependent attenuation values to finally determine the fractional bone volume indicated as bone volume to total volume (BV/TV) [25].

Materials with k-edge energies in the range of the X-ray spectrum (such as iodine) can be easily distinguished from biological materials with relatively low k-edge energies. The K-edge effect refers to an abrupt increase in the photoelectric absorption of X-ray photons observed at an energy just above the binding energy of the K-shell electrons of the absorbing atom. This K-edge effect is thus specific to each element according to its atomic number. However, the most abundant elements in human tissue, namely hydrogen, carbon, oxygen and nitrogen, have low K-shell binding energies, and therefore the K-edge effect is of interest not for these elements but for elements with larger K-shell binding energies that lie in the useful region of the X-ray energy spectrum, such as the element iodine. Contrast agents containing iodine are typically used in CT examinations. Therefore, some materials may have very similar HU values at a single selected X-ray energy spectrum, while differing much better at another X-ray energy spectrum when the K-edge effect leads to an increase in photoelectric absorption for one of the materials. For this reason, the dual energy index (u) was introduced for material differentiation in clinical CT examinations, in which the normalized ratio of CT numbers calculated from two energy spectra of different kVp levels is calculated:

$$u = \frac{x_{E1} - x_{E2}}{x_{E1} + x_{E2} + 2000}$$

where x_{E1} is the CT number obtained from the low energy CT scan and x_{E2} from the high energy CT scan. Their difference is normalized by their sum in the denominator. The summand 2000 results from the attenuation difference between air and water for x_{E1} and x_{E2} (and cancels out in the numerator) and corresponds to material not dissolved in water. More about this method of material differentiation is described in the chapter on bullet identification or can be found in the literature [26–29].

4.8 WHAT IS A VOXEL?

As is well known, a two-dimensional image is composed of picture elements (pixels). However, CT images are composed of volume elements (voxels), since a CT data set is three-dimensional. The voxels contain the information about the attenuation values in the form of CT numbers (given in HU), which are displayed as gray values in the CT image. CT reconstruction typically uses 12-bit reconstruction, resulting in 4096 CT numbers, usually ranging from −1024 HU to +3071 HU. For the special examination of metallic implants or foreign bodies, some manufacturers also allow 16-bit reconstructions or an extended CT scale (or extended HU scale) algorithm to expand the number of gray values by a factor of 10, which enables the more detailed representation of metal objects (Figure 4.9). For more information on the extended CT scale reconstructions and the study of metallic objects in the body, the reader is referred to the chapter on bullet profiling in this book.

Since the CT numbers of different types of soft tissues, for example, are in a range of only about 100 HU, hardly any contrast between soft tissues is visible when the gray values of the entire HU ranges are displayed. Therefore, the gray scale is adjustable by choosing a desired window width and center, which is then called windowing. Changing the windowing does not change the CT number information within the voxel, whereas choosing a different kernel does. Although in a measured region of interest (ROI), the mean of the HU values changes only slightly, their range of values within the ROI and consequently the standard deviation changes considerably (Figure 4.10).

But not only the kernel has an influence on the HU values but also the size of a reconstructed voxel. The size of the voxel is defined by the slice thickness (d_{slice}), e.g. 1 mm. the selected size of the reconstruction field of view (FOV), e.g. 500 × 500 mm² for a thorax scan, together with the system-dependent reconstruction matrix, i.e. the number of pixels (N), which is typically 512 × 512 or 1024 × 1024:

$$\text{voxel} = d_{\text{slice}} \cdot \prod_{i=x,y} \frac{\text{FOV}_i}{N_i}$$

Thus, in-plane resolution, i.e., voxel size along the x- and y-axes, is defined by the FOV and matrix, while resolution along the z-axis, i.e., parallel to the scan direction, is defined by the slice thickness (Figure 4.11).

At this point it should be noted that the scanned FOV is circular, while the reconstruction FOV and thus the reconstructed CT image is square. Therefore, if the corners of the reconstruction FOV are larger than the diameter of the scanned FOV (the center of which is at the isocenter, i.e., the center of the CT gantry), there is no information in it; i.e., these voxels are given a CT number of −1024 HU in the CT image.

Especially for the visualization of CT data sets in other orientations (multi-planar reformations) or the three-dimensional representation (e.g. volume renderings), small isotropic voxel sizes are desired, so voxels with almost the same extension in all three dimensions (Figure 4.12).

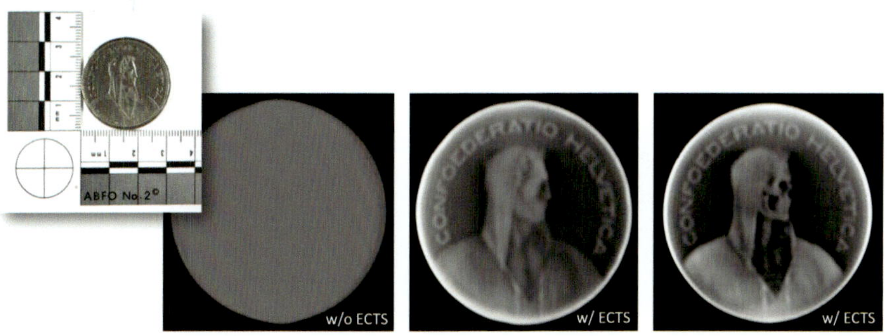

Figure 4.9 Multi-planar reconstructions of the high-resolution CT scan of a Fünfliber (the highest circulation coin of the Swiss currency) without extended CT scale and with extended CT scale (scan parameters: 140 kV, 666 mAs, pitch: 0.3, rotation time: 1 second; reconstruction parameters: 0.4 mm slice thickness, field-of-view: 50 × 50 mm², reconstruction matrix: 512 × 512, kernel: U68). While the standard CT scale is limited to a maximum value of 3071 HU, the extended CT scale allows windowing into HU ranges up to a maximum value of 30710 HU to image metallic objects in more detail and visualize, for example, the mintage of a coin [29]. This makes it possible, for example, to distinguish a coin from a button battery in the digestive tract. Going even further, one may even visualize the skeleton of the alpine shepherd in the *Sennechutteli* as illustrated in the image far right (just kidding, this is a special coin imitation).

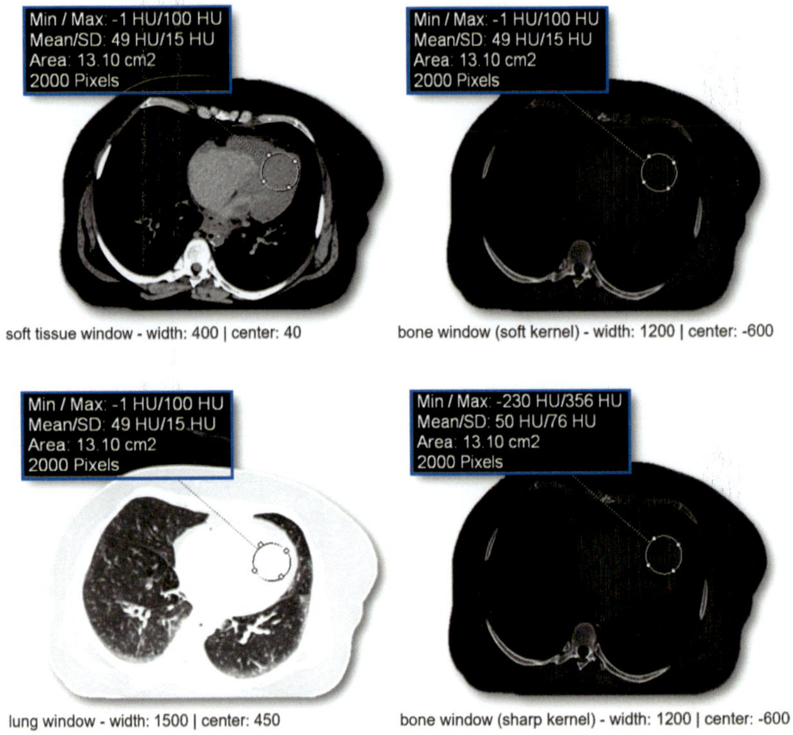

Figure 4.10 CT images with different windowing, a typical soft tissue window (top left image), a typical bone image (top right image), and a typical lung window (bottom left image). The mean attenuation value (49 HU) within the region-of-interest (ROI) measurements does not change between the different windows of these three CT images. The same reconstruction kernel was used for these three CT images with different windows. In practice, a different kernel is used for soft tissue imaging (soft kernel) and for bone and lung imaging (sharp kernel), resulting in slight variations of the mean CT numbers (50 HU) and a higher standard deviation (SD) within the ROI mainly due to increased noise (bottom left image).

The size of the reconstruction FOV or the in-plane voxel size affects not only the representation of structures in the cross-sectional images but also three-dimensional reconstructions, which is commonly referred to as a "pixelated image" (Figure 4.13).

To minimize noise in cross-sectional images, the slice thickness can be arbitrarily increased during visualization using appropriate software, which is usually done when assessing soft tissue. The data set reconstructed from the raw data should always be calculated with a low slice thickness, especially because reconstructions with the soft tissue kernel are often also used for 3D reconstructions (volume renderings) and an increased slice thickness leads to artifacts in the three-dimensional view (Figure 4.14). For more detailed information on 3D visualization of CT datasets, the reader is referred to the corresponding chapter on visualization.

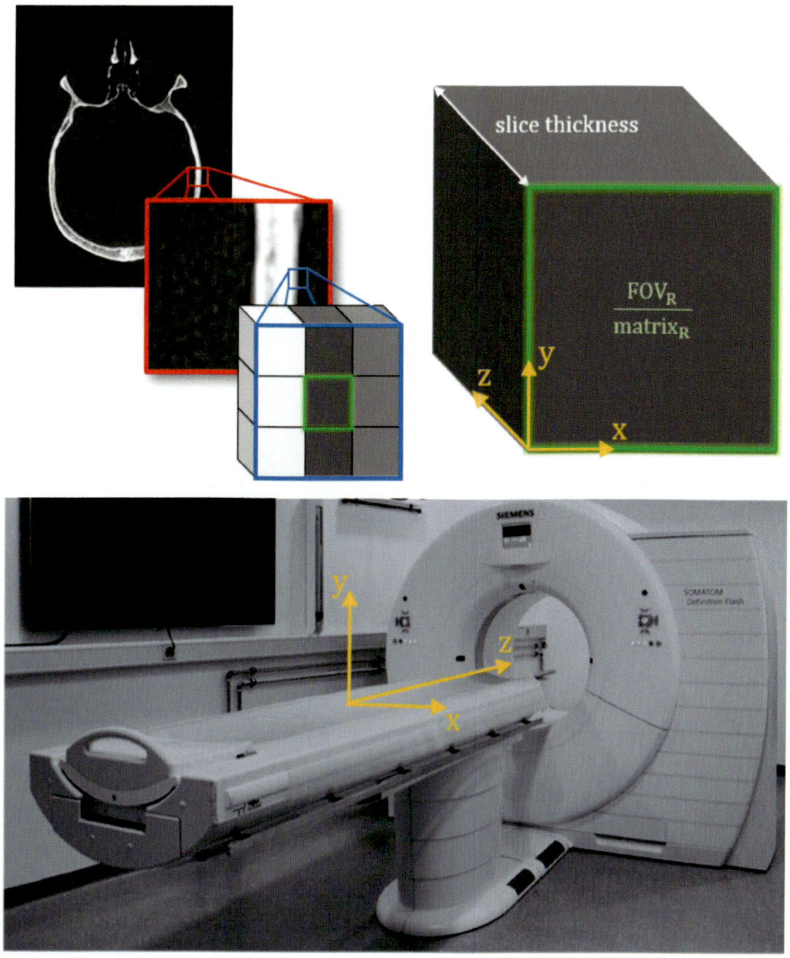

Figure 4.11 The three-dimensional CT data consists of up to several hundred slices (depending on the degree of slice overlap) and is therefore also referred to as a slice stack. Each individual CT slice consists of "three-dimensional pixels" called voxels, which contain the calculated gray value. The size of the voxel in the plane of view (along x- and y-axis) is defined by the reconstruction field of view (FOV$_R$) and the reconstruction matrix (matrix$_R$), while the size of the voxel in the third dimension (the z-axis, in the direction of the table feed) is determined by the slice thickness (SL).

Figure 4.12 The reconstruction FOV is adjusted to the subject to keep the in-plane voxel size (x and y) as small as possible. The FOV for head and neck scans is approximately 300 × 300 mm^2, that for thorax and abdomen scans is approximately 500 × 500 mm^2, and for full-body scans, an extended FOV of up to 780 × 780 mm^2 is commonly used. To generate isotropic voxels, slice thicknesses of 0.6 mm (head and neck scan), 1.0 mm (thorax and abdomen scan), and 1.5 mm (whole-body scan) are selected for the reconstructions from the raw data.

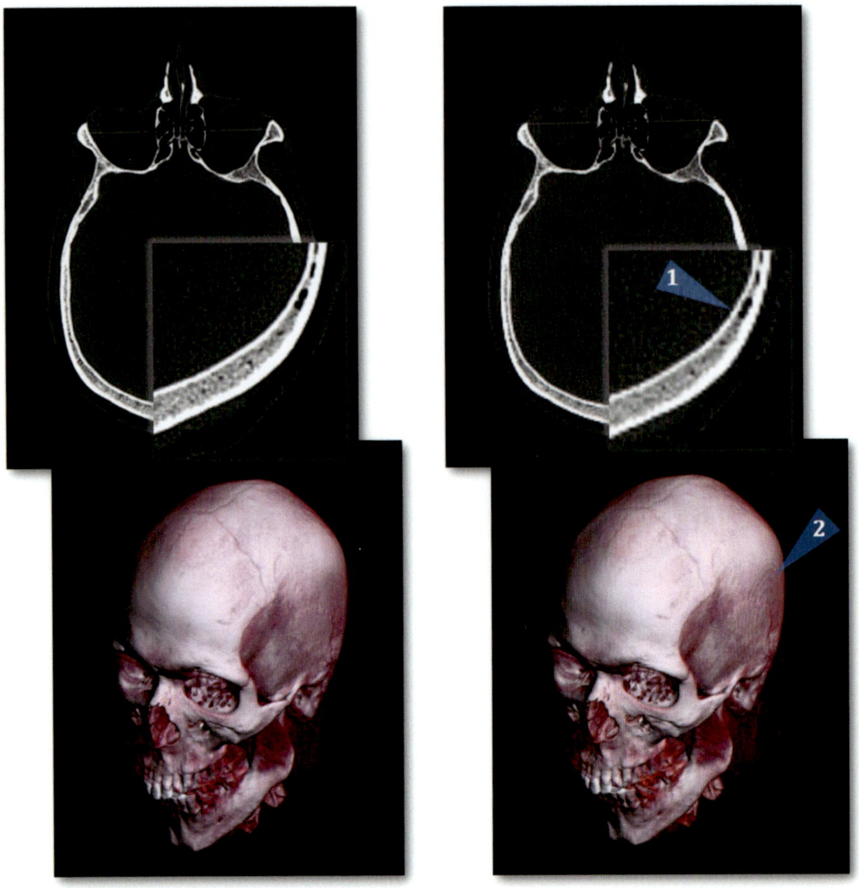

Figure 4.13 The effect of in-plane voxel size on cross-sectional images and 3D reconstructions. On the left we see the data set of a CT scan of the head reconstructed with a slice thickness of 0.6 mm and a reconstruction field-of-view (FOV) of 300 × 300 mm^2 (thus an isotropic voxel) and on the right the data set of the same CT scan reconstructed with a slice thickness of 0.6 mm and a reconstructed FOV of 500 × 500 mm^2. The disadvantage of a large FOV is clearly visible in the cross-sectional images due to the pixelated step artifacts at the bone edges (arrowhead 1). Therefore, it is always recommended to reconstruct the head and neck area with a small FOV in addition to the reconstruction of a full body scan, which requires a large FOV. In the 3D reconstruction of the dataset with a large FOV, ring-shaped artifacts are visible (arrowhead 2), as in the 3D reconstructions of the thick-slice dataset, which now appear perpendicular to the scan plane. These artifacts are less pronounced here, since the difference in size of the individual voxels in the in-plane plane is smaller than the difference in slice thickness in the corresponding example. Note also that increasing the FOV did not create an artificial finding here as it did in the 3D reconstruction of the thick-slice dataset before (arrowhead 3).

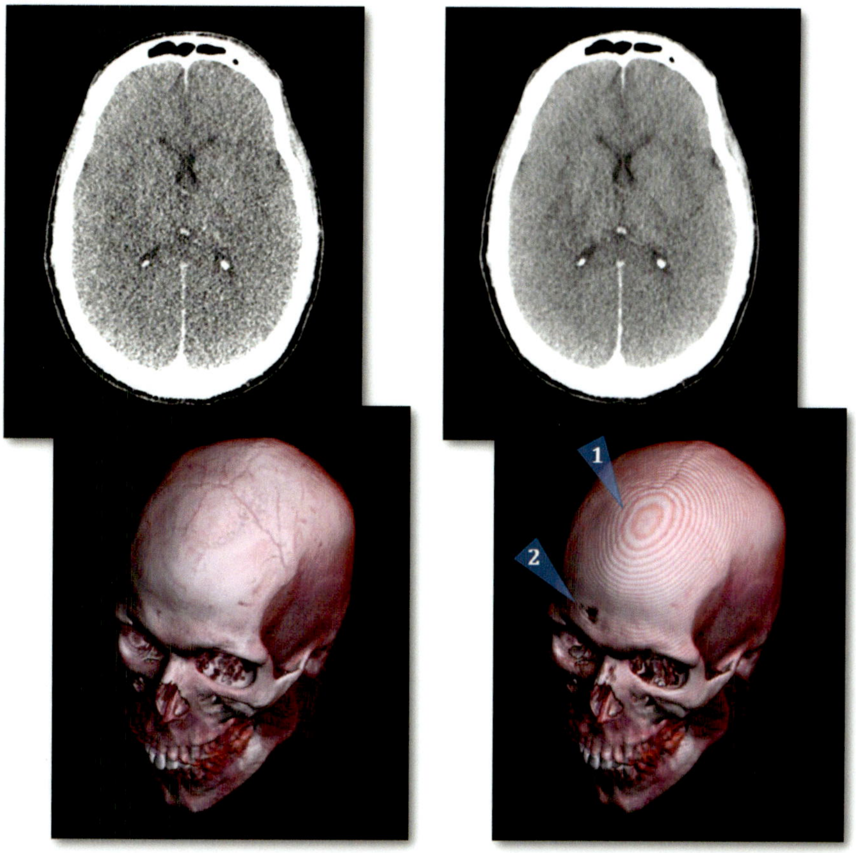

Figure 4.14 The effect of slice thickness on cross-sectional images and 3D reconstructions. On the left we see the data set of a CT scan of the head reconstructed with a slice thickness of 0.6 mm and on the right the data set of the same CT scan reconstructed with a slice thickness of 2 mm. The left image is noisier due to the thinner slice thickness. However, the slice thickness can be readjusted at any time during the visualization of the data sets and thus the noise for the assessment of the soft tissues can be reduced by increasing the slice thickness so that the left image also has comparably low noise as the right image. In the 3D reconstructions, the disadvantage of thick-slice reconstructions is clearly visible from the slice artifacts (arrowhead 1). The 3D reconstruction also impressively demonstrates the pitfall of creating artificial findings (arrowhead 2). This apparent finding is due to the fact that at this point the bone is less radiopaque because of a vessel. Note that in the 3D reconstruction, this apparent finding is further enhanced.

4.9 WHAT DEFINES THE IMAGE QUALITY?

The image quality can be described with objective quality criteria, which include contrast, spatial resolution, and noise. The contrast of the CT image is given by the linear attenuation coefficient (μ), which is energy dependent and thus influenced by the tube voltage. The resolution of our CT image, the image sharpness, is determined by the point spread function PSF(x,y,z), the reconstruction kernel. The PSF(x,y,z) is convolved with the three-dimensional object function μ(x,y,z). Accordingly, the image function I(x,y,z) is given by [30]:

$$I(x,y,z) = C(E) \cdot \mu(x,y,z) * PSF(x,y,z) + \eta + A$$

where C(E) is the energy-dependent contrast, η is the noise and A the artifacts.

Thus, noise and artifacts further affect the image quality.

The noise (η) is given by the applied dose, which is mainly affected by the tube current. The noise sources are different, as there is shot noise with a Poisson distribution, which has a physical origin; then there is thermal noise with a Gaussian distribution, which is caused by electronics; and then there is conversion noise with a binomial distribution, which is caused by the conversion from analog to digital signal. This section only briefly discusses shot noise, which refers to uncertainty in the detected intensities. A major advantage of postmortem CT is the ability to neglect radiation exposure with respect to patient dose considerations. Increasing the radiation dose reduces image noise caused by fluctuations in the number of photons per square millimeter. These fluctuations result in a statistical fluctuation in the signal intensity of each voxel called shot noise, which follows a Poisson distribution. Since shot noise (SN) follows a Poisson distribution, its standard deviation (SD_{SN}), which is the square root of the variance, is proportional to the square root of the mean number of photons detected:

$$SD_{SN} \propto \sqrt{\bar{N}}$$

Thus, the variance increases when more photons are detected. However, since the strength of the signal itself increases more rapidly when more photons are detected, the relative noise decreases; i.e. the signal-to-noise ratio (SNR) increases:

$$SNR = \frac{\bar{N}}{\sqrt{\bar{N}}} = \sqrt{\bar{N}}$$

Thus, the noise in the image can be reduced by increasing the number of intensities detected, which can be achieved by increasing the tube current and scan time (Figure 4.15). For medical CT data, SNR is often calculated from the values of region-of-interest measurements in the tissue of interest and in the background. The ratio of the mean HU value in the tissue of interest to the standard deviation of the HU values in the background (SD_{HU}), and thus the background noise, is

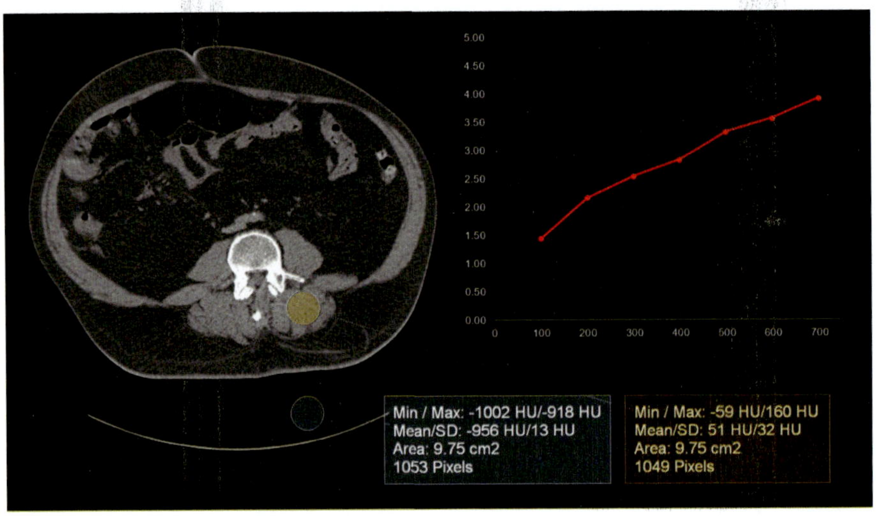

Figure 4.15 CT scans were performed repeatedly in the lumbar spine region, increasing the tube current-scan time product (unit: milliampere-seconds [mAs]) for each individual scan; i.e., CT scans were performed in increments of hundreds from 100 to 700 mAs. Subsequently, ROI measurements were performed in muscle tissue and background, respectively. Here in the image on the left the measurements in the muscle tissue (mean: 51 HU) and in the background (SD: 13 HU) of the data set with 700 mAs are shown. The SNR was calculated from the measured values and is shown in the graph as a function of the tube current. It is clear to see how the SNR increases with increase in tube current-scan time product.

then reported as SNR. The contrast to noise ratio (CNR) is the difference in mean HU values between two tissue types of interest and its ratio to background noise.

However, for the evaluation of CT imaging with different reconstruction techniques, the use of these criteria as quality criteria is less useful, and methods based on statistical decision theory such as mathematical model observers should be used instead to describe the image quality [31]. In addition to such objective image quality calculations, the subjective perception of image quality should also be mentioned here, which depends primarily on individual preferences and previous experience [13]. Since this influences the individual assessment of images, at least two experienced evaluators are required for the scientific assessment of image quality based on visual assessment. Notwithstanding this, the two basic scan parameters for noise and contrast are tube current (or tube current-time product) and tube voltage, respectively (Figure 4.16).

Most manufacturers also offer the option of automatic exposure control (AEC) (Figure 4.17). This measures the body's X-ray attenuation along the z-axis during the initial radiographic projections, known as the localizer, the topogram, or the scout view. The AEC aims to provide consistent image quality for each body composition, which means that the volume CT dose index ($CTDI_{vol}$) in helical acquisition is increased for more obese bodies and decreased for leaner bodies compared to a reference body of about 70–80 kg. The major manufacturers use different methods for modulation.

The final adaptation of the X-ray exposure according to the rotation angle at the actual CT scan can be based entirely on the information of the radiographic projections or partly additionally on the actual helical scan. Since the modulation refers to a reference body with a defined weight, the actual weight of the scanned body can also be calculated from the exposure finally administered [32]. By using the AEC, for example, the X-ray exposure for full-body postmortem CT scan can be increased for the torso region and reduced in the region of the lower extremities, where less X-ray attenuation occurs due to the lower body volume (Figure 4.18). Furthermore, a more uniform image quality may be preferred for certain scientific investigations.

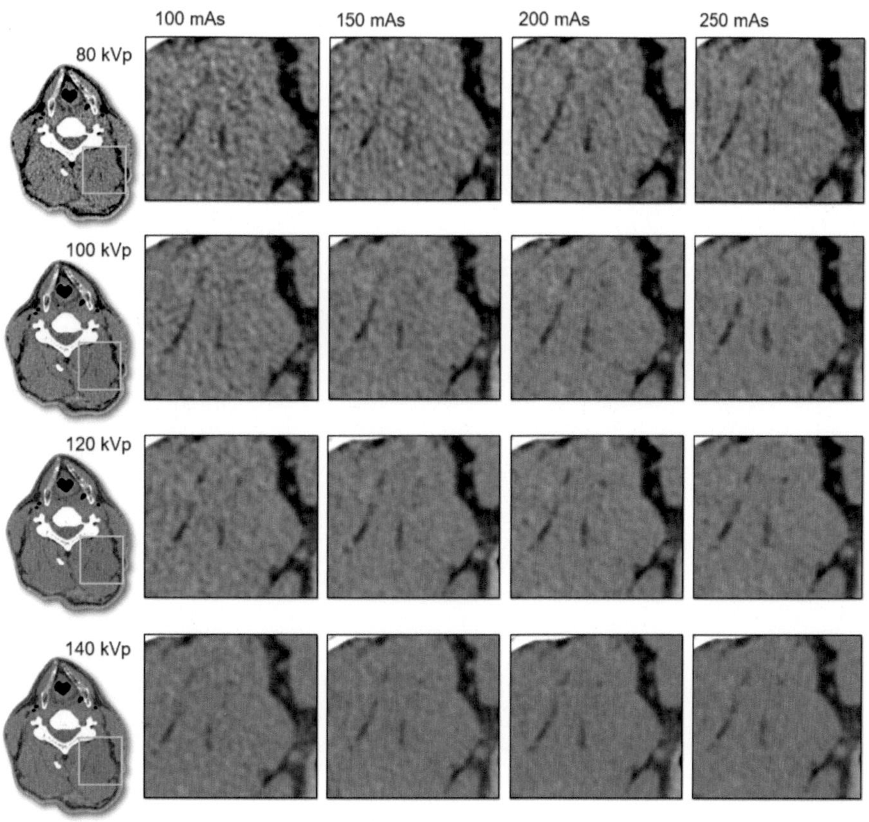

Figure 4.16 Axial CT images of a decedent at the anatomical level of the neck from repeated CT scans acquired with tube voltages of 80, 100, 120, and 140 kVp (rows) and with tube currents of 100, 150, 200, and 250 mAs. It is impressive to see how the noise decreases with the current (from left to right) and how the contrast between different tissue types decreases with the tube voltage (from top to bottom).

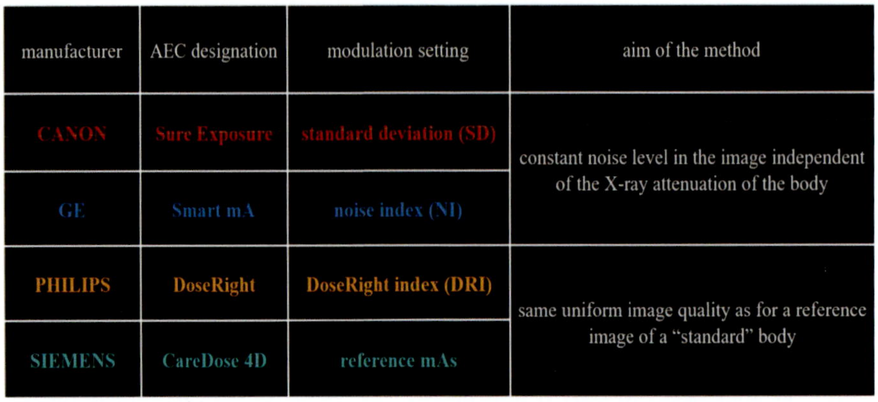

manufacturer	AEC designation	modulation setting	aim of the method
CANON	Sure Exposure	standard deviation (SD)	constant noise level in the image independent of the X-ray attenuation of the body
GE	Smart mA	noise index (NI)	
PHILIPS	DoseRight	DoseRight index (DRI)	same uniform image quality as for a reference image of a "standard" body
SIEMENS	CareDose 4D	reference mAs	

Figure 4.17 The major manufacturers have different names for their method of automatic exposure control (AEC) and different setting parameters. The aim of modulation is primarily to achieve constant noise and uniform image quality over the entire scan area.

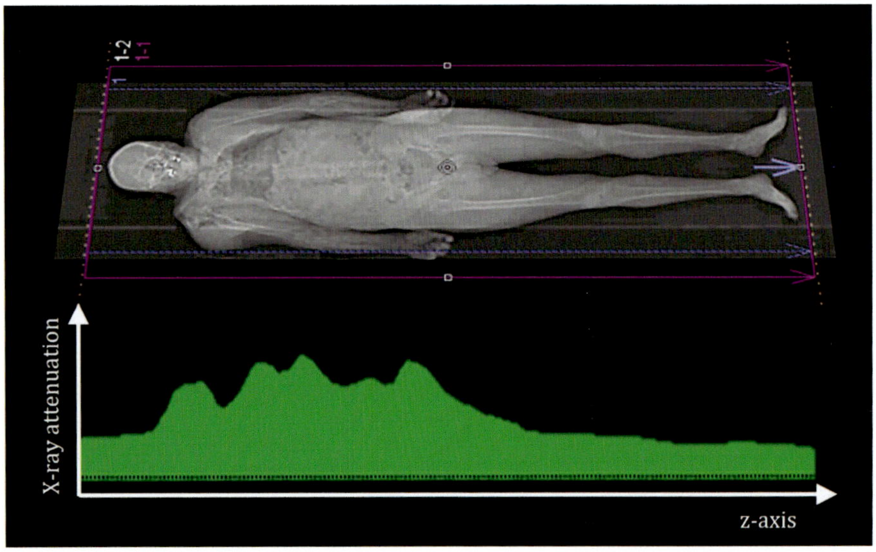

Figure 4.18 The X-ray attenuation profile is obtained from radiographic projections acquired during the generation of the topogram. For the subsequent CT scan, the radiation exposure is adjusted according to the measured profile. Thus, the torso region, which due to its larger volume leads to a stronger X-ray attenuation, is finally displayed with a better image quality and the image quality over the whole body becomes more uniform.

4.10 WHAT SETTINGS SHOULD BE CONSIDERED FOR POSTMORTEM CT SCANNING?

Postmortem CT examinations are already performed in several forensic institutes as standard in highly relevant forensic cases. The role of CT examination as a substitute for autopsy is still a controversial issue [33,34]. However, the attitude towards this probably relates more to the relevant specialty and experience of the person making the judgment than to the imaging technique, which has become an indispensable part of clinical radiology.

For the diagnostic evaluation of postmortem CT images, the reader is referred to the literature and the corresponding chapters in this book, which demonstrate the use of postmortem CT examination for the detection of relevant findings in different causes of death. In the following, we will briefly discuss what needs to be considered when performing a postmortem CT scan and then summarize the most important parameter adjustments for postmortem CT scans based on more than 10 years of experience with our in-house CT scanner (Figure 4.19). Note: The parameter options described in the following may differ not only among manufacturers but also due to new hardware and software releases.

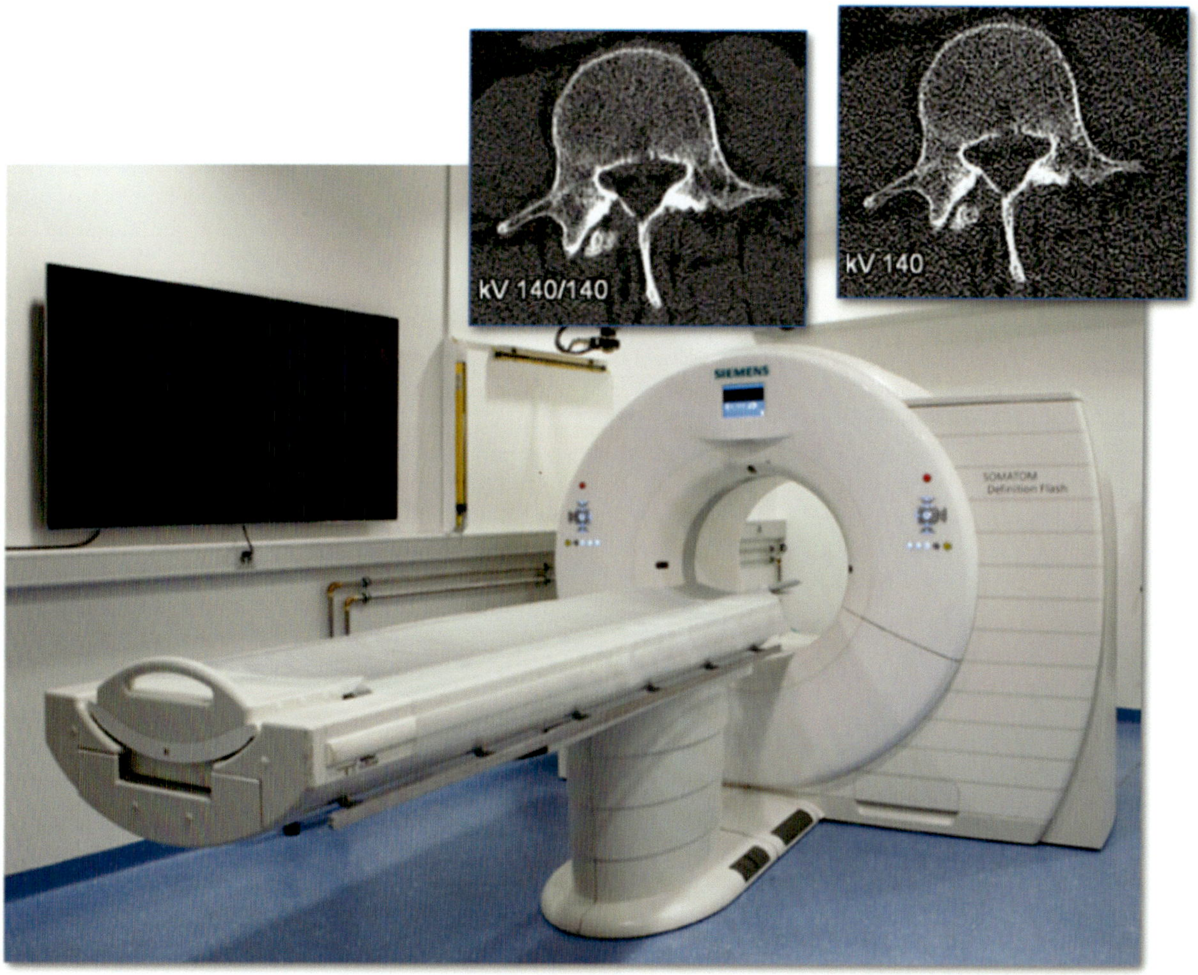

Figure 4.19 Dual-source CT scanner at the Zürich Institute of Forensic Medicine (SOMATOM Definition Flash, Siemens Healthineers). Apart from dual-energy protocols, the advantage of the two-source system is that both tubes can be used simultaneously for the CT scan of an obese corpse, which means that a higher radiation dose can be achieved and the noise in the image can be significantly reduced, albeit only in a limited FOV (330 × 330 mm^2), since the detector of the B-tube is smaller. In the upper right images, this is illustrated with a two-source scan (left image) compared to a single-source scan (right image) of the same lumbar vertebra. Such a CT scanner acquires data continuously as the X-ray tube rotates and the CT table moves through the CT gantry, which is referred to as helical or spiral CT scanning.

But before it comes to scanning and reconstruction, the preparation and appropriate positioning of the body are relevant. In postmortem CT scanning, care must be taken to ensure that the deceased is scanned on a suitable base (e.g. a plastic sheet) or in a body bag to prevent contamination of the CT equipment through fluid leakage. If possible, the body should be scanned in the supine position with the hands crossed over the lower abdomen ("praying position"). The head should be positioned straight with the nose in the sagittal axis of the body. No increased force should be used to position the body so that no additional postmortem findings occur. This is especially true for an additional CT scan of the torso with the arms raised to reduce streak and dark band artifacts due to beam hardening [13,35]. Transferring from the transport table to the CT table and vice versa after the scan can be facilitated by a flat CT table support (as used in CT-based radiotherapy treatment planning). To take advantage of the scan length of approximately 2000 mm for a full body scan, the body on the CT table must often still be pulled cranially or caudally (depending on whether head first or foot first is scanned). For the person standing next to the CT table pulling the body, this movement is harmful to the spine, which is why special table supports for CT scanners are recommended (Figure 4.20).

A separate scanning protocol should be established for CT examinations of deceased persons in order to take advantage of the fact that in postmortem imaging, radiation protection can be neglected with respect to the person being examined. In the following, scan parameter settings are described, which should be adjusted by the user for postmortem imaging. These setting options may differ slightly depending on the manufacturer. In case of deviations in the setting options, the reader is referred to consult the corresponding manufacturer. In general, a small collimation should be selected unless it is already selected by default. A small collimation has a positive effect on spatial resolution, although it also increases noise, which is, however, reduced by an increased tube current. In the spiral or helical CT scan method, i.e. with continuous table feed during the rotation of the X-ray tube in the gantry, the pitch factor is another relevant parameter. The pitch factor, which is defined by the ratio between the CT table feed in millimeter per 360° rotation and the total slice collimation, is usually set between 1 and 2 in clinical CT scanning to quickly cover a scan volume and reduce the applied dose [36]. However, for postmortem CT, the pitch factor is set to a value <1 to increase the resolution along the z-axis (the direction in which the CT table moves) and to allow an increase in the tube current-time product, which is expressed in milliampere-seconds (mAs). Therefore, a small collimation, low pitch, and high mAs value are the most important scan parameters that should be selected for postmortem CT scanning to increase spatial resolution and decrease noise [13]. In addition, a medium to slow (tube) rotation time is selected (Figure 4.21).

The X-ray beam energy, which is adjustable by the applied tube voltage, should be chosen with respect to the desired tissue contrast. A tube voltage of 120 kV is usually recommended because it provides the best image contrast for a normal adult decedent. For skeletal remains, a tube voltage of 140 kV; for iodine contrast-enhanced angiography, a tube voltage of 100 kV; and for infants, a tube voltage of 80 kV can slightly increase contrast between different tissue types.

At our institute, a three-step scanning protocol has proved effective. Step 1 is a whole-body scan (basic scan) from head to toe (maximum scan length: almost 2000 mm). Step 2 is a separate head and neck scan with higher tube current setting for a higher signal to noise ratio. For both, step 1 and step 2, separate reconstructions are later calculated for the assessment of the bones and soft tissues. If relevant, a special high-resolution scan (slice thickness: 0.4 mm) over the maxilla and mandible (dental scan) for identification or over the laryngeal region (laryngeal scan) in case of strangulation and hanging is also performed before step 3 (Figure 4.22).

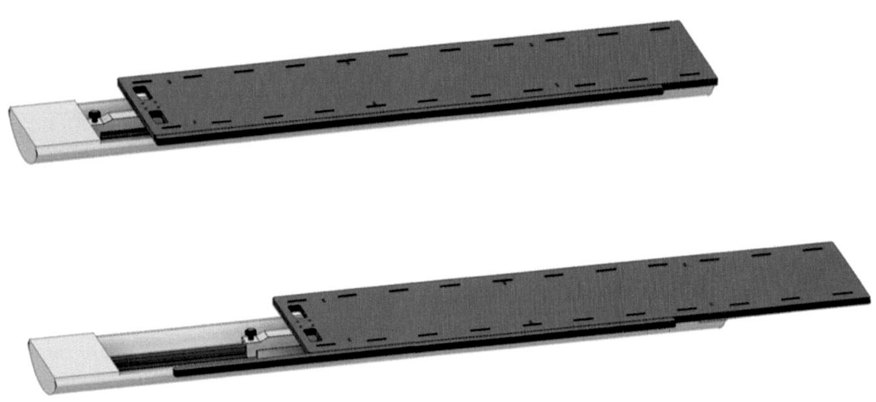

Figure 4.20 A sliding CT table support that can be fixed at any position over a length of 500 mm was specially designed by us (Dominic Gascho and Patrik Wili) together with Jonas Vonaesch (Neurad GmbH) for easy positioning of deceased persons along the z-axis of the CT scanner (table feed direction).

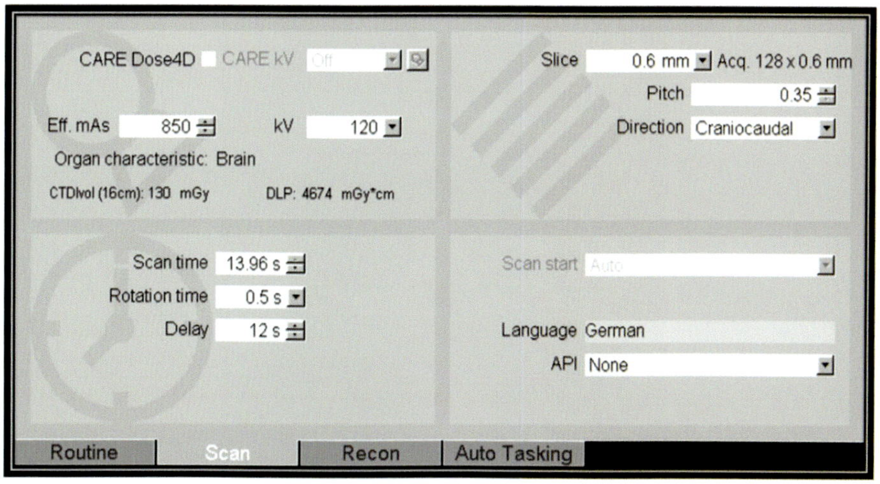

Figure 4.21 Example of the user interface of a CT scanner on which the scan parameters for postmortem examinations have been set. Small collimation (in this case with 128 × 0.6 mm by default), low pitch, high tube current (or high effective tube current-time product), and medium (0.5 seconds) to slow (1 second) rotation time should be set for postmortem CT scans.

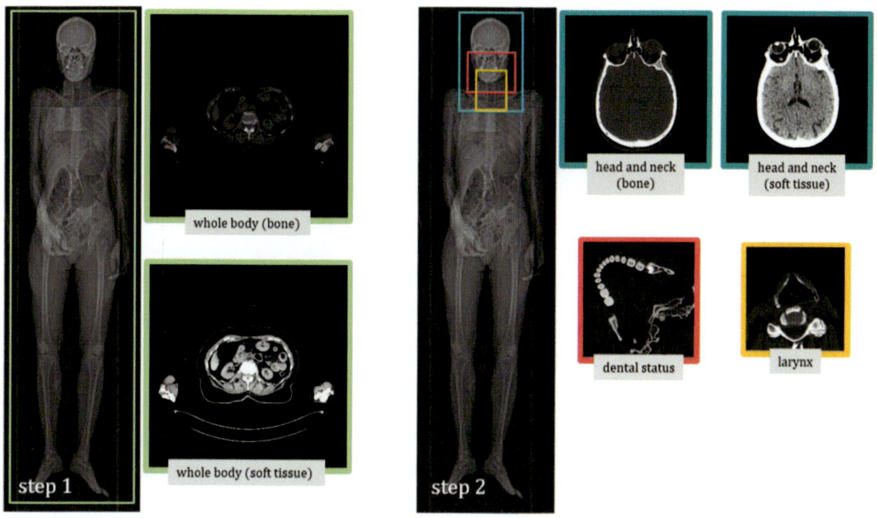

Figure 4.22 Postmortem CT scan protocol consisting of a whole-body scan (step 1) and a head and neck scan (step 2). In the case of identification, an additional dental scan is performed, and in the case of strangulation and hanging, an additional laryngeal scan is performed.

Step 3 is a separate thorax and abdomen scan with arms over the head if body condition and rigor mortis permit. When the arms are above the head, image artifacts in the torso area can be reduced (Figure 4.23), but there is a risk of shoulder dislocation or fractures, especially in older osteoporotic subjects. In addition, it should be noted that postmortem hemorrhage is often caused in the shoulder joint area. Therefore, it is not recommended to lift the arms above the head in case of foul play, i.e. criminal violence or activity that resulted in the person's death, and accidents.

After the acquisition of the raw data, the parameters for the reconstructions must now be defined. Since the CT raw data takes up a lot of storage space, it is usually deleted after 1 week. Therefore, it is even more important to define a suitable reconstruction protocol. An important reconstruction parameter is the reconstruction FOV, which should be manually adjusted for the corresponding body regions for each scanned case. The FOV is reduced according to the respective body region (head and neck scan: $\leq 300 \times 300$ mm^2; thorax and abdomen scan: $\leq 500 \times 500$ mm^2) or is extended (whole-body scan: maximum 780×780 mm^2). The slice thickness is selected according to an isotropic voxel in relation to the size of the FOV (head and neck scan: 0.6 mm; thorax and abdomen scan: 1.0 mm; whole-body scan: 1.5 mm; dental scan and laryngeal scan: 0.4 mm). In addition, an increment of about two-thirds of the slice thickness is usually chosen to generate an image stack with overlapping CT slices, creating smoother transitions when scrolling through the image stack. As mentioned before, separate reconstructions are calculated to evaluate the bones and soft tissues, but neither the FOV nor the slice thickness and increment are changed for this purpose, only the kernel

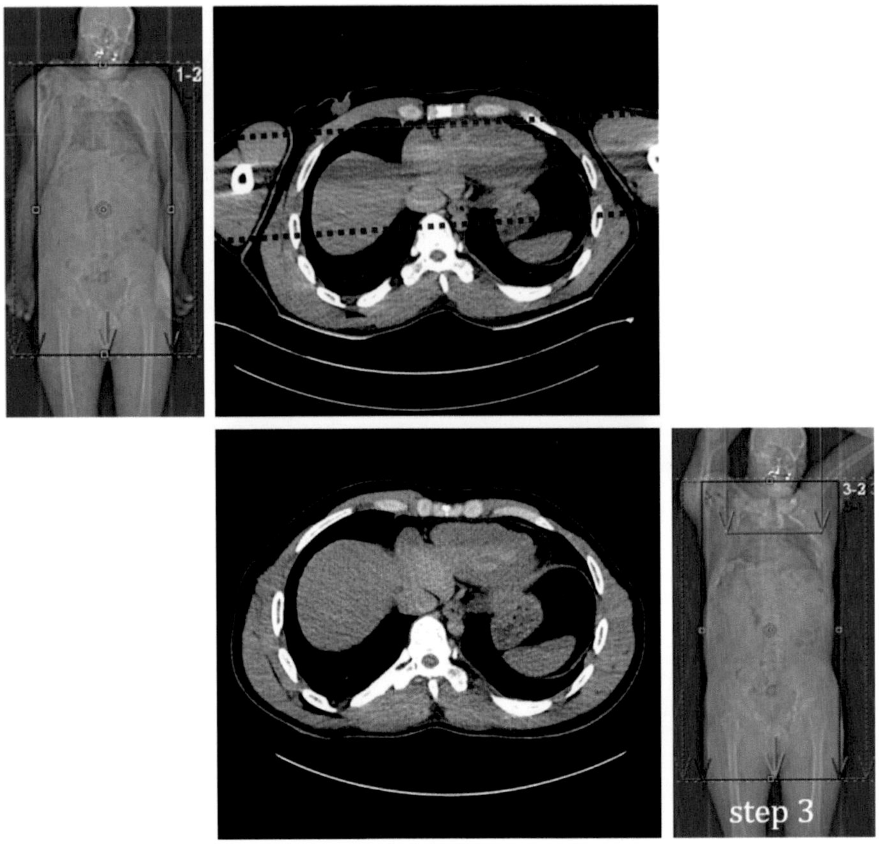

Figure 4.23 Thorax and abdomen scan with the arms next to the body (upper row) and the arms above the head (lower row), considerably reducing radiation hardening artifacts. However, lifting the arms above the head can also cause shoulder dislocations, fractures, and postmortem hemorrhages.

(whole-body scan: B38 for soft tissue and B60 for bones; head and neck scan: H38 for soft tissue and H60 for bones; thorax and abdomen scan: B38 for soft tissue; B60 for bones and lungs). For the assessment of the lung structures, the same hard kernel (B60) is used as for the bones; thus no additional reconstruction is calculated, but only the HU window is adjusted for the assessment. For the laryngeal scan, a medium-hard kernel (H45) is used to visualize the cartilaginous structure well. For the dental scan, a very hard kernel (U68) is used and additionally the extended CT scale algorithm is applied, which allows one to visualize metallic implants and foreign bodies more accurately based on appropriate HU windowing. The individual reconstruction parameters should be saved as standard protocols.

If relevant, further reconstruction jobs can be defined from the raw data of the individual CT scans, for instance to adjust the FOV to a small area in the case of an unclear fracture or to additionally calculate with extended CT scale in the case of a foreign body. In this case, parameters such as the slice thickness and increment should also be adjusted on the operator interface (Figure 4.24).

By adapting the examination protocols for postmortem CT examinations, high-resolution images with low noise can be generated, which is the basis for detecting smaller but still relevant findings in forensic medicine. Defining and saving appropriate protocols not only helps to establish an examination standard, but also facilitates the operation of the CT device and speeds up the CT examination considerably. If necessary, the application specialist of the manufacturer can help to select the appropriate scan and reconstruction parameters. In summary, the appropriate protocol can help to detect findings, although the quality of reporting ultimately depends on the training and experience of the person who diagnostically visualizes the CT images.

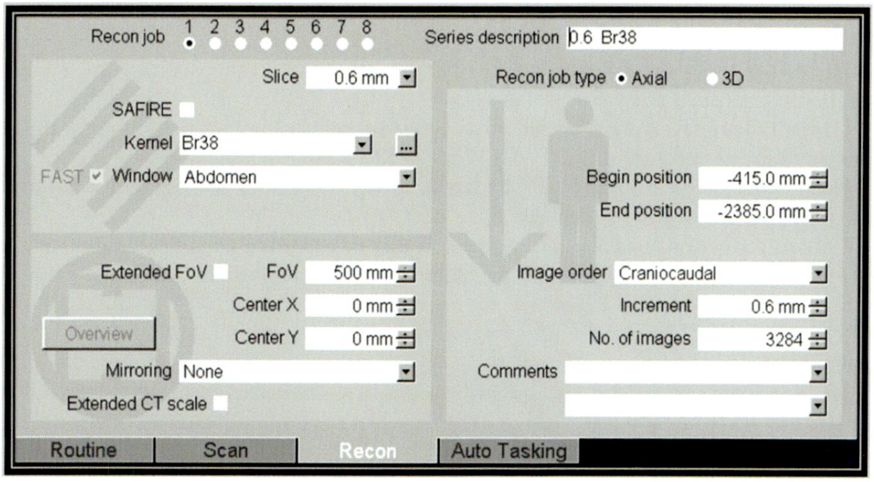

Figure 4.24 Example of the user interface of a CT scanner, where the reconstruction parameters should be set. The slice thickness should be adapted to the FOV in order to create isotropic voxels. Thus, in this case, with a reconstruction FOV of 500 × 500 mm^2 (and a matrix of 512 × 512), a slice thickness of 1 mm (with an increment of 0.6) would be appropriate. The kernel as well as the window preset can be selected from a list (note: the window can be changed afterwards on the viewing software). The extended FOV (501 × 501 mm^2 to 780 × 780 mm^2 in diameter) and extended CT scale can be activated by clicking in the white box. A FOV < 500 × 500 mm^2 (or < 300 × 300 mm^2 for head and neck scans) can be shifted as required for specific additional reconstructions within the area of 500 × 500 mm^2 (or 300 × 300 mm^2) with the isocenter of the gantry as the center, i.e. the center of the smaller FOV shifts accordingly from the isocenter (center X and center Y). Sinogram-affirmed iterative reconstruction (SAFIRE) can also be activated for additional reconstructions and five different strength levels are then available.

References

[1] W.C. Röntgen, On a new kind of rays, Science. 3 (1896) 227–231.

[2] B.G. Brogdon, J.E. Lichtenstein, Brogdon's forensic radiology—Chapter 2. Forensic radiology in historical perspective. In: Thali MJ, Viner MD, Brogdon BG, editors. Brogdon's forensic radiology, Second Edition 2011, CRC Press—Taylor & Francis Group, p. 9–24, n.d.

[3] A.M. Cormack, Representation of a function by its line integrals, with some radiological applications, J. Appl. Phys. 34 (1963) 2722–2727. https://doi.org/10.1063/1.1729798.

[4] A.M. Cormack, Representation of a function by its line integrals, with some radiological applications. II, J. Appl. Phys. 35 (1964) 2908–2913. https://doi.org/10.1063/1.1713127.

[5] G.N. Hounsfield, Computerized transverse axial scanning (tomography): Part 1. Description of system, Br. J. Radiol. 46 (1973) 1016–1022. https://doi.org/10.1259/0007-1285-46-552-1016.

[6] J. Ambrose, Computerized transverse axial scanning (tomography): Part 2. Clinical application, Br. J. Radiol. 46 (1973) 1023–1047. https://doi.org/10.1259/0007-1285-46-552-1023.

[7] B.J. Perry, C. Bridges, Computerized transverse axial scanning (tomography): Part 3. Radiation dose considerations, Br. J. Radiol. 46 (1973) 1048–1051. https://doi.org/10.1259/0007-1285-46-552-1048.

[8] M. Schumacher, M. Oehmichen, H.G. König, H. Einighammer, Intravital and postmortal CT examinations in cerebral gunshot injuries, ROFO. Fortschr. Geb. Rontgenstr. Nuklearmed. 139 (1983) 58–62. https://doi.org/10.1055/s-2008-1055840.

[9] P. Krantz, S. Holtås, Postmortem computed tomography in a diving fatality, J. Comput. Assist. Tomogr. 7 (1983) 132–134.

[10] M.J. Thali, K. Yen, W. Schweitzer, P. Vock, C. Boesch, C. Ozdoba, G. Schroth, M. Ith, M. Sonnenschein, T. Doernhoefer, E. Scheurer, T. Plattner, R. Dirnhofer, Virtopsy, a new imaging horizon in forensic pathology: Virtual autopsy by postmortem multislice computed tomography (MSCT) and magnetic resonance imaging (MRI): A feasibility study, J. Forensic Sci. 48 (2003) 386–403.

[11] R. Dirnhofer, C. Jackowski, P. Vock, K. Potter, M.J. Thali, VIRTOPSY: Minimally invasive, imaging-guided virtual autopsy, RadioGraphics. 26 (2006) 1305–1333. https://doi.org/10.1148/rg.265065001.

[12] D. Lira, A. Padole, M.K. Kalra, S. Singh, Tube potential and CT radiation dose optimization, Am. J. Roentgenol. 204 (2015) W4–W10. https://doi.org/10.2214/AJR.14.13281.

[13] D. Gascho, M.J. Thali, T. Niemann, Postmortem computed tomography: Technical principles and recommended parameter settings for high-resolution imaging, Med. Sci. Law. 58 (2018) 70–82. https://doi.org/10.1177/0025802417747167.

[14] W.A. Kalender, Computed tomography: Fundamentals, system technology, image quality, applications—Chapter 1: Principles of computed tomography, John Wiley & Sons, 2011.

[15] J.H. Hubbell, S.M. Seltzer, X-ray mass attenuation coefficients, NIST Stand. Ref. Database. 126 (2004) 20899. https://doi.org/10.18434/T4D01F.

[16] M.J. Willemink, M. Persson, A. Pourmorteza, N.J. Pelc, D. Fleischmann, Photon-counting CT: Technical principles and clinical prospects, Radiology. 289 (2018) 293–312. https://doi.org/10.1148/radiol.2018172656.

[17] W.A. Kalender, Computed tomography: Fundamentals, system technology, image quality, applications—Chapter 9: Mathematical aspects of image reconstruction, John Wiley & Sons, 2011.

[18] P. Suetens, Fundamentals of medical imaging—Chapter 2: X-ray computed tomography, Cambridge University Press, 2017.

[19] N.B. Smith, A. Webb, Introduction to medical imaging: Physics, engineering and clinical applications—Chapter 1: General image characteristics, data acquisition and image reconstruction, Cambridge University Press, 2010.

[20] L.L. Geyer, U.J. Schoepf, F.G. Meinel, J.W. Nance, G. Bastarrika, J.A. Leipsic, N.S. Paul, M. Rengo, A. Laghi, C.N. De Cecco, State of the art: Iterative CT reconstruction techniques, Radiology. 276 (2015) 339–357. https://doi.org/10.1148/radiol.2015132766.

[21] C. Ghetti, F. Palleri, G. Serreli, O. Ortenzia, L. Ruffini, Physical characterization of a new CT iterative reconstruction method operating in sinogram space, J. Appl. Clin. Med. Phys. 14 (2013) 263–271. https://doi.org/10.1120/jacmp.v14i4.4347.

[22] G.E. Moore, Cramming more components onto integrated circuits. Electronics. 38 (8), Sidsel Lond Grosen Agnete Meldgaard Hansen Jo Krøjer. (1965).

[23] W.-D. Zech, C. Jackowski, Y. Buetikofer, L. Kara, Characterization and differentiation of body fluids, putrefaction fluid, and blood using Hounsfield unit in postmortem CT, Int. J. Legal Med. 128 (2014) 795–802. https://doi.org/10.1007/s00414-014-1030-0.

[24] C. Booz, P.C. Hofmann, M. Sedlmair, T.G. Flohr, B. Schmidt, T. D'Angelo, S.S. Martin, L. Lenga, D. Leithner, T.J. Vogl, J.L. Wichmann, Evaluation of bone mineral density of the lumbar spine using a novel phantomless dual-energy CT post-processing algorithm in comparison with dual-energy X-ray absorptiometry, Eur. Radiol. Exp. 1 (2017) 11. https://doi.org/10.1186/s41747-017-0017-2.

[25] E.L. Nickoloff, F. Feldman, J.V. Atherton, Bone mineral assessment: New dual-energy CT approach, Radiology. 168 (1988) 223–228. https://doi.org/10.1148/radiology.168.1.3380964.

[26] M.U. Sedlmair, Dual-Energy-CT: physikalische Modelle und Anwendungen, Shaker, 2010.

[27] T. Johnson, C. Fink, S.O. Schönberg, M.F. Reiser, Dual energy CT in clinical practice, Springer Science & Business Media, 2011.

[28] D. Gascho, N. Zoelch, H. Richter, A. Buehlmann, P. Wyss, S. Schaerli, Identification of bullets based on their metallic components and X-Ray attenuation characteristics at different energy levels on CT, Am. J. Roentgenol. (2019) 1–9. https://doi.org/10.2214/AJR.19.21229.

[29] D. Gascho, N. Zoelch, H. Richter, A. Buehlmann, P. Wyss, M.J. Thali, S. Schaerli, Heavy metal in radiology: How to reliably differentiate between lodged copper and lead bullets using CT numbers, Eur. Radiol. Exp. 4 (2020) 43. https://doi.org/10.1186/s41747-020-00168-z.

[30] W.A. Kalender, Computed tomography: Fundamentals, system technology, image quality, applications—Chapter 4: Image quality, John Wiley & Sons, 2011.

[31] F.R. Verdun, D. Racine, J.G. Ott, M.J. Tapiovaara, P. Toroi, F.O. Bochud, W.J.H. Veldkamp, A. Schegerer, R.W. Bouwman, I.H. Giron, N.W. Marshall, S. Edyvean, Image quality in CT: From physical measurements to model observers, Phys. Med. 31 (2015) 823–843. https://doi.org/10.1016/j.ejmp.2015.08.007.

[32] D. Gascho, L. Ganzoni, P. Kolly, N. Zoelch, G.M. Hatch, M.J. Thali, T.D. Ruder, A new method for estimating patient body weight using CT dose modulation data, Eur. Radiol. Exp. 1 (2017) 23. https://doi.org/10.1186/s41747-017-0028-z.

[33] J.C. Herath, S.O. Herath, Is it time for targeted and minimally invasive postmortem examination using total body computed tomography in a medicolegal autopsy?, Forensic Sci. Med. Pathol. 17 (2021) 175–176.

[34] R. Tse, P. Morrow, M. Sage, S. Stables, Targeted and minimally invasive postmortem examination with total body computed tomography not recommended in New Zealand, Forensic Sci. Med. Pathol. 17 (2021) 373–374.

[35] P.M. Flach, D. Gascho, W. Schweitzer, T.D. Ruder, N. Berger, S.G. Ross, M.J. Thali, G. Ampanozi, Imaging in forensic radiology: An illustrated guide for postmortem computed tomography technique and protocols, Forensic Sci. Med. Pathol. 10 (2014) 583–606. https://doi.org/10.1007/s12024-014-9555-6.

[36] W.A. Kalender, Computed tomography: Fundamentals, system technology, image quality, applications—Chapter 3: Spiral CT, John Wiley & Sons, 2011.

CHAPTER 5

INTRODUCTION TO MAGNETIC RESONANCE

Niklaus Zoelch and Dominic Gascho

DOI: 10.4324/9781003413998-6

5.1 INTRODUCTION

The development from the description of the phenomenon of magnetic resonance to magnetic resonance imaging (MRI) and magnetic resonance spectroscopy (MRS) as we know them today is an incredible intellectual achievement by countless scientists and engineers that deserves our greatest admiration[1]. It is presumptuous to attempt to cover the basics of MRS and MRI in a short section within the scope of this book. The aim of this text is therefore only to give a simple approach to the subject in order to understand the following chapters. Whenever possible, we refer to articles and books for further reading, which we warmly recommend.

The fascinating insights into the body provided by magnetic resonance imaging and magnetic resonance spectroscopy are due to a special property of atomic nuclei known as spin or spin angular momentum[2]. Whether a certain atomic nucleus has a spin or not depends on the mass and the number of neutrons and protons in the nucleus[2]. The nuclear spin is a fixed property of the nucleus, which can change direction but cannot become larger or smaller for a given nucleus. So even at zero Kelvin, nuclei have spin. Closely associated with the spin is a magnetic moment which leads to interactions with electromagnetic fields and, surprisingly, measurable effects over a large number of nuclei. The spin is characterized by the nuclear spin quantum number I, and for a hydrogen nucleus (which we consider in this book exclusively), ^{1}H is I=1/2. Also, other nuclei such as ^{13}C, ^{19}F, and ^{31}P have spin I =1/2 and can be used for spectroscopy or imaging[2].

If we bring such a nuclear spin into a strong magnetic field, we can in principle distinguish different states of the spin because the different states have distinguishable energy levels. The differences of these energies are co-determined by the strength of the field at the location of the nucleus. Therefore, if we succeed in somehow "picking up" the energy between different spin states, we can use them to measure the strength of the field at the site of the nucleus. This very fact is used for MRI and MRS, albeit in a slightly different way. It is important to emphasize that we do not measure individual spins in MR experiments; the sensitivity of the devices is far from sufficient for this. For a real MR experiment, therefore, an ensemble of spins must always be considered.

In MRI, the applied magnetic field (and thus the field at the location of the nuclei) is specifically varied by means of field gradients across the measured object. The measured signals then carry the information of the magnetic field at the location of the nuclei, which allows one to reconstruct the origin of the signal and thus to obtain an image. In MRS, we can determine which molecule a specific nucleus (or better, a specific group of nuclei) belongs to based on the measured magnetic field at the location of the nucleus, since the chemical environment influences this.

In the following, we will describe a simple magnetic resonance experiment to illustrate how the magnetic field at the location of the nucleus is measured.

The nuclear spin is per se a quantum mechanical quantity, and in general, only quantum mechanics can correctly quantitatively describe the interaction of the nucleus with magnetic fields. Therefore, for the description of the magnetic resonance experiment, the expectation value of the magnetization (sum of all magnetic moments in a sample) operator would have to be calculated using the Schrödinger equation and the density operator[2]. However, for non-interacting nuclear spins (such as for hydrogen nuclei in water) in a sufficiently large sample, the expectation value of the magnetization during the MR experiment has been shown to behave as classically expected for the magnetic moment of a particle with angular momentum, so e.g. as for a bar magnet rotating around its own axis in an external field. The motion of the magnetization M in this case can be described by,

$$\frac{d}{dt}\boldsymbol{M} = -\gamma \boldsymbol{M} \times \boldsymbol{B} \tag{1}$$

where γ is the proton gyromagnetic ratio and B the magnetic field.

Thus, if we somehow accept that in a certain amount of water in an external magnetic field B_0 a certain magnetization is created and with this magnetization an angular momentum is associated, we can describe the behavior of this magnetization in an MR experiment with classical equations without having to use quantum mechanics[3]. This not only gives us a more intuitive picture of the experiment but also allows us to make correct statements about the observable quantities. To better understand certain aspects of the MR experiment like relaxation or RF excitation, it can sometimes be useful to look at the magnetization in more detail. The "spin-orientation sphere"[3,4] is a very good way to do this without having to use quantum mechanics.[1] Here you look at a multitude of magnetic moments with the same angular momentum pointing in all possible directions. This description results in a useful and intuitive insight into the origin of the magnetization.

For now, we recommend to just take it for granted: If we put a milliliter of water into a strong external magnetic field, a macroscopic magnetization over the sample is built up whose motion can be described with equation 1. This magnetization is relatively small and points in the direction of the external magnetic field (conventionally aligned with the z-axis). The magnetization that builds up is greater at higher external fields and lower at higher temperatures. We cannot directly measure this static magnetization and therefore some manipulation is necessary[5]. If we now take alternating current with a suitable frequency, in our case the same frequency used to generate radio waves, and pass it through a coil, a rotating magnetic field B_1 (near-magnetic field) can be generated perpendicular to B_0. Such a field leads to a deflection of the magnetization from the z-axis, as can be shown by solving the equation 1 given above[6]. As soon as the magnetization is deflected, it starts to process in the xy-plane with the frequency ν_0

$$\nu_0 = \frac{\omega_0}{2\pi} = \frac{\gamma}{2\pi} B_0 \tag{2}$$

This also makes it clear why B_1 must have a suitable frequency. Figuratively, only if the B_1 field follows the motion of the magnetization is a deflection possible. For $\frac{\gamma}{2\pi} \approx 42.577$ MHz T^{-1} and common field strengths, ν_0 is in the range of frequencies used for radio transmission[7].

The mentioned precession can also be observed with a gyroscope in the gravitational field of the earth. If the gyroscope has no angular momentum (it does not rotate around its own axis), it simply falls over when deflected. However, if it rotates fast enough around its own axis, it starts to rotate around the axis of the gravity field.

The resulting rotating magnetization finally induces an electromotive force in a receiving coil, which needs to be brought close to the water sample[8]. This is a similar situation as in a bicycle dynamo, where the rotating magnet induces a current in the coils of wire. The induced electromotive force depends on the sensitivity of the coil at the location of the water sample. In the so-called receive chain, this electromotive force is finally converted into a measurable AC voltage. As a result, we get a voltage whose frequency depends on the strength of the magnetic field and whose amplitude is directly proportional to the number of nuclear spins in the sample.

With equation (1), the signals would remain observable for any length of time after deflection by B_1. However, this is clearly not the case, and the magnetization will return to its initial state prior to deflection M_0, which is denoted as the equilibrium state. The return to the equilibrium state is denoted as relaxation[9]. By adding the relaxation process to equation 1, we arrive at a set of differential equations known as the Bloch equations[6,10]. In the Bloch equations, the effects of relaxation are considered by introducing relaxation times T_1 and T_2[11]. The constant T_1 is the longitudinal or spin-lattice relaxation time and describes the evolution of M_z (magnetization in the z direction) back to its equilibrium value M_0. A change in M_z involves energy exchange of ^1H with their atomic and molecular environment. T_2 describes the loss of the magnetization in the x-y plane (denoted as transversal magnetization) and is mainly due to interaction between the proton spin itself and involves no energy exchange with the surroundings. If the relaxation process is characterized by a simple single exponential decay, as in the Bloch equations, the transverse magnetization will decay so that it loses 63% of its initial value after time T_2. B_0 inhomogeneities can accelerate the decay of the transverse magnetization, which then is described by an effective decay time T_{2^*}, but this loss can be refocused in echo experiments and does not describe a true relaxation process.

Note

1 Quantum mechanics provides us with rules, which result in the formation of magnetization, but is not very intuitive and somehow you never get beyond just applying rules without a lot of practice. Most explanations based on classical mechanics allow an intuitive idea and a correct prediction about the magnetization that occurs. Nevertheless, unfortunately these are always simplifications, which reach their limits in other contexts and let you down, because they disregard rules of quantum mechanics. It is a bit like a fairy tale. The fairy tale may explain an important and correct moral, but a wolf will simply never be able to speak.

References

1. Becker ED. A brief history of nuclear magnetic resonance. *Analytical Chemistry*. 1993;65(6):295A–302A.
2. Levitt MH. *Spin dynamics: Basics of nuclear magnetic resonance*. John Wiley & Sons; 2013.
3. Hanson LG. Is quantum mechanics necessary for understanding magnetic resonance? *Concepts in Magnetic Resonance Part A*. 2008;32A(5):329–340.
4. De Graaf RA. *In vivo NMR spectroscopy: Principles and techniques*. John Wiley & Sons; 2019.
5. Mlynarik V. Introduction to nuclear magnetic resonance. *Analytical Biochemistry*. 2017;529:4–9.
6. Worthoff WA, Yun SD, Shah NJ. Chapter 1 Introduction to magnetic resonance imaging. *In: Hybrid MR-PET imaging: Systems, methods and applications*. The Royal Society of Chemistry; 2019:1–44.
7. Physical Measurement Laboratory of NIST. The NIST Reference on Constants, Units, and Uncertainty. https://physics.nist.gov/cgi-bin/cuu/Value?gammapbar. Published 2018. Accessed January 17, 2022.

8. Hoult DI. The origins and present status of the radio wave controversy in NMR. *Concepts in Magnetic Resonance Part A*. 2009;34A(4):193–216.
9. Keeler J. *Understanding NMR spectroscopy*. John Wiley & Sons; 2011.
10. Bernstein MA, King KF, Zhou XJ. *Handbook of MRI pulse sequences*. Elsevier; 2004.
11. Johnston ER. Solution of the bloch equations including relaxation. *Concepts in Magnetic Resonance Part A*. 2020; 2020:1–5.

CHAPTER 6

MAGNETIC RESONANCE IMAGING—BASICS AND FORENSIC POSTMORTEM APPLICATIONS

Dominic Gascho and Niklaus Zoelch

DOI: 10.4324/9781003413998-7

6.1 INTRODUCTION

Magnetic resonance imaging (MRI) is essentially the use of magnetic resonance (MR) to produce cross-sectional images of the human (or animal) body in analogy to X-ray computed tomography, which is why this imaging technique is sometimes referred to as magnetic resonance tomography. For this purpose, the basic property of MR is used in a clever way to measure the field at the location of the hydrogen nucleus. By making the magnetic field location dependent, the spatial origin of the measured signals can be determined. Contrast-specific sequences for acquiring the MR signals can then be used to provoke differences in the signals between different locations in the object, thus collecting spatially resolved information about the measured object. Common contrasts are generated by proton density (PD), relaxation times (T1, T2, and T2*), susceptibility, or diffusion, all of which are also described in this chapter. But it should also be mentioned here that there are many more contrasts in MRI, and some may not have been discovered yet. However, before the spatial assignment of MR signals and the basic principles of image contrasts are explained in more detail, a brief excursion into MR basics and the history of MRI will serve as an introduction to this chapter.

MRI is based on the excitation of net magnetization, which results from the distribution of the nuclear spin orientations of hydrogen protons in a strong magnetic field. The resonant frequency necessary to excite net magnetization is the characteristic Larmor frequency, which is determined by the gyromagnetic ratio times the static magnetic field (B_0 field). The value of the gyromagnetic ratio (γ)for hydrogen, usually given as $\gamma/2\pi$, is 42.58 MHz per Tesla (T). The frequency used in MRI is therefore in the radio frequency range (around 20 kHz to around 300 GHz). Accordingly, for 3.0 T MRI the wavelength in free space is about 235 cm, and taking into account the dielectric constant of tissue, the wavelength in tissue is about 30 cm [1]. Since the resolution of an imaging system is theoretically limited by half the wavelength of the applied radiation due to the diffraction limit, direct MR signal acquisition for imaging is insufficient even for imaging of full-grown elephants, as noted by Richard Robert Ernst [2] before describing the imaging encoding concept in MRI, which is based on spectroscopic encoding. But let's start at the beginning. In 1973, Paul Christian Lauterbur introduced the idea of using linear magnetic field gradients in different directions for image acquisition [3], followed by the article "Magnetic Resonance Zeugmatography" [4] in 1974, which described the use of gradients and projection techniques to create a "zeugmatograph" of an animal. Then in 1975, Anil Kumar, Dieter Welti, and Richard Robert Ernst of the Eidgenössische Technische Hochschule (ETH) Zürich published the article "NMR Fourier Zeugmatography" [5], in which they proposed to apply a sequence of pulsed orthogonal linear field gradients to the sample during the free induction decay (FID) and to use the Fourier transform to reconstruct the spatial proton density function, which is still the basis of image reconstruction in magnetic resonance tomography. It is also worth mentioning the line scanning and selective irradiation method developed by Peter Mansfield et al. [6,7], which led to the first in vivo MR image of a human in 1977, a cross-sectional image of the finger of Andrew Maudsley, a student of Peter Mansfield [8]. From then on a steady increase in the use of MR imaging in clinical medicine can be observed, and whole-body MR scanners with increasingly strong static magnetic fields were built in parallel with the further development of superconducting magnets. Thus, already in 1985, the first clinical 1.5 T MR scanner was introduced, which is still widely used today, and finally in 1998, the first 3.0 T MR scanner was implemented in clinical medicine [9]. In 2017, the first 7.0 Tesla MR scanner was approved for clinical diagnostics, and shortly thereafter, postmortem 7.0 Tesla MRI examinations were performed and scientifically evaluated for forensic purposes [10,11]. Regarding postmortem 7.0 Tesla MRI in forensic medicine, the reader is referred to the corresponding further chapter on these topics in this book.

This chapter briefly discusses the process of spectroscopic encoding and describes the principles of image contrast in MRI. In addition, standard T_1-weighted and T_2-weighted sequences as well as advanced sequences such as susceptibility-weighted and diffusion-weighted sequences are described. The final part of this chapter addresses postmortem MR scanning in practice and highlights the differences between clinical and postmortem MRI.

6.2 HOW DOES IMAGE FORMATION IN MRI WORK?

The main task for creating an MR-based image is to make the magnetic field spatially dependent so that the measured signals of the rotating magnetization can be assigned to a certain location within the measured object. This can be achieved by gradients. A gradient creates a spatial variation of the static magnetic field by superimposing a gradient field (in $T{\cdot}m^{-1}$) over the B_0 field [12]. Therefore, the Larmor frequency (ω), the frequency of the precession of the magnetization, becomes a function of the position along the gradient that is applied, for example, along the z-axis. For a gradient with strength G_z along the z-axis, the Larmor frequency is described by [12]:

$$\omega(z) = \gamma \cdot B_0 + \gamma \cdot G_z \cdot z$$

In a coordinate system that is rotating with $\omega_0 = \gamma \cdot B_0$ (rotating frame of reference), which is commonly used to simplify the description of MR experiments, the equation simplifies to:

$$\omega(z) = \gamma \cdot G_z \cdot z$$

By the way, switching on a gradient field causes loud noises during MR scanning. Rapidly switched electrical currents pass through the gradient coils, which, in the presence of a strong magnetic field, lead to strong Lorentz forces. This leads to vibration of the gradient coils, which then leads to acoustic noises.

To describe image formation, a homogeneous head-shaped object is assumed. Homogenous means that the object has a homogeneous density of hydrogen nuclei of water and thus initially a homogeneous distribution of magnetization, which is formed when the object is placed in the static magnetic field. To explain the principle of spectroscopic encoding in a simple way, a slice-selective gradient echo sequence with one excitation pulse and five gradients is illustrated (Figure 6.1).

The following describes the function of gradients 1 to 5 for image formation. For more detailed information on the RF pulse and how it acts on the proton spins and how finally a detectable signal is generated, the reader is referred to the corresponding chapter on the general introduction to magnetic resonance.

In the first step, a particular slice is selected by a slice-selecting gradient (gradient 1), which is applied in combination with the RF pulse. This gradient determines in which slice through the object the RF pulse leads to an effective

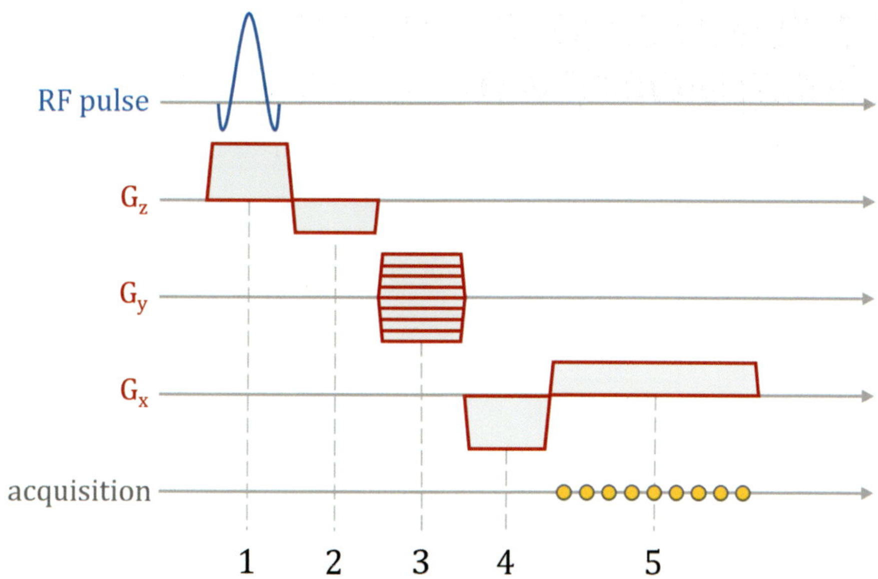

Figure 6.1 Gradient echo sequence. The slice selection gradient (1) is switched on during the transmission of the radiofrequency (RF) pulse. Then the rephaser gradient (2) is applied to rewind the phase accumulated during the duration of the pulse. After that the phase-encoding gradient (3) is applied. Gradient 4 and the first half of gradient 5 are then used to create the gradient echo. Gradient 5 also generates the frequency encoding during the acquisition of the signal, represented by the yellow dots.

excitation of the magnetization, and the excitation bandwidth of the RF pulse in combination with the gradient strength determines the thickness of the selected slice (Figure 6.2).

Since such an excitation pulse is relatively long, with a typical duration of about 1 ms, the protons acquire a certain phase during excitation (due to the gradient along z). This phase is rewound by the rephaser gradient (gradient 2) to avoid signal loss.

In the second step, further gradients are used to vary the field within the slice so that signals can be assigned to the corresponding position from which they originate. For this, it needs phase-encoding gradients and frequency-encoding gradients. The frequency-encoding gradient (gradient 5) is switched on during the actual signal detection. This gradient leads to a local variation of the frequency at which the magnetization rotates after excitation. In our case, this is done along the x-axis. Thus, during image reconstruction, the local origin of the signal along the x-axis can be determined based on the detected frequency (Figure 6.3.).

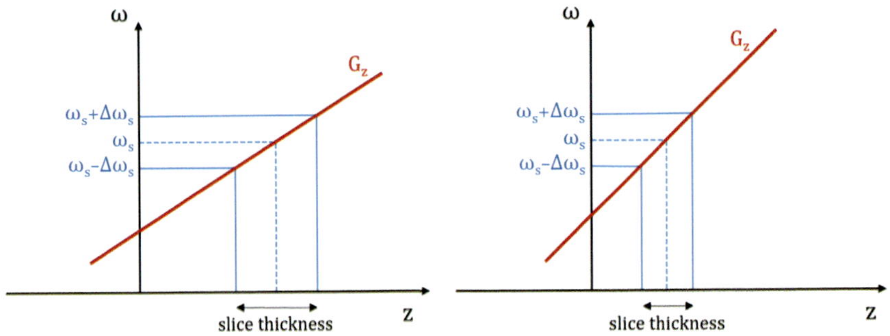

Figure 6.2 The selected center frequency of the excitation pulse (ωs) defines the position along the z-axis, and the excitation bandwidth of the pulse (±Δωs) together with strength of the gradient (Gz) defines the slice thickness.

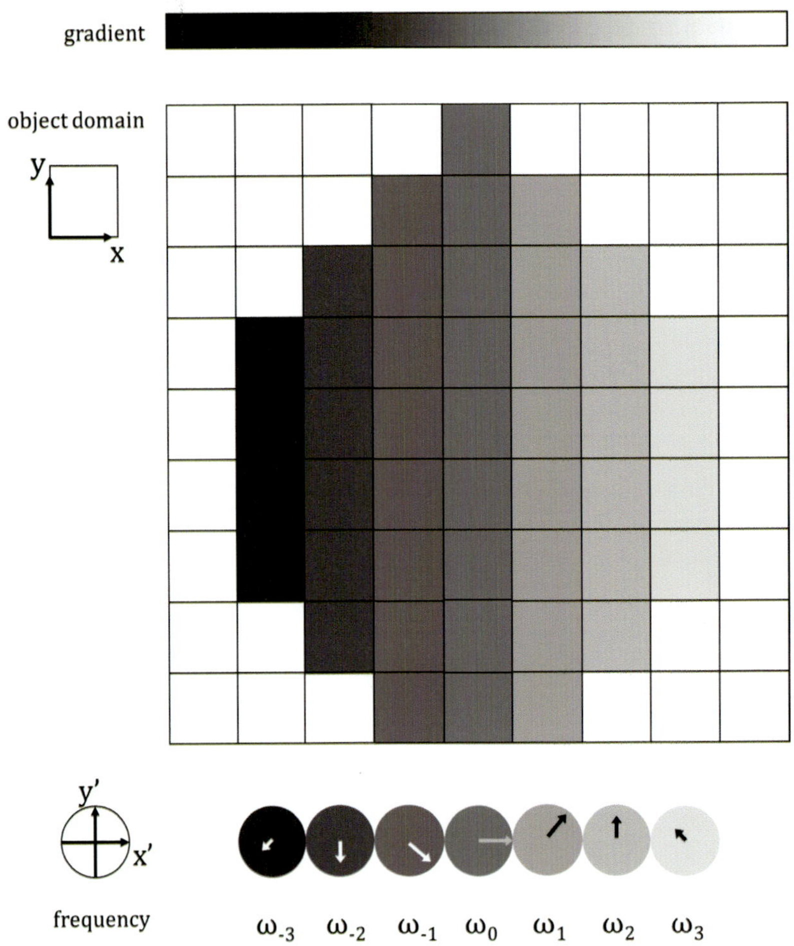

Figure 6.3 The gradient along the x-axis causes a different frequency (ω) in each row along the x-axis. The sum of magnetizations along each row is represented by the size of the vector arrow. There is no tissue in the leftmost and rightmost columns, and thus no magnetization is detected.

Until now, the positions have only been encoded in frequencies along one spatial axis. Since frequency encoding cannot be used simultaneously on two spatial axes, the position along the y-axis is encoded into the signal phase. This is done by the phase-encoding gradient (gradient 3), which in our case is applied along the y-axis for a certain period of time (τ). After that, the signal phase (φ) depends on the position along the y-axis (Figure 6.4) [12]:

$$\varphi\left(G_y,\tau\right) = \omega\left(y\right)\cdot\tau = \gamma\cdot G_y\cdot y\cdot\tau$$

However, by applying only one gradient strength along y, it is not possible to encode the positions in the individual rows along the y-axis within the selected slice. While with frequency encoding, a signal from a specific location can be obtained directly from the sum of the signals via a change from the time domain to the frequency domain by means of Fourier transformation, this is not directly possible with phase encoding, so that several manipulations are necessary to finally be able to separate the signals from different locations. In fact, there needs to be one phase-encoding gradient step for each location or voxel in the phase-encoding gradient direction. In our case we applied nine phase-encoding gradients with different strength for each repetition to encode the origin along y-axis correctly. The gradient strength is changed in equal steps between the maximum amplitude of the gradient and the minimum value, this is illustrated by the different lines in gradient 3. The time between repetitions of the sequence is called the repetition time (TR), which will be discussed in more detail later in this chapter.

Gradient number 4 in the sequence (and also the first half of gradient 5) are there to form the so-called gradient echo. This is in principle not mandatory for image formation. In another typical sequence, the spin echo sequence, which will be discussed in more detail later in this chapter, gradient 4 is replaced by a 180° RF pulse. Or, in principle, it would be possible to measure directly the signal generated by the 90° pulse without the need to generate an echo. However, the gradient echo method has proved a very clever signal manipulation for MRI, since it can be repeated quickly (short TR) and provides a good signal-to-noise ratio despite the time that is needed for phase and frequency encoding.

In MRI, the signal is given by the transverse magnetization and detected with receiver coils, which in principle consists of conductor loops. The transverse magnetization performs a precession, i.e. a rotation about the z-axis, as it relaxes, resulting in a magnetic field that changes over time while it permits through the conductor loop. This induces

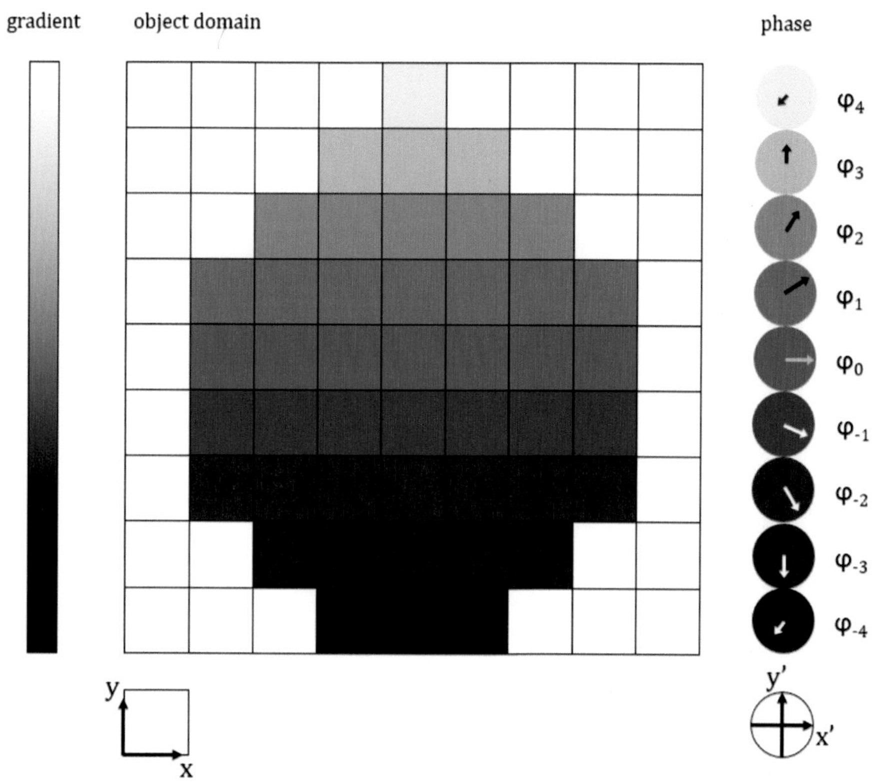

Figure 6.4 The gradient along the y-axis causes a different phase (φ) in each line along the y-axis. The sum of magnetizations along each line is represented by the size of the vector arrow.

a voltage that is just proportional to the rate of change of the magnetic field through the loop, as described by Faraday induction. The magnetic moment generated within the range of such a conductor loop is thus converted into a voltage, which is then converted into digital data for numerical processing. Therefore, the receiver coil should be positioned close to the region of interest, and is usually located close to the body surface. The digital data is sampled as complex values in the so-called k-space, where "k" refers to the spatial frequency of waves. The concept of k-space is situated in the spatial frequency domain, i.e. the Fourier transform of the image measured. Thus, the k-space contains the raw data before reconstruction. The object domain and the k-space are reciprocity domains. The width of the columns (Δx) in the object domain is given by the ratio between the field-of-view (FOV_x) and the number of sampling points, while the distance between two points sampled in k-space (Δk_x) is given by the ratio of the bandwidth (BW) to the number of sampling points. The FOV_x is inversely proportional to Δk, while the BW_x is inversely proportional to Δx, so the BW is inversely proportional to the resolution (Figure 6.5). For a larger FOV with the same resolution, more sampling points are required, resulting in a longer scan time. At this point it should be noted that the time for sampling the signal is limited, because the signal decreases due to relaxation during sampling. Thus, the relaxation sets a limit to the resolution.

In MRI, the resolution depends not only on the gradients but also on the signal and relaxation times, since the resolution is also determined by the contrast between tissue structures. The relaxation times will be described in more detail later. In the following, signal sampling in the k-space by switching the gradients will be discussed. Signal sampling happens while the frequency gradient is applied, while the phase gradient determines which line is read in k-space, depending on its strength: Positive gradients for lines above the center line and negative for lines below the center line (as illustrated for the phase gradient in Figure 6.1). According to frequency and phase encoding, the k-space coordinates k_x and k_y are given by [12]:

$$k_x = \frac{\gamma}{2\pi} \cdot G_x \cdot t$$

$$k_y = \frac{\gamma}{2\pi} \cdot G_y \cdot t$$

where t is the duration of the gradient.

The MR signal (s) that is sampled as a function of the k-space positions is proportional to the double integral or, as discrete points are sampled, to the sums of the proton density (ρ) at each position along the x- and y-axis times, the exponential corresponding to frequency encoding and phase encoding [12]:

$$s(k_x, k_y) \propto \sum\sum \rho(x,y) \cdot e^{-j \cdot 2\pi \cdot k_x \cdot x} \cdot e^{-j \cdot 2\pi \cdot k_y \cdot y}$$

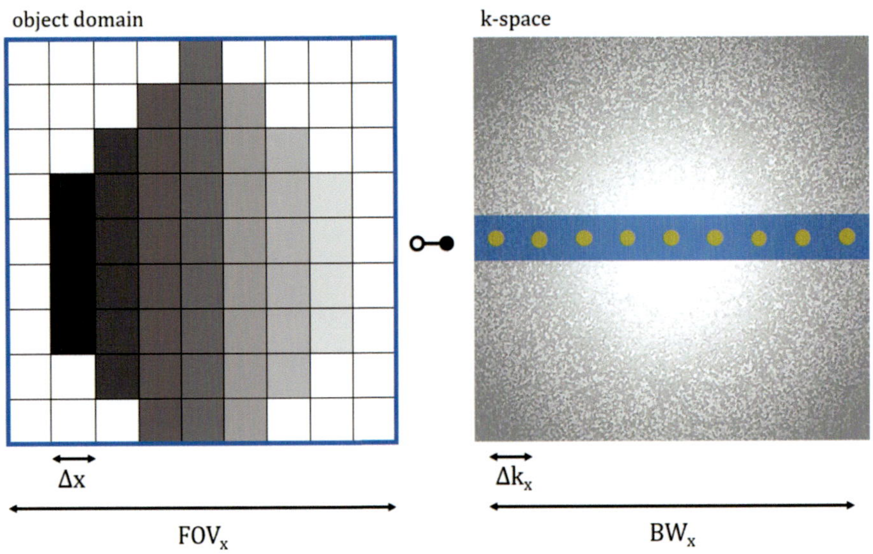

Figure 6.5 The sum of magnetizations of the individual columns in the object domain becomes a projection onto the center line in k-space. The object domain and k-space are reciprocity domains, with FOV_x inversely proportional to Δk_x and BW_x inversely proportional to Δx.

Finally, the proton density as a function of the position is the desired information for the image formation, where the image represents the signal intensity as a function of the position (x,y), which can be obtained by applying the inverse Fourier transform:

$$\rho(x,y) = i(x,y) = \Sigma\Sigma s(k_x, k_y) \cdot e^{+j \cdot 2\pi \cdot (k_x \cdot x + k_y \cdot y)}$$

As mentioned before, the digital data are sampled as complex values and so the inverse Fourier transform also provides a complex image with real and imaginary components and can be expressed in terms of its magnitude and phase. Typical MR images such as T1-weighted and T2-weighted images are usually displayed in magnitude mode, i.e., the magnitude of the real and imaginary components.

6.3 HOW DO T_1 AND T_2 RELAXATION CONTRIBUTE TO IMAGE CONTRAST?

So far, it has been explained how an image is formed based on spectroscopic encoding. Now let's look at the potential information that can be represented in these images. Or, in other words, what contrasts there are. The most obvious contrast in MR images is certainly the difference due to the different relaxation times, which will be discussed in the following.

After the excitation pulse is switched off, the net magnetization vector returns to its original state, the equilibrium; i.e. the net magnetization vector aligns with B_0 again. This recovery of the longitudinal magnetization along the z-axis is described by T_1 relaxation and mainly attributed to "spin-lattice" interaction, although the term "lattice" actually refers only to solids and is therefore not really appropriate here. It can be said that T_1 relaxation depends on how the water molecules are embedded in their environment or how mobile they are in the tissue. In fact, molecules are in constant random motion; they tumble at a certain frequency, which is described by the molecular tumbling rate. The magnetic fields of the spins that reside on this molecule fluctuate at the same frequency. When the tumbling rate of molecules corresponds well to the Larmor frequency, as is the case for lipids, and thus a long correlation time is given, good conditions prevail for energy exchange between the protons, the "spins", and the environment, the "lattice", leading to rapid relaxation and thus short T_1 values [13]. Thus, the same principle applies to relaxation as to excitation, that when frequencies match, energy transfer is promoted. "Free" water, in turn, has a very high molecular tumbling rate. Since water molecules have a very small molecular size, they can tumble very rapidly in their free form, such as in low viscosity liquids like cerebrospinal fluid (CSF). Consequently, CSF has long T_1 values. In biological tissue, the water molecules are restricted in their movement, and, depending on the tissue, the molecular tumbling rate gets closer to the Larmor frequency, resulting in shorter T_1 values. A T_1 value describes the duration (in ms) until the longitudinal magnetization (M_z) reaches 63% of its initial value.

Besides the increase of magnetization along the z-axis (longitudinal magnetization), the magnetization in the xy-plane (transverse magnetization) decreases when the spins lose coherence and eventually is no longer present. The decrease in transverse magnetization is commonly referred to as T_2 relaxation due to "spin-spin" interaction. T_2 relaxation proceeds much faster than T_1 relaxation and accompanies all T_1 relaxation processes, except when the molecular tumbling rate falls below the Larmor frequency, in which case T_2 relaxation occurs independently of T_1 relaxation (Figure 6.6). A slow molecular tumbling rate leads to a faster decay of the transverse magnetization, resulting in short T_2 values. Consequently, very low T_2 values are observed in solids, such as bones, whereas for CSF, almost no T_1-independent T_2 relaxation is observed, and T_2 becomes almost as long as T_1. [13]. The T_2 value describes the duration (in ms) until the transverse magnetization (M_{xy}) has been reduced to 37% of its initial value.

Thus, water molecules (and other molecules) in different tissue and fluid types have different T_1 and T_2 relaxation times. By setting the sequence parameters, differences in signal intensity can be evoked due to the different relaxation times, and thus the contrast of the MR image can be generated.

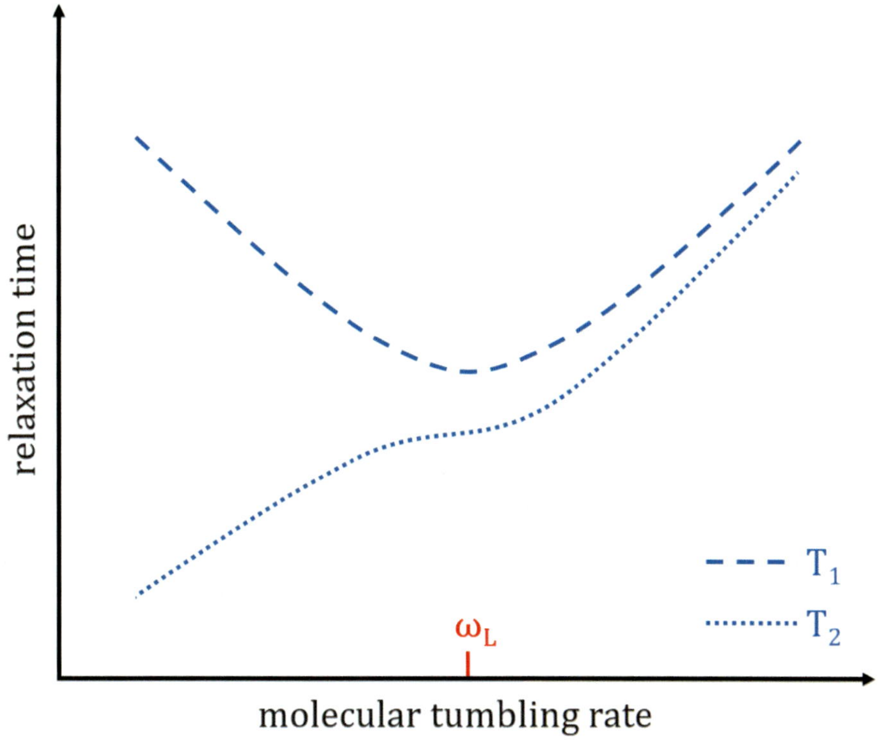

Figure 6.6 The plot of relaxation time as a function of molecular tumbling rate illustrates the dependence of T_1 relaxation times (dashed line) on the coincidence of molecular tumbling rate with Larmor frequency (ω_L), as well as the nearly linear increase in T_2 relaxation times (dotted line) with the increase in molecular tumbling rate.

6.4 WHAT DO REPETITION TIME, INVERSION TIME, AND ECHO TIME MEAN?

Having described the processes of T_1 and T_2 relaxations, it will now be discussed how the desired contrast can be selected depending on the time interval between repeated excitation pulses, whether additional pulses are applied in between, and the time interval between the excitation pulse and the formation of an echo signal. In this context, three times are defined: the repetition time (TR), the inversion time (TI), and the echo time (TE). Using three standard methods, the influence of these times on the image contrast is described in the following, assuming only 90° excitation and 180° inversion pulses for simplicity.

First the partial saturation method. Here a 90° excitation pulse is switched on; i.e. the net magnetization vector along the z-axis (M_z) is tilted into the xy-plane, and after a certain TR, another 90° pulse is switched on; thus the TR is the time interval between excitation pulses. Since the T_1 relaxations of different tissues or fluids pass at different rates, they thus have different amounts of M_z by the second 90° pulse depending on the duration of the TR (Figure 6.7). If the TR is short, this leads to different signal intensities resulting from these differences in T1 relaxation. If the TR is long, tissues and fluids with long T_1 values also have sufficient time to build up M_z, and the different signal intensities result only from the number of protons in the corresponding volume, i.e. the proton density. Thus, T_1-weighted images are characterized by a short TR. M_z recovery is governed by the Bloch equations. With a 90° excitation pulse, M_z recovery can be described with the following exponential increase:

$$M_z \text{ recovery} = 1 - e^{-TR/T_{1(\text{tissue})}}$$

However, to finally generate T_1-weighted images or proton density weighted images, the influence of T_2 relaxation must also be avoided; i.e. a short TE must be chosen, which will be discussed later.

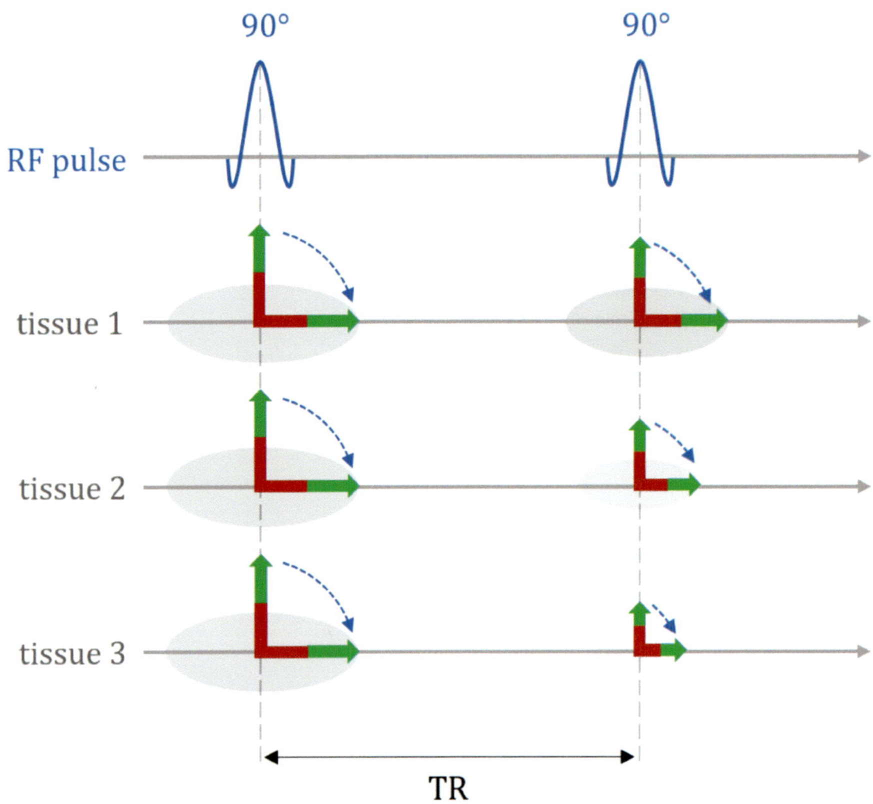

Figure 6.7 After an initial 90° excitation pulse and a certain repetition time, tissue types with different fast T_1 relaxations have recovered a different amount of M_z. When these are now flipped by switching the second 90° pulse, different signal intensities are thus finally displayed in the so-called T_1-weighted images. To get a strong T_1 contrast, the TR must be chosen to be short. This method of contrast generation is called partial saturation.

Second is the inversion recovery method. Instead of starting with a 90° excitation pulse as in the partial saturation method, a 180° inversion pulse is switched on first, which inverts M_z. Now after a certain TI, the 90° excitation pulse is applied. TI is thus the time between a 180° inversion pulse and a 90° excitation pulse (Figure 6.8). Depending on how long the TI is, tissue types with different T_1 relaxations will have different amounts of M_z. If the TI is chosen so that a tissue type has no M_z at the time of the 90° excitation pulse, it will be displayed in the image without signal intensity (Figure 6.9). This allows, for example, the signal from fat to be suppressed in the image. In the inversion recovery method, the TR refers to the duration until the next 180° excitation pulse. The M_z recovery after a 180° inversion pulse is given by the exponential equation:

$$M_z \text{ inversion recovery} = 1 - 2 \cdot e^{-TI/T_{1(\text{tissue})}}$$

Third is the spin echo method. In addition to the TR, the TE is particularly decisive for the contrast between different tissue types. TE is the time between the maximum amplitude of the signal immediately after the excitation pulse is switched off and the maximum amplitude of the echo signal. In the spin echo sequence, a signal echo is formed by an additional 180° inversion pulse after the 90° excitation pulse. The signal at the echo top, the maximum amplitude of the echo signal, is then T_2-weighted. In contrast, the signal echo formed with prephaser gradients, as described for the gradient echo sequence, is weighted by T_2* relaxation. The signal resulting from the excitation is actually a sinusoidal wave, referred to as the free induction decay (FID), and its oscillation is damped by the T_2* decay. The T_2* decay is composed of the pure T_2 relaxation and the effects of local B_0 field inhomogeneities. With the spin-echo sequence, the signal part that is lost due to local B_0 field inhomogeneities is recovered, since the phase of the spins is recovered with the 180° inversion pulse, and therefore, the signal at the echo top is T_2 weighted:

$$M_{xy} \text{ spin echo} \propto e^{-TE/T_{2(\text{tissue})}}$$

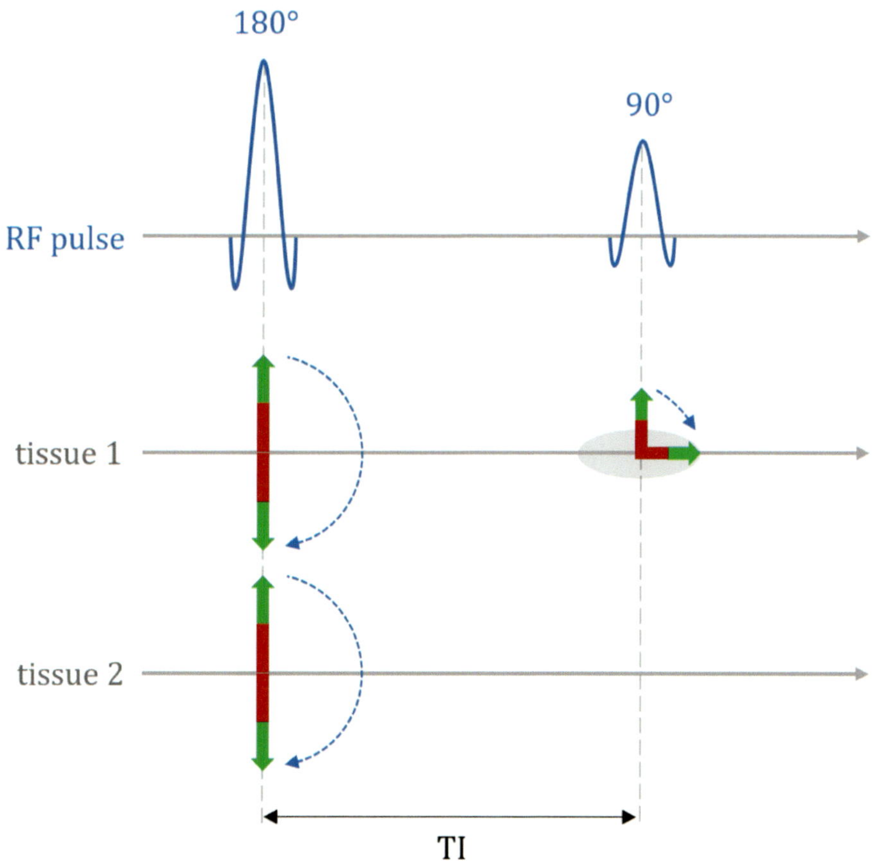

Figure 6.8 After a 180° inversion pulse and a certain TI, tissue types with different T_1 relaxations times recover to a different amount of M_z before excitation. Tissue types with zero M_z after the selected TI appear dark in the resulting images. The inversion recovery method can be used for T1-weighted images as well as for the T2-weighted images, for example, to suppress the CSF signal.

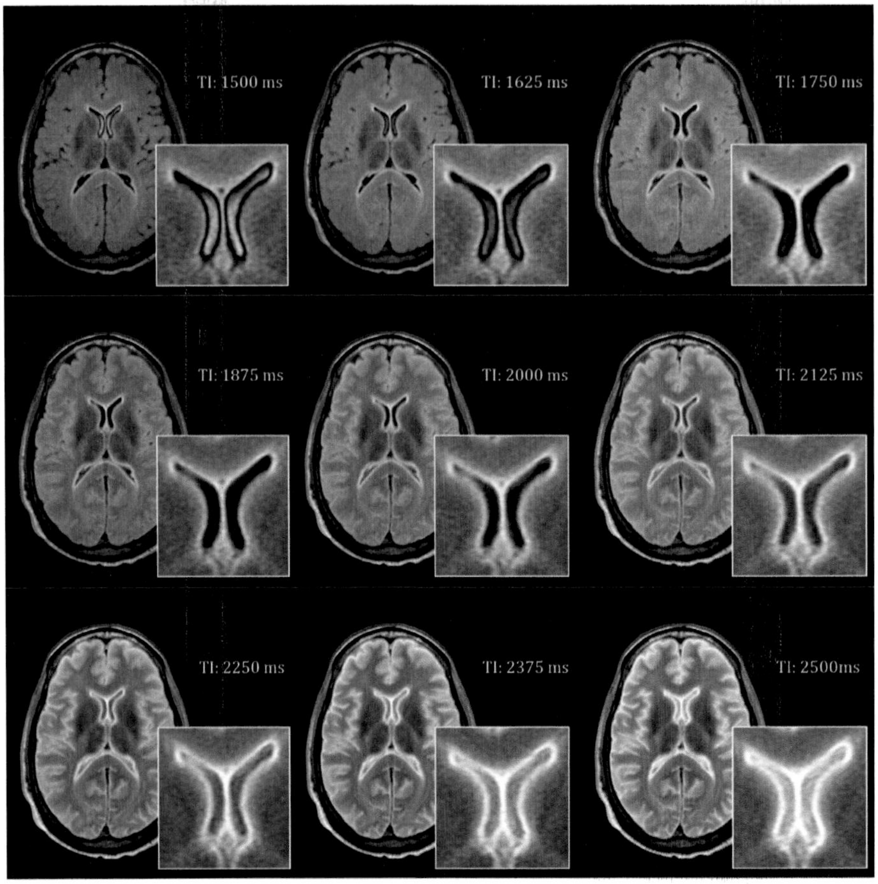

Figure 6.9 3 Tesla MRI of the head with different TI. Each of the nine images has a TE of 135 ms and a TR of 8000 ms and differ only by their TI. In the first image on the upper left, the TI is 1500 ms and increases from image to image by 125 ms so that the TI in the last image on the lower right is 2500 ms. Note how the representation of the CSF changes. A section of the anterior lateral ventricles has been enlarged for better visualization. In these images, the image at the bottom right with a TI of 2500 ms would actually correspond to a standard fluid-attenuated inversion recovery (FLAIR) sequence used for in vivo application, where the CSF should be appear dark (hypointense). However, it is clear that the signal of the CSF in the left image in the middle row is suppressed at a TI of 1875 ms. The reason for this is the lower temperature of the deceased compared to the body temperature of a living patient.

To obtain T_2-weighted images, long TE is required to generate strong contrast differences. With short TE, T_2 differences result in only small signal intensity differences for tissue types with different T_2 relaxation times. To avoid T_1 weighting, long TR times are required for T_2-weighted images. The spin echo sequence can also be used to acquire T_1-weighted images. This can be achieved by choosing a short TE and a short TR (as learned before). An advantage of the spin echo sequence over the gradient echo sequence is that phase shifts due to local B_0 inhomogeneities are also refocused. However, because two pulses (excitation pulse and inversion pulse) must be applied to generate the signal echo, the minimum TE and TR values for the spin echo sequence are generally longer than for the gradient echo sequence. If a long TR but a short TE is set, a proton-weighted image can be generated (as learned before).

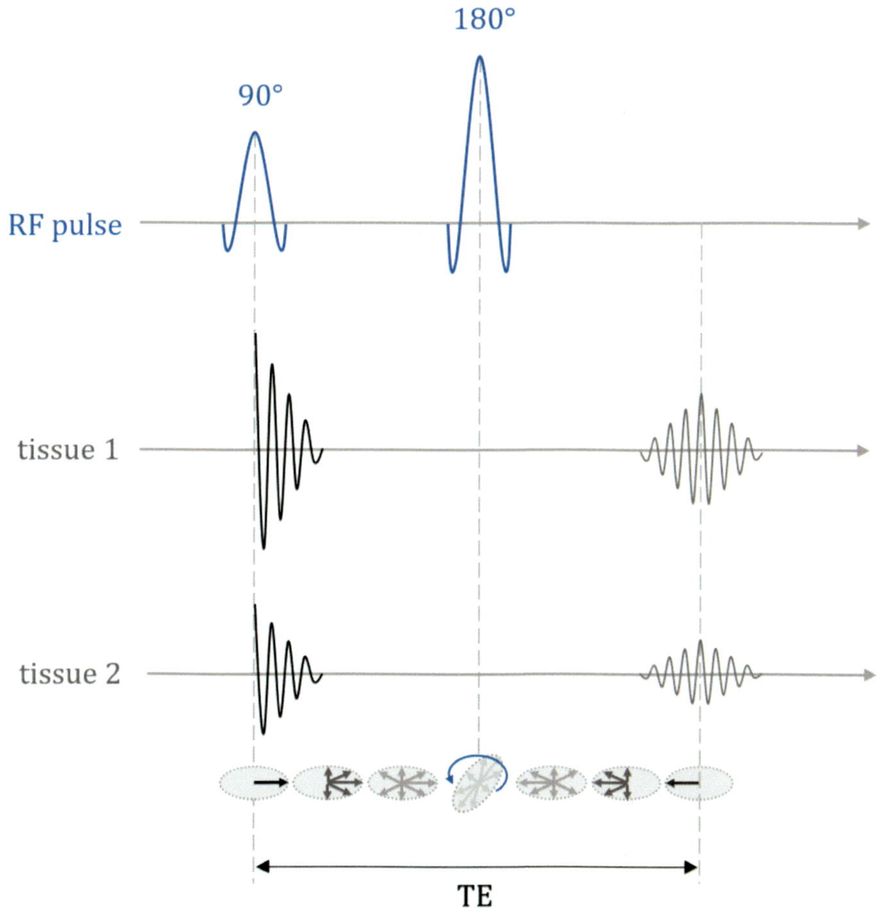

Figure 6.10 With a 90° excitation pulse, the distribution of magnetic moments forming the (macroscopic) magnetization is brought to the transversal plane resulting in a detectable signal. However, this initial distribution is quickly lost due random molecular processes described by T_2 relaxation and due different Larmor frequencies caused by local B_0 inhomogeneities and therefore the signal decays rapidly. The resulting signal that can be observed is called free induction decay (FID). The dephasing of magnetic moments due to B_0 inhomogeneities can be reversed by a 180° inversion pulse, and thus a part of the signal is be recovered in the formed spin echo. This is illustrated for a small selection of magnetic moments in the images at the bottom. Consequently, the signal amplitude at the echo top is weighted solely by pure T_2 relaxation during the time TE.

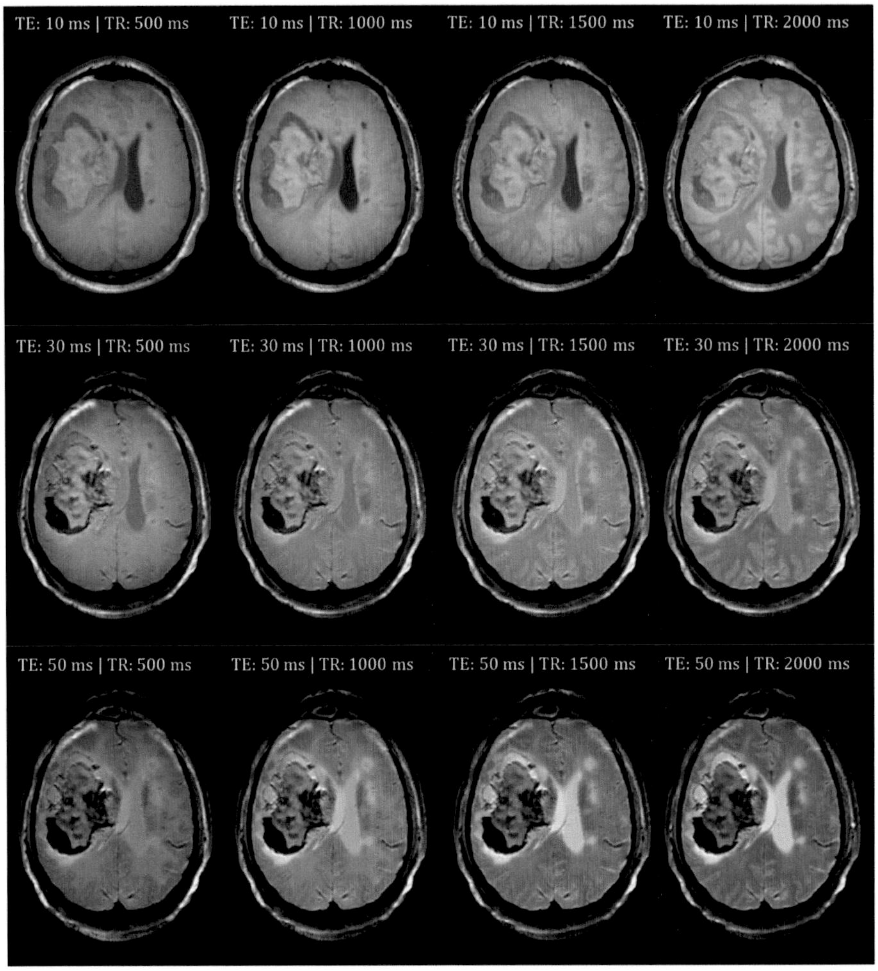

Figure 6.11 Postmortem 3 Tesla MRI of the head with cerebral hemorrhage acquired using spin-echo sequences with a TE of 10 ms (first row), 30 ms (second row), and 50 ms (third row) with a TR of 500 ms (first column), 1000 ms (second column), 1500 ms (third column), and 2000 ms (fourth column). The image on the top left with a short TE and short TR represents a standard T_1-weighted image, where low viscosity liquids like CSF are displayed dark (hypointense). Note also that the T_1 contrast between gray and white matter is not as strong as on T1-weighted images of living patients (not shown here), where the gray matter is clearly darker than the white matter. The lower postmortem contrast is due to the lower body temperature. If the TE is kept short but the TR becomes longer, more and more a proton-weighted image is created, as shown in the upper right picture. If the TE also becomes longer, more and more a T_2-weighted contrast is created, as shown in the lower right image.

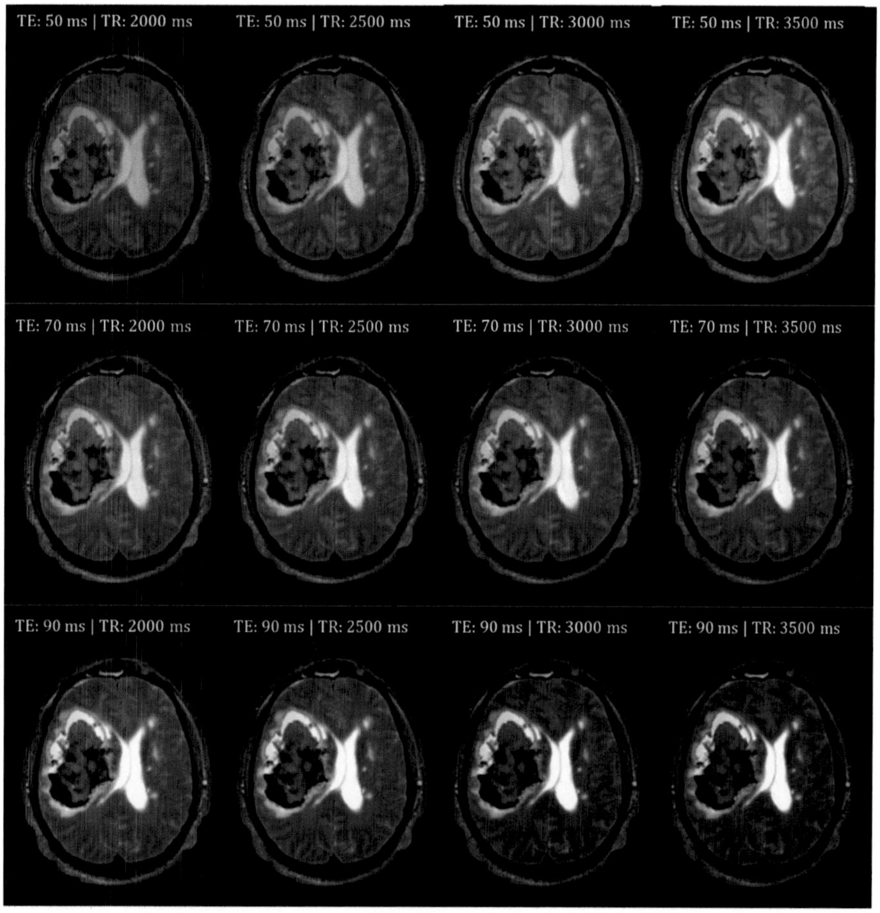

Figure 6.12 Postmortem 3 Tesla MRI of the head with cerebral hemorrhage acquired using Turbo-spin-echo sequences with a TE of 50 ms (first row), 70 ms (second row), and 90 ms (third row) with a TR of 2000 ms (first column), 2500 ms (second column), 3000 ms (third column), and 3500 ms (fourth column). In this sequence of images, it becomes clear how the signal in CSF increases with longer TE and longer TR until the lower right image, where CSF is displayed very bright (hyperintense). Although all images here are dominated by T_2 contrast, a typical T_2-weighted image would be at a TE of 70–90 ms and a TR of 3000–3500 ms. The postmortem lower body temperature seems to have less influence in T_2 imaging than in T_1 imaging.

6.5 WHAT IS MEANT BY T_2*-WEIGHTED AND SUSCEPTIBILITY-WEIGHTED IMAGING?

As mentioned earlier, T_2* relaxation is composed of local B_0 inhomogeneities in addition to "pure" T_2 relaxation. Whenever an object, in our case tissue, is introduced into the static magnetic field, it distorts the magnetic field, causing B_0 inhomogeneities, and the reason for this is magnetic susceptibility. In other words, magnetic susceptibility is a measure of the magnetizability of tissue which is placed in a strong external magnetic field. It is a unitless physical quantity that describes the ratio of tissue magnetization to the applied magnetic field strength. In general, tissue can be considered diamagnetic and therefore has a susceptibility of less than zero. However, there are also components in the body which are paramagnetic, where the susceptibility is greater than zero. Such a paramagnetic component is deoxy-hemoglobin due to the iron atom in the center of the heme group, whereas oxy-hemoglobin is diamagnetic because the attached oxygen neutralizes the paramagnetic effect of the iron atom. Since the paramagnetic effect accelerates T_2* relaxation, different relaxation times for deoxygenated blood compared to oxygenated blood result. This effect is used for blood oxygen level–dependent (BOLD) imaging for example for functional MRI examinations. Postmortem MRI is limited to the paramagnetic effect of blood, which is why T_2*-weighted imaging is primarily used to detect smaller demarcated hemorrhages, which are highlighted by susceptibility-related signal dropouts. While T_2*-weighted imaging uses only the magnitude information of the signal, susceptibility-weighted (SW) imaging uses the magnitude and phase information of the complex-valued signal to generate the SW image, making this method even more sensitive to susceptibility effects (Figure 6.13).

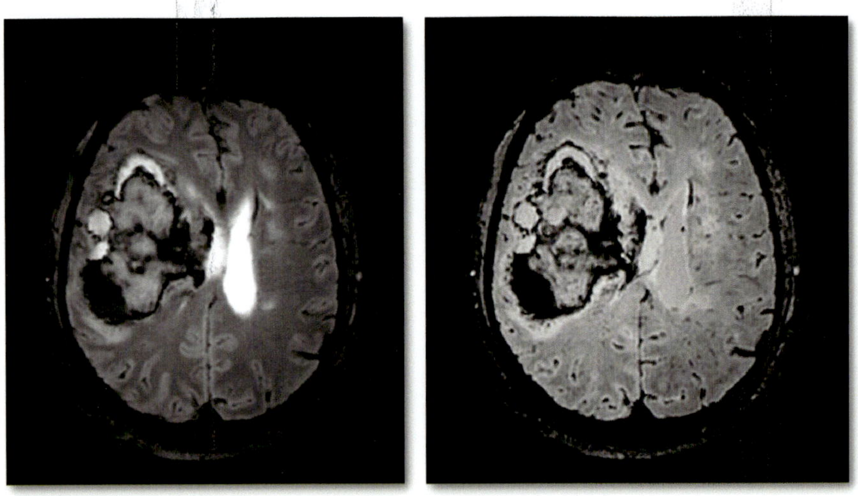

Figure 6.13 Postmortem MRI of a cerebral hemorrhage with a T_2*-weighted standard sequence (left image) and a susceptibility-weighted (SW) sequence (right image). In T_2*-weighted imaging, only the magnitude information of the signal is used. In contrast, SW imaging uses the magnitude and phase information of the complex-valued signal to generate the image. Therefore, SW imaging is more sensitive to susceptibility effects than standard T_2*-weighted imaging.

6.6 WHAT ARE DIFFUSION-WEIGHTED IMAGING AND DIFFUSION-TENSOR IMAGING?

As the name implies, diffusion plays a role in this imaging methods. More precisely, it is about the self-diffusion of water molecules related to thermal energy, which is also known as Brownian motion. How far a water molecule moves by self-diffusion is described by the average molecular displacement ⟨r⟩, which is given by the Einstein–Smoluchowski equation [14]:

$$r = \sqrt{6 \cdot t_D \cdot D}$$

where t_D is the diffusion time and D is the diffusion coefficient in three-dimensional space.

For the diffusion of spherical particles through a liquid with a low Reynolds number, D is given by the Stokes–Einstein equation [14]:

$$D = \frac{k_B \cdot T}{6 \cdot \pi \cdot \eta \cdot r}$$

where k_B is the Boltzmann constant, T is the absolute temperature, η is the viscosity of the medium, and r is the radius of the spherical particle.

The diffusion of water molecules in aqueous solution at 37°C is approximately 3.10^{-3} mm²/s. In biological tissue, however, the diffusion of water molecules is restricted by many mechanisms, such as the cellular microstructure, so that the diffusion coefficient is smaller by a factor of 2–10. Therefore, the term apparent diffusion coefficient (ADC) is used in diffusion-weighted imaging (DWI) to estimate the diffusion of water molecules in the tissue [15].

To map the displacement of a water molecule over a period of time, diffusion-sensitizing bipolar gradients are applied [14]. By switching on such bipolar gradients, molecules that are moving acquire a different phase according to the duration of the two gradients, while molecules at rest do not accumulate phase. However, since the displacement is of the order of µm, while the reconstruction voxel has a size of the order of mm, phase changes occur within a voxel. This means that although DWI is based on the principle of phase contrast, the effect seen in the images at the end is associated with the decrease of transverse magnetization, the change of magnitude, due to the changed phases within a voxel. This signal loss is described by an exponential decay:

$$e^{-b \cdot ADC}$$

where b is the measure of diffusion sensitivity with the unit s/mm².

The b-value is given by:

$$b = \gamma^2 \cdot G^2 \cdot \delta^2 \cdot \left(\Delta - \frac{\delta}{3}\right)$$

where γ is the gyromagnetic ratio, G is the gradient strength, δ is the duration of the gradient, and Δ is the time between the gradients (Figure 6.14).

The diffusion-weighted signal (S_{DWI}) is then given by [16]:

$$S_{DWI} = S_0 \cdot e^{-b \cdot ADC}$$

where S_0 is the signal that is detected without applying diffusion-sensitizing gradients.

Therefore, the ADC can be derived by:

$$ADC = \frac{-\ln\left(\frac{S_{DWI}}{S_0}\right)}{b}$$

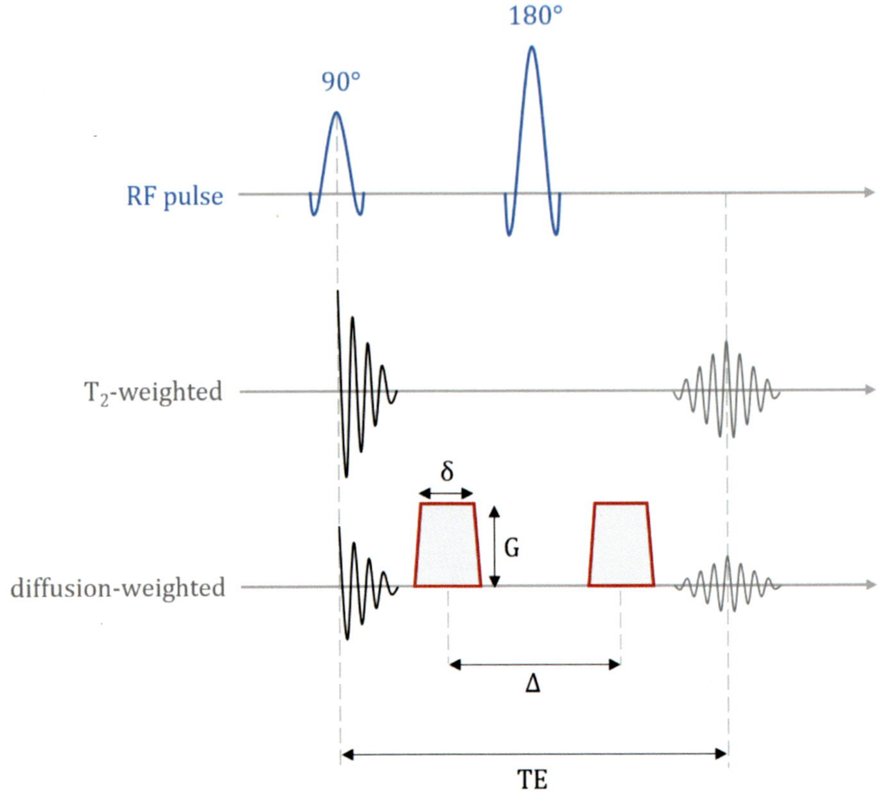

Figure 6.14 Compared to a T_2-weighted spin echo sequence, diffusion-weighted MRI uses diffusion-sensitizing gradients. These gradients are bipolar (note the 180° inversion pulse in the graphical representation). The abbreviation G stands for the gradient strength, δ for the duration of the gradient, and Δ for the time between gradients.

This means that the ADC map is finally computed from T_2-weighted baseline images, the b_0 images, and the DWI-weighted images sensitized to diffusion in several different directions. In ADC imaging, multiple conventional DWI images are acquired with different DWI weights, and the signal change is proportional to the diffusion rate.

For example, CSF that is bright on T_2-weighted images becomes darker and darker on images as the b-value increases because such fluid allows for a lot of motion and motion in turn leads to phase contrast within a voxel, which then causes a decrease in transverse magnetization, i.e., a change of magnitude. Such diffusion-weighted MRI is sensitive to changes in water diffusion, for example, due to cell swelling as seen in cytotoxic brain edema [12]. When failure of the Na+/K+-ATPase pump results in a net translocation of water from the extracellular to the intracellular space, as occurs in ischemic stroke, this alters the distribution of water molecules between the intracellular and extracellular compartments, ultimately leading to impeded diffusion [17]. This means that the corresponding region with impeded diffusion remains bright as the B-value increases, while diffusion-sensitive tissue becomes darker.

Differences in the diffusion of water molecules are also caused by the myelin sheaths of the axons. The displacement of water molecules is greater along the myelin sheaths of the axons than perpendicular to them because the myelin sheaths act as a kind of barrier. However, the b-value and the ADC only provide information about the diffusion sensitivity of the tissue, the magnitude, but not about the differences in the diffusion direction. DWI is thus a measurement for diffusion scalars. To display also directional information, the diffusion tensor is used, which is why this technique is called diffusion tensor imaging (DTI). Since the reconstruction problem is now a tensor problem, we must have a sufficient number of measurements to obtain the values for the unknowns. Diffusion (D) can occur along the three spatial axes (D_{xx}, D_{yy}, D_{zz}) or the diagonals (D_{xy}, D_{xy}, D_{xz}, D_{yx}, D_{zx}, D_{zy}), leading to a nine-dimensional problem. However, since diffusion is a random process, it cannot be predicted whether it proceeds in a positive or negative direction, so it can be reduced to a six-dimensional problem. Thus, after taking at least six measurements, eigenvalues (λ_1, λ_2, and λ_3) and eigenvectors (ε_1, ε_2, and ε_3) can be derived from the collected data. These eigenvalues can now be viewed as axis of an ellipsoid, with the lengths of these eigenvectors corresponding to the eigenvalues (Figure 6.15).

The eigenvalues can also be used to calculate the fractional anisotropic map, which is an index of the amount of diffusion asymmetry in a voxel. More commonly, however, the impressive fiber tracking map is created. For this, axonal tracts are usually mapped using a deterministic method known as FACT (fiber assignment by continuous tracking). In this method, the user can select a region in a certain plane from which the software then calculates the fiber trajectories that

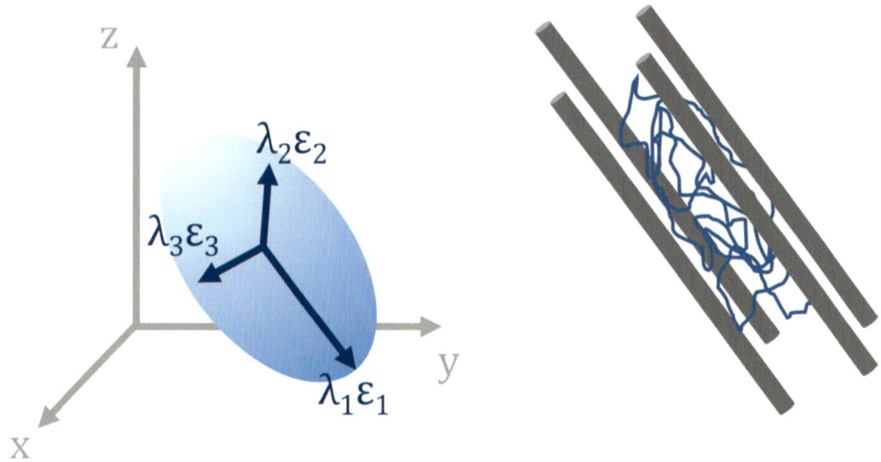

Figure 6.15 The eigenvectors (ε) with a length corresponding to the eigenvalues (λ) can be represented as the axis of an ellipsoid, and the shape of this ellipsoid corresponds to the slightly restricted diffusion. In the case of isotropic diffusion, the ellipsoid becomes a sphere. In the case of anisotropic diffusion, it becomes a cigar, with the long axis indicating the diffusion direction with the higher likelihood, which is e.g. parallel to the axonal tracts. If the third eigenvalue is zero, it becomes a pancake.

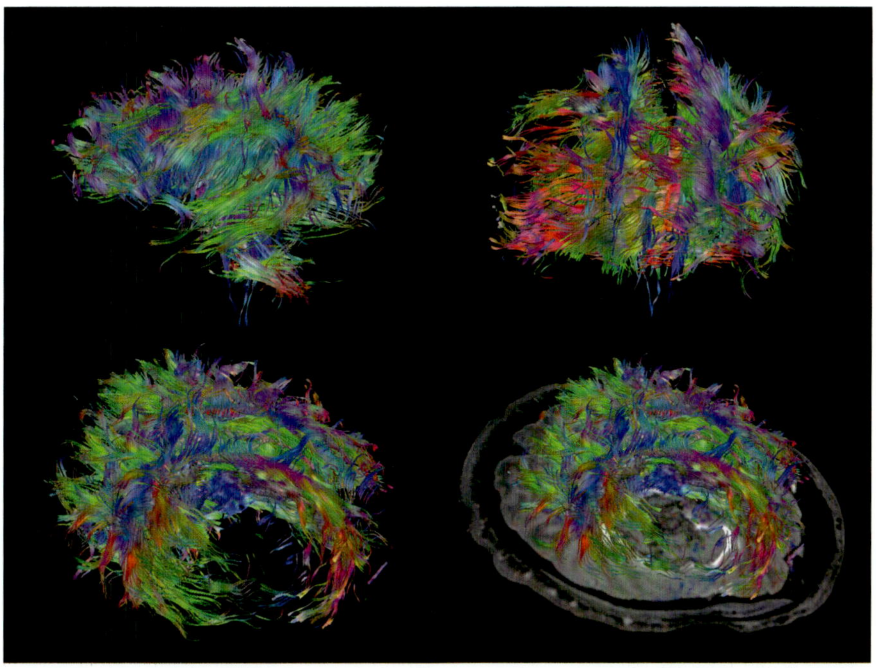

Figure 6.16 Postmortem representation of the fiber tracts with color coding according to their direction of trajectory. The three-dimensional data set is shown here in lateral view (upper left) and coronal view (upper right), as well as in oblique axial view and oblique axial view with fused T2-weighted cross-sectional view. Note here the recess of the fibers in the area of the cerebral hemorrhage, which is clearly visible in the lower left image.

pass through that plane. In this process, the primary eigenvector in each voxel is tracked until it encounters a neighboring voxel, whereupon the trajectory is changed in the direction of the new eigenvector. This method thus impressively demonstrates fiber tracts, for example, in tumorous regions or in hemorrhagic regions, provided that the corresponding region is large enough. The fibers displayed in a three-dimensional data set can additionally be colored differently according to the direction of their trajectory (Figure 6.16).

6.7 IS THERE A RISK THAT A CADAVER WILL WARM UP DURING A POSTMORTEM EXAMINATION?

Now that it has been described how an image is created and how common contrasts are generated, the body temperature or the warming of bodies in the practical application of MRI will be briefly discussed before finally focusing on postmortem imaging. For usual body sizes, the body of the deceased does not completely fill the gantry diameter, and no substantial heating (leading to denaturation) is expected, especially at lower field strengths less than or equal to 1.5 T [18]. However, the E-fields in the MR scanner bore become larger with distance from the isocenter, so heating can occur in obese bodies, especially if sequences with long pulses and short TE are used repeatedly at higher static magnetic fields of greater than or equal to 3 T. The specific absorption rate (SAR), which roughly corresponds to the power deposition, increases fourfold at 3 T, since the power depositions are proportional to B_0^2. Thus, superficial heating of voluminous bodies should not be disregarded during examinations of the thorax and abdominal regions (Figure 6.17).

Regarding examinations of the head, there is hardly any risk of heating during examinations of the head. This also applies to projectiles that are stuck centrally in the brain, since they hardly heat up due to their shape and position near the isocenter [18]. However, there is a possibility that ferromagnetic projectiles in the brain will rotate and align along the B_0 field depending on the tissue destruction [18]. In addition, migration of a ferromagnetic projectile in the head, while unlikely, cannot be completely ruled out depending on the destruction of the tissue and its degree of decomposition [18]. It should be noted here that no case is known to date in which a projectile in the body has changed its position significantly during a postmortem MRI examination, thus causing another artificial wound channel. However, a change in position of a projectile due to the preparation of a deceased for a surface scan, including head shaving and cleaning, is described in the literature [19].

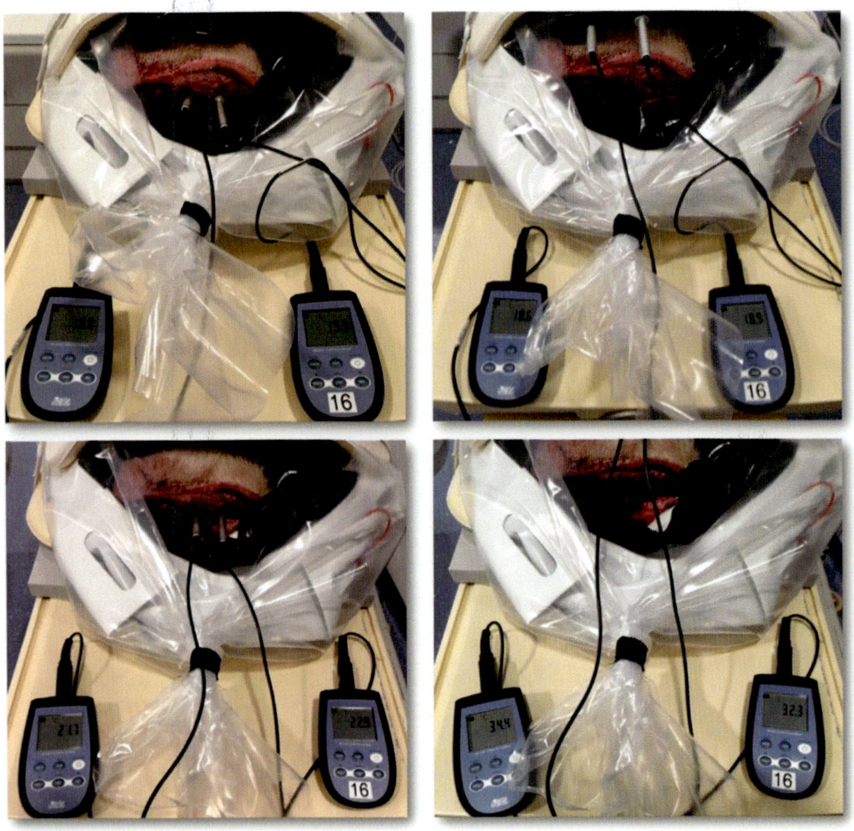

Figure 6.17 Shown here are the temperature measurements in a pig cadaver before the MR examination (top row: 18.0°C | 19.3°C and 18.6°C | 18.9°C) and after the MR examination (bottom row: 21.7°C | 22.9°C and 34.4°C | 32.3°C) in the center of the body (left column) and on the surface (right column) respectively. Warming during postmortem MR examination was provoked using a T1-weighted sequences with short TE and TR. After two hours, the surface showed an increase in temperature from an initial 18°C to almost 35°C, while the interior of the body showed a much lower temperature increase to about 22°C. The diameter of the pig carcass was only half the diameter of the gantry. Even stronger warming can be assumed if a body completely fills the gantry.

6.8 WHAT DO WE KNOW ABOUT POSTMORTEM MRI EXAMS?

Postmortem MRI examinations of different body regions have been presented in various studies, usually to complement CT findings but also for advanced forensic questions related to soft tissue injuries that cannot be answered by CT examination. MRI examinations of deceased persons allow, for example, the detection of cardiovascular pathologies in cases of sudden cardiac death (Figure 6.18) [20–23]. Furthermore, quantitative MRI studies of the heart with respect to myocardial infarction can be found in the literature [24–26]. As early as the early 2000s, the use of whole-body MRI examinations in comparison to pathological autopsies was described and the postmortem changes in the body were visualized radiologically [27].

With regard to traumatic injuries, the sensitivity and specificity of whole-body MRI examinations for the detection of traumatic injuries in multiple trauma have also been described [28]. Furthermore, targeted MRI examinations of subcutaneous fatty tissue and graduations of fatty tissue injury based on these examinations in the context of forensic traumatology have also been presented in the literature [29]. In cases of hanging and strangulation, high-contrast visualization of soft tissues enables assessment of, for example, hemorrhage into muscle [30–32]. Therefore, this non-invasive examination modality is also recommended in live victims of strangulation, especially in choke holds which often leave hardly any external signs of injury [33]. The utility of postmortem MRI examinations is also demonstrated for the study of cerebral injuries in fatal trauma [34–36]. If child abuse is suspected, an MRI exam may reveal relevant findings (Figure 6.19). The diagnostic value of postmortem MRI examination in fatal gunshot wounds is also pointed out in the literature under the term radiologic wound ballistics [18,37–39]. In addition, the literature also highlights the advantages and disadvantages of postmortem imaging with regard to pediatric examinations and recommends appropriate protocols for postmortem MRI examinations [40,41].

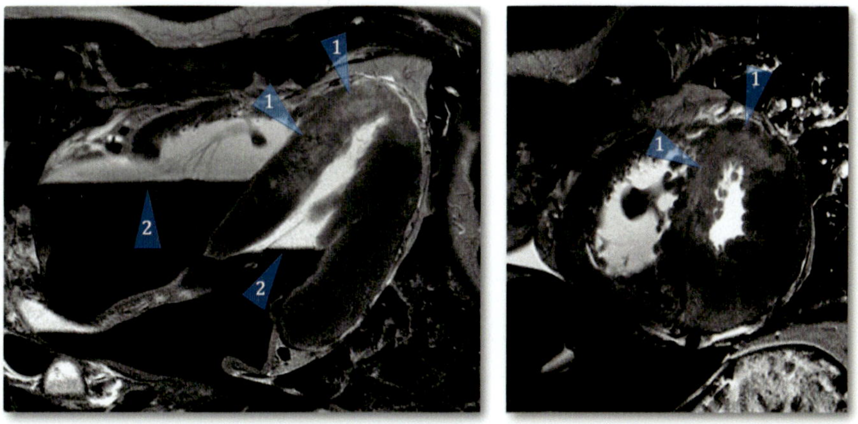

Figure 6.18 Postmortem MRI examination of the heart. The T2-weighted images show a hyperintense (bright) region in the septum and parts of the anterior wall of the left ventricle indicating an myocardial edema (arrowheads 1). The images also clearly show sedimentation of blood in the right and left ventricles due to non-circulation of blood (arrowheads 2).

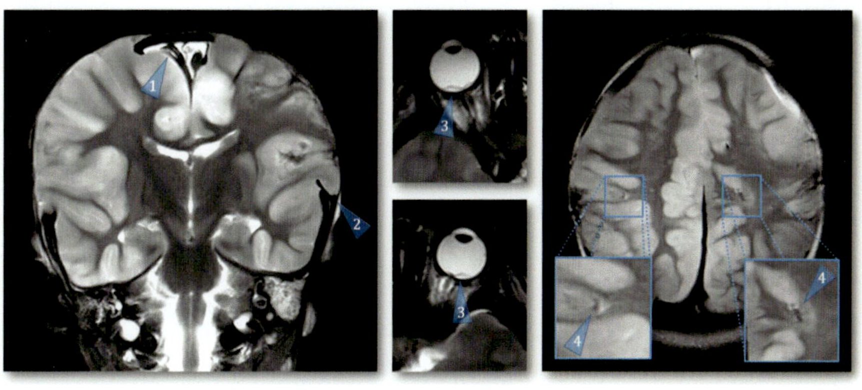

Figure 6.19 Postmortem MRI examination in suspected shaken-baby syndrome. Condition after extensive removal of the skull bones on both sides. The coronal T_2-weighted image shows hemorrhage along the cerebral falx (left image: arrowhead 1) and brain swelling with cerebral tissue swelling over the edges of the remaining skull bones (left image: arrowhead 2). On the transverse T2-weighted image retinal hemorrhages are indicated (images in the middle: arrowheads 3). Furthermore, shearing injuries at the gray–white matter boundary are visible (left image: arrowheads 4).

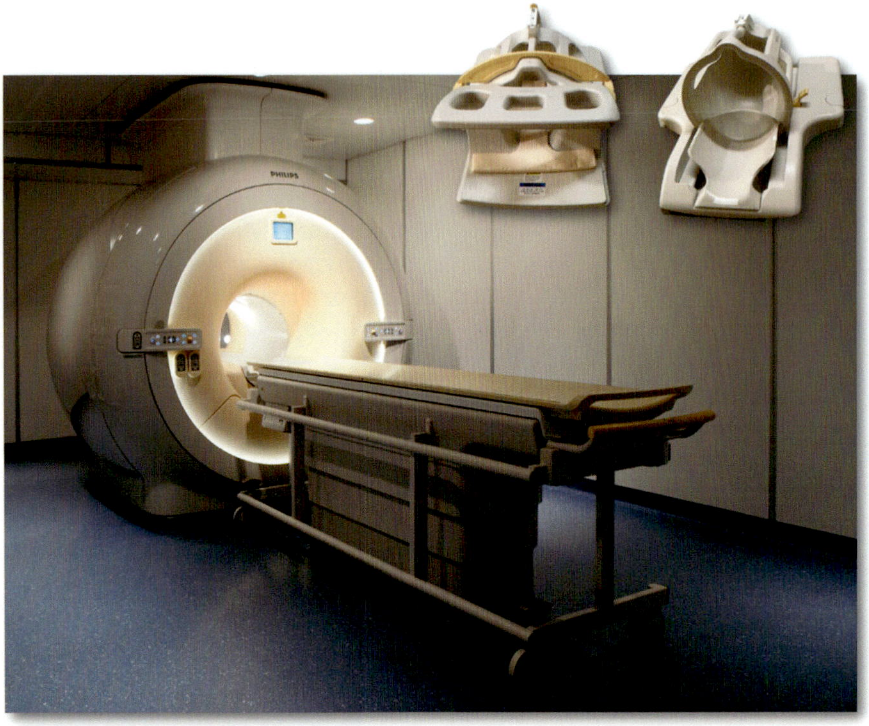

Figure 6.20 3 Tesla Philips Achieva 3.0T TX at the Zürich Institute of Forensic Medicine. Such a MR scanner allows whole-body examinations with a receiving coil built into the gantry, as well as detailed examinations of various body regions using the appropriate coil. At the upper right corner of the image, a (head and) 16-channel neck coil (left image) and a 8-channel head coil (right image) are shown.

Nevertheless, postmortem MRI examinations have not yet made the leap from research to practice in forensic medicine. This is often attributed to three reasons: first, limited access to MR scanners; second, the duration of an MRI examination; and third, the complexity of MR technology and the assessment of images with different weightings. As MRI becomes more and more established in clinical radiology, the first reason is probably only a matter of time since more and more institutes are gaining access to this examination technique. The interest of the manufacturers to reach the worldwide market with cheaper low-field MR scanners also facilitates access for forensic institutes. The second reason is probably based on postmortem MRI studies of earlier days with long examination times. This problem has become less important in recent years, as the acquisition of MR data has accelerated considerably over the years. The desire for shorter examination times from clinical radiology has driven the development of advanced fast imaging techniques such as compressed sensing [42]. This leaves the third reason, which relates to the complexity of this research modality. An additional problem here may be the temperature of the deceased, which leads to altered contrast in the images. Although several studies on the diagnostic evaluation of postmortem MRI examination in the forensic context can be found, hardly any studies on parameter adjustment in relation to temperature-related contrast changes can be found [43–48]. Although these temperature-related changes are known, there is still a lack of a rapid practical solution to quickly adapt them for individual examination. However, because access to MRI scanners currently varies globally [49] (Figure 6.20), it will certainly take some time until postmortem MRI is widely used in forensic medicine; extensive postmortem studies on its use are already desired to evaluate the versatile potential of this imaging modality for forensic questions.

References

[1] K.M. Franklin, B.M. Dale, E.M. Merkle, Improvement in B1-inhomogeneity artifacts in the abdomen at 3T MR imaging using a radiofrequency cushion, Journal of Magnetic Resonance Imaging. 27 (2008) 1443–1447. https://doi.org/10.1002/jmri.21164.

[2] R.R. Ernst, Methodology of magnetic resonance imaging, Quarterly Reviews of Biophysics. 19 (1987) 183–220. https://doi.org/10.1017/S0033583500004121.

[3] P.C. Lauterbur, Image formation by induced local interactions: Examples employing nuclear magnetic resonance, Nature. 242 (1973) 190–191. https://doi.org/10.1038/242190a0.

[4] P.C. Lauterbur, Magnetic resonance zeugmatography, Pure and Applied Chemistry. 40 (1974) 149–157. https://doi.org/10.1351/pac197440010149.
[5] A. Kumar, D. Welti, R.R. Ernst, NMR Fourier zeugmatography, Journal of Magnetic Resonance (1969). 18 (1975) 69–83. https://doi.org/10.1016/0022-2364(75)90224-3.
[6] P. Mansfield, A.A. Maudsley, T. Bains, Fast scan proton density imaging by NMR, Journal of Physics E: Scientific Instruments. 9 (1976) 271–278. https://doi.org/10.1088/0022-3735/9/4/011.
[7] P. Mansfield, A.A. Maudsley, Planar spin imaging by NMR, Journal of Magnetic Resonance (1969). 27 (1977) 101–119. https://doi.org/10.1016/0022-2364(77)90197-4.
[8] In memoriam: Sir Peter Mansfield (1933–2017), Magnetic Resonance in Medicine. 78 (2017) 826–827. https://doi.org/10.1002/mrm.26855.
[9] R.R. Edelman, The history of MR imaging as seen through the pages of radiology, Radiology. 273 (2014) S181–S200. https://doi.org/10.1148/radiol.14140706.
[10] D. Gascho, E. Deininger-Czermak, N. Zoelch, C. Tappero, S. Sommer, N. Hinterholzer, M.J. Thali, Noninvasive 7 tesla MRI of fatal craniocerebral gunshots: A glance into the future of radiologic wound ballistics, Forensic Science, Medicine and Pathology. 16 (2020) 595–604. https://doi.org/10.1007/s12024-020-00300-w.
[11] D. Gascho, N. Zoelch, S. Sommer, C. Tappero, M.J. Thali, E. Deininger-Czermak, 7-T MRI for brain virtual autopsy: A proof of concept in comparison to 3-T MRI and CT, European Radiology Experimental. 5 (2021) 3. https://doi.org/10.1186/s41747-020-00198-7.
[12] N.B. Smith, A. Webb, Introduction to medical imaging: Physics, engineering and clinical applications—Chapter 5: Magnetic resonance imaging (MRI), Cambridge University Press, 2010.
[13] E. Berry, A.J. Bulpitt, Fundamentals of MRI: An interactive learning approach—Chapter 4: Relaxation mechanism, CRC Press, Boca Raton, 2013. https://doi.org/10.1201/b15767.
[14] S. Morozov, K. Sergunova, A. Petraikin, E. Akhmad, S. Kivasev, D. Semenov, I. Blokhin, I. Karpov, A. Vladzymyrskyy, A. Morozov, Diffusion processes modeling in magnetic resonance imaging, Insights into Imaging. 11 (2020) 60. https://doi.org/10.1186/s13244-020-00863-w.
[15] D.L. Bihan, H. Johansen-Berg, Diffusion MRI at 25: Exploring brain tissue structure and function, Neuroimage. 61 (2012) 324–341. https://doi.org/10.1016/j.neuroimage.2011.11.006.
[16] M.Y. Park, J.Y. Byun, Understanding the mathematics involved in calculating apparent diffusion coefficient maps, American Journal of Roentgenology. 199 (2012) W784–W784. https://doi.org/10.2214/AJR.12.9231.
[17] T.C. Kwee, T. Takahara, R. Ochiai, K. Katahira, M. Van Cauteren, Y. Imai, R.A.J. Nievelstein, P.R. Luijten, Whole-body diffusion-weighted magnetic resonance imaging, European Journal of Radiology. 70 (2009) 409–417. https://doi.org/10.1016/j.ejrad.2009.03.054.
[18] D. Gascho, C. Tappero, N. Zoelch, E. Deininger-Czermak, H. Richter, M.J. Thali, S. Schaerli, Synergy of CT and MRI in detecting trajectories of lodged bullets in decedents and potential hazards concerning the heating and movement of bullets during MRI, Forensic Science, Medicine and Pathology. 16 (2020) 20–31. https://doi.org/10.1007/s12024-019-00199-y.
[19] D. Gascho, G. Ampanozi, L.C. Ebert, S.A. Bolliger, S. Eggert, S. Franckenberg, M.J. Thali, P.M. Flach, A moot point! A homicide case report on ambiguous projectile movement on postmortem MR, Journal of Forensic Radiology and Imaging. 5 (2016) 62–67. https://doi.org/10.1016/j.jofri.2015.08.004.
[20] C. Jackowski, N. Schwendener, S. Grabherr, A. Persson, Post-mortem cardiac 3-T magnetic resonance imaging, Journal of the American College of Cardiology. 62 (2013) 617–629. https://doi.org/10.1016/j.jacc.2013.01.089.
[21] B. Guidi, G.D. Aquaro, M. Gesi, M. Emdin, M. Di Paolo, Postmortem cardiac magnetic resonance in sudden cardiac death, Heart Failure Reviews. 23 (2018) 651–665. https://doi.org/10.1007/s10741-018-9705-0.
[22] I.M. Wagensveld, B.M. Blokker, A. Pezzato, P.A. Wielopolski, N.S. Renken, J.H. von der Thüsen, G.P. Krestin, M.G.M. Hunink, J.W. Oosterhuis, A.C. Weustink, Diagnostic accuracy of postmortem computed tomography, magnetic resonance imaging, and computed tomography-guided biopsies for the detection of ischaemic heart disease in a hospital setting, European Heart Journal—Cardiovascular Imaging. 19 (2018) 739–748. https://doi.org/10.1093/ehjci/jey015.
[23] D. Gascho, A. von Allmen, A. Landsmann, T. Hünermund, C. Tappero, M.J. Thali, E. Deininger-Czermak, Diagnostic value of T1- and T2-weighted 3-Tesla MRI for postmortem detection and age stage classification of myocardial infarction, Forensic Science, Medicine and Pathology. (2023). https://doi.org/10.1007/s12024-023-00592-8.
[24] H.J.A. Crooijmans, T.D. Ruder, W.-D. Zech, S. Somaini, K. Scheffler, M.J. Thali, O. Bieri, Feasibility of quantitative diffusion imaging of the heart in postmortem MR, Journal of Forensic Radiology and Imaging. 1 (2013) 124–128. https://doi.org/10.1016/j.jofri.2013.05.009.
[25] W.-D. Zech, N. Schwendener, A. Persson, M.J. Warntjes, C. Jackowski, Postmortem MR quantification of the heart for characterization and differentiation of ischaemic myocardial lesions, European Radiology. 25 (2015) 2067–2073. https://doi.org/10.1007/s00330-014-3582-2.
[26] N. Schwendener, C. Jackowski, A. Persson, M.J. Warntjes, F. Schuster, F. Riva, W.-D. Zech, Detection and differentiation of early acute and following age stages of myocardial infarction with quantitative postmortem cardiac 1.5T MR, Forensic Science International. 270 (2017) 248–254. https://doi.org/10.1016/j.forsciint.2016.10.014.
[27] L. Patriquin, A. Kassarjian, M. O'Brien, C. Andry, S. Eustace, Postmortem whole-body magnetic resonance imaging as an adjunct to autopsy: Preliminary clinical experience, Journal of Magnetics Resonance Imaging. 13 (2001) 277–287. https://doi.org/10.1002/1522-2586(200102)13:2<277::AID-JMRI1040>3.0.CO;2-W.

[28] S. Ross, L. Ebner, P. Flach, R. Brodhage, S.A. Bolliger, A. Christe, M.J. Thali, Postmortem whole-body MRI in traumatic causes of death, American Journal of Roentgenology. 199 (2012) 1186–1192. https://doi.org/10.2214/AJR.12.8767.

[29] K. Yen, P. Vock, B. Tiefenthaler, G. Ranner, E. Scheurer, M.J. Thali, K. Zwygart, M. Sonnenschein, M. Wiltgen, R. Dirnhofer, Virtopsy: Forensic traumatology of the subcutaneous fatty tissue; multislice computed tomography (MSCT) and magnetic resonance imaging (MRI) as diagnostic tools, Journal of Forensic Science. 49 (2004) 1–8.

[30] D. Gascho, J. Heimer, C. Tappero, S. Schaerli, Relevant findings on postmortem CT and postmortem MRI in hanging, ligature strangulation and manual strangulation and their additional value compared to autopsy: A systematic review, Forensic Science, Medicine and Pathology. 15 (2019) 84–92.

[31] E. Deininger-Czermak, J. Heimer, C. Tappero, M.J. Thali, D. Gascho, The added value of postmortem magnetic resonance imaging in cases of hanging compared to postmortem computed tomography and autopsy, Forensic Science, Medicine and Pathology. (2020). https://doi.org/10.1007/s12024-020-00233-4.

[32] E. Deininger-Czermak, J. Heimer, C. Tappero, M.J. Thali, D. Gascho, Postmortem magnetic resonance imaging and postmortem computed tomography in ligature and manual strangulation, The American Journal of Forensic Medicine and Pathology. Publish Ahead of Print (2020). https://doi.org/10.1097/PAF.0000000000000542.

[33] D. Gascho, J. Heimer, M.J. Thali, P.M. Flach, The value of MRI for assessing danger to life in nonfatal strangulation, Forensic Imaging. 22 (2020) 200398. https://doi.org/10.1016/j.fri.2020.200398.

[34] K. Yen, K.-O. Lövblad, E. Scheurer, C. Ozdoba, M.J. Thali, E. Aghayev, C. Jackowski, J. Anon, N. Frickey, K. Zwygart, J. Weis, R. Dirnhofer, Postmortem forensic neuroimaging: Correlation of MSCT and MRI findings with autopsy results, Forensic Science International. 173 (2007) 21–35. https://doi.org/10.1016/j.forsciint.2007.01.027.

[35] J. Añon, L. Remonda, A. Spreng, E. Scheurer, G. Schroth, C. Boesch, M. Thali, R. Dirnhofer, K. Yen, Traumatic extra-axial hemorrhage: Correlation of postmortem MSCT, MRI, and forensic-pathological findings, Journal of Magnetic Resonance Imaging. 28 (2008) 823–836. https://doi.org/10.1002/jmri.21495.

[36] E. Scheurer, K.-O. Lovblad, R. Kreis, S.E. Maier, C. Boesch, R. Dirnhofer, K. Yen, Forensic application of postmortem diffusion-weighted and diffusion tensor MR imaging of the human brain in situ, American Journal of Neuroradiology. 32 (2011) 1518–1524. https://doi.org/10.3174/ajnr.A2508.

[37] D. Gascho, N. Zoelch, C. Tappero, S. Kottner, E. Bruellmann, M.J. Thali, E. Deininger-Czermak, Fracture MRI: Optimized 3D multi-echo in-phase sequence for bone damage assessment in craniocerebral gunshot injuries, Diagnostic and Interventional Imaging. 101 (2020) 611–615. https://doi.org/10.1016/j.diii.2020.02.010.

[38] D. Gascho, M. Marosi, M.J. Thali, E. Deininger-Czermak, Postmortem computed tomography and magnetic resonance imaging of gunshot wounds to the neck, Journal of Forensic Sciences. 65 (2020) 1360–1364. https://doi.org/10.1111/1556-4029.14311.

[39] D. Gascho, S.A. Bolliger, M.J. Thali, C. Tappero, Postmortem computed tomography and magnetic resonance imaging of an abdominal gunshot wound, The American Journal of Forensic Medicine and Pathology. 41 (2020) 119–123. https://doi.org/10.1097/PAF.0000000000000547.

[40] O.J. Arthurs, J.C. Hutchinson, N.J. Sebire, Current issues in postmortem imaging of perinatal and forensic childhood deaths, Forensic Science, Medicine, and Pathology. 13 (2017) 58–66.

[41] E. Whitby, A.C. Offiah, S.C. Shelmerdine, R.R. van Rijn, M. Aertsen, W.M. Klein, D. Perry, S.K. Goergen, C. Abel, A. Taranath, D. Gascho, E. Miller, O.J. Arthurs, Current state of perinatal postmortem magnetic resonance imaging: European Society of Paediatric Radiology questionnaire-based survey and recommendations, Pediatric Radioloy. 51 (2021) 792–799. https://doi.org/10.1007/s00247-020-04905-9.

[42] L. Feng, T. Benkert, K.T. Block, D.K. Sodickson, R. Otazo, H. Chandarana, Compressed sensing for body MRI, Journal of Magnetic Resonance Imaging. 45 (2017) 966–987. https://doi.org/10.1002/jmri.25547.

[43] T. Kobayashi, T. Isobe, S. Shiotani, H. Saito, K. Saotome, K. Kaga, K. Miyamoto, K. Kikuchi, H. Hayakawa, H. Akutsu, K. Homma, Postmortem magnetic resonance imaging dealing with low temperature objects, MRMS. 9 (2010) 101–108. https://doi.org/10.2463/mrms.9.101.

[44] T.D. Ruder, G.M. Hatch, L. Siegenthaler, G. Ampanozi, S. Mathier, M.J. Thali, O.M. Weber, The influence of body temperature on image contrast in post mortem MRI, European Journal of Radiology. 81 (2012) 1366–1370. https://doi.org/10.1016/j.ejrad.2011.02.062.

[45] C. Birkl, C. Langkammer, J. Haybaeck, C. Ernst, R. Stollberger, F. Fazekas, S. Ropele, Temperature-induced changes of magnetic resonance relaxation times in the human brain: A postmortem study, Magnetic Resonance in Medicine. 71 (2014) 1575–1580. https://doi.org/10.1002/mrm.24799.

[46] K. Tashiro, S. Shiotani, T. Kobayashi, K. Kaga, H. Saito, S. Someya, K. Miyamoto, H. Hayakawa, Cerebral relaxation times from postmortem MR imaging of adults, MRMS. 14 (2015) 51–56. https://doi.org/10.2463/mrms.2013-0126.

[47] W.-D. Zech, N. Schwendener, A. Persson, M.J. Warntjes, C. Jackowski, Temperature dependence of postmortem MR quantification for soft tissue discrimination, European Radiology. 25 (2015) 2381–2389.

[48] C. Berger, M. Bauer, H. Wittig, E. Scheurer, C. Lenz, Post mortem brain temperature and its influence on quantitative MRI of the brain, Magnetic Resonance Materials in Physics, Biology and Medicine. (2021) 1–13.

[49] S. Geethanath, J.T. Vaughan Jr., Accessible magnetic resonance imaging: A review, Journal of Magnetic Resonance Imaging. 49 (2019) e65–e77. https://doi.org/10.1002/jmri.26638.

CHAPTER 7

MAGNETIC RESONANCE SPECTROSCOPY

Basics and Forensic Postmortem Challenges

Niklaus Zoelch and Dominic Gascho

DOI: 10.4324/9781003413998-8

7.1 BASICS

With magnetic resonance spectroscopy (MRS), we indirectly measure the magnetic field at the location of the nucleus and use this information to determine to which molecule[1] the hydrogen nucleus belongs. This is possible because although an external magnetic field B_0 is applied, the field B that the nucleus senses depends strongly on the local electronic environment of the nucleus[1]. The electrons surrounding the nucleus minimally change the field at the nucleus. This fact is called chemical shift and was discovered in 1950[2,3]. The main effect that will lead to the change is as follows: As soon as the electrons are brought in the B_0 field, they start to rotate in circular orbits around the direction of B_0. Since electrons are charged particles, a magnetic moment is generated, analogous to a magnetic field generated by current in a conductor loop. The direction of the generated field is inversely proportional to the applied field and thus reduces the field at the location of the nucleus[4].

$$B = B_0 (1 - \sigma)$$

where σ is a dimensionless quantity and is called shielding constant. The resonance condition can be modified to

$$v = \frac{\gamma}{2\pi} B_0 (1 - \sigma)$$

The observed shieldings are small, about one millionth of the applied field. At 3T with a Larmor frequency ν_0 of 127 MHz, the shielding by the electrons causes changes of a few 100 Hz in the resonance frequency. The strength of the shielding is determined by the number of electrons and the probability that the electrons are at the location of the nucleus. And this, in turn, is determined by the type of bond that the hydrogen proton is part of. Since the shielding therefore varies for different molecules, individual signals can be assigned to the molecules with the help of tabulated values,[5] and thus it can be determined which molecules/metabolites are present in the examined tissue.

For this comparison, it is not convenient to give the resonance frequency in Hz as in the equations above, as the peaks appear at different positions when measured at different field strength. Resonance frequencies are therefore mostly given relative to a reference frequency[6].

$$\delta_{ppm} = \frac{v - v_{ref}}{v_{ref}} \times 10^6$$

where δ is denoted as chemical shift (in ppm), and v and v_{ref} are the Larmor frequencies (in Hz) of the compound under investigation and of a reference compound, respectively. Different reference compounds are available, which needs to be kept in mind when tabulated values are used. For aqueous solutions, 2,2-dimethyl-2-silapentane-5-sulfonate (DSS) is usually recommended as a reference compound, since the chemical shift of DSS is largely independent of pH and temperature. However, DSS does not occur in the body, and therefore it is recommended to use the methyl peak of creatine for calibration of the chemical shift scale when available with enough SNR. In accordance with the widely used spin system parameters tabulated elsewhere, it should be set to 3.027 ppm[7] (or 3.024 ppm)[8].

In a typical in vivo spectrum (see Figure 7.1), the signals to the right of the water peak (at ~4.7 ppm) belong to nuclei with less chemical shift. The further right you go, the smaller the chemical shift in ppm, i.e. more shielding by the electrons and thus a lower resonance frequency (less negative) is measurable. DSS, if it could be measured in vivo, would appear at 0 ppm.

However, the field at the location of the nucleus can also differ within a molecule. We will look at this using ethanol as an example. Ethanol has two primary hydrogen nuclei groups, which can be measured: The hydrogen nuclei of the methyl group (CH_3) and the hydrogen nuclei of the methylene group (CH_2). All nuclei of a group give the same signal here; i.e. they are magnetically equivalent. Due to the proximity to the oxygen atom, with its high electronegativity, the probability that the electrons are at the location of the hydrogen nuclei of the methylene group is smaller compared to methyl group. Therefore, for the CH_2 nuclei, the resonance frequency is higher (more negative), and the chemical shift is larger (~3.7 ppm) than for CH_3 (~1.17 ppm)[5].

The example of ethanol shows that we can measure several peaks per molecule/metabolite with sometimes only slightly different frequencies. This leads to the fact that the signals of different metabolites overlap when measuring in human tissue at usual field strengths (1.5 to 3T).

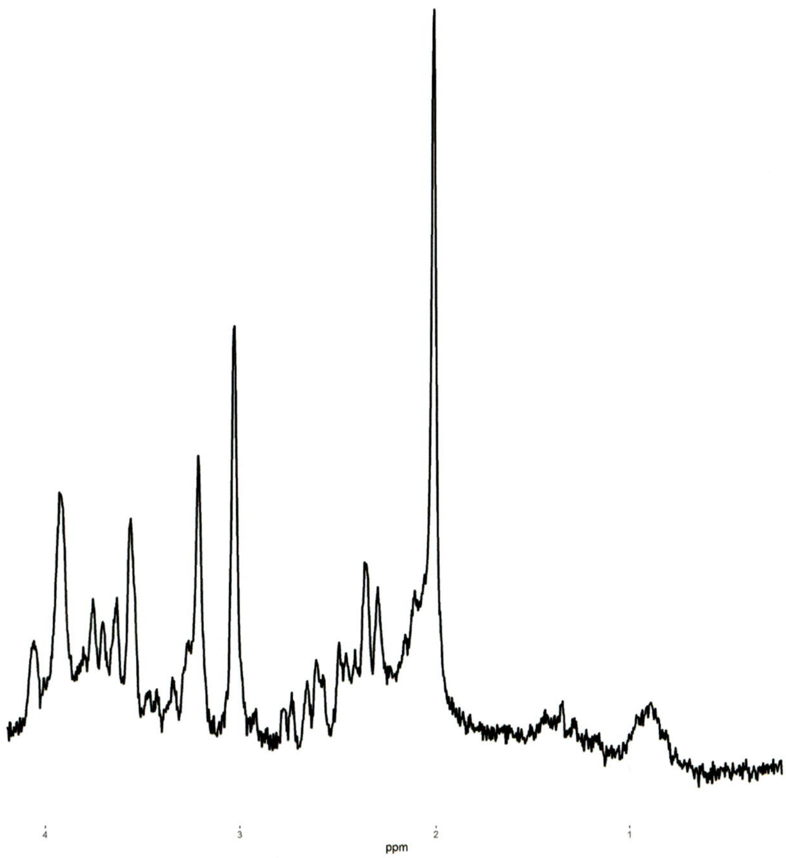

Figure 7.1 In vivo MRS spectrum acquired in a 12-year-old healthy girl at 3 T. The voxel with a size of 8 mL (20 × 20 × 20 mm3) was placed centrally in the brain above the lateral ventricle, containing a mixture of mainly GM and WM tissue. The localization was performed with a sLASER sequence with a TE of 30 ms and a TR of 5 s. The acquisition of the whole spectrum took 5 min and 30 sec.

7.1.1 J-Coupling or Scalar Coupling

Furthermore, there are also interactions between the nuclei in a molecule that change the energy levels of the nuclei and thus influence the measurable frequencies. These interactions lead to a splitting of the peaks. Most relevant for in vivo and in situ measurements is the indirect spin-spin coupling, also denoted as J-coupling or scalar coupling. Figuratively speaking, the energy of a hydrogen nucleus nuclear spin depends on the orientation of the other nuclear spin connected by bonds. Since the orientation is quantized in B_0, only very specific energies are possible. For example, for two coupling nuclei with spin quantum number I=1/2, this leads to two peaks instead of one per nucleus. For protons, only nuclei, which are connected over maximal three bond lengths, interact visibly. This interaction is independent of the field strength. A detailed explanation of the J coupling can be found elsewhere[9].

In general, J-coupling complicates the spectrum by leading to further overlap. In addition, the available signal is distributed over different peaks, which, especially in the case of low concentrated metabolites with J-coupling, makes the signals difficult to distinguish from noise. This becomes evident, for example, in the case of gamma-aminobutyric acid (GABA), which has a low concentration in brain tissue. The signals from GABA overlap with the signals of other metabolites and the signal is distributed among different peaks, which makes the detection of GABA generally difficult. On the other hand, such couplings can also be used to improve the detection of coupled metabolites (via spectral editing[10]) or when previously unobserved signals in the spectrum need to be assigned to a metabolite[11].

7.2 QUANTIFICATION

Quantification aims to determine a meaningful measure of the amount of metabolite in the measured tissue from the MRS signal of a metabolite. The basis for this must always be a qualitatively good measurement, which is therefore referred to by various authors as the first step of quantification[4]. How to obtain and recognize a high quality measurement will be discussed later in the text. The actual quantification can be roughly divided into two steps: fitting and data calibration. Fitting is the assignment of the measured signals (more precisely the measured peak areas) to the corresponding metabolites. Data calibration is then the conversion of the peak areas into a suitable measure of the amount of metabolite in the examined tissue. The latter step is possible because the MRS signal strength of a metabolite is directly proportional with the number of protons contributing to the signal. Therefore, in principle, it is even possible to express the amount of a metabolite in standard concentration units (biochemical units) as mole per liter (mol/l, molar) or mole per kilogram (mol/kg, molal).

7.2.1 Fitting

As outlined above, many peaks can be expected even with a small number of metabolites, and since the differences between the various metabolites at clinical field strengths (1.5 to 3 T) are relatively small, the peaks of different metabolites overlap at least partially, which makes an exact assignment difficult. Since the signal strength and thus the area under the peak is proportional to the number of nuclear spins contributing to this signal, the area of the peak must be determined as precisely as possible to make an estimation of the amount of compound in the measured tissue. Simple approaches such as integrating the peak areas are no longer valid for overlapping peaks (Figure 7.2).

For the peak area estimation and the correct assignment linear combination model fitting is used today. Software packages such as LCModel, Tarquin, INSPECTOR, or Osprey are available for this purpose. In linear combination model fitting, the measured signal is modeled as a linear combination of model spectra of individual metabolites and thus the contribution of the individual metabolite is estimated[12]. The model spectra are called basis spectra and are either measured in a phantom or simulated. Basis spectra must be generated for all metabolites expected in the spectrum, otherwise the metabolite cannot be identified in the spectrum. Once the full set of metabolite basis spectra (the basis set) is produced, a constrained non-linear least-squares analysis can be used to fit a linear combination of the basis spectra to the acquired MR spectrum by adjusting their individual amplitudes and frequencies. This process is called fitting and the result is called the fit. The difference between measured signal and fit, the fit residual, can be taken to estimate the goodness of the used model. Cramér–Rao minimum variance bound (CRMVB) estimates (the most commonly used measure of the uncertainty of the model fit parameter estimates) give the minimum error one has in the obtained parameters and consequently also in the variables derived from them, such as the concentration.

7.2.2 Data Calibration

Unfortunately, the measured signal/peak area does not only scale with the number of protons but is influenced by factors of the measurement. Important influencing factors are among others the achieved flip angle within the localized volume, the metabolites relaxation times and the timing of the localization method, the reception sensitivity, the loading of the coil and other possible gains in the receive chain[13,14]. Only if the influence of all these factors is eliminated or at least controlled, the results can be expressed in biochemical units and thus directly compared with results of other measurements. The elimination of influencing factors is associated with a relatively large effort and in most cases with additional measurements. In practice, therefore, complete elimination is never fully achieved and some form of estimation will be incorporated into the measured metabolite quantity. Consequently, in most cases, MRS only provides an estimate of concentrations in biochemical units. In MRS studies in the living brain, molar or molar concentrations are usually estimated. An overview of common concentration measures for brain measurements can be found here[15]. Depending on the given problem, a complete elimination of the influencing factors is not necessary and semi-quantitative measures (expressed in internal or institutional units)[16] are sufficient. Semi-quantitative measures are for example still dependent on the chosen localization sequence or the scanner used for the measurement and therefore may not directly be compared with results obtained with another localization sequence or at another scanner. It is therefore an important part of the preparation for MRS investigations to determine the necessary amount of corrections and to report them accurately together with the results[12].

The conversion of the metabolite signal (peak areas) into the desired measure of the quantity (e.g. concentration) happens finally via some form of reference signal. A good overview of used reference signals with a discussion of the advantages and disadvantages can be found here[12]. In vivo most commonly, internal reference signals are used, such as creatine or water signals. Since these signals are recorded from the same position and in the same measurement, influencing

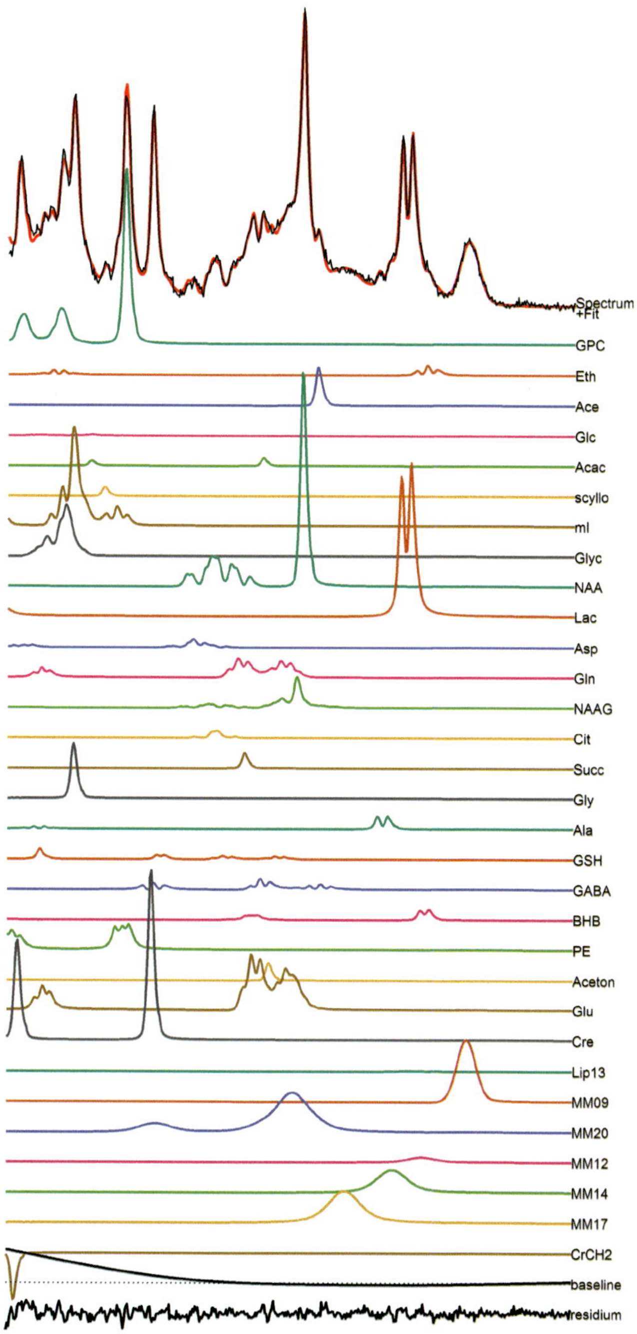

Figure 7.2 Postmortem MRS spectrum acquired at 3T in a 34-year-old male decedent. The measured spectrum and the obtained fit are shown together on top. Below, the estimated signal contributions from all metabolites present in the basis set are listed. Primarily, increases in lactate, acetate and choline as well as an increase in the region of myo-inositol is observed compared to in vivo measurements. Using the given basis set, the increased signal in the myo-inositol region is attributed to high levels of glycine (Gly), glycerol (Glyc), and myo-inositol. Details of the acquisition: PRESS localization, TE=30 ms, TR 2300ms, NSA=320, voxel size of 10.5 mL (35x20x15 mm3) placed centrally in the brain above the lateral ventricle, containing a mixture of mainly GM and WM tissue.

factors such as receiver sensitivity, coil loading and effective flip angles can be easily eliminated. For example, an estimate of moles of metabolite per mass of brain water (excluding CSF) can be calculated using the following equation

$$[M]_{molal} = \frac{S_M}{S_{H_2O}} \frac{2}{N_M} \frac{1}{(1 - f_{csf})} [H_2O]_{molal}$$

where S_M and S_{H_2O} are the metabolite and water signals, respectively. Two protons are contributing to S_{H_2O}, N_M is the number of metabolite protons contributing to S_M. f_{csf} is the CSF water mole fraction and $[H_2O]_{molal}$ is the molal concentration of pure water which is 55.51 moles/kg, the inverse of its molecular weight (18.015×10^3 kg/mole)[12,17]. However, this equation is only true for fully relaxed signals, i.e. measured at a very long repetition time (TR) (>30 s) and an echo time (TE) of 0 s. Practically, this is never fully achieved due to time limitations (TR) or technical limitations (TE) and the relax attenuation needs to be corrected. In vivo this can be achieved by using tabulated relaxation times, especially for the water reference signal[17]. In a postmortem setting however, correction based on literature values is not recommended due to sparse data and the potential influence of temperature and postmortem interval (PMI). If accurate concentrations are of interest, the settings should be chosen such that potentially different relaxation times have as little effect as possible.

As internal alternative to water, total creatine (tCr) can be used as a reference. Using tCr as a reference, is even more uncertain postmortem than in vivo. Depending on the tissue, it is likely that tCr changes with PMI or already deviates from normal concentrations due to disease ante mortem. Ratios to tCr (intensity or concentration) can be used as markers of metabolic differences. Their interpretation is subject to assumptions regarding which part of the ratio is changing, if not both, and in which direction.

Alternatively, external reference signals can be used, which are measured in phantom solution with exactly known concentration of a reference substance[13,18]. Since the concentration in the phantom solution is precisely known and the relaxation times in the phantom can be measured with accuracy, such assumptions are omitted compared to internal references. However, because the reference measurements are not taken from the same tissue and, for example, at a different temperature, other influencing factors such as coil loading, temperature and reception sensitivity must be adjusted to the actual measurement. Especially for postmortem MRS applications such external references can be advantageous. However, a problem remains the relaxation time of the investigated metabolite, which must also be taken into account here.

7.3 TECHNICAL ASPECTS OF MRS ACQUISITION

Technically, it is possible to measure the chemical composition in a single voxel as well as in a 2 to 3 dimensional grid of voxels. The former is called single voxel (SV) MRS, the latter is called magnetic resonance spectroscopic imaging (MRSI). MRSI allows the acquisition of metabolite maps e.g. over the whole head. MRSI is measured over a much larger area than SV MRS and thus has higher demands on the homogeneity of B_1 (both transmit and receive) and B_0. It is relatively difficult to achieve sufficient spectral quality everywhere at the same time. In addition, at high spatial resolution, measurements take a relatively long time. Only through acceleration techniques and the use of higher field strengths, meaningful metabolite maps have become possible. At rather low resolution, folding of fat signals is also a major problem. An overview of fast MRSI sequences and promising applications in the field of MRSI can be found in ref. [19]. Due to our expertise, we focus here on SV MRS applications. However, we think that measuring the distribution of metabolites could also provide new important insights for forensic medicine.

7.3.1 SV MRS/Localization

In SV MRS measurements, the signals of the nuclei are measured only from a specific volume. For this purpose, so-called single-shot localization sequences named as PRESS, sLASER (LASER) and STEAM are used. They are all based on the combination of gradients and frequency-selective pulses in three orthogonal directions. By applying the gradient, a variation of the resonance frequencies over the measured object is achieved (see also the chapter about MR imaging). The frequency selective pulse then excites only a certain range of frequencies and thus only a certain region in the object. Strictly speaking, one layer can be excited at a time. By repeating this in three orthogonal directions, only the signal from the intersection of the three layers is excited. The available sequences differ in subtleties that can be crucial depending on the application. For example, STEAM can be used to obtain short echo times, which may allow detection of metabolites with short T_2. Unfortunately, however, the signal is only half as large as with PRESS or sLASER (with equal TR and TE). sLASER uses adiabatic pulses and is thus very robust against B_1 inhomogeneities (during transmit) within the volume. In addition, the pulses used have a wide bandwidth, which reduces the chemical shift displacement artefact problem.[2] With PRESS, the CSD artifact is relatively large, which is especially noticeable at field strengths >3T and should therefore not be used there. At 3T the CSD can be handled by using saturation bands (OVERPRESS). In general, the PRESS sequence is very robust and often used. For standard brain studies, sLASER is increasingly recommended as the sequence of choice. Care should be taken to ensure that the sequence available from the vendor is stably implemented[20].

7.3.2 Water Suppression

Besides the localizations, the handling of the water signal is an important step for each MRS measurement. The water signal is about 10,000 times higher than the metabolite signal, depending on the measured tissue. The concentration of water is about 46 mol per liter volume of gray matter, whereas metabolites have a concentration of a few mmol per liter. When measured without special precautions, the detection of metabolites is difficult, although most have a resonance frequency different from water. The edges of the large water signal peak affect the metabolite peak and make it difficult to determine whether the signal is from the metabolite or from the water. Furthermore, additional peaks, so-called gradient sideband artifacts, appear which can be misinterpreted as a metabolite signals. Therefore, the water is usually suppressed by adding a water suppression element before the localization sequence. Various implementations (also vendor-dependent) are available, whereby VAriable Power radiofrequency pulses with Optimized Relaxation delays (VAPOR)[21] achieves the best and most robust suppression of the water signal. The basic element of most water suppression techniques is a frequency selective excitation with which only the nuclei of the water are excited followed by a gradient which dephases the generated transverse magnetization. Optimally, at the beginning of the localization, no polarization is present for the water and no signal is measurable. By repeating the basic element and optimizing the times between them, flip angles, etc., VAPOR allows good water suppression even in the case of B_1 and B_0 inhomogeneities and in the case of different relaxation times of the water signal, as it is often the case with voxels containing CSF, WM and GM.

Alternatively, a sequence known as metabolite cycling[22] can be used. Here, the water signal remains untouched. In two successive scans, the metabolites on the left and the metabolites on the right of the water are inverted. If the scans are subtracted from each other, water and gradient sidebands can be eliminated and the metabolite signal becomes visible. This is particularly interesting for metabolites that exchange protons with water and are thus partly suppressed with VAPOR[23]. Postmortem, we observe that, for example, the signal of acetate can be significantly improved with metabolite cycling.

7.3.3 B_0 Shimming

The placement of a human in the bore of an MR scanner results in distortions of the magnetic B_0 field.[3] These distortions result from large-scale susceptibility differences between different tissue types and air[24]. At interfaces of different tissues, or tissue and air, therefore magnetic field inhomogeneities form, whose amplitude is proportional to the applied magnetic field and the susceptibility difference. These field inhomogeneities affect the quality of the MRS spectrum in several ways, such as localization errors when the voxel is localized with frequency selective pulses, reduced water suppression and especially degraded spectral resolution. By a process called active B_0 shimming, the inhomogeneities in B_0 can be counteracted. In principle, the B_0 is measured in a fast way and then an additional field is overlayed with the help of shim coils to compensate the inhomogeneities. B_0 shimming is indispensable for a successful MRS measurement and good B_0 shimming methods for SV MRS are available on common MR scanners. It can also help to plan the voxel based on an acquired B_0 map in addition to an anatomical image, to improve the spectral quality[24]. This is especially useful postmortem, where the field may be disturbed by postsurgical metal clips, deoxygenated blood or similar.

7.3.4 Quality

The assessment of the quality of the spectrum is an important step in any spectroscopic investigation[25]. Only spectra of sufficient quality allow a reliable fit of the measured signal and thus a valid estimation of the concentration. A detailed discussion of the different aspects of quality can be found in the recent consensus paper[26]. In this section, we will only briefly highlight the most important aspects, which should allow a MRS beginner to recognize bad spectra.

Spectral resolution: Narrow linewidths of the metabolite peaks are necessary to be able to assign correctly the signals in the fitting process. On the one hand, the linewidth depends on the T_2 relaxation time and on smallest B_0 variations on molecule and cell level (and therefore on the measured tissue). On the other hand, the linewidth also depends on the macroscopic B_0 homogeneity and can therefore be influenced by B_0 shimming. An estimation of the spectral resolution can be made by the full width at half maximum (FWHM) of a selected peak in Hz (or ppm) at half maximum height. The FWHM is best determined from the unsuppressed water, which is usually measured in a separate measurement. Guide values for acceptable FWHM are given for different field strengths and tissues[26,27]. In the brain, linewidths up to approx. 13 Hz are considered acceptable at 3T. A quick visual evaluation of the shim quality can also be done based on the dominant creatine and choline peaks. If the point where these peaks meet is at the level of the baseline, the shim is very good, the higher the worse.

SNR: The signal to noise ratio is defined as signal divided by the standard deviation of the noise[28]. The energies in MRS are generally small and therefore the signals in single measurements are difficult to separate from noise. This means that multiple measurements must be collected for a useful SNR or large volumes need to be measured instead. General SNR limits are not found in the literature.

Remaining water signal: If the suppressed water signal has a visible influence on the baseline in the spectrum, this spectrum should be excluded from further analysis. If necessary, a filter can be applied to remove the remaining water during pre-processing.

Artifacts: There are several possible artifacts that can appear in a spectrum. A gallery of artifacts can be found here[29]. Common are e.g. lipid signals that do not originate from the measured location but fold in from the adjacent scalp tissue. Another issue are unwanted signals that occur during localization and appear as high frequency pattern in the final spectrum. Such artifacts should result in an exclusion of the spectrum, if the artifact cannot be eliminated with pre-processing.

Some of the above parameters can be improved by suitable pre-processing. The quality should therefore be assessed after preprocessing before the actual fitting process. See also the work from Near et al.[12] for further information about the possible pre-processing steps.

7.4 TECHNICAL ASPECTS OF POSTMORTEM MRS MEASUREMENTS

7.4.1 Metabolites

In postmortem MRS, new metabolites emerge that are not commonly observed in in vivo measurements. A primary investigation of postmortem metabolites was done by Ith et al. with a longitudinal analysis of eight sheep heads for up to 18 days[30]. The basis set was extended twice: First, metabolites known to occur under pathologic conditions were included for the early PMI (3 days). In the late PMI (15 days), unknown peaks were observed, analyzed in vitro by NMR experiments and identified to be free trimethyl ammonium, butyrate, isobutyrate, and propionic acid. In accordance, Banaschak et al. repeatedly measured five pig heads for up to 21 days and observed the formation of succinate and free trimethyl ammonium at a PMI greater than 100 hours[31]. Musshoff et al. also detected free trimethyl ammonium, propionate, and acetate 9 days postmortem in a severed sheep head and a complete sheep[32]. A list of relevant metabolites identifiable in postmortem MRS of the brain and their reported postmortem changes are summarized in Table 7.1.

7.4.2 Temperature

Unlike measurements in the living, temperature varies over a relatively large range in postmortem measurements. Depending on passive cooling (finding situation, clothing, climatic region, PMI) and active cooling (storage temperature, duration), the core body temperatures of cadavers vary considerably. At our institute, we measure cadavers in the range of approximately 5–37°C body temperature. Temperatures in the brain can differ considerably from the core body temperature (measured rectally) due to the different cooling curves[33]. Temperature differences have various effects on the MRS measurement which are briefly discussed below.

7.4.2.1 Relaxation Times

The effects of temperature on the relaxation parameters of the fresh (unfixated) human brain has not been extensively studied[34]. Only the temperature dependence of the relaxation times of water has been studied to a small extent, while no such studies exist in situ for the metabolites. The main reason for this is certainly that the temperature in the living brain is very stable. Another reason is the difficulty to measure temperature directly in the head with MR compatible devices. Previous studies have mainly found an increase in T_1 with increasing temperature, with a greater increase in GM compared to WM[34–36]. The temperature dependencies found vary between the different publications. For T_2 times of water, most authors have found no significant temperature dependence[35,37]. However, the only group that measured the temperature directly in the fresh brain has found a significant decrease in T_2 with increasing temperature of −0.5 ms/°C[37].

Deviations in the relaxation times T_1 and T_2 influence the MRS results, especially via data calibration. Relaxation times determined in vivo used for correction of relaxation attenuation can lead to incorrect concentrations estimates. For example, when internal water is used as a reference and in vivo T_1 times are used for the relaxation attenuation correction of the water signal. This will lead to an overestimation of the fully relaxed water signal and thus an underestimation of the concentration of the metabolite (assuming temperature stable T_1 of the metabolite) in postmortem measurements with temperatures below 37°C. This is best avoided during acquisition by choosing the TR much longer than the expected T_1 (4–5 times longer), at least for the measurement of the water reference. Unfortunately, we cannot choose TE infinitely short, so it is all the more important to correct for T_2 influences. Based on the available data, corrections should best be based on individually measured T_2 times, which is possible for water in a relatively short time.

The temperature dependence of water relaxation times can also affect MRS measurements in a completely different way. Water suppression implementations are optimized for in vivo applications and the corresponding relaxation times. Therefore, the suppression may need to be optimized for use at lower temperatures. Based on previous experience, VAPOR provides relatively good water suppression even at lower temperatures, but other sequences may be more susceptible.

7.4.2.2 Initial Magnetization

As can be deduced from equation 1, the MR signal is inversely proportional to the temperature of the tissue under investigation. For lower temperatures, there is a larger initial magnetization and thus a larger signal[16]. Metabolites and water are equally affected and therefore this effect is corrected for when signals are referenced to internal references such as water or creatine. However, if external reference signals are used to estimate concentrations, the difference in initial magnetization must be corrected for. A table with the correction factors over the temperature range of 4–37°C can be found in the book by Tofts (Table 9.1)[36].

The temperature dependence of water density has a small effect (<1%) on the calculated concentrations when calculating molar concentrations with internal water.

7.4.2.3 Chemical Shift

The chemical shifts of some protons in certain metabolites are temperature dependent. Measurements at 7T between 1 and 43°C showed temperature dependencies from $-6.7 \cdot 10^{-4}$ up to $+7.6 \cdot 10^{-4}$ ppm/K[38]. These temperature dependencies can lead to wrong assignments in the fitting due to wrong prior knowledge (i.e. chemical shifts measured at 37°C) and partly due to increased overlap between metabolite peaks and thus influence the results considerably. To minimize these influences, different basis sets can be simulated for different temperatures ranges[38]. However, while many metabolite peaks are only slightly affected by temperature, the chemical shift of water is strongly temperature dependent with ~0.01 ppm/°C[39]. Since the resonance frequency of water is used as a reference in clinical MR scanners and is automatically shifted to the same position of ~4.7 ppm, most metabolites are displayed at strongly shifted positions directly on the scanner and in many pre-processing tools. This can be corrected by shifting the spectrum back to the correct position based on temperature stable metabolites such as creatine or lactate. Note that many MR sequences use the water peak as reference for the frequency of the applied RF pulses. For example, inversion pulses whose frequency is shifted by 300 Hz relative to water. This must be taken into account when using measurements that have been successfully used in vivo in postmortem applications.

7.4.3 Acidity

The pH under physiological conditions is maintained in small ranges slightly above 7. In postmortem tissue however, the pH can be considerably lower. Takahashi measured pH values in postmortem brain tissue ranging from 7.01 down to 5.58 including various causes of death[40]. The main reason for the pH decrease is postmortem anaerobic glycolysis and the associated accumulation of lactic acid (present as lactate in this pH range). Anaerobic glycolysis proceeds at a high rate during the first 10 hours after death and then slows considerably until it stops, e.g. due to depletion of the substrates. According to Henssge and Madea, anaerobic glycolysis also stops at a pH of 6.3[41]. The pH seems to be determined not only by postmortem changes, but also by the cause of death and the duration of the agonal phase[40,42,43]. Due to the small pH differences observed in vivo, there are few studies available on the influence of pH on ^{1}H-MRS. It is known from metabolomic analysis of biofluids that chemical shifts and linewidth are sensitive to changes in pH[44,45]. However, the pH dependence of chemical shifts is strongly dependent on the actual pH range, e.g. alanine is pH dependent in the range of pH ~11–9, while changes are observed for acetate in the range of ~6.8–2[45,46]. Chemical shift changes might complicate the fitting process. But the possible changes are relatively small, e.g. ~ 0.17 ppm over the entire range from 6.8 to 2 in the case of acetate. On the other hand, the change in chemical shift can also be used to measure pH[47]. We recently tried to identify very low pH values in postmortem brain tissue by MRS using the acetate position[48]. Besides chemical shifts, relaxation times can also be subject to changes due to pH variations[49].

Table 7.1 List of Relevant Metabolites Identifiable in Postmortem MRS of the Brain and Their Reported Postmortem Changes. Some metabolites are commonly grouped under an umbrella term (class), as these metabolites are hardly distinguishable at common field strengths. The proton peak patterns are described by type[4]: s: singlet, d: doublet; t: triplet; q: quartet; m: multiplet. The number of metabolite peaks and their location was referenced from Govindaraju/Govind et al.[7,52], the Madison Metabolomics Consortium Database[51], and the library of parameterized compounds by GISSMO[5,54] in this order if not found in the prior resource.

Metabolite	Class	Chemical Shift (ppm)	Type	Ref	Reported Trends
Acetate		1.90	s	7	Sigmoid increase[50]
Acetoacetate		2.27; 3.44	s; s	51	
Acetone		2.22	s	51	
Alanine		1.47; 3.77	d; q	7	Quadratic increase[50]
Aspartate		2.65; 2.80; 3.89	dd; dd; dd	7	
γ-Aminobutyrate		1.89; 2.28; 3.01	q; t; t	52	Slow linear Increase[53]
Butyrate		0.88; 1.54; 2.14	t; m; t		Quadratic increase[50]
β-Hydroxybutyrate		1.19; 2.29; 2.40; 4.14	d; dd; dd; m	5,54	
Iso-butyrate		1.05; 2.37	d; m	5,54	

Table 7.1 (Continued)

Metabolite	Class	Chemical Shift (ppm)	Type	Ref	Reported Trends
Choline	Cho	3.19; 3.50; 4.05	s; m; m	7	Initial decrease, then slow linear increase[53]
Phosphocholine		3.21; 3.64; 4.28	s; m; m	7	Initial decrease, then slow linear increase[53]
Glycerophosphocholine		3.21; 3.61; 3.66; 3.67; 3.87; 3.90; 3.95; 4.31	s; dd; m; dd; m; m; m; m	7	Initial decrease, then slow linear increase[53]
Creatine	tCr	3.03; 3.91; 6.65	s; s; s	7	initial decrease, then slow linear increase[53]
Phosphocreatine		3.03; 3.93; 6.58; 7.30	s; s; s; s	7	initial decrease, then slow linear increase[53]
Ethanol		1.17; 3.64	t; q	5,54	
Free trimethylammonium		2.86	s	55	Linear increase after 50 hours[50]
Glucose (α)		3.40; 3.52; 3.70; 3.82; 3.83; 5.22	t; dd; t; m; dd; d	7	
Glucose (β)		3.23; 3.39; 3:45; 3.47; 3:71; 3.88	dd; t; m; t; dd; dd	7	
Glutamine	Glx	2.13; 2.43; 3.75	m; m; t	7	Initial decrease, then slow linear increase[53]
Glutamate		2.04; 2.34; 3.74	m; m; dd	7	Initial decrease, then linear increase[53]
Glutathione		2.16; 2.51; 2.93; 3.77, 3.77, 4.56; 7.15; 8.18	m; m; dd; t; s; dd; t; d	7	Initial decrease, then slow linear increase[53]
Glycerol		3.55; 3.55; 3.77	dd; dd; m	7	linear increase with PMI of 3–195h[56]
Lactate		1.31; 4.10	d; q	7	Initial increase[31,53] then slow linear increase[53], disappearing after 200h[29]
Myo-Inositol	mI&Gly	3.27; 3.34; 3.52; 3.52; 3.61; 3.61; 4.05	t; s; dd; dd; t; t; t	7	Initial decrease, then linear increase[53]
Glycine		3.55	s	7	
N-Acetylaspartate	tNAA	2.01; 2.67; 4.38; 7.82	s; dd; dd; d	7	Exponential decrease[50], decreasing and disappearing[29], initial decrease, then stable with slow linear decrease[53]
N-Acetylaspartylglutamate		1.88; 2.04; 2.19; 2.72; 4.13; 4.61	m; s; m; dd; dd; dd	7	
Phosphoethanolamine		3.22; 3.98	m; m	7	
Propionate		1.04; 2.17	t; q	5,54	Linear increase after 20 hours[50]
Pyruvate		2.36	s	7	
Scyllo-inositol		3.34	s	7	
Succinate		2.39; 2.39	s; s	7	
Taurine		3.42	t	7	Initial decrease, then slow linear increase[53]
Valine		0.98; 1.03; 2.26; 3.60	d; d; m; d	7	

Notes

1 Since we mostly study biochemical metabolic processes with MRS, we usually use the term metabolite for the molecules we study with MRS.

2 Chemical shift displacement artifact: Slightly different volumes are excited during localization with gradients and frequency selective RF pulses for nuclei with different chemical shift i.e. the chemical composition is measured in slightly offset volumes.

3 For materials with a rather low susceptibility, such as tissues, the assumption $B \sim B_0$ is commonly made, where B_0 describes the magnetic B field in vacuum. The changes caused by the measured object and also the inhomogeneities are in the range of a few ppm. Strictly speaking, one should not speak of changes in the B_0 and rather use B, but this is how it is generally done in MR literature.

References

1. Levitt MH. *Spin dynamics: Basics of nuclear magnetic resonance*. John Wiley & Sons; 2013.
2. Proctor W, Yu F. The dependence of a nuclear magnetic resonance frequency upon chemical compound. *Physical Review*. 1950;77(5):717.
3. Dickinson W. Dependence of the F 19 nuclear resonance position on chemical compound. *Physical Review*. 1950;77(5):736.
4. De Graaf RA. *In vivo NMR spectroscopy: Principles and techniques*. John Wiley & Sons; 2019.
5. Dashti H, Westler WM, Tonelli M, Wedell JR, Markley JL, Eghbalnia HR. Spin system modeling of nuclear magnetic resonance spectra for applications in metabolomics and small molecule screening. *Analytical Chemistry*. 2017;89(22):12201–12208.
6. Keeler J. *Understanding NMR spectroscopy*. John Wiley & Sons; 2011.
7. Govindaraju V, Young K, Maudsley AA. Proton NMR chemical shifts and coupling constants for brain metabolites. *NMR in Biomedicine: An International Journal Devoted to the Development and Application of Magnetic Resonance In Vivo*. 2000;13(3):129–153.
8. Ulrich EL, Akutsu H, Doreleijers JF, et al. BioMagResBank. *Nucleic Acids Research*. 2007;36(suppl_1):D402–D408.
9. de Graaf RA, Rothman DL. In vivo detection and quantification of scalar coupled1H NMR resonances. *Concepts in Magnetic Resonance*. 2001;13(1):32–76.
10. Choi IY, Andronesi OC, Barker P, et al. Spectral editing in (1) H magnetic resonance spectroscopy: Experts' consensus recommendations. *NMR Biomedicine*. 2020:e4411.
11. Pouwels PJ, van de Lagemaat M, van de Pol LA, Witjes BC, Zonnenberg IA. Spectroscopic detection of brain propylene glycol in neonates: Effects of different pharmaceutical formulations of phenobarbital. *Journal of Magnetic Resonance Imaging*. 2019;49(4):1062–1068.
12. Near J, Harris AD, Juchem C, et al. Preprocessing, analysis and quantification in single-voxel magnetic resonance spectroscopy: Experts' consensus recommendations. *NMR Biomed*. 2020:e4257.
13. Zoelch N, Hock A, Henning A. Quantitative magnetic resonance spectroscopy at 3T based on the principle of reciprocity. *NMR in Biomedicine*. 2018;31(5):e3875.
14. Zoelch, Niklaus, et al. "Accurate determination of brain metabolite concentrations using ERETIC as external reference." *NMR in Biomedicine* 2017;30(8):e3731.
15. Kreis R, Ernst T, Ross B. Absolute quantitation of water and metabolites in the human brain: II. Metabolite concentrations. *Journal of Magnetic Resonance, Series B*. 1993;102(1):9–19.
16. Kreis R. Quantitative localized 1H MR spectroscopy for clinical use. *Progress in Nuclear Magnetic Resonance Spectroscopy*. 1997;31(2):155–196.
17. Gasparovic C, Song T, Devier D, et al. Use of tissue water as a concentration reference for proton spectroscopic imaging. *Magnetic Resonance in Medicine*. 2006;55(6):1219–1226.
18. Zoelch N, Hock A, Heinzer-Schweizer S, Avdievitch N, Henning A. Accurate determination of brain metabolite concentrations using ERETIC as external reference. *NMR in Biomedicine*. 2017:e3731-n/a.
19. Maudsley AA, Andronesi OC, Barker PB, et al. Advanced magnetic resonance spectroscopic neuroimaging: Experts' consensus recommendations. *NMR in Biomedicine*. 2021;34(5):e4309.
20. Deelchand DK, Berrington A, Noeske R, et al. Across-vendor standardization of semi-LASER for single-voxel MRS at 3T. *NMR in Biomedicine*. 2021;34(5):e4218.
21. Tk\'a v, I, Star\vc, uk, Z, Choi I, Gruetter R. In vivo 1H NMR spectroscopy of rat brain at 1 ms echo time. *Magnetic Resonance in Medicine*. 1999;41(EPFL-ARTICLE-177519):649–656.
22. Dreher W, Leibfritz D. New method for the simultaneous detection of metabolites and water in localized in vivo 1H nuclear magnetic resonance spectroscopy. *Magnetic Resonance in Medicine*. 2005;54(1):190–195.
23. de Graaf RA, van Kranenburg A, Nicolay K. Off-resonance metabolite magnetization transfer measurements on rat brain in situ. *Magnetic Resonance in Medicine: An Official Journal of the International Society for Magnetic Resonance in Medicine*. 1999;41(6):1136–1144.
24. Juchem C, Cudalbu C, de Graaf RA, et al. B0 shimming for in vivo magnetic resonance spectroscopy: Experts' consensus recommendations. *NMR in Biomedicine*. 2021;34(5):e4350.
25. de Barros NP, Slotboom J. Quality management in in vivo proton MRS. *Analytical Biochemistry*. 2017;529:98–116.
26. Wilson M, Andronesi O, Barker PB, et al. Methodological consensus on clinical proton MRS of the brain: Review and recommendations. *Magnetic Resonance in Medicine*. 2019;82(2):527–550.
27. Oz G, Deelchand DK, Wijnen JP, et al. Advanced single voxel (1) H magnetic resonance spectroscopy techniques in humans: Experts' consensus recommendations. *NMR in Biomedicine*. 2020:e4236.
28. Kreis R, Boer V, Choi IY, et al. Terminology and concepts for the characterization of in vivo MR spectroscopy methods and MR spectra: Background and experts' consensus recommendations. *NMR in Biomedicine*. 2020:e4347.
29. Kreis R. Issues of spectral quality in clinical 1H-magnetic resonance spectroscopy and a gallery of artifacts. *NMR in Biomedicine*. 2004;17(6):361–381.
30. Ith M, Bigler P, Scheurer E, et al. Observation and identification of metabolites emerging during postmortem decomposition of brain tissue by means of in situ 1H-magnetic resonance spectroscopy. *Magnetic Resonance in Medicine: An Official Journal of the International Society for Magnetic Resonance in Medicine*. 2002;48(5):915–920.

31. Banaschak S, Rzanny R, Reichenbach J, Kaiser W, Klein A. Estimation of postmortem metabolic changes in porcine brain tissue using 1 H-MR spectroscopy—preliminary results. *International Journal of Legal Medicine.* 2005;119(2):77–79.
32. Musshoff F, Klotzbach H, Block W, Traeber F, Schild H, Madea B. Comparison of postmortem metabolic changes in sheep brain tissue in isolated heads and whole animals using 1H-MR spectroscopy—preliminary results. *International Journal of Legal Medicine.* 2011;125(5):741–744.
33. Henssge C, Frekers R, Beckmann E. Determination of the time of death based on simultaneous measurement of brain and rectal temperatures. *Zeitschrift fur Rechtsmedizin Journal of Legal Medicine.* 1984;93(2):123–133.
34. Berger C, Bauer M, Wittig H, Scheurer E, Lenz C. Post mortem brain temperature and its influence on quantitative MRI of the brain. *MAGMA.* 2021.
35. Birkl C, Langkammer C, Haybaeck J, et al. Temperature-induced changes of magnetic resonance relaxation times in the human brain: A postmortem study. *Magnetic Resonance in Medicine.* 2014;71(4):1575–1580.
36. Tofts P. *Quantitative MRI of the brain: Measuring changes caused by disease.* John Wiley & Sons; 2005.
37. Tashiro K, Shiotani S, Kobayashi T, et al. Cerebral relaxation times from postmortem MR imaging of adults. *Magnetic Resonance in Medical Sciences.* 2015;14(1):51–56.
38. Wermter FC, Mitschke N, Bock C, Dreher W. Temperature dependence of (1)H NMR chemical shifts and its influence on estimated metabolite concentrations. *MAGMA.* 2017;30(6):579–590.
39. Kuroda K. Non-invasive MR thermography using the water proton chemical shift. *International Journal of Hyperthermia.* 2005;21(6):547–560.
40. Takahashi K. Relationship between acidity and swelling in the brain. *The Tohoku Journal of Experimental Medicine.* 1966;90(3):261–268.
41. Brinkmann B, Madea B. *Handbuch gerichtliche Medizin.* Vol 1: Springer; 2003.
42. Monoranu CM, Apfelbacher M, Grunblatt E, et al. pH measurement as quality control on human post mortem brain tissue: A study of the BrainNet Europe consortium. *Neuropathology and Applied Neurobiology.* 2009;35(3):329–337.
43. Hardy J, Wester P, Winblad B, Gezelius C, Bring G, Eriksson A. The patients dying after long terminal phase have acidotic brains; implications for biochemical measurements on autopsy tissue. *Journal of Neural Transmission.* 1985;61:253–264.
44. Weljie AM, Newton J, Mercier P, Carlson E, Slupsky CM. Targeted profiling: Quantitative analysis of 1H NMR metabolomics data. *Analytical Chemistry.* 2006;78(13):4430–4442.
45. Tredwell GD, Bundy JG, De Iorio M, Ebbels TM. Modelling the acid/base (1)H NMR chemical shift limits of metabolites in human urine. *Metabolomics.* 2016;12(10):152.
46. Marti M, Labouesse J, Canioni P, Merle M. N-acetyl-L-aspartate and acetate 1H NMR signal overlapping under mild acidic pH conditions. *Magnetic Resonance in Medicine.* 1993;29(5):692–694.
47. Ackerman JJ, Soto GE, Spees WM, Zhu Z, Evelhoch JL. The NMR chemical shift pH measurement revisited: Analysis of error and modeling of a pH dependent reference. *Magnetic Resonance in Medicine.* 1996;36(5):674–683.
48. Frese S, Gascho D, Thali MJ, Kozerke S, Zoelch N. In situ determination of pH in brain tissue using postmortem 1H-MRS. ISMRM Annual Meeting; 2023; Toronto, Canada.
49. Bertram HC, Whittaker AK, Andersen HJ, Karlsson AH. pH Dependence of the progression in NMR T(2) relaxation times in postmortem muscle. *Journal of Agriculture Food Chemistry.* 2003;51(14):4072–4078.
50. Scheurer E, Ith M, Dietrich D, et al. Statistical evaluation of time-dependent metabolite concentrations: Estimation of postmortem intervals based on in situ1H-MRS of the brain. *NMR in Biomedicine.* 2005;18(3):163–172.
51. Wishart DS, Feunang YD, Marcu A, et al. HMDB 4.0: The human metabolome database for 2018. *Nucleic Acids Research.* 2018;46(D1):D608–D617.
52. Govind V, Young K, Maudsley AA. Corrigendum: Proton NMR chemical shifts and coupling constants for brain metabolites. Govindaraju V, Young K, Maudsley AA, NMR Biomed. 2000; 13: 129–153. *NMR in Biomedicine.* 2015;28(7):923–924.
53. Hsu CH, Lin S, Ho AC, et al. Comparison of in vivo and in situ detection of hippocampal metabolites in mouse brain using (1) H-MRS. *NMR in Biomedicine.* 2021;34(2):e4451.
54. Dashti H, Wedell JR, Westler WM, et al. Applications of parametrized NMR spin systems of small molecules. *Analytical Chemistry.* 2018;90(18):10646–10649.
55. Ith M, Bigler P, Scheurer E, et al. Observation and identification of metabolites emerging during postmortem decomposition of brain tissue by means of in situ 1H-magnetic resonance spectroscopy. *Magnetic Resonance in Medicine.* 2002;48(5):915–920.
56. Michaelis T, Helms G, Frahm J. Metabolic alterations in brain autopsies: Proton NMR identification of free glycerol. *NMR in Biomedicine: An International Journal Devoted to the Development and Application of Magnetic Resonance In Vivo.* 1996;9(3):121–124.

CHAPTER 8

SONOGRAPHY

Lars Ebert and Jakob Heimer

Unlike computed tomography (CT) or magnetic resonance imaging (MRI), sonography does not utilize electromagnetic waves for imaging but rather mechanical ones. Typically, medical ultrasound uses sound waves with frequencies of 2–12 MHz. The basis for ultrasound technology as it is used clinically is the piezoelectric effect. Some materials, when put under mechanical stress, generate an electrical charge (piezoelectric effect) and when exposed to an electric field will mechanically deform (inverse piezoelectric effect). Using both effects in conjunction, devices, so-called transducers, can be built that are able to emit mechanical waves using the inverse piezoelectric effect, but can also record incoming waves by converting mechanical movement to electrical signals. Transducers send ultrasound waves into the tissue, where they are reflected by structures they hit. The time required by the sound waves to be reflected can be used to calculate the distance of the reflective surface, which in turn is used to generate images. Lower frequencies penetrate the tissue deeper at the cost of resolution, whereas higher frequencies produce higher resolution images, at the cost of penetration depth. For different purposes, different shapes for ultrasound probes exist. Sector scanners have a small contact surface, which allows access to otherwise difficult to reach areas. Structures that are closer to the scanner are displayed in higher resolution than structures that have a larger distance. Linear probes are useful for structures that are closer to the surface. Convex probes allow the creation of overview scans, i.e. of the abdomen due to the large field of view.

Sonography in general has some clear advantages, many of which mainly play out clinically. Because sonography does not utilize ionizing radiation, it poses little to no health risk. The technique in general is well understood and routinely used clinically, as in Figure 8.1a–c. Compared to CT or MRI scanners, the devices are relatively affordable. In recent years, mobile transducers have been developed that can be operated with mobile devices such as smartphones. This makes sonography very flexible and mobile. On the other hand, sonography has strong limitations, as it cannot penetrate bone or gas accumulations, which means some anatomical structures are difficult if not impossible to depict. This is especially true in a postmortem setting, where decompositional gas can be present. Positioning the transducer and interpreting the resulting images correctly requires extensive experience and training. Because of this and the fact only single ultrasound images can be recorded (as opposed to volumes in CT and MRI imaging), it can be difficult to communicate findings to medical laypeople or get a second opinion.

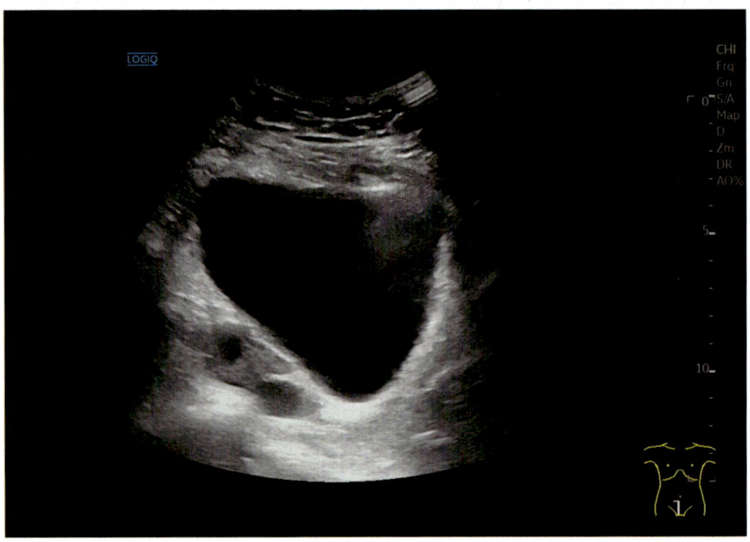

(a)

Figure 8.1 Sonography of a bladder. a: sagittal view. b: transversal view.

DOI: 10.4324/9781003413998-9

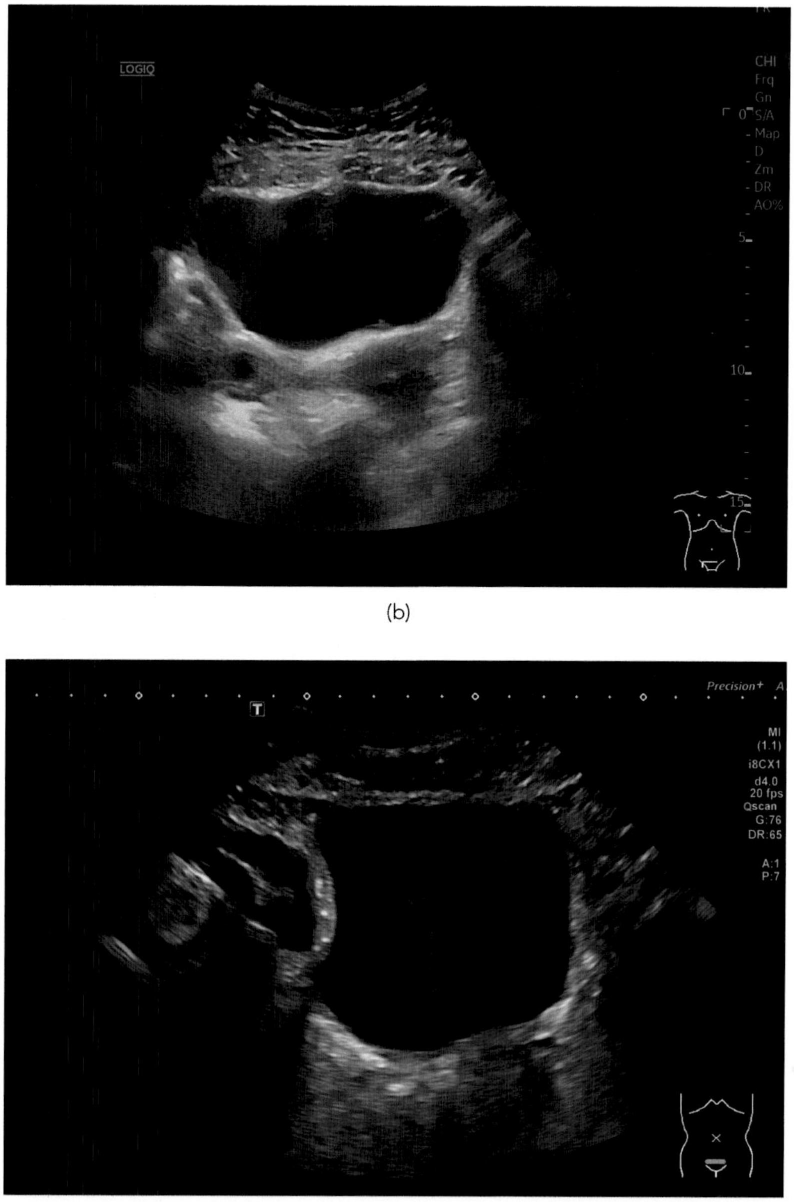

(b)

(c)

Figure 8.1 (Continued)

Nonetheless, some applications for sonography in forensic medicine have been discussed in the literature.

For postmortem imaging, the application of sonography has so far been limited, considering superior techniques are available. However, its potential for forensic medicine should not be underestimated. Sonography could become the modality of choice when its advantages (low cost, mobility, immediacy) outweigh its disadvantages [1]. First, the affordability of sonography makes it a substitute for MRI in fetuses and neonates and is currently researched thoroughly for this cause. A practical guide for perinatal postmortem ultrasound examinations was published by Shelmerdine et al. [2]. Sonography reached an overall sensitivity of 74.5% and a specificity of 83.3% in comparison to autopsy in a recent study [3]. Sonography has shown to be significantly more often "non-diagnostic" (insufficient for making a diagnosis) than MRI. In a recent study, sonography was non-diagnostic in 26.9% of cases, while MRI was non-diagnostic in only 4.4% in a study by Kang et al. investigating 160 fetal cases [4]. However, if both methods were diagnostic, no significant differences in diagnostic accuracy were discovered (67.8% for sonography, 78.0% for MRI).

More research is needed for the application of sonography in further settings, which are theoretical concepts for the moment: The unique advantage of being a highly mobile device could be exploited in transporting a sonography device to the external examination. On-site sonography may lead to improved decision making involving the necessity of further imaging or a full autopsy. However, the low expected diagnosticity and sensitivity for different causes of death in adult examinations limit the extent of this improvement. The combination of high mobility, non-ionizing and immediacy of results makes sonography a prime candidate for examinations in cases of non-fatal strangulation. As non-fatal strangulation often presents with few visible external findings, sonography could improve the diagnosis in such cases. As findings in these cases often represent subcutaneous and intramuscular hematoma and edema, sonography could perform well in such cases.

To conclude, sonography currently plays an important role in perinatal postmortem investigations. Apart from that, sonography has not found a relevant forensic application so far. Further applications are imaginable, but research must show possible additional value and/or practicability.

References

[1] Fariña J, Millana C, Fdez-Aceñero JM, Furió V, Aragoncillo P, Martín VG, et al. Ultrasonographic autopsy (echopsy): A new autopsy technique. Virchows Arch 2002;440:635–9. https://doi.org/10.1007/s00428-002-0607-z.

[2] Shelmerdine SC, Sebire NJ, Arthurs OJ. Perinatal post mortem ultrasound (PMUS): A practical approach. Insights Imaging 2019;10:35. https://doi.org/10.1186/s13244-019-0723-9.

[3] Kang X, Shelmerdine SC, Hurtado I, Bevilacqua E, Hutchinson C, Mandalia U, et al. Postmortem examination of human fetuses: Comparison of two-dimensional ultrasound with invasive autopsy. Ultrasound Obstet Gynecol Off J Int Soc Ultrasound Obstet Gynecol 2019;53:229–38. https://doi.org/10.1002/uog.18828.

[4] Kang X, Sanchez TC, Arthurs OJ, Bevilacqua E, Cannie MM, Segers V, et al. Postmortem fetal imaging: Prospective blinded comparison of two-dimensional ultrasound with magnetic resonance imaging. Ultrasound Obstet Gynecol Off J Int Soc Ultrasound Obstet Gynecol 2019;54:791–9. https://doi.org/10.1002/uog.20217.

CHAPTER 9

IMAGING-GUIDED TISSUE (BIOPSY) OR FLUID RETRIEVAL

Carlo Tappero and Lars Ebert

DOI: 10.4324/9781003413998-10

9.1 INTRODUCTION

Biopsy (in Greek, *bios* = life and *opsis* = view) is a word whose etymology means view and analyze the life.

This procedure is well established in the medical community. Already in the latter half of the tenth century, an Arabian physician, Albucasim, described the needle puncture of the thyroid to distinguish different types of goiter (Anderson & Webb, 1987). However, it was the French dermatologist Ernest Besnier who created this neologism in 1879 (DD., 1994).

Nowadays this technique has evolved from blind or surgical sampling, which could be quite invasive, to a minimally invasive image-guided procedure.

The most used imaging techniques at present are ultrasound, CT and MR, the latter two also present at some forensic institutes.

The use of CT and MR imaging in the forensic field is well known and very useful. They are powerful non-destructive techniques that save a virtual copy of the subject and are useful to mostly identify forensic relevant macropathologies such as trauma, internal organ injuries or abnormal gas collection. Although the resolution of these techniques is now submillimetric, we are still far from performing a virtual histology, even if promising results are coming from micro-CT and micro-MRI.

However, the possibility of conducting minimally invasive tissue sampling through image-guided biopsy technique is very useful to the pathologist as it allows investigation even if an autopsy is not permitted due to objections of the family or religious beliefs.

With a postmortem image-guided biopsy, it is possible to sample liquids such as intracardiac blood, pleural or pericardial effusion, cerebro-spinal fluid, intrabdominal collection and urine, as well as tissue sampling from every organ and tissue. All the samples can be processed with the same anatomopathological techniques used in normal autopsy. Furthermore, performing a postmortem biopsy is far easier than in living people because of lack of moving artefacts or risk of health complications. Performing a biopsy by a trained physician is a quick and safe procedure that can decrease the examination time and moreover protects the operators from being exposed to potential infectious situations due to contamination.

9.2 TECHNICAL EVOLUTION

9.2.1 Non–Image-Guided Biopsy and Needle Autopsy

Tissue sampling in medicine has a long history, including its use in the postmortem field. However, non–image-guided biopsy, as an alternative to autopsy has not been described until the second half of the twentieth century. One of the first applications was focused on the diagnosis of infectious diseases in South America (Manson-Bahr, 1950). Another one compared the results of liver biopsy with those of the autopsy, but the range of delay between the two was up to 202 days and it was more focused on proving whether a minute specimen could be representative of the structural changes of the liver as a whole, rather than its postmortem accuracy (Wagoner, Ulevitch, Gall, & Schiff, 1951). The first detailed report of "needle necropsy" is the work of Terry (1955) in which he proposed a modified version of his liver biopsy needle to perform biopsy of different organs, including the brain through a small hole in the skull near the suspected area. His series had a high success rate, with 22 diagnoses out of 24 cases.

From this moment onwards, other papers came out discussing the accuracy of blind biopsy versus autopsy.

Already in 1957, 50 consecutive needle biopsies were compared with conventional autopsy, finding discrepancies in 26 cases, judging the technique limited in its value (West & Chomet, 1957). The most common causes of discrepancy were myocardial infarctions, pulmonary infarctions, cerebrovascular accidents, heart failures and gastrointestinal tract hemorrhages. In 1969, Wellman reported a series of 394 consecutive needle biopsies, 37 of which were limited to only one or two organs, while in the other cases he sampled liver, heart, lung and kidney with sampling success rates of 92%, 55%, 46% and 34%, respectively. For these authors, needle autopsy was an inadequate substitute to conventional autopsy but valuable when an autopsy was not possible for different reasons. Other interesting papers were published in 1983 by Benbow and Underwood (Underwood, Slater, & Parsons, 1983). The latter noted that small-tissue samples tended to be more carefully analyzed than the larger samples coming from conventional autopsy, where a wider range of examinations are usually performed to obtain any possible information. Other studies were published in the 90s presenting biopsies as an alternative to reduce operator risk in case of HIV-positive subjects (Lucas, 1993) and showing how they were useful to identify the cause of death, including infections (Baumgart KW, 1994), ischemic heart disease or a large pulmonary thromboembolism (Foroudi F, 1995). An interesting development of this technique was the microbiological culture of tissue samples. In the case of Huston et al. (1996) postmortem needle lung cultures correlated with the complete autopsy in 17 (85%) of 20 patients and 16 (80%) of 20 spleen cultures. Cultures of the brain (one patient), cerebrospinal fluid (two patients), peritoneum (two patients) and serum (one patient) correlated 100% when compared to the complete autopsy. Furthermore, in 1998 Aranda et al. compared microbiological cultures of samples from autopsies with those from fine-needle aspirations (FNAs) in 92 adult cases. As a result, they found that FNA performed in the immediate postmortem period adds relevant microbiological information to the clinicopathologic picture and provides higher specificity than autopsy cultures. At the beginning of the 2000s the focus was again on HIV-positive patients (Guerra, et al., 2001) (Satyanarayana, et al., 2003) with identification of other infective diseases that were not clinically suspected (Guerra, et al., 2001).

Summing up, these studies showed that tissue sampling without any guidance is not of equal value in all cases compared to a full autopsy, but it has some useful peculiarities. Most of the authors agree that with blind biopsies, it is difficult to obtain samples of the desired organs, and this made the diagnostic reliability of the technique worse than traditional autopsy. (West & Chomet, 1957) (Underwood, Slater, & Parsons, 1983) (Baumgart KW, 1994) (Guerra, et al., 2001)

9.2.2 Image-Guided Biopsy

Under imaging guidance nowadays two types of biopsies are performed: cytology and histology. In cytology only some cells are sampled and examined while in histology a piece of tissue is sampled and examined. In the first case, normally a 22G (0.643 mm) needle is inserted in the lesion and a vacuum is applied during a rocking movement to obtain a small amount of liquid with cells. In the second case a special needle of 14 to 20G diameter adapted to cut the tissue is sited near or inside the lesion and a sample is cut out.

There are three image modalities used mostly to guide the position of a biopsy needle: ultrasound, MRI and CT.

9.2.2.1 Ultrasound (US)

US is a low-cost, easily available image modality. It has the advantage of not using ionizing radiation and of producing images in real time, but it requires a skilled operator to perform a biopsy. Because ultrasound cannot penetrate gas and air, its use is limited in a postmortem setting due to decompositional gas. Because of these reasons, very few examples for use in a postmortem setting are present in the forensic literature (Duarte-Neto AN, 2019) (Brook OR, 2021).

The needle can be controlled either free hand or with the help of a special needle guide fixed on the probe (see Figure 9.1). In this case it is possible to visualize the hypothetical path of the needle on the screen even if it is not visible with the US.

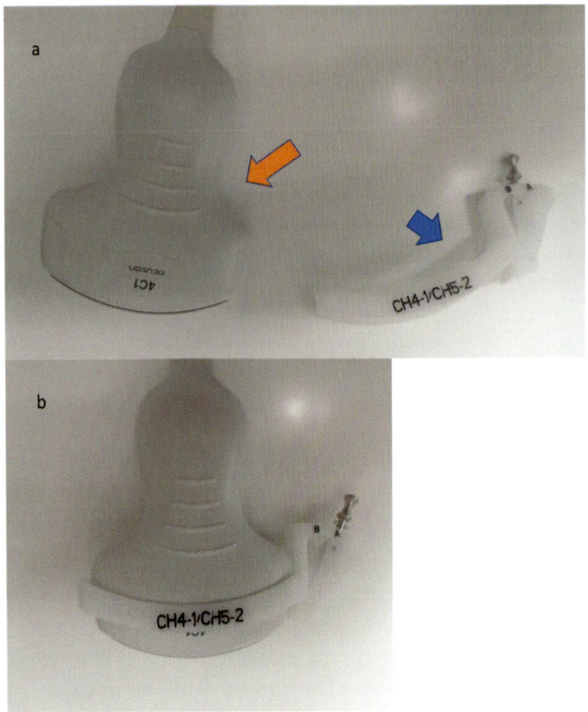

Figure 9.1 (a) Ultrasound probe (orange arrow). Needle guide (blue arrow) with two possible insertion angles for the needle (A and B) (b) Ultrasound probe with installed needle guide.

9.2.2.2 Magnetic Resonance Imaging (MRI)

MRI is a well-established instrument in the forensic field and it offers high-quality anatomical information about the soft tissue in the target area. When compared to CT, this comes at the cost of a slightly lower spatial resolution. It has also the advantage of not exposing the operator to ionizing radiation. In clinical radiology, MR-guided biopsies are still an emerging field. Because of the strong magnetic field, handling the standard ferromagnetic needles is almost impossible and special non ferromagnetic needles are required.

Using special ultrafast sequences, it is possible to follow the insertion of these special coaxial needles in real time like needle placement under US. Because of the monoplanar acquisition, insertion point and target lesion must be on the same plane, which can limit the accuracy for oblique insertion angles. Once the coaxial needle is in place, the corpse is moved out from the magnetic field to take the required samples with a standard biopsy needle.

Because the coaxial needle must be in the middle of the magnetic field for the best visualization, the person conducting the biopsy has to partly enter the narrow MR gantry almost lying on the subject and this may be inconvenient, especially in a forensic setting.

9.2.2.3 Computed Tomography (CT)

CT is now the standard diagnostic tool in modern forensic pathology due to its speed and high spatial resolution and relatively inexpensive cost compared to MRI, at the cost of lower soft tissue contrast.

There are several techniques to perform CT-guided biopsy. Here we will present three of them: step and shot, CT-fluoroscopy and helical acquisition.

9.2.3 Step and Shot

A standard conventional planning CT is performed to plan the approach to the lesion and the needle position. Then, after having planned the trajectory on the first acquisition, the skin entry point is routinely marked and the needle is put in place. Standardized control scans can be performed using a foot switch. For every control, normally three slices are acquired—one centered on the lesion, one above and one below. The slice thickness can vary from 1 to 10 mm depending on the CT producer and operator preferences. With this technique even if the acquisition is in low dose, the operator can be exposed to a small amount of ionizing radiation. Furthermore, an out-of-plane approach is difficult to accomplish and requires a skilled operator. Older CT scanners allowed for a gantry tilt to compensate for this limitation but are becoming rare.

9.2.4 CT Fluoroscopy

Thanks to the evolution in informatics, it is possible to reconstruct the scanned image almost in real time. This technique combines cross-sectional imaging with the real-time imaging, tracking and motion perception typical of fluoroscopy.

The planning of the procedure is always done with standard CT. After having verified the right entrance point, it is then possible to follow the progression of the needle inside the body in real time. Even in this case, the entrance point and the target have to be strictly on the same plane. While performing the needle positioning, the operator needs to be near the gantry and so he/she will be exposed to radiation, which is the biggest disadvantage. This technique permits real-time correction of the needle placement and therefore allows a faster procedure.

9.2.5 Conventional CT

The third technique is just a repetition of the planning CT, but with the needle in place. It has the advantage of allowing oblique planes in a cranio-caudal approach, but is normally slower. The operator exits the room during every acquisition to avoid radiation exposure and to evaluate the needle trajectory, especially in case of oblique approaches.

9.2.5.1 Computer-Aided Biopsy

Surgical navigation systems can guide the operator to put the needle in the right position without performing control during the needle positioning.

There are different systems on the market, but they work with roughly the same principle. A CT acquisition is done with markers on the skin of the patient. The acquisition is then sent to the navigation system that recognizes the markers and calculates the relation between patient and CT data, a step called registration. The desired trajectory is planned and stored in the system. Then a tracked aiming device, recognized by the system by using magnetic or active/passive infrared trackers, is used to guide the user to the planned trajectory. When the aiming device is in the right position, the needle is advanced on the trajectory to the desired depth, which is continuously monitored by the system. Most of these systems can fuse the planning acquisition with previous CT or MRI scans to be sure to reach the target, even if it is not visible in the planning CT. The biggest problem of these systems is that if the target moves with respiration, the accuracy decreases because the skin markers are moving and the CT scan is static. In a clinical setting, this makes it necessary to reduce such moving artifacts with different techniques, but it is not an issue in a postmortem environment (Aghayev, et al., 2008). The biggest advantages of these systems are that even less skilled operators can perform accurate biopsies with just a bit of training. The technique is fast, and there is no radiation exposure to the operator.

9.2.5.2 Robotic Biopsy

The use of surgical navigation systems still requires the user to manually place the needle, which can lead to errors. If a large number of image-guided samples have to be taken, this procedure can also be time consuming. In addition, the handling of needles, tissue samples and body fluids can come with contamination issues and be a potential health hazard. This led to the development of systems that use automation techniques from robotics. In a postmortem setting, automation can roughly be separated into two groups: Robotic systems with needle guide and systems with automatic needle placement. Robotic systems with needle guide tend to be smaller and mobile. After planning (see Figure 9.2), the system places a needle guide, which can be used to manually push a coaxial introducer needle into the final position, which can then be used for tissue sampling as described above (Martinez, et al., 2014). Robotic systems with automatic needle placement either place the needles directly through robot motion (Gromniak, Neidhart, Heinemann, Püschel, & Schlaefer, 2020) (Neidhar, et al., 2022) or by using additional pneumatic actuators, which has the advantage of higher speed, resulting in lower soft tissue movement (Bert, et al., 2014).

In general, the use of robotic systems allows for accurate and fast tissue sampling with a low risk of contamination during needle placement. Because of the required specialized equipment, however, it can be costly. More details about the Virtobot biopsy system can be found in the corresponding chapter.

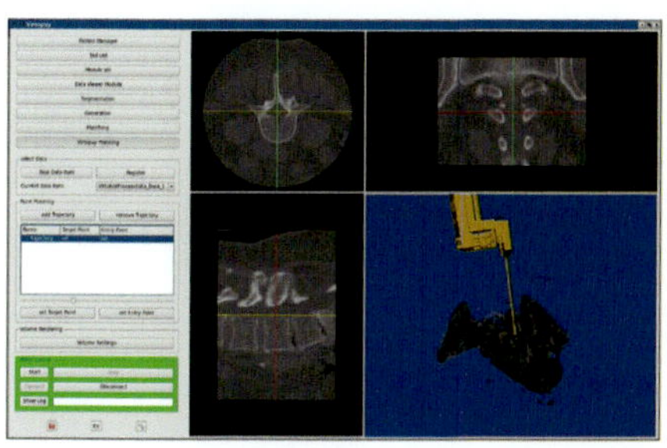

Figure 9.2 Planning of a needle biopsy using the first prototype of the Virtobot system. After an initial registration step, target point, entry point and the orientation of the robot end effector can be defined.

9.3 BIOPSY NEEDLES AND TECHNIQUES

As mentioned before, there are two main biopsy classifications: fine needle aspiration (FNAB) and tissue biopsy.

9.3.1 Fine Needle Aspiration Biopsy

With this technique, a fine Chiba needle (20 to 25G) is inserted in the lesion, and using a vacuum and a rocking movement, some cells are taken and then deployed either on a microscope slide or inside a special medium. This method only allows for limited cytological analysis because of the thinness of the needle. Because of the required movement, it is easy to displace the needle during biopsy. Due to these limitations, this technique is not suitable for a comprehensive postmortem forensic examination.

9.3.2 Tissue Biopsy

To take a tissue sample there are nowadays two types of needles—the Trucut (see Figure 9.3) and full core needles. Both types of needle have a diameter between 14 and 20 G.

Trucut biopsy guns can be automatic or semiautomatic. Trucut needles are special needles with a compartment to store the sample. The needle is inserted near the lesion. The part of the needle with the slot is pushed inside the lesion, and finally a sheath slides over the slot to cut the sample.

In the semiautomatic version, the part of the needle with the slot is manually pushed inside the lesion, and only the cutting sheath is fired by a spring. In this case it is possible to confirm that the slot is really taking the targeted sample. In the automatic version, when the needle is near the lesion, both the slot and the sheath are moved by a spring. According to some studies (Krokidis, 2019), the full automatic version produces better samples than the semiautomatic one.

In case of full core biopsies, the needle tip is put near the lesion and then a hollow cutting sheath slides inside the lesion with the help of a spring. This type of needle is normally slightly bigger and heavier than Trucut and so is a bit more difficult to control the position of the needle and if the lesion is small and hard sometimes it slides on it.

Normally full core needles produce up to 30% larger samples than those of Trucut needles of the same caliber.

9.3.3 Coaxial Technique

To avoid replacing the biopsy needle after every sample, it is possible to place a hollow needle with an inner trocar near the lesion, and after removing the trocar insert the biopsy needle. These coaxial needles (see Figure 9.4) are smaller, stiffer and easier to handle than biopsy needles. It's possible to insert them even when space in the CT or MRI gantry is limited. Furthermore, it is possible to take multiple samples with only one needle positioning, substantially shortening the working time.

Coaxial needles also allow for additional flexibility, as they can directly be used for tissue and liquid sampling but also for other tasks such as the injection of contrast agents or the placement of guide wires.

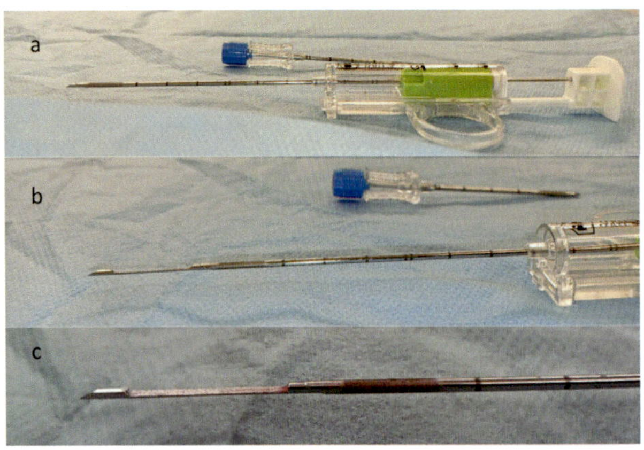

Figure 9.3 Trucut needle. (a) Closed slot, (b) opened slot and (c) magnification of the opened slot

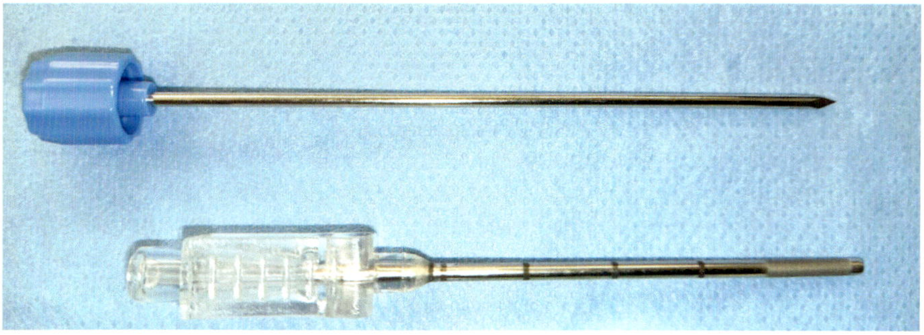

Figure 9.4 Coaxial needle.

9.4 FLUID RETRIEVAL

Most of the previously mentioned techniques can also be used for liquid sampling. For this purpose, a needle is placed and a syringe is connected to draw liquid. It should be kept in mind that more viscous liquids (i.e. pus) require larger diameter needles. In contrast to planning a biopsy, where the needle tip is planned on top of the target, because the tissue sample is taken 1–2 cm under the needle tip, for liquid sampling, the needle tip is planed directly into the center of liquid collection.

CHAPTER 10

USE OF CONTRAST AGENT IN PMCT

Wolf Schweitzer and Carlo Tappero

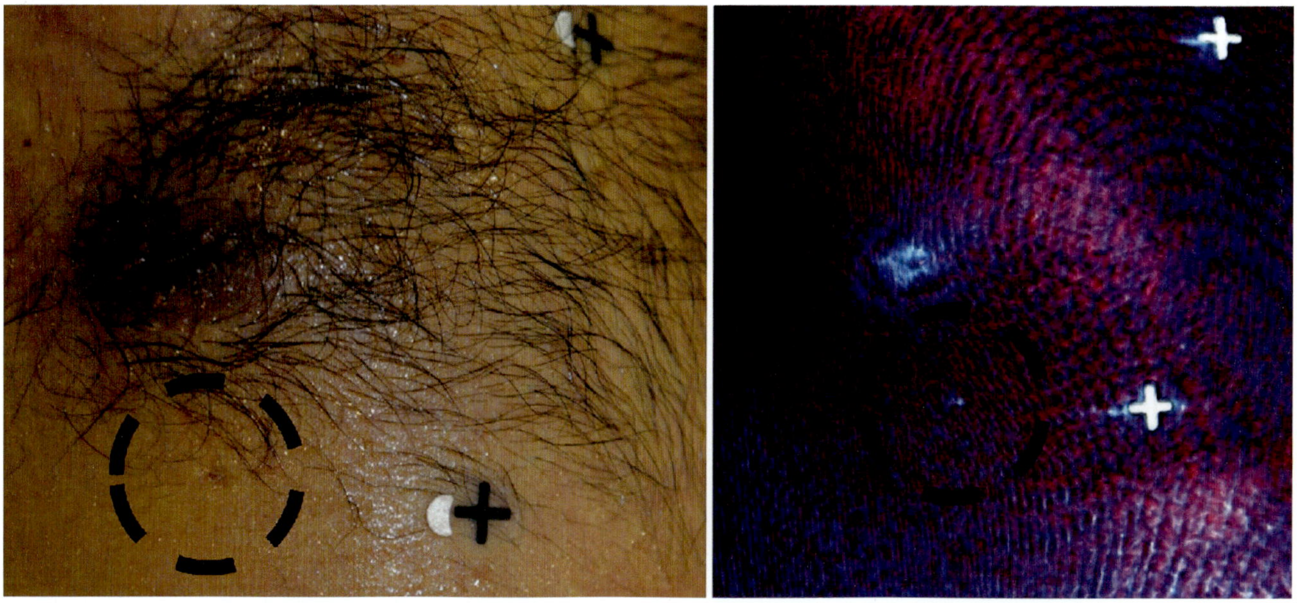

Figure 10.1 Contrast enhancement of skin injury 1/3. These images show contrast medium (watery solution of Optiray, 1:20) applied to the skin, with a needle mark (marked with circle) both in photograph (left) and VRT of PMCT (right). The contrast agent effectively highlights the needle mark, as evidenced by the correlated point type finding in the VRT reconstruction with regard to the photograph.

DOI: 10.4324/9781003413998-11

10.1 BACKGROUND

While soft tissue and organ contrast in native CT and native PMCT may turn out to be relatively low, it may be increased either by maximizing CT dose or by adding contrast agent. After death, a wider range of contrast agents come into consideration than in living subjects.

The typical aspects to consider generally are of a common nature, regardless of how exactly a contrast agent is applied.

An issue affecting the targeted CT density of any contrast admixture is that of trading off the partial volume effect against beam-hardening artifacts. The partial volume effect plays out at the effective resolution of the scan: The smallest structures or dimensions that are contrasted may well fall under the partial volume effect, whereby the CT density of a small structure in the actual CT scan that result from that examination will be lower than the usual density of the contrast agent. This is an inherent aspect of this technology. So, increasing the density of a contrast agent fluid to be very high may cause beam-hardening artifacts in larger structures but yield better contrast in structures that (still) succumb to the partial volume effect. This may also illustrate how the partial volume effect is not to be understood as an artifact of CT scanning but as a correctly obtained result that manifests a mismatch between object size and scan resolution and material CT density versus achieved object or structure CT density. Conversely, if one already increases the viscosity of the carrier fluid to avoid contrast in smaller structures such as vessels, there seems to be no justification for overly increasing contrast density. In that context, it is generally advised to obtain higher-resolution scans and reconstructions and attempt the acquisition of isotropic data [1].

For contrast agent, a preference in our team is water-based contrast. This includes iodine dilutions, also for environmental reasons [2]. The cost of installing and maintaining an oil/water waste water separator both for purchase/installation as well as required checks, maintenance, and supervision may be regarded as prohibitive [3]. A particularly popular application of contrast agent is full-body forensic postmortem CT angiography (PMCTA), but it is not the only way to use contrast agent in conjunction with PMCT.

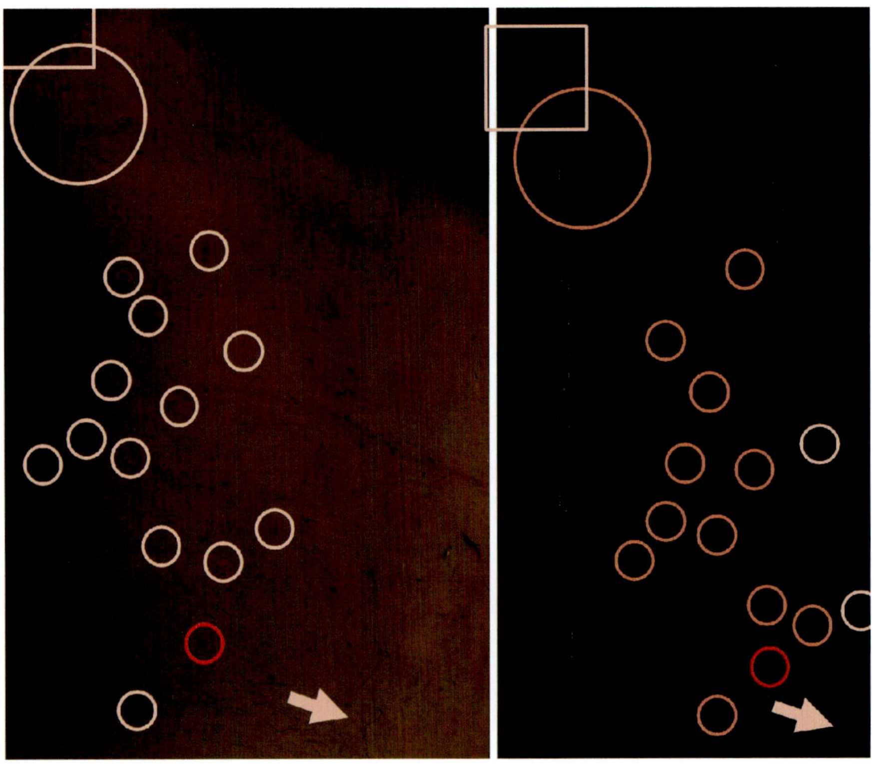

Figure 10.2 Contrast enhancement of skin injury 2/3. These images show contrast medium (admixture of Optiray and Ponal white glue, 1:5) applied to the skin and then wiped gently with paper towels, with superficial skin defect injuries that presumably were caused by small water animals such as duck fleas in a body retrieved from a lake (red small irregularly shaped skin bleeding: marked with circles) both in photograph (left) and VRT of PMCT (right). If the right image is just a dark rectangle with circles then the quality of the rendering (screen, print) is too low. Contrast is also enhanced in areas of body hair (square, arrow, red circle), so not all contrasted findings match a skin defect here. It is therefore essential to interpret the images carefully, recognizing that contrast enhancement does not exclusively indicate skin defects, to also correlate contrast enhancement findings with other observations, such as the actual location in its context, specific shape and appearance of the finding.

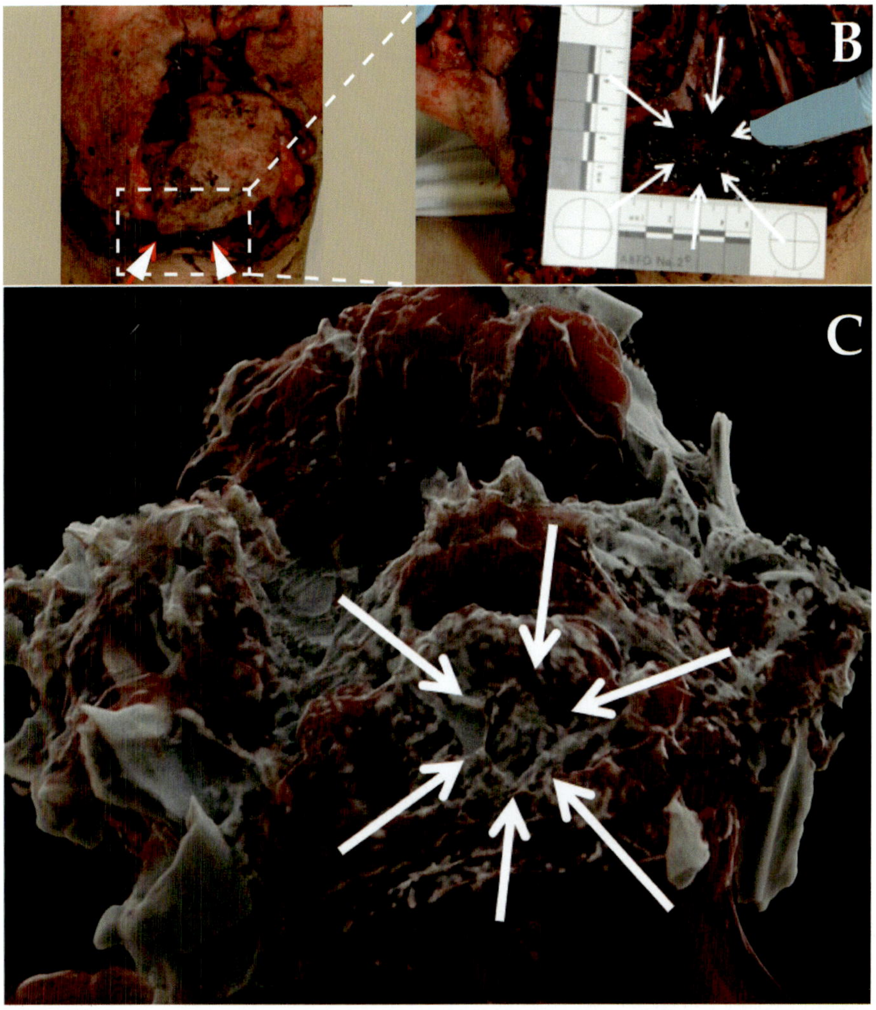

Figure 10.3 Contrast enhancement of skin injury 3/3. These images show contrast medium (Optiray) applied to the exposed inner body structures after a rifle shot to the head. What remained of the entrance wound in the chin region (A, B, arrows) was identified macroscopically, but in the PMCT data, surface contrast was helpful to allow for easier 3D visualization of what otherwise appeared to be a poorly delineated soft tissue defect shape (C, arrows).

In that context, use of the term "artifact" may imply a perceptual misrepresentation or a factual error in imaging or reconstruction such as from defective parts, as it is usually used in such a context. For example, metal "artifacts" are streaky visualizations presented in axial slice image data reconstructed from CT raw data that tend to occur due to the assumptions made by the backprojection algorithm that then are not met by the combination of a polyenergetic X-ray and

In postmortem CT angiography (PMCTA), when low-viscosity (watery) contrast agents are used, the observed leaks and extravasations—although sometimes producing streaky or heterogeneously dense appearances—generally reflect true postmortem vascular conditions and authentic contrast distribution, representing an expected and technically appropriate imaging outcome for the given specimen. Likewise, blurred CT images of a beating heart when really a sharp focused image was expected [5] typically do not represent a wrong calculation or defect in a sensor but the application of a use of technology that is at odds with the goal, and attributing such a motion-related image blur to the domain of "artifacts" may not do the user error (that provides a mismatch between technology and specimen, subject or object characteristics) justice. So, to a degree, the term "artifact" tends to get used for anything that does not conform to the user's hopes and expectations, while shifting the blame to the technology that is used.

So if there are image aspects that one does not immediately understand, the suggested line of investigation is to ask whether there are natural science-based explanations for these aspects, if they can be reproduced and what their significance may be.

A watery viscosity of the contrast fluid combined with a low osmolarity and mechanical hemolysis allows for a particularly extensive contrast fluid distribution also into skin vessels, possibly resulting in apparent edema and thus possibly altering the post-procedural appearance of the skin and, notably, the face and neck by swelling and reddish discoloration. While that is an authentic correlate of how the technology works and could be termed side effect, there are aspects of this—such as contrast admixture—that still may benefit from further improvements.

Postmortem CT angiography and other contrast agent use thus may have to be treated as research methods and applied responsibly and with caution.

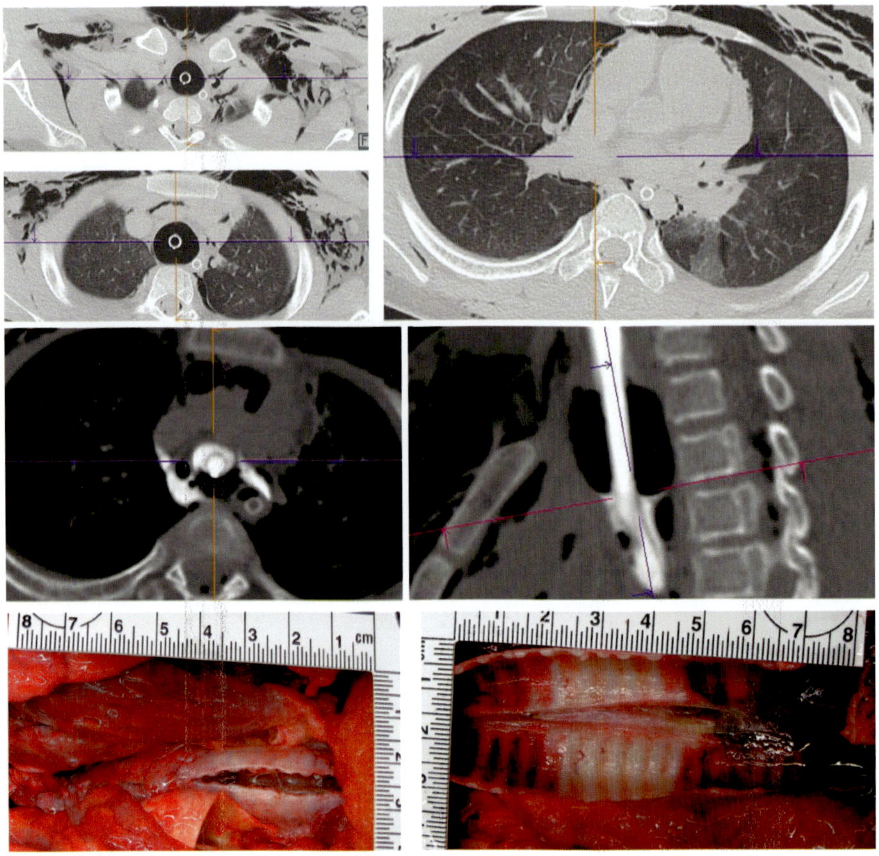

Figure 10.4 Contrast enhancement of trachea. In this image, native PMCT showed extensive soft tissue emphysema, so tracheography with contrast medium (iodine) administration to the trachea; the contrast medium indicated a leak from the trachea into surrounding tissues, which was matched by an autopsy finding of a longitudinal tear overlying the balloon of the endotracheal tube, where the tissue also showed hemorrhage.

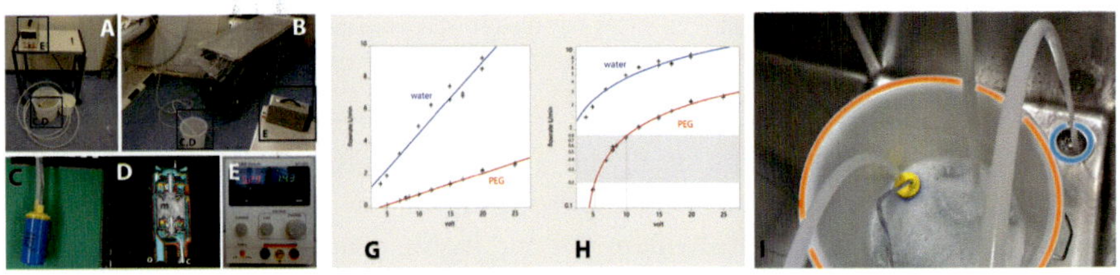

Figure 10.5 Immersion pump set up for usage in postmortem CT angiography. A-E: Setup: Pump (C, D: CT of pump, view on front facing longitudinal cut surface with inflow (i), outflow and conical tube connector (o), cable (c), motor (m), shaft (s) and paddle wheel (f)) is immersed in contrast agent solution contained in plastic container (see A and B). Pump draws electricity from voltage controllable power supply (E whole device, also see A, B). Display of entire setup (B), silicone tube connects pump that is immersed into contrast agent to femoral region (F) of body that is positioned on table of CT scanner.—**G, H: calibration:** voltage-dependent flowrates for both water (diagram: +) and more viscous PEG (diagram: **x**) using immersion pump (model 0444, Barwig, method section) (linear regression; water R^2 = 0.96, PEG R^2 = 0.99). Linear regression (same data in both diagrams—G: linear flowrate scale; H: logarithmic flowrate scale) shows linear relation between pump voltage and both watery (higher) and viscous (lower) fluid's flow rates. For PMCTA targeted flow rates between 0.2 and 0.8L/min, PEG is best pumped with pump voltages between 5 and 10 volts, which is more easy to identify on logarithmic voltage-flow rate plot (right diagram). This coincides with this particular pump model's permanent usage's comfort zone of 5–10V. **I: cleaning:** Cleaning very low cost immersion pump is achieved by immersing it in detergent containing water and pumping that fluid, thus both cleaning pump from outside and inside while transporting detergent/water from fluid container (orange portal) to sink (blue portal). More details see [4]. [Image source, with permission: Schweitzer W, Flach PM, Thali M, Laberke P, Gascho D. Very economical immersion pump feasibility for postmortem CT angiography. Journal of Forensic Radiology and Imaging. 2016 Jun 1;5:8–14.].

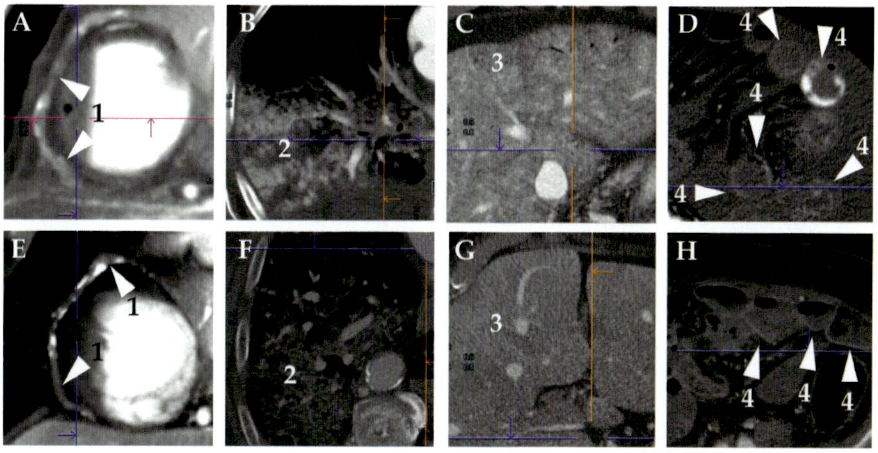

Figure 10.6 Contrast enhancement of vasculature—viscosity of contrast admixture. These images allow one to compare a watery contrast agent viscosity (A–D: a case where we used Optiray and water) with a contrast agent admixture with elevated viscosity (E-H: a case where we used Optiray and PEG-200 [6]). In both instances, a whole-body PMCTA was performed using an immersion pump [4], the body was in a supine position on the CT table in both examinations. The right coronary artery (1) is well contrasted regardless of the viscosity differences. This shows that relatively large vessels are not affected by that viscosity difference. The lungs (2) show more contrast extravasation into the alveoli in the lower viscosity case (B) than in the case contrasted with the Optiray/PEG admixture (F). The liver (3) shows greater texturized contrast after watery filling (C) than after using higher-viscosity fluid (G). The small intestine (4) showed more contrast staining of the wall and in one instance a contrast leak without any macroscopic finding or correlate at all when examined at autopsy (D), whereas the higher viscosity fluid raised the intestinal wall contrast less (H).

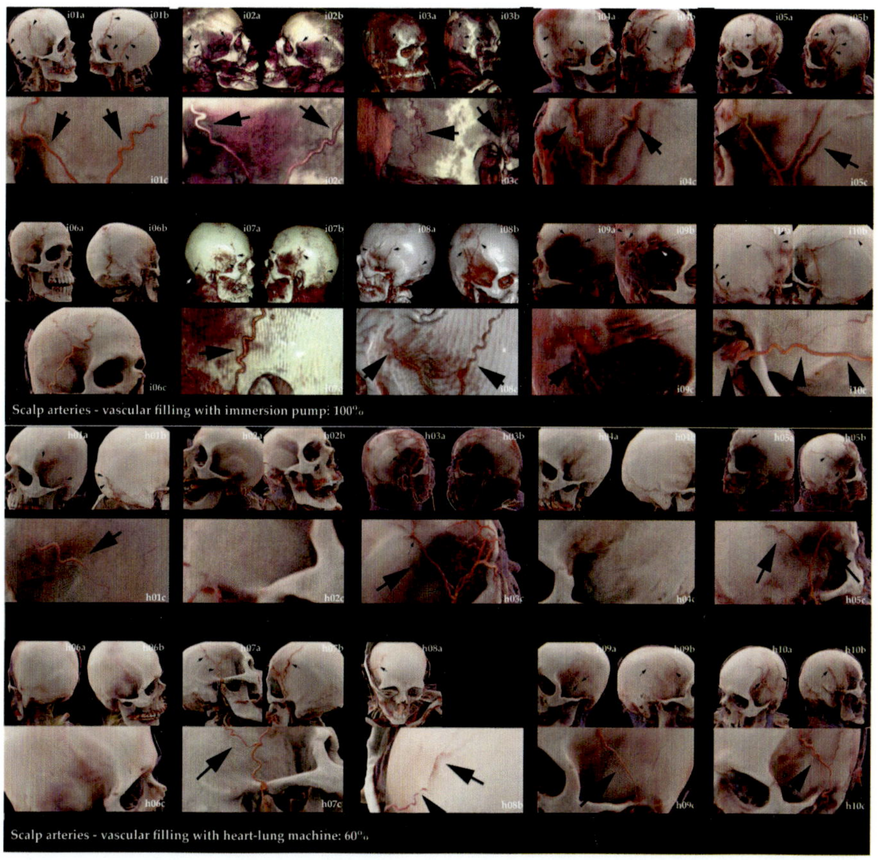

Figure 10.7 PMCTA of the head arteries via femoral access. Scalp arterial perfusion was observed in all instances with the immersion pump (images i01-i10). Conversely, the heart-lung machine (images h01-h10) facilitated perfusion of the temporal arteries in 70% and occipital arteries in 50% of the cases, resulting in an overall mean arterial filling rate of 60% for the scalp. Image derived from [7].

10.2 USE CASE EXAMPLES

10.2.1 Contrast Enhancement of Wound Trajectories

Instilling contrast medium into wound trajectories and into penetrating injuries such as stab wounds has been shown to aid the depiction of stab wound tracks in PMCT, but the determination of the depth or length of the penetration was not found to be particularly accurate [8]. Traditionally, autopsies have been performed using a method where tissues are dissected and examined layer by layer, particularly when examining penetrating injuries such as stab wounds. This approach allows for a detailed examination of the injuries, and while it may be time consuming, it may be time well invested. When stab wounds are instilled with a fluid contrast agent, PMCT can provide enhanced images of the injury. Alternatively, one might be tempted to mechanically probe anything that appears to be a penetrating injury. However, neither PMCT with wound contrast nor probing (which is not recommended, as it may alter any dissection findings) are sufficiently accurate when it comes to examining the depth of wounds, particularly also those caused by less sharp and more blunt objects, such as screwdrivers. This could be due to irregular wound paths, whereby less sharp objects may cause more complex wound paths with torn edges rather than sharp objects like knives or ice picks that tend to deliver a more cut-like tissue injury. Trajectography was reported to be relatively specific (96%–98%) but not very sensitive (42%–46%) for colorectal lesions [9], whereas different specificities (50% to 92%) and sensitivities (73% to 100%) resulted for penetrating diaphragmatic injuries [10].

10.2.2 Contrast Enhancement of Skin or Tissue Surfaces

The skin surface, as well as exposed inner bodily surfaces after injury, may be radiologically stained using contrast agent. Contrast agent may be tested for its usability systematically, whereas staining may also be acquired by immersing the samples or tissues [11] or applied in form of various, possibly sticky, admixtures. See Figures 10.1, 10.2, and 10.3 for a demonstration of this application of contrast agent, here in a combination of iodine-based contrast and a watery dilution of water-soluble white glue [12]. As can be shown, also relatively small injuries may be high-lighted reasonably well within the imagery of PMCT while the contrast agent may also adhere to other structures, such as skin areas with body hair.

10.2.3 Contrast Enhancement of Vasculature

Both vasculature-related aspects—anatomical variations of vasculature and vessel leaks or injuries in general—have attracted attention in research and practice. A way to examine the structure and intactness of vessels may be the administration or use of contrast agent. A first step towards better vascular contrast was the introduction of postmortem CT-angiography (PM-CTA) [13–15].

Using contrast universally delineates any contrasted anatomical structures in a more intense fashion, though the clarity of any contrasted structures tends to vary, especially in instances with a high degree of injury and in cases with decomposition [16] or depending on other specifics.

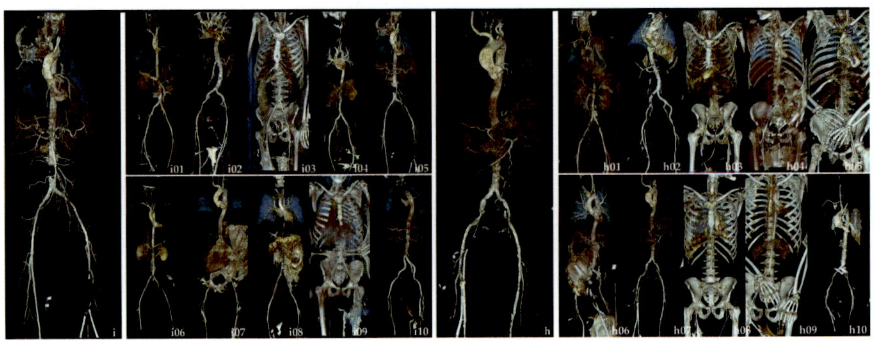

Figure 10.8 PMCTA of the thorax and abdomen via femoral access. Each pump variant that we used (i: immersion pump, h: heart-lung machine with roller pump) allows for what appears to be adequate perfusion across thoracic and abdominal organs when checked in an overview, with variances observed in coronary artery delineation. These imaging results generally facilitate detailed visualization of the vascular architecture and possible leaks. Image derived from [7].

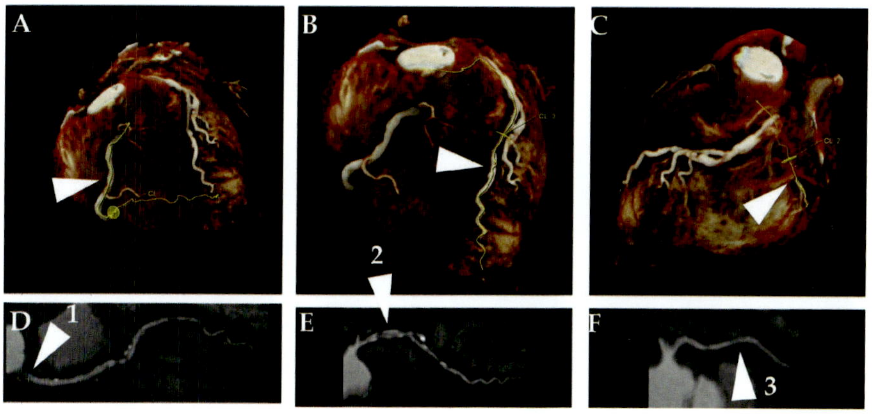

Figure 10.9 Contrast enhancement of coronary arteries. In this case, iodine-based watery contrast [6] was administered in context of a whole-body PMCTA via femoral access, whereas the body had been placed in supine position on CT table. The contrast transport into the vasculature was achieved using an immersion pump [4]. A-C: 3D/VRT reconstructions, highlighting a manually placed yellow marker line for curved multiplanar reconstruction (cMPR) along the lumen center of the respective coronary artery. D-F: cMPR.—A, D: right coronary artery. A, D: left anterior descending coronary artery. C, F: left circumflex coronary artery.—The contrast filling exhibits a relatively good quality. While there were several locations with relative narrowing of the lumen (1, 2: autopsy correlate of ~ 80% lumen narrowing; 3: autopsy correlate of ~ 50% lume narrowing), no complete blockage was identified.

Technical considerations surrounding the use of contrast agents are paramount, given their fluid properties, and the particular viscosity that can be achieved through specific formulations is a relevant aspect.

By incorporating iodine contrast with water, a solution of watery viscosity is created, enhancing its ability to reach finer vascular branches, including capillaries. This property may be interesting for detecting and documenting potential leaks or extravasation also of small-caliber vasculature, which may be revealed in subsequent computed tomography scans. Not all findings might hold what could appear to be significant immediate forensic value, as they may just underscore physiological alterations occurring postmortem. For instance, gastrointestinal mucosa damage could originate from ischemia or from postmortem decomposition in due course of the natural progression of bodily changes after death.

Increasing the viscosity of the contrast agent admixture is a strategic attempt to controlling the diagnostic value of PMCTA findings: increasing viscosity limits and restricts the contrast agent's penetration to larger vessels while at the same time preventing it from reaching the finest vascular networks, depending on viscosity, diameter of vessels, leaks or defects, and pressure. This modification can reduce the detection of minor leaks and curtail the likelihood of observing extensive extravasation from decomposing tissues, particularly in the gastrointestinal tract. A side-by-side comparison is explained in Figure 10.6.

The decision to utilize a contrast agent of lower or higher viscosity is ultimately a diagnostic consideration that may be balanced against the specifics of each case. The examiner is tasked with evaluating the benefits and drawbacks of each approach, making an informed choice that best suits the objectives of the postmortem examination and the nuances of the case at hand.

A pump used to administer contrast agent rather than manual administration has the advantage that aspects of fluid perfusion may be better documented and thus controlled. This may be a relevant step towards repeatable or standardized method development, also in any quality-controlled environment. In our experience, a very affordable immersion pump may exhibit better flow and filling characteristics than an expensive roller pump [4, 7].

10.2.3.1 Contrast Enhancement of Vasculature: Selective PMCTA

Blood vessels may be contrasted in a selective way. One example for this method is selective coronary artery [18] or pulmonary artery [19] angiography. The advantage here is clearly a very good contrast for arteries. That may be relevant in instances where one does not command a pump device for whole-body PMCTA that also contrasts the coronary arteries (see, e.g., Figure 10.9).

Another example is the manual contrasting of pulmonary veins [17], as shown in Figure 10.11.

10.2.3.2 Contrast Enhancement of Vasculature: Whole-Body PMCTA

Almost 20 years after roller pumps were introduced for PMCTA [20], it remains unclear exactly why such a relatively expensive pump mechanism was even chosen as pump mechanism in the first place for PMCTA [21, 22]. Technically, roller pumps are typically applied for non-Newtonian fluids, like blood, where also the risk of mechanical hemolysis is a reason to use that type of mechanism (or, for example, for other non-Newtonian fluids such as paint or ketchup [23, 24]).

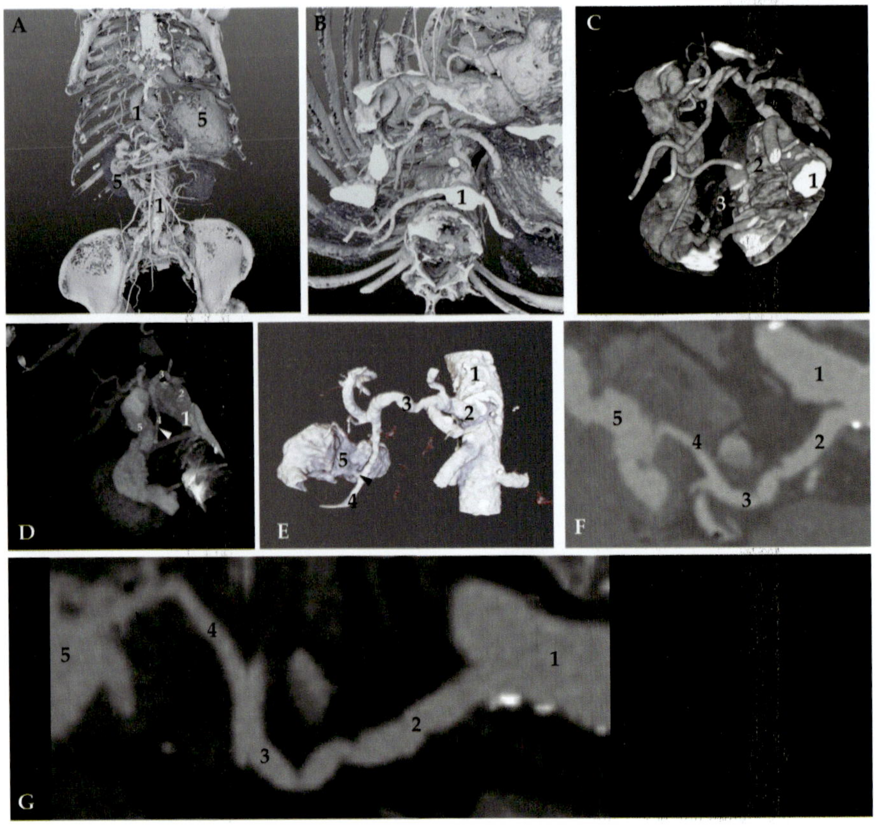

Figure 10.10 Contrast enhancement of duodenal ulcer hemorrhage. Here, a PMCTA shows findings of a leak pertaining to a gastrointestinal hemorrhage, after arterial contrast filling via femoral access.—Contrasted structures are aorta (1), truncus coeliacus (2), arteria hepatica communis (3), arteria gastroduodenalis (4) and contrasted duodenal (and gastric etc.) lumen (5). Small connecting artery to ulcer and hemorrhage indicated with arrow. Images are VRT (A, B, C, D, E) and curved MPR (F, G).

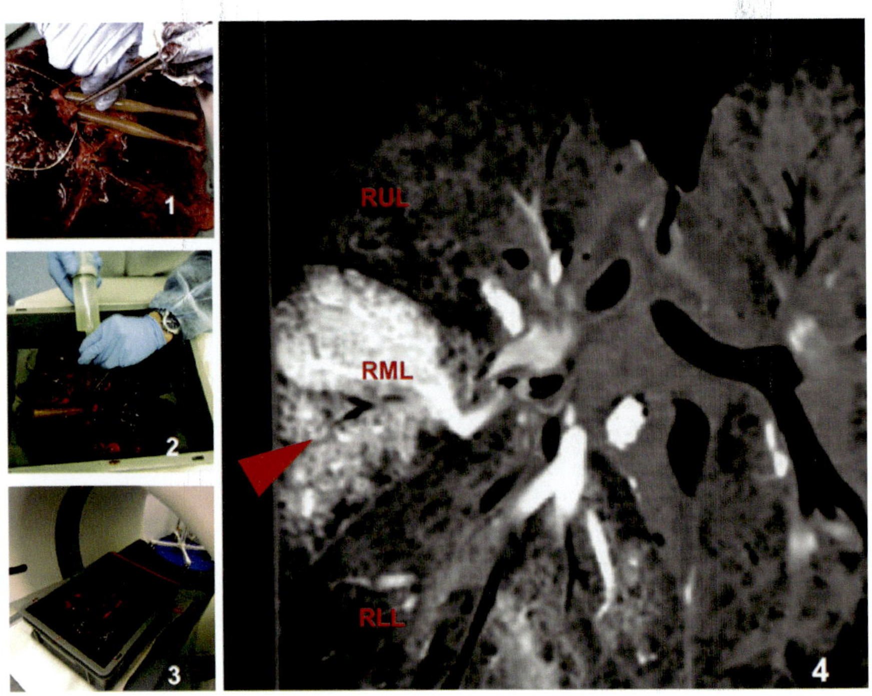

Figure 10.11 Contrast enhancement, selectively, of pulmonary veins. This shows the diffuse vascular damage in the right middle lobe of the lung (RML) after selective manual contrast enhancement of the pulmonary veins (see photos for visual demonstration). Image source [17], with permission.

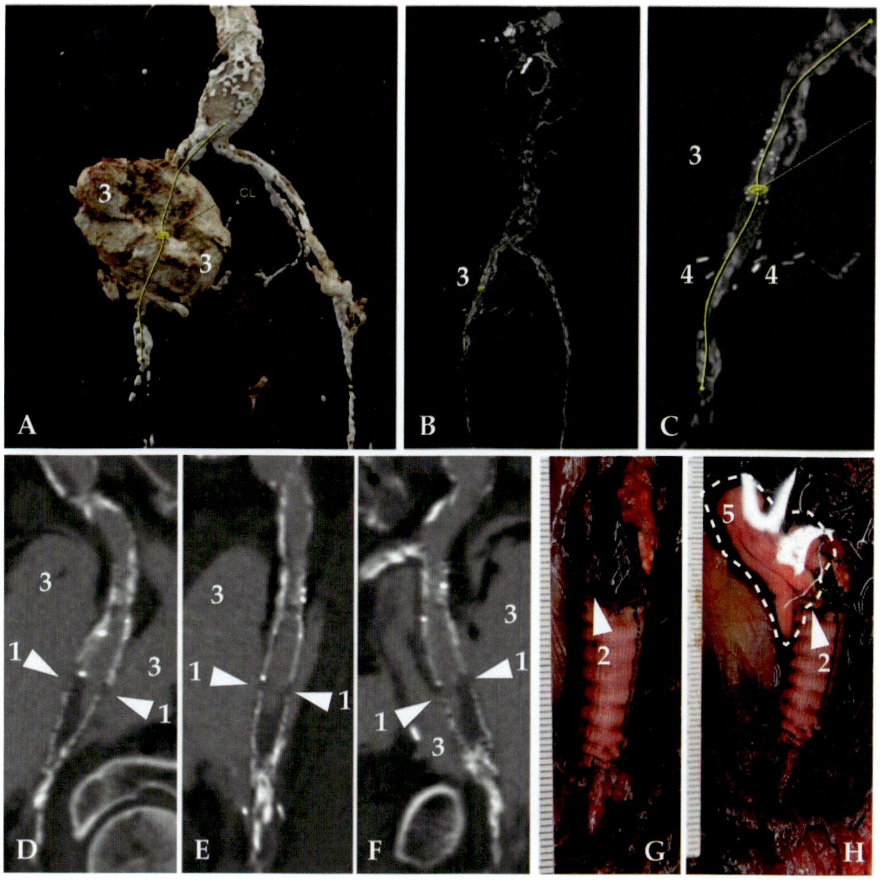

Figure 10.12 Contrast enhancement of a right external iliac artery as part of whole-body PMCTA via contralateral femoral access. A: PMCTA/cinematic rendering: extravasation of contrast medium (3) at the level of the right external iliac artery. B: MIP of PMCTA showing contrast agent (3) surrounding right iliac artery and a relatively high degree of what appeared to be atherosclerotic calcification. C: Detail/enlarged section of B, also exhibiting CT-dense clips (4). D E F: curved MPR (various angles across D, E and F to allow for circumference inspection of vascular wall) of artery and extravasated contrast agent surrounding it (3), blurred boundary between vessel lumen and extravasated contrast with absent demarcated wall structure, indicating possible area of leak (1). G, H: autopsy finding with leak (G before and H after manual injection of white emulsion at autopsy to demonstrate leak 2) at stent–graft junction of the iliac artery (1). Location of 1 and 2 similar.

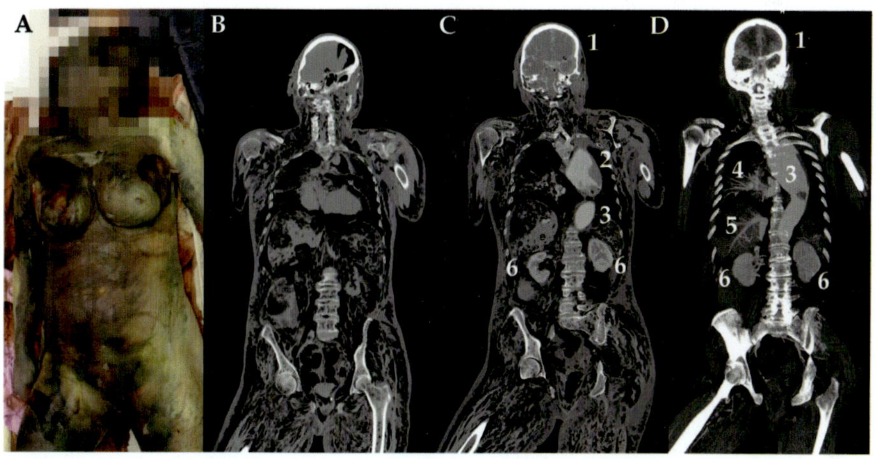

Figure 10.13 Contrast enhancement of aortic rupture, in a body with putrefaction, whole-body PMCTA. A: photograph to show general state of the body; B-D: coronal MPR. B: native PMCT. C: thin MPR. D: thick MPR (30 mm). C, D: Arterial contrast filling using iodine-based contrast [6] in context of a whole-body PMCTA via femoral access, in supine position on CT table, using immersion pump [4]. Images show extensive gas in tissues and organs alongside loss of organ structure typical for postmortem decomposition. While the region of heart and aorta exhibits a large homogenous gray finding highly suggestive of a rupture, no further details can be delineated in native PMCT. PMCTA yields what appears to be a complete contrast filling up to the intracranial arteries. (1) However, there is considerable diffuse extravasation typical for postmortem decomposition. Technically good contrast filling is identified for the heart (2) and aorta (3) as well as lungs (4). Liver vessels (5) and kidneys (6) with contrast enhancement. Details of circulatory system findings: see Figure 10.14.

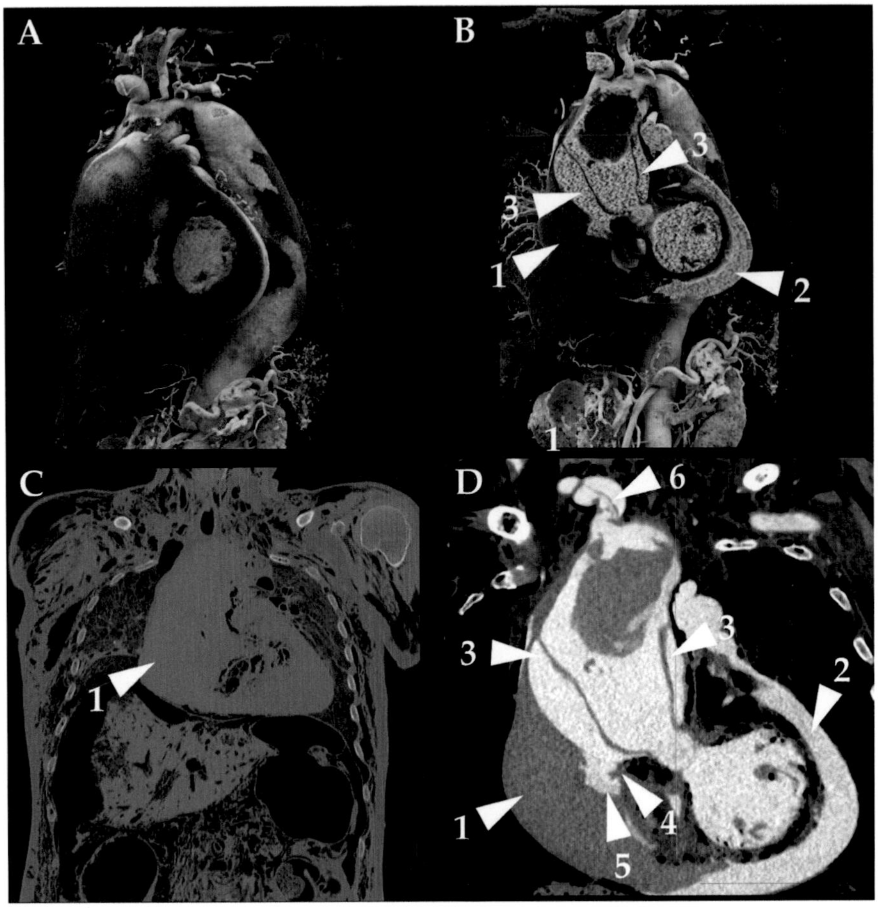

Figure 10.14 Contrast enhancement of aortic rupture, in a body with putrefaction, whole-body PMCTA. Arterial contrast filling using iodine-based contrast [6] in context of a whole-body PMCTA via femoral access, in supine position on CT table, using immersion pump [4]. A, B: 3D/VRT; C, D: coronal MPR. Hemopericardium with both non-stained (1) and contrast-stained (2) parts after PMCTA, indicating impediment of contrast distribution by blood in pericardium. Dissection flap (3) apparent in aorta. The (outer) aortic wall shows an irregularly jagged aspect of what thusly appears to be a disruption (4) with blurred continuity between intraluminal and pericardial contrast (5), likely marking rupture site. Dissection extends also into brachiocephalic artery (6). Overview of state of body in Figure 10.13.

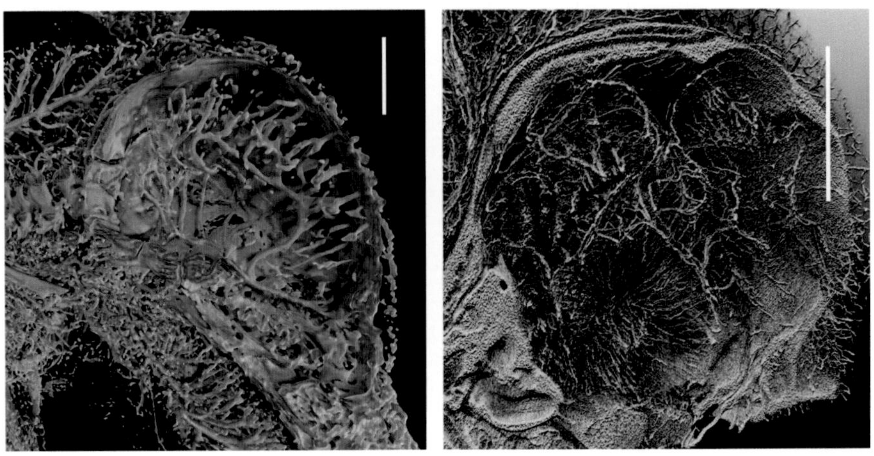

Figure 10.15 Contrast enhancement of a pig neonate using 1-mm silicone catheter access line and immersion pump. Postmortem angiography of the veins of a piglet, both images offering a view into the virtually/digitally exposed skull cavity. Left image: piglet, scanned with conventional clinical CT scanner (Somatom Definition Flash). Right image: same piglet, micro-CT, isotropic resolution of 35 micrometers. Bar ~10 mm.

In PMCTA, however, non-Newtonian fluids are not the type of fluid or material that is being pumped. There, *homogeneous* oily [25] or watery [6, 20] liquid is being pumped into the vascular system of a body [7]. We had previously established in a feasibility study [4, 26] that a simple centrifugal type immersion pump (with the pump itself priced as low as ~15–20 USD) can be calibrated to obtain a linear voltage–flowrate relationship for the iodine and PEG-based contrast agent solution used (Figure 10.5).

While roller pumps seem to work well for most thoracicoabdominal examinations (see the case of bleeding from ulcer via gastroduodenal artery; Figure 10.10), the comparison of ten immersion pump (IP) contrasted cases in a case-control (same sex, similar age) comparison with a heart-lung machine (HLM) contrasted series showed a better result for the more affordable device (IP) (Figures 10.7 and 10.8). Other cases examined with an immersion pump include that of an intraoperative hemorrhage due to a defect in the right iliac artery (Figure 10.12) [27]. We also successfully contrasted vasculature in cases with putrefaction (see Figures 10.13 and 10.14).

10.2.4 Contrast Enhancement of Vasculature: Towards Fetuses and Newborns

A particular problem may be posed when performing PMCTA on newborns or fetuses. There, the diameter of the vasculature is so small that the pressure to administer the contrast agent tends to be substantially higher than for adult PMCTA [28]. In a dedicated project 2, we started to address a number of prior shortcomings of PMCTA. A first result is shown in Figure 10.15, where a side-by-side comparison of conventional CT scanner and dedicated micro-CT scanner is demonstrated.

10.2.5 Contrast Enhancement of Trachea and Other Tubular Structures

Tubular structures such as the trachea may be contrasted in forensic death examination directly by applying contrast agent before performing contrasted PMCT (see Figure 10.4). Tracheography had been performed clinically with tantalum powder, whereas the staining or contrast reportedly was of good quality in all cases [29]. Other selective tubular structure enhancements are retrograde cystourography techniques [30] or retrograde contrast enhancement of ureters [31].

10.2.6 General Aspects of Interpretation and Side Effects in Postmortem Imaging Contrast Enhancement

An ideal imaging finding is one that provides a good contrast of trauma or pathology, which then finds a correlation at autopsy. Leaks or paravasation as well as so-called false negative findings (i.e., false negative compared with an expectation of pathology or trauma due to a variety of reasons), however, are also as much a reality of contrast use as a possibly profuse staining of a region with considerable extravasation without delineation of a macroscopically evident vascular trauma or pathology. The examiners also have to consider what one may term side effects, such as skin discoloration or haptic changes of tissues and organs. Last but not least, it is up to the examiners to properly sequence their steps of securing any surface related or other evidence, depending on area of contrast use, as well as in the case of PMCTA the securing of tissues or fluids for further tests such as toxicology before possibly impeding such by use of a contrast agent.

Notes

1 From [32]: Ring artifacts are caused by miscalibrated or defective detector elements, leading to rings centered on the center of rotation. These can often be fixed by recalibrating the detector.—Poisson noise arises from the statistical error of low photon counts, resulting in random thin bright and dark streaks along the direction of greatest attenuation. This noise can be reduced using iterative reconstruction or by combining data from multiple scans. Such noise reduction techniques enable diagnostic scans at much lower radiation doses. Iterative reconstruction at low doses decreases resolution slightly but only increases noise marginally. Model-based iterative reconstruction (MBIR), for example, smooths out noise while pre-serving edges, giving a plastic-like appearance with small clusters of pixels having similar Hounsfield units.—Beam hardening and scatter both produce dark streaks between two high-attenuation objects (such as metal or bone), with surrounding bright streaks. These artifacts can be reduced using iterative reconstruction. While dual energy CT can reduce beam hardening, it does not address scatter. Beam hardening and scatter also cause pseudoenhancement of renal cysts.—Metal streak artifacts result from multiple mechanisms, including beam hardening, scatter, Poisson noise, motion, and edge effects. The Metal Deletion Technique (MDT) is an iterative method that reduces artifacts from all these causes, potentially improving diagnostic image quality.—Out of field artifacts stem from suboptimal reconstruction algorithms and can be corrected with better algorithms. Using improved algorithms, images can be acquired with a field of view much smaller than the object being scanned, thus reducing the radiation dose. Higher resolution scanners will likely require iterative reconstruction or limited field of view scans to achieve an acceptable level of noise with reduced radiation doses.

2 KTI-contract 28314.1 PFLS-LS—Institut fuer Rechtsmedizin UZH.

ACKNOWLEDGMENTS

The authors express their gratitude to Emma Louise Kessler, MD, for her generous donation to the Zürich Institute of Forensic Medicine, University of Zürich, Switzerland. Released under CC BY 4.0. Output of AI tools was critically considered for use in the generation of this chapter [Instatext, Chat Generative Pre-trained Transformer]. Figure 10.5 reprinted from [4] with explicit permission from Elsevier [Elsevier license 5723670896350]. Figure 10.11 reprinted from [17], with explicit permission from Elsevier [Elsevier license 5751341243800]. Publication [7] was distributed under a CC BY 4.0 license where re-use does not require extra permission. While the authors have made efforts to ensure clarity and structure, the development of this manuscript occurred under the constraints of available time and timelines. The retrospective use of anonymized data was submitted to and evaluated by the Ethics Committee of the Canton of Zürich (reference number 15–0686).

References

[1] Monnin P, Sfameni N, Gianoli A, Ding S. Optimal slice thickness for object detection with longitudinal partial volume effects in computed tomography. Journal of Applied Clinical Medical Physics. 2017;18(1):251–259.

[2] Visser S, Lee B, Hall J, Krieger D. Bioremediation Study of olefins, mineral oils, iso-paraffin fluids and diesel oils used for land-based drilling. In: SPE International Conference and Exhibition on Health, Safety, Environment, and Sustainability? SPE; 2002. p. SPE-73921.

[3] Gerdes GL, Hanson MJ, Cosper SD, Bevelheimer SJ, Overbay LW. Reducing Motorpool Oil/Water Separator Management Costs. US Army Corps of Engineers, Engineer Re-search and Development Center; 2001.

[4] Schweitzer W, Flach PM, Thali M, Laberke P, Gascho D. Very economical immersion pump feasibility for postmortem CT angiography. Journal of Forensic Radiology and Imaging. 2016;5:8–14.

[5] Kalisz K, Buethe J, Saboo SS, Abbara S, Halliburton S, Rajiah P. Artifacts at cardiac CT: Physics and solutions. Radiographics. 2016;36(7):2064–2083.

[6] Ross S, Spendlove D, Bolliger S, Christe A, Oesterhelweg L, Grabherr S, et al. Postmortem whole-body CT angiography: Evaluation of two contrast media solutions. American Journal of Roentgenology. 2008;190(5):1380–1389.

[7] Schweitzer W, Ebert L, Thali MJ. Comparison of very affordable immersion pump and heart-lung-machine for post mortem CT angiography in forensic pathology: A case-control study. Forensic Imaging. 2022;30:200496.

[8] Bolliger S, Preiss U, Gläser N, Thali M, Ross S. Radiological stab wound channel depiction with instillation of contrast medium. Legal Medicine. 2010;12(1):39–41.

[9] Dreizin D, Boscak AR, Anstadt MJ, Tirada N, Chiu WC, Munera F, et al. Penetrating colorec-tal injuries: Diagnostic performance of multidetector CT with trajectography. Radiology. 2016;281(3):749–762.

[10] Dreizin D, Borja MJ, Danton GH, Kadakia K, Caban K, Rivas LA, et al. Penetrating di-aphragmatic injury: Accuracy of 64-section multidetector CT with trajectography. Radiology. 2013;268(3):729–7310.

[11] Pauwels E, Van Loo D, Cornillie P, Brabant L, Van Hoorebeke L. An exploratory study of contrast agents for soft tissue visualization by means of high resolution X-ray computed tomography imaging. Journal of Microscopy. 2013;250(1):21–31.

[12] Schweitzer W, Thali M, Gascho D. Rechtsmedizinische Bildgebung: Verbesserung des Haut-und Weichteiloberflächenkontrasts in der PMCT durch oberflächliche Anwendung von Kontrastmittel. Rechtsmedizin; 2016.

[13] Jackowski C, Sonnenschein M, Thali MJ, Aghayev E, von Allmen G, Yen K, et al. Virtopsy: Postmortem minimally invasive angiography using cross section techniques—implementation and preliminary results. Journal of Forensic Sciences. 2005;50(5):JFS2005023-12.

[14] Grabherr S, Djonov V, Friess A, Thali MJ, Ranner G, Vock P, et al. Postmortem angiography after vascular perfusion with diesel oil and a lipophilic contrast agent. American Journal of Roentgenology. 2006;187(5):W515–W523.

[15] Ross SG, Bolliger SA, Ampanozi G, Oesterhelweg L, Thali MJ, Flach PM. Postmortem CT angiography: Capabilities and limitations in traumatic and natural causes of death. Radio-Graphics. 2014;34(3):830–846.

[16] Franckenberg S, Flach PM, Gascho D, Thali MJ, Ross SG. Postmortem computed tomography-angiography (PMCTA) in decomposed bodies: A feasibility study. Journal of Forensic Radi-Ology and Imaging. 2015;3(4):226–234.

[17] Schweitzer W, Marti M, Majcen R, Steinmann P, Thali M, Fornaro J, et al. Lethal lung tear and coronary artery gas embolism in shallow water apnea diving exercise–PMCT, PMMRI, autopsy and histology. *Journal of Forensic Radiology and Imaging.* 2014;2(4):199–204.

[18] Saunders SL, Morgan B, Raj V, Robinson CE, Rutty GN. Targeted postmortem computed tomography cardiac angiography: Proof of concept. International Journal of Legal Medicine. 2011;125(4):609–616.

[19] Franckenberg S, Sieberth T, Ptacek W, Fürst M, Colacicco G, Ebert L. Semiautomated tar-geted postmortem computed tomography angiography of the pulmonary arteries using a robotic system. Forensic Science International. 2023;348:111712.

[20] Jackowski C, Sonnenschein M, Thali MJ, Aghayev E, von Allmen G, Yen K, et al. Virtopsy: Postmortem minimally invasive angiography using cross section techniques—implementa-tion and preliminary results. Journal of Forensic Science. 2005;50(5):1175–1186.

[21] Grabherr S, Grimm JM, Heinemann A. Atlas of Postmortem Angiography. Springer; 2016.

[22] Hislop D, Bosua R, Helms R. Knowledge Management in Organizations: A Critical Intro-duction. Oxford University Press; 2018.

[23] Walker C, Goulas A. Performance characteristics of centrifugal pumps when handling non-Newtonian homogeneous slurries. Proceedings of the Institution of Mechanical Engineers, Part A: Journal of Power and Energy. 1984;198(1):41–49.

[24] Alfa Laval. All You Need to Know . . . Alfa Laval Pump Handbook. Aartselaar, Belgium: Gillain & Co; 2002.

[25] Grabherr S, Doenz F, Steger B, Dirnhofer R, Dominguez A, Sollberger B, et al. Multi-phase postmortem CT angiography: Development of a standardized protocol. International Journal of Legal Medicine. 2011;125(6):791–802.

[26] Schweitzer W, Enders M, Thali M. Very affordable post mortem CT angiography kit: Feasibility study using immersion pump and 3D printed parts. Journal of Forensic Radiology and Imaging. 2019;16:11–18.

[27] Knaute DF, Thali MJ, Gascho D, Schweitzer W. Post mortem CT angiography of iliacofemoral arteries after cardiovascular surgery. Journal of Forensic Radiology and Imaging. 2017;9:40–43.

[28] Schweitzer W, Thali M, Martinez RM, Ebert L. Post mortem CT angiography in fetuses or newborn: Very affordable pump and catheter solution. Journal of Forensic Radiology and Imaging. 2018;13:17–22.

[29] Stitik FP, Proctor DF. Tracheography with the experimental contrast agent tantalum. Annals of Otology, Rhinology & Laryngology. 1973;82(6):838–843.

[30] Darge K. Voiding urosonography with ultrasound contrast agents for the diagnosis of vesi-coureteric reflux in children: I. Procedure. Pediatric Radiology. 2008;38:40–53.

[31] Usawachintachit M, Tzou DT, Mongan J, Taguchi K, Weinstein S, Chi T. Feasibility of ret-rograde ureteral contrast injection to guide ultrasonographic percutaneous renal access in the nondilated collecting system. Journal of Endourology. 2017;31(2):129–134.

[32] Boas FE, Fleischmann D, et al. CT artifacts: Causes and reduction techniques. Imaging Medicine. 2012;4(2):229–240.

CHAPTER 11

DICOM/PACS

Lars Ebert and Wolf Schweitzer

Standardization is an important means to ensure interoperability between different devices and software. For radiological data, the Digital Imaging and Communications in Medicine (DICOM) standard is commonly used. The DICOM standard is the successor to the earlier American College of Radiology–National Electrical Manufacturers Association (ACR-NEMA) standard. It is an open standard that encompasses a file format for medical data (i.e. image data, segmentation information, image registration data and others) but also network protocols for data exchange, printing, reporting and others [1]. The DICOM standard is an open standard, which means that any manufacturer can implement it without limitations.

Any current generation device that generates medical image data, such as computed tomography (CT) or magnetic resonance imaging (MRI) scanners, uses the Digital Imaging and Communications in Medicine standard to store and exchange the generated datasets. DICOM is also the basis for radiological image database systems, the so-called Picture Archiving and Communications Systems (PACS).

DOI: 10.4324/9781003413998-12

11.1 DICOM FILE FORMAT

The DICOM file format is a container format, which means that it can contain a variety of different data and thus is not limited to image data alone. Even though the DICOM file format is a container format and could contain all sorts of non-medical data such as point clouds or polygon data, many of these are usually stored in other formats and thus are not compatible with systems based on the DICOM standard. An example would be standard file formats for 3D models such as STL or OBJ that are prepared for 3D printing contain formats are not DICOM compatible, such as STL or OBJ.

Examples of data routinely stored as DICOM files are images such as scanned CT or MRI images, ultrasound images, photos, audio data, reports or printing information. When used for medical image data, DICOM files typically consist of two parts—the image itself and meta information that usually precedes the image data within any DICOM file; hence, DICOM meta information may be referred to as the "DICOM header". Meta information that is embedded in medical DICOM images usually contains information about the patient, the scan location and time as well as details about the scan parameters, such as the modality, the scanner model used or the device settings.

As DICOM headers usually contain patient-related data, all image files of a study may have to be anonymized before they may be used for certain types of applications, such as teaching or research.

Volumetric data such as CT or MRI data is usually stored as a stack of single files. There, each file contains its own meta information; a typical difference between each slice image of a single series is the table position of each image.

How the image data itself is stored within the DICOM file also depends on the type of the image data. An image compression may be omitted or used, including the use of lossy compression formats.

11.2 DICOM NETWORK COMMUNICATION STANDARD

The DICOM standard is not only a file standard but also contains standards and protocols for networking and communication.

This aims to achieve interoperability between different devices and data storage systems from different manufacturers and makes it possible to integrate DICOM-based systems with other IT systems such as hospital information systems (HISs) or radiological information systems (RISs).

Through the DICOM standard, imaging devices and PACS can communicate with each other and exchange data. Within a PACS, data has a structure that reflects the actual scanning workflows. One patient can have multiple appointments with scans using different imaging devices (i.e. one MRI scan and an ultrasound).

With the same image modality, different studies (usually a study defines an examination of one or several scans at a particular date and time) and different series (i.e. different sequences for MRI or different regions of the body or reconstruction kernels using CT) may be performed. So, in other words, the PACS representation of this is that each patient can have multiple studies, which symbolize different appointments over time, and different series, which symbolize technical variation inside a study for a given specific image modality. So, for each study, multiple series may be present, with each series being a separate scan.

This may play a particular role particularly when exporting, exchanging, importing or reading data: is all data of a patient, only all data pertaining to a particular study or only all data of a particular series (inside a study) exchanged, exported, imported or read?

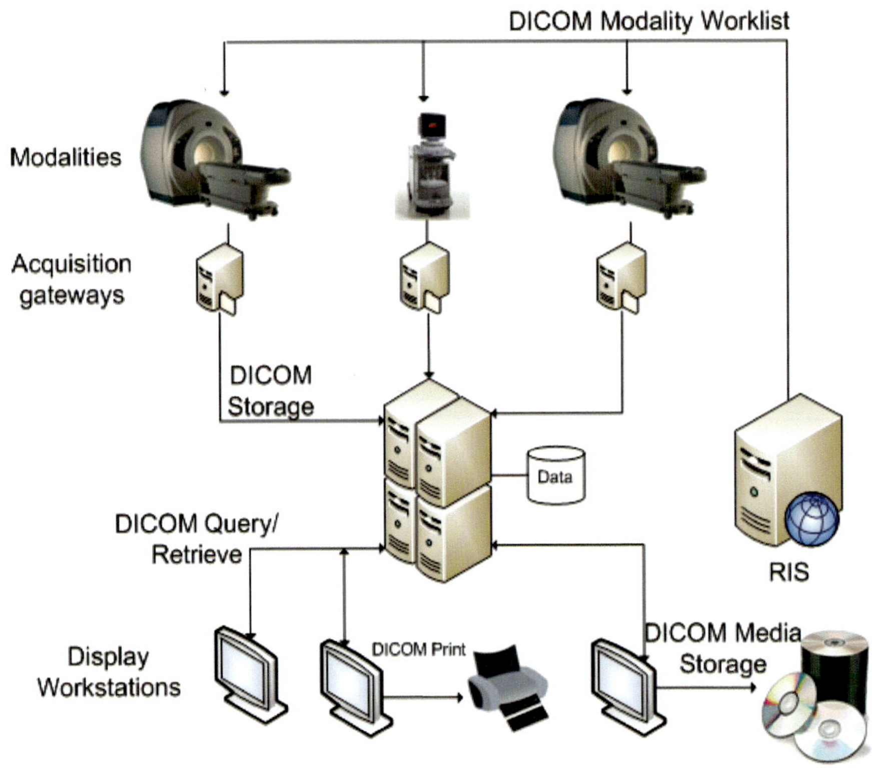

Figure 11.1 Setup of a PACS. Data from different image modalities are stored in a PACS server. PACS workstations can access and display these datasets. Image Credit: Valente F, Costa C, Silva A (2013) Dicoogle, a Pacs Featuring Profiled Content Based Image Retrieval. PLoS ONE 8(5): e61888. License under Creative Commons Attribution 3.0 Unsupported https://doi.org/10.1371/journal.pone.0061888/ File:PACS overview.png—Wikimedia Commons.

11.3 PACS SERVER AND WORKSTATION

In order to store and visualize medical image datasets, PACS are used. PACS usually consist of two components—a PACS server that is responsible for data storage and backup and PACS workstations that allow a user to access the image data and offer a variety of visualization techniques (Figure 11.1). It may be possible to use mobile devices such as laptops, tablets or even cell phones as PACS workstations. The workload of storing, reading and rendering data may be delivered in different ways, depending on the way a particular system works or is set: a particular workstation device may render all data locally, on the workstation device itself. Then, that computer has to provide the necessary resources such as sufficient memory and graphics display options. Alternatively, the server may provide streaming of rendered visualizations, with the data processing for visualization being conducted on the server.

PACS workstations often possess functionality for particular examination aspects such as tools for measuring (distance, Hounsfield units, angles etc.) or for volume segmentation, for instance, for 3D printing or volumetry [2]. PACS workstations additionally can have functionality regarding reporting, data transfer using DICOM network protocols and backup capabilities.

References

[1] Bidgood WD, Horii SC. Introduction to the ACR-NEMA DICOM standard. Radiogr Rev Publ Radiol Soc N Am Inc 1992;12:345–55. https://doi.org/10.1148/radiographics.12.2.1561424.

[2] Lemke HU, Berliner L. PACS for surgery and interventional radiology: Features of a Therapy Imaging and Model Management System (TIMMS). Eur J Radiol 2011;78:239–42. https://doi.org/10.1016/j.ejrad.2010.05.030.

CHAPTER 12

VISUALIZATION

Lars Ebert and Wolf Schweitzer

DOI: 10.4324/9781003413998-13

12.1 BACKGROUND

Good visualization demonstrates the relevant findings so that the readers understand where and what they are looking at. Medicolegal findings may have to be reported, preferably in a visual and structured way [1, 2]. What is relevant in forensic pathology and in clinical forensic medicine may be read as a trick question whose answer may at least in part be inferred from the relevant textbooks and scientific literature. Relevance in forensic medicine is a term usually narrowed by the case circumstances and by other case related evidence. Relevance also bases on the hypotheses to be tested, but it is to no small part also defined by what others may overlook and/or what has general relevance across all examinations, and/or what allows a reader to see that this case has been checked meticulously and carefully. One might surmise that what is not in the files is not in the world.

The target audience in forensic medicine may be different from a clinical medical audience that many doctors including radiologists are accustomed to; forensic imaging routinely delivers results to forensic pathologists, police, forensic scientists, state attorneys, or judges as well as the bereaved, society in general, and there in particular with regard to aspects related to reportable disease and injury or disease prevention.

The requirements for adequate visualization in forensic medicine furthermore are that the images are accurate, sufficiently detailed, and generated in sufficient speed. If the time between a postmortem computed tomography scan (PMCT) and the autopsy is only 2 hours, then the relevant visuals required to inform the pathologist about the PMCT findings may have to be generated in that time.

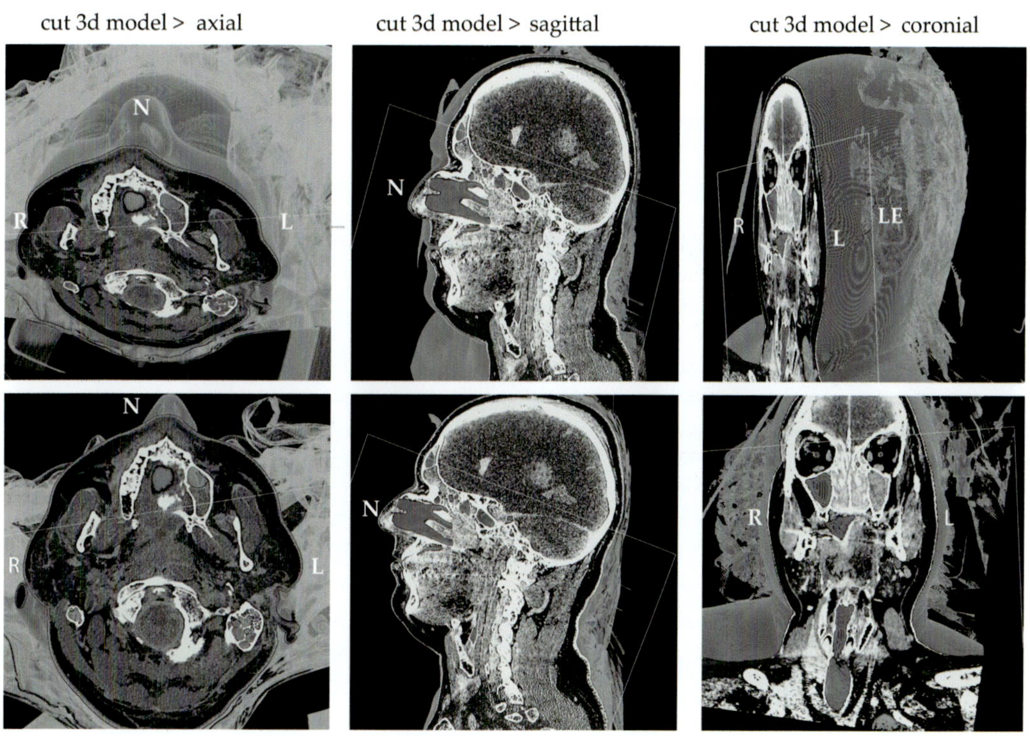

Figure 12.1 Cut planes of a PMCT data set. These images show the relation of standard orientation planes when oriented correctly (bottom row) to the 3D appearance of a head (top row). Left: an axial image typically shows a section that is perpendicular to the scan table and also perpendicular to the z-axis as the longitudinal axis of the CT-table; if the body (here: the head) is oriented so that the person lies on the back and head looks upwards from the table on a typical clinical CT scanner configuration, then the axial images will show anatomically left structures (L) towards the right margin of the image, anatomically right structures (R) on the left side of the image, and anatomically anterior structures (N: nose) towards the top margin. Middle: a sagittal view shows the rostral/cranial aspects towards the top margin of the image, the caudal direction is towards the bottom image margin, the anatomically anterior aspects (N: nose) may be shown towards the left (or right) image margin, and the anatomically dorsal parts may be towards the right (or left). Right: a typical coronal view exposes the cranial/rostral anatomy towards the image top margin, the anatomically right direction (R) is pictured towards the image's left margin, and the anatomically left structures (L) are on the right side of the image (for orientation: LE left ear).

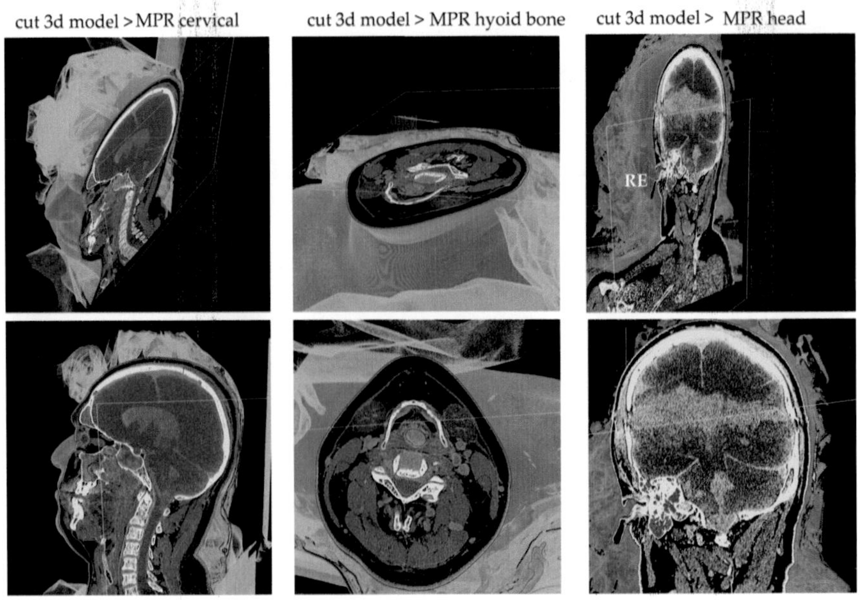

Figure 12.2 Cut planes of a PMCT data set. As opposed to strictly axial, sagittal, or coronal images as shown in Figure 12.1, these are free-angle oriented MPR images. Left: the cervical spine is shown in a quasi-sagittal view; as PMCT often contains an oblique body orientation; however, MPR is used to provide a midline oriented cut view on the cervical spine. Middle: the hyoid bone is shown in cut that is horizontal relative to the hyoid bone. That cut plane appears to be slightly tilted in relation to a strict axial plane. The angle of a hyoid bone may deviate considerably from an axial image plane in PMCT, as the head and neck position after death may not be oriented as precisely to gantry and table of a CT-scanner compared to clinical CT. Right: this MPR shows a quasi-coronal view containing both skull entrance and exit defect of a firearm injury to the head; the power of MPR plane orientation lies in that at the same time, also the CT-correlate of a brain injury with hemorrhage—obviously the firearm related brain injury and subsequent hemorrhage as vital reaction—are shown in relation to the bone defects. Please note that none of these MPR cut planes are 0 or 90° to table.

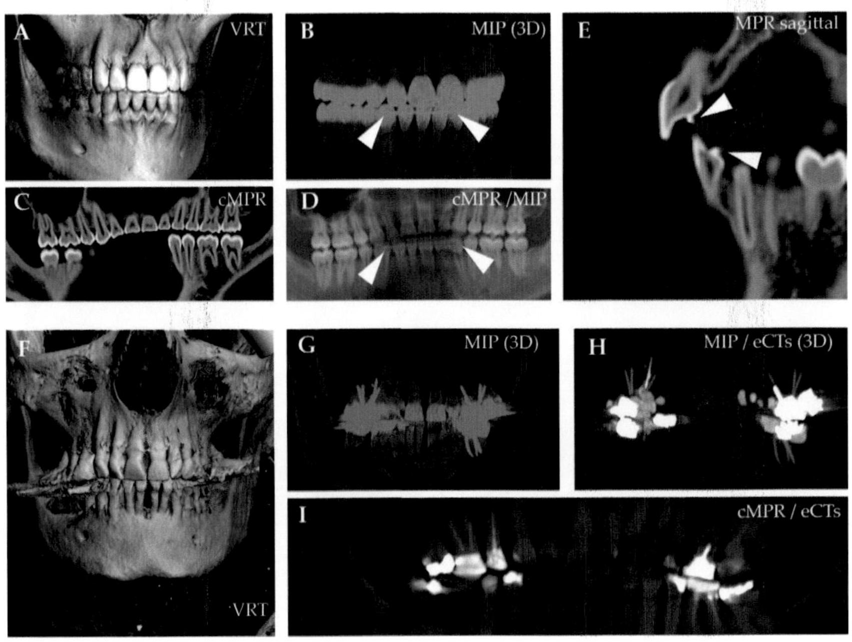

Figure 12.3 Maximum intensity projection (MIP). Maximum intensity projection methods will show the highest signal intensity or CT density across any number of slices. This can be seen here with two cases (case 1: A-E, case 2: F-I), where in 3D using VRT (A, F), the bone and teeth are set to appear as visually impenetrable, whereas a MIP of the same 3D-rendering (B, G) will allow the reader to see dental restorations with greater ease. Use of an extended CT-scale (eCTs) allows the software to differentiate higher densities, in that the normal density range of clinical CT scanners (around −1000 to +3000 Hounsfield units is extended to around +30,000 Hounsfield units) with the effect that higher density dental work is shown in greater contrast detail in the MIP rendering of that data (H). When a curved multiplanar reconstruction is obtained for a thin slice (cMPR, in C and I), a display can be obtained that some may find easier to approach comparisons with panorama radiographs as performed by dentists. However, clinical CT scanners still may lack some of the contrast and spatial resolution of dental radiographs. For example, the retainer wires at the inside of the incisivi (sagittal view, E) are shown only marginally well in the MIP (B, arrows) and cMPR/MIP (D, arrows).

12.2 RADIOLOGICAL/TECHNICAL BACKGROUND

In forensic imaging, a variety of different data types are generated and have to be processed. Data types may depend on the image modality used [3, 4].

The forensic analysis of PMCT tends to be the application of forensic medicine and forensic pathology principles using the tools of radiology and imaging, so typically, DICOM data is processed.

When the bulk of clinical radiology may be sufficiently assessed using axial image viewing only, the examiner will study a case with hundreds of single images by constantly scrolling back and forth, creating a mental 3D-model of the findings. It seems to be good practice not to look at the whole CT image at once but focus on a particular area—so it is required to go back a few times to view other image regions.

With reading and interpreting the PMCT data in a forensic or medicolegal context, different approaches also to data presentation may be helpful than those used in clinical radiology, and 3d visualisation may take a ,ore predominant role depending on case specifics, while axial and multiplanar slice images remain a predominant focus (see the image presentation for example in [5]).

A particular emphasis in forensic radiology may be on detection of even small (and therapeutically irrelevant) lesions that still may have reconstructive significance, whereas on the whole, MPR and 3D reconstruction often seems to constitute a far larger aspect than in clinical radiology. That particular difference between forensic and clinical radiology was discussed in depth with the involved clinical radiologist in a case published by Schweitzer et al. [6].

12.2.1 Two-Dimensional Images and Volume Data

The simplest digital "two-dimensional" image data in forensic imaging in general appear to be digital photographs, as they are produced by digital photo cameras, for example. Numerically, so-called 2D images may have further dimensions, such as a different value per pixel for red, green and blue, possibly also for transparency. So generally, 2D images consist of an array of picture elements (pixels), with each pixel having three color channels and sometimes a transparency value [7]. A particular aspect or focus may be given to the specifics of image processing that is performed on the camera device, where the user has often little or no control. This has been found to affect the quality of photogrammetry, for example, where less elaborate and older cameras lacking extensive on-camera image processing may yield better photos for photogrammetry than modern consumer devices.

Medical cross-sectional scanners such as computed tomography (CT) and magnetic resonance imaging (MRI) scanners produce volumetric data sets that typically are stored in single stacks or series of two-dimensional images; occasionally an actual volume format is used [8]. That data contains single image point elements in three dimensions, so-called volume elements (voxels). A three-dimensional stack of scanned serial slides from histology may well contain three color-values per voxel [9]. But when the volume results from a radiological examination such as a CT or MRI scan, the voxels typically

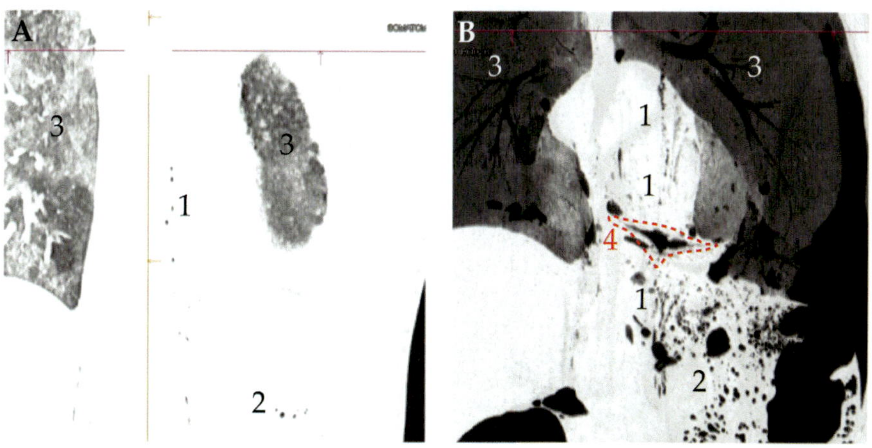

Figure 12.4 Minimal intensity projection (MINIP). Minimal intensity projection will show the lowest signal intensity or CT density across any number of slices. In the left image (A), a regular coronal reconstruction of the mediastinum and the lungs is shown (3), where there are a few scant para-aortic gas bubbles (3) as well as a few scant para-gastric gas bubbles (2). In fact, this was a case investigated as Boerhaave's syndrome with esophageal rupture. The right image (B) is a MINIP reconstruction: it now shows the same case and same coronal projection, but now with a visual summary of minimal densities, there are many (and not just a few single) gas bubbles para-aortically (1) and even more around the stomach (2), whereas there was no gas in the liver; bronchi (3) labelled for reference. The anatomic pattern of the gas distribution is shown more adequately in the MINIP projection ("MINIP projection": note the RAS [redundant acronym syndrome] syndrome). Even the clinically known "V-sign" referenced in Boerhaave's syndrome-related radiology literature may be investigated in these projection shapes (4).

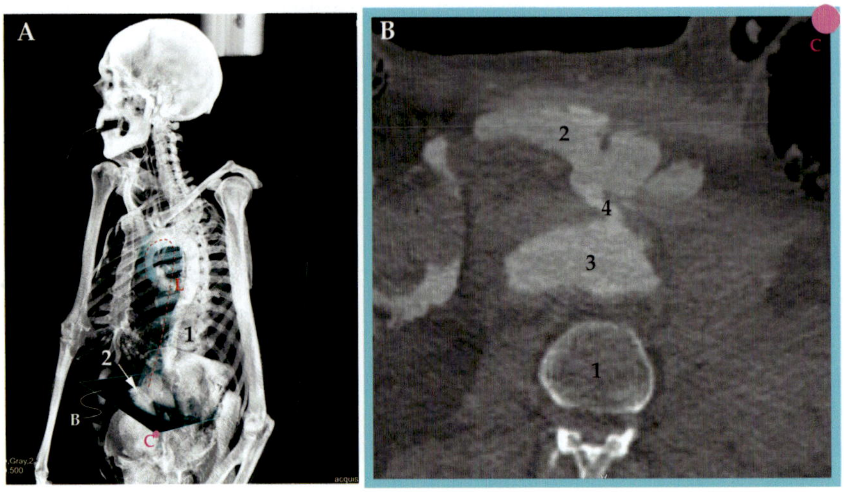

Figure 12.5 Sequential MPRs orthogonal to a curve. A marker line was placed in the data in order to define a center-line in the aortic lumen (L: cyan colored image vectors extending from red-dashed curve, following the aorta) in order to allow the automatic read-out of images that are orthogonal to that line; the relevant MPR-image in this case of a postmortem angiogram of a ruptured infrarenal aortic aneurysm is shown in B (pink dot C for reference) where in front of the spine (1), the aneurysmatically dilated aorta (3) exhibits a contrasted connection (4) to the extravasated contrast agent (2) that is surrounded by the native contrast of blood after aortic aneurysm rupture.

do not possess a color value (as typical digital photography image files have) but rather a value connected to the image modality, such as the X-ray attenuation in Hounsfield units (HUs) in the case of CT [10] or T1- or T2-relaxation times in MRI. For medical CT, the single images of the stack are usually square with 512 × 512, 1024 × 1024, or 2048 × 2048 voxels per image, typically containing 12–16-bit information or one out of 4096–65,536 different values. Metainformation such as the voxel size, spacing, or scanning parameters can be extracted from the DICOM header if required.

12.2.2 Polygon Data

Surface documentation scans using photogrammetry and structured light scanning usually result in polygon meshes that approximate the outside shape of the scanned object but not the inside of the object [11]. That data typically lists all vertex coordinates and a description how these vertices are connected to form the surface polygon. This may include color information of texture images that are mapped onto the polygon mesh. Figure 12.7 shows data from a PMCT scan of a left foot. While subtle, the effect of postprocessing such as smoothing can help to obtain a more realistic 3D-model. The format of isosurfaces is not commonly used for medical visualization but is the typical format for 3D printing. Capturing surface data has its own challenges (e.g., [12–14]) that to a degree may be approached by attempting a remote turnkey device [15] but that one really may overcome only with better camera use technique, which may also include extra equipment such as a tripod or additional light and, ultimately, subjectively immersing oneself into the case and patience [16]. Recent advances of forensic photogrammetry exploit massive flashlight setups and very fast serial photography options, using a hand held consumer camera yielding mesh qualities far beyond the current standard [Lux C, Ebert L, Schweitzer W, Ampanozi G, Chatzaraki V, Bösch M, Kreitmeier FF, Kettner M. Photography techniques for 3D surface reconstruction during autopsy. Forensic Science International. 2026 Feb 1;378:112702.]. So this field may be still seen as subject to ongoing research and development.

Some surface documentation techniques such as laser scanning produce so-called point clouds to describe a surface. A point cloud consists of a set of single, sometimes colored point in virtual space whose position and color describe the scanned surface. Point cloud data of crime scenes documented by laser scanners can be fairly large, often consisting of tens of millions of points, making real time processing on consumer computers challenging.

12.2.3 Data Formats

Digital data come (and go [17]) in a variety of formats. It is, to a certain extent possible, to convert between different 2D and 3D formats.

2D images can be stored uncompressed (i.e. TIFF) resulting in larger file sizes, with lossless compression (i.e. PNG) or lossy compression formats (i.e. jpg) resulting in smaller file sizes, in the case of lossy formats through loss of detail. A list of 2D file formats in use in forensic radiology has been compiled by Graham et al. [18].

Volumetric data can be stored as DICOM stacks, which is the standard for medical images, or as a volume data file [8, 19]. A detailed compilation of volume data formats can be found under [20]. Stakeholders and users of DICOM image formats must consider that in any image series, each and every file header might contain personal or identifying information about the individual. DICOM image series may have to be anonymized before publication.

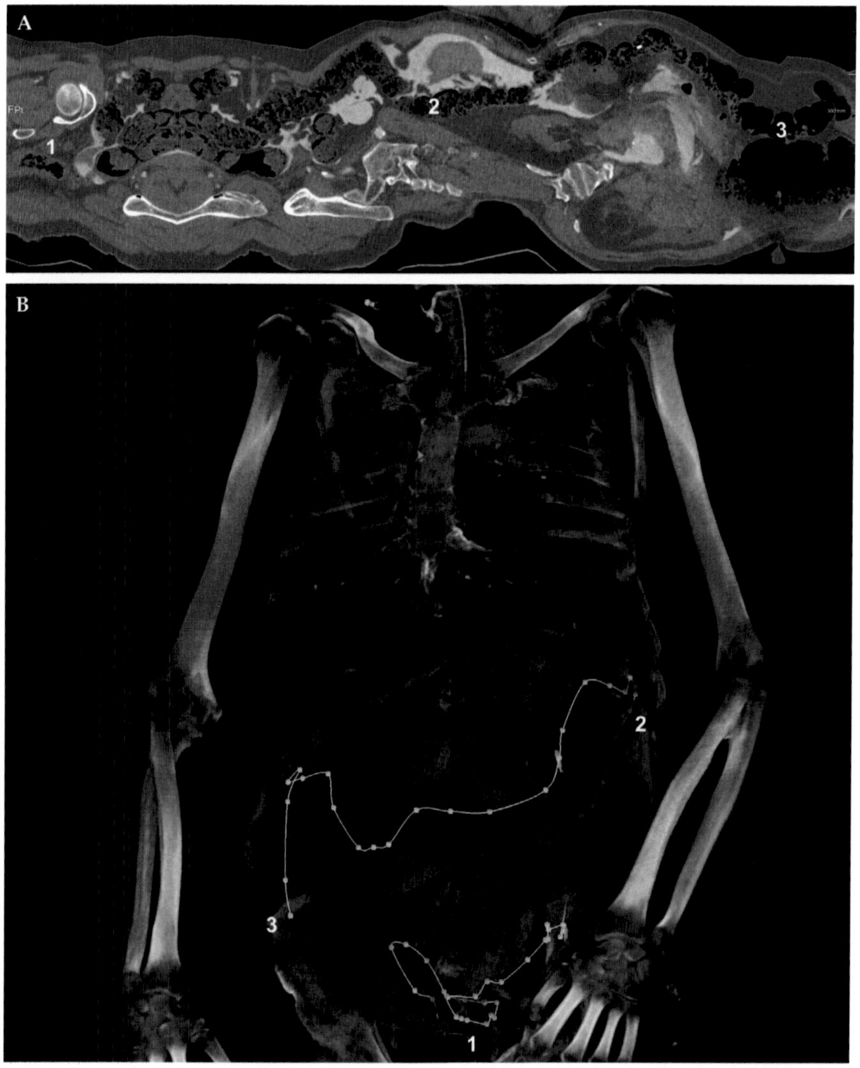

Figure 12.6 Longitudinal curved MPRs. The curved multiplanar reconstruction (A) follows a relatively complicated curve, as shown by the connected manually placed marker points (B, cyan line); the MPR itself (A) then is a flat thin-sliced projection following the marker points; in this case of a postmortem angiography of a ruptured infrarenal aortic aneurysm, the colon was examined for suspicious lesions (1: region of rectum, 2: region of left flexure; 3: region of coecum).

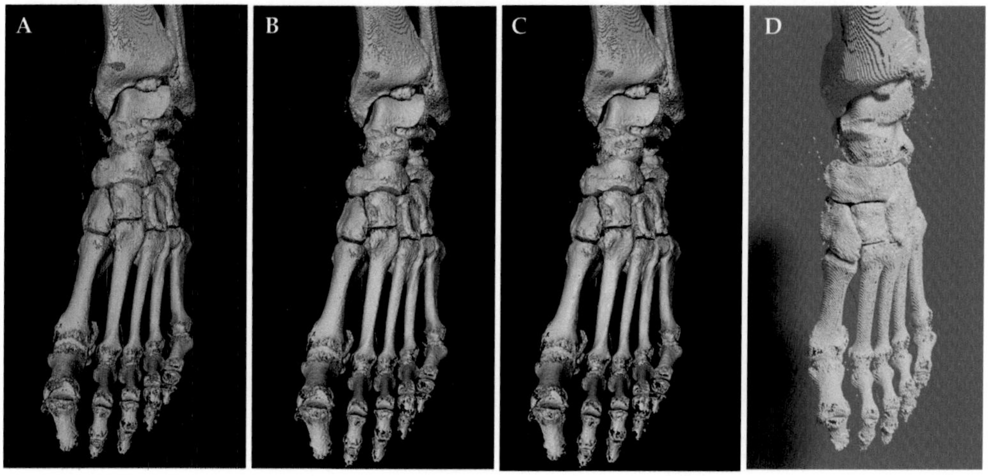

Figure 12.7 3D-rendering of an iso-surface representing the skeletal structures of a left foot. The effect of postprocessing such as smoothing can help to obtain a more realistic 3D model. The format of isosurfaces is the typical format for 3D printing, see also the chapter about 3D printing in this book. A: unsmoothed, B: smoothed (factor 3), C: smoothed (factor 10). A differently rendered view of this isosurface model is in D where artificial lighting and added shadowing may help the viewer to understand depth cues better.

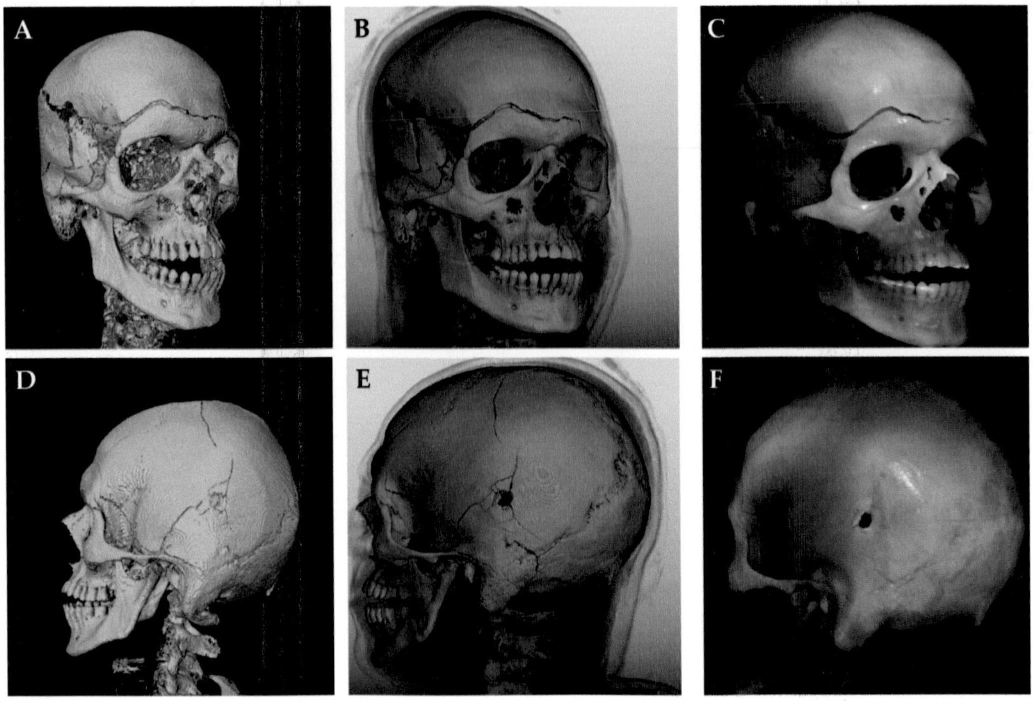

Figure 12.8 3D rendering of a skull as scanned with computed tomography. A, D: as iso-surface or shaded surface display renderings show a single non-transparent surface that the user places at a particular CT density value (approximated here: bone); they are technically relatively simple and were thus faster to render, and they were the standard for real-time rendering, at a time some 20 years ago when volume renderings (VRTs) took a lot longer on normal desktop computers. To this date, however, isosurface data is relevant as it is required for 3D printing; after it is exported from CT data, it may be thus visualized for validation. B, E: volume rendering allows one to produce more sophisticated renderings whereas all voxels are assigned different colors and opacities depending on their CT densities; that assignment is also termed transfer function. C, F: This shows VRT using a different transfer function where it remains unclear just why this rendering was given an appearance of a reflective if not wet surface.

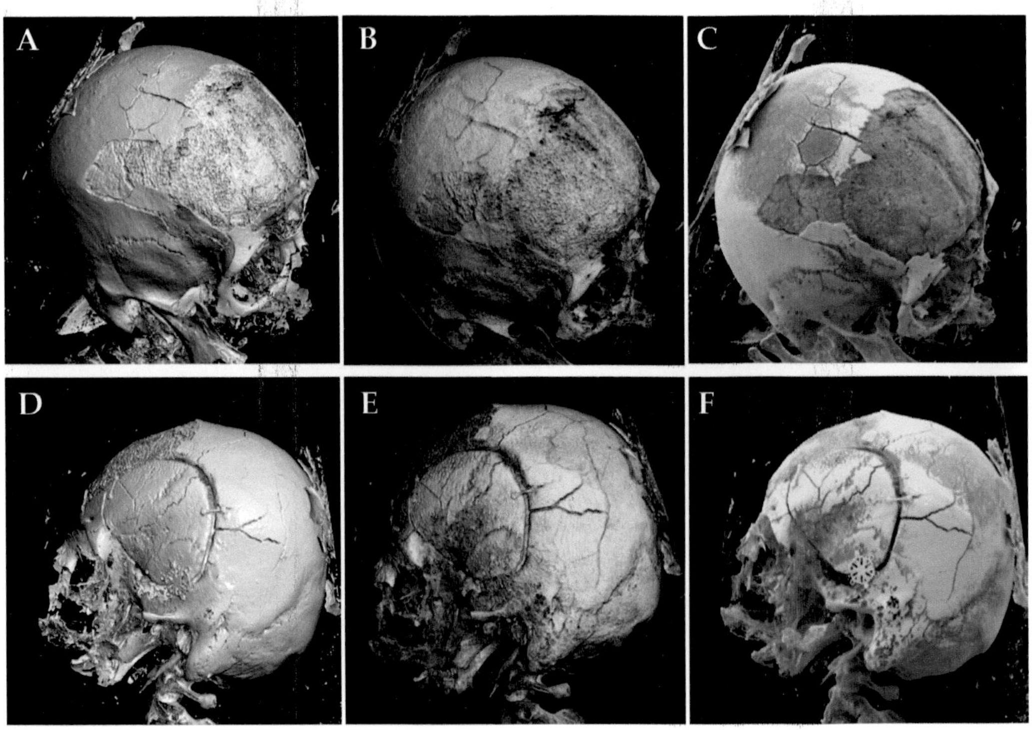

Figure 12.9 Aspects of different 3D-renderings of a skull as scanned with computed tomography. Skull of a burnt body. Both the large outer layer defects of the skull and the burst fractures are of interest here. A, D: A factory-issued standard preset of the VRT software may yield a first survey of the findings. B, E: A different VRT transfer function allows one to identify some smaller findings, but at the expense of visual clutter. C, F: Using a different VRT rendering may present the relevant finding more clearly.

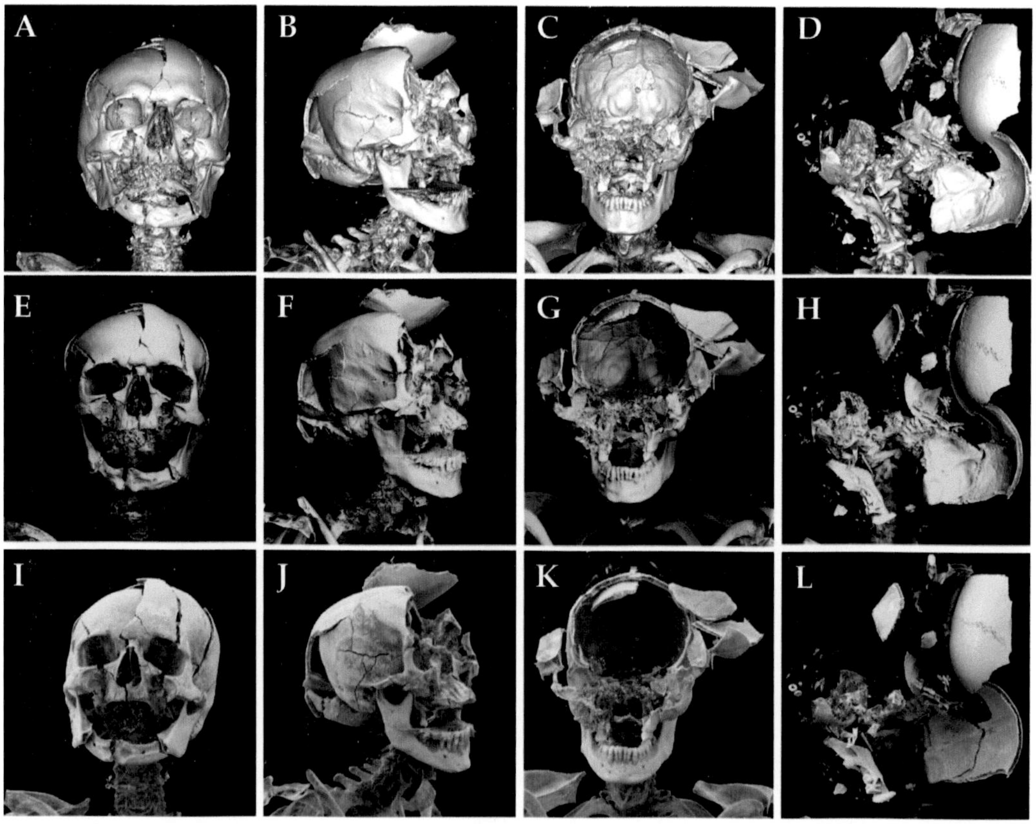

Figure 12.10 Aspects of different 3D-rendering of skulls as scanned with computed tomography. Various degrees of trauma (blunt trauma, firearm injury) with emphasis of visualization to allow the viewer to consider the fracture patterns. Case 1 (shotgun): A, E, I. Case 2 (blunt trauma, train): B, F, J. Case 3 (shotgun): C, G, K. Case 4 (blunt trauma, train): D, H, L. Conventional factory issue software VRT: A-D. Customized transfer function with VRT: E-H. Customized transfer function with so-called photon-mapping VRT: I-L. The user may choose whatever they see fit to deliver the visual message they regard as adequate and relevant. Where the visual presentation lacks clarity, a written report or an in-person verbal description could be considered as explanatory adjunct.

For point clouds, file formats are often proprietary and specific to a manufacturer of laser scanners. The Point Cloud Library (PCL) format is a free data format that tries to overcome the interoperability issues of the proprietary formats. A similar issue exists for polygon data, which multiple vendor specific file formats. Polygon formats are often specific to a use, i.e. in the case of the STL format, which has been developed for the purpose of 3D printing. Some formats are ASCII based and editable with a text editor (usually larger), some are binary formats. Interoperability should also be considered, as not all software is able to read all files. A review of 3D formats was published by McHenry et al. [21].

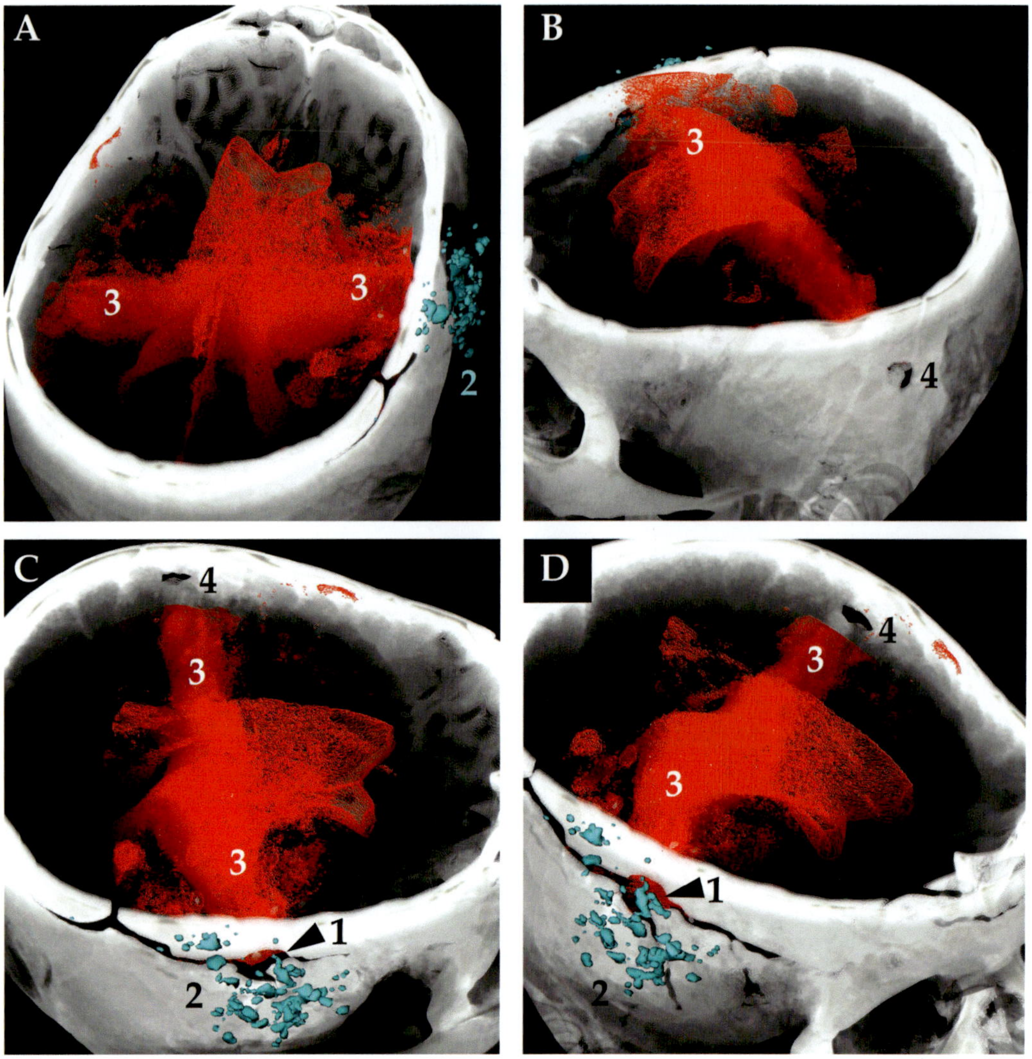

Figure 12.11 Complex reconstruction: presumed gunshot residue and skull PMCT. I. This series of 3D-reconstructions combine a VRT model of the skull as well as polygon models of data substructures after combined manual and CT-density based segmentation. In this case of a pistol shot injury to the head, there was an entrance defect of the skull (1), there was a correlate for presumed gunshot residue (2, rendered as isosurface), a correlate for what based on its CT-density is consistent with blood (3, rendered as isosurface) and the exit defect of the skull (4). Skull rendered as volume rendering/VRT, not as isosurface. These images may not exhibit the same refined appearance as when advanced volume rendering is used (as in Figure 12.12), but the details still are visualized in an effective way.

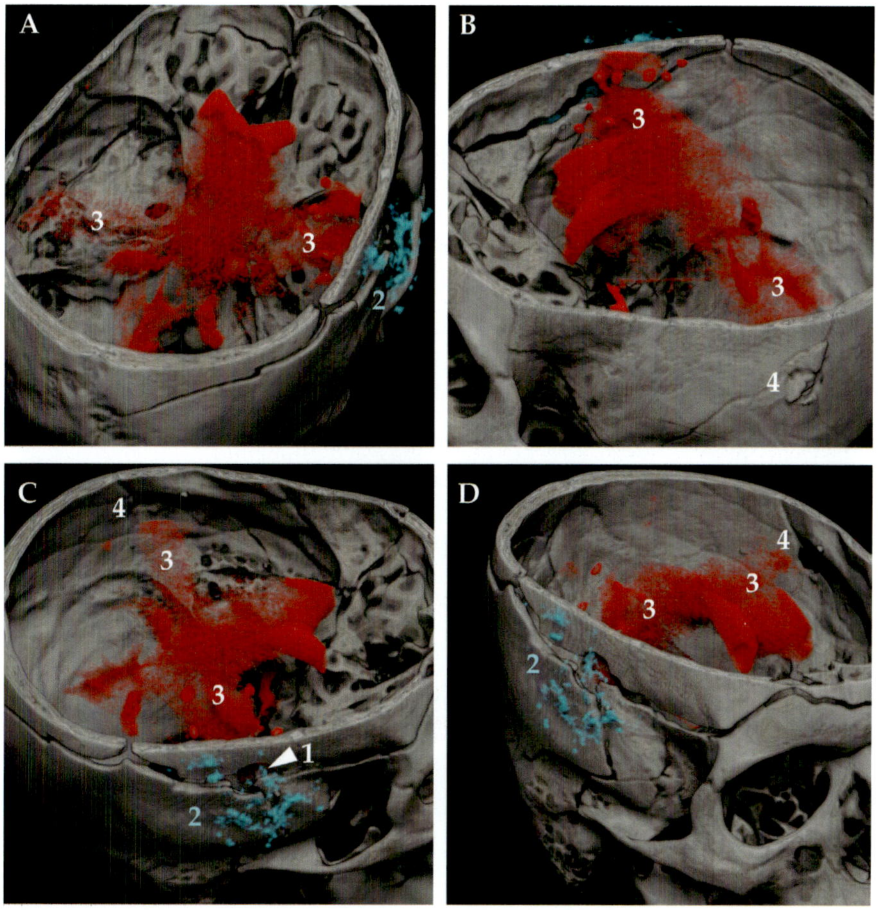

Figure 12.12 Complex reconstruction: presumed gunshot residue and skull PMCT. II. This image series shows how a manually and CT-density-based segmentation (same as in Figure 12.11), combined with advanced volume rendering, as provided by modern rendering software with various options related to what is termed photon-mapping. In this case of a pistol shot injury to the head, there was an entrance defect of the skull (1), there was a correlate for presumed gunshot residue (2), a correlate for what based on its CT-density is consistent with blood (3), and the exit defect of the skull (4). The overall appearance of this rendering may seem to convey a higher degree of realism inasmuch as the skull is concerned, whereas the added models for presumed gunshot residue and presumed blood clearly add an abstract and artificial aspect to it. The medicolegal aspect here is reduced to the extent of the injury in as much as native PMCT capabilities are concerned.

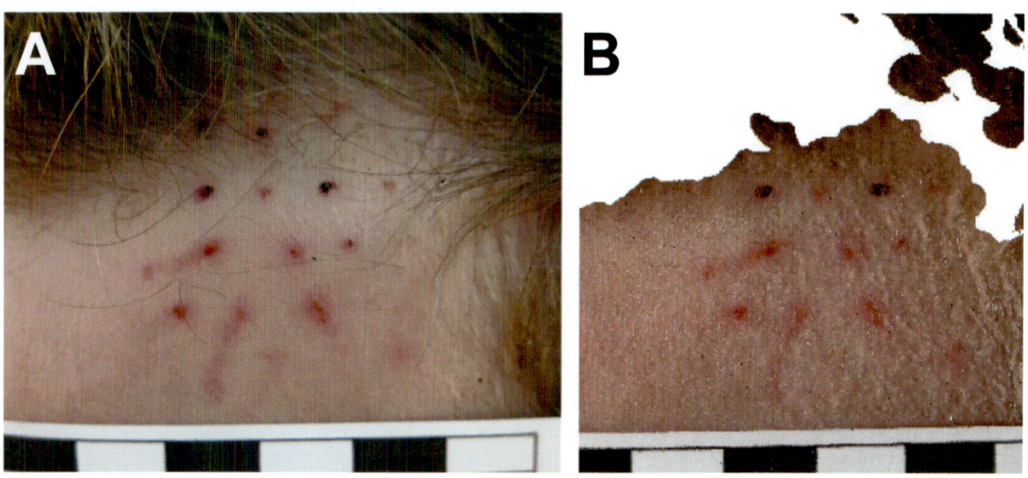

Figure 12.13 Polygon surface data visualization. Skin abrasions at the left forehead. The injury contains regularly spaced and arranged punctate abraded round lesions. Less deep or more superficial are a number of red striate skin discolorations immediately attaching or connected to the punctate deeper lesions. A: Photograph. B: 3D-model derived from photogrammetry. Scale bars 10 mm. Figure republished from publication [22], reuse licensed as 6040211304915.

12.3 USES OF VISUALIZATION IN FORENSIC IMAGING

There are a large variety of different visualization methods, all with respective advantages and disadvantages [23]. For the visualization of forensic medical 3D data, PMCT in particular, a variety of 2D and 3D methods may be considered for use [24].

12.3.1 Two-Dimensional Visualization Techniques

Axial image slices are routinely used for diagnostic purposes. A well-positioned body will result in axial slices showing good alignment with the table and gantry angles, allowing perpendicular planes to correspond to sagittal and coronal planes, as they are defined anatomically (Figure 12.1). With software that allows for that, one may freely define a differently angled cross-section through the volume by arbitrarily tilting the reslice plane in order to yield so-called multiplanar reconstructions (MPRs) (Figure 12.2).

With more refined requirements to rendering thinly sliced data, following a curved path to generate a series of cross sectional images (Figure 12.5) or performing MPR along a curve path (Figure 12.6) may visualize any curved structure or any findings not arranged in any single plane, such as gastrointestinal or vascular structures.

Because of an apparent higher spatial resolution of 2D visualizations over 3D renderings, subtle details may be better displayed, making these types of visualizations the preferred mode if not standard for medical examinations also of PMCT and PMMR.

Combining information of several adjacent slices can be done by either averaging, picking the highest value voxel (maximum intensity projection; Figure 12.3), or the lowest intensity voxel (minimum intensity projection [minIP; Figure 12.4]). Typical applications are the visualization of pulmonary calcifications (MIP) or gas accumulations (minIP).

12.3.2 Three-Dimensional Visualization Techniques

Volume data can be visualized over all three spatial dimensions, mapping voxel values to color and transparency values. Volume rendering and 3D-visualisation can be exploited for a range of specific questions (e.g., [25–31]). For medicolegal use, also 3D surface data may be exploited for forensic interpretation [22, 26, 32].

Volumetric data may be used to generate 3D polygons to represent 3D surfaces. When extracting them from CT data, these polygons usually are conveniently placed at a particular CT density, hence their descriptor "iso-surface". The simplest and fastest method for 3D volume visualization will be isosurface rendering and surface shaded display (SSD), (see Figures 12.7 and 12.8). From the view of the data structure, any of these 3D polygons may be regarded as a standard for visualizing objects and persons in video games and computer-generated images for movies, and they are a standard for 3D printing [33]. Segmentation with polygon generation may be used for measurements [34–36]. For the computational identification of particular tissues or materials, a variety of different tools exist to manually, semi-automatically, or automatically select structures in the dataset based on their Hounsfield value, anatomical structure, or other criteria [37]. These selection can then be either used for visualization or to automatically calculate a polygon mesh that describes the border of the selected structure. Many dedicated gaming engines and animation software on the market as well as rendering hardware are optimized to handle large amounts of polygon data. For crime and incident reconstructions as well as virtual reality applications, polygon rendering pipelines represent a convenient industrial de-facto standard as they may employ available high performance software and hardware.

The so-called volume rendering technique (VRT) is fast enough to be used in real time on many consumer computers. There, transparency and colors are combined and mapped to CT attenuation values. Advanced modern VRT can allow for even more realistic looking visualizations (Figure 12.8). Particular applications such as burnt bodies (Figure 12.9) or an increasing degree of fragmentation in high-grade trauma cases (Figure 12.10) may be presented with acceptable visual clarity using VRT methods.

In principle, different types of visualization can be mixed, i.e. volume and polygon renderings. While this approach makes it possible to combine the advantages of different visualization techniques, they may require other software than standard CT vendor PACS viewing facilities and may thus be more time consuming to create. Nevertheless, these visualizations may become routine. In relation to more recent techniques, such as robot-coordinated needle placements, needle placement visualization may be part of a baseline case documentation (e.g., [38]).

An example for a mixed visualization of both volume and polygon renderings is shown in Figure 12.11, whereas the very same case of a pistol contact shot injury to the head was visualized with three different transfer functions (Figure 12.12). There, visualization attempts to display presumed gunshot residue in its actual configuration at the

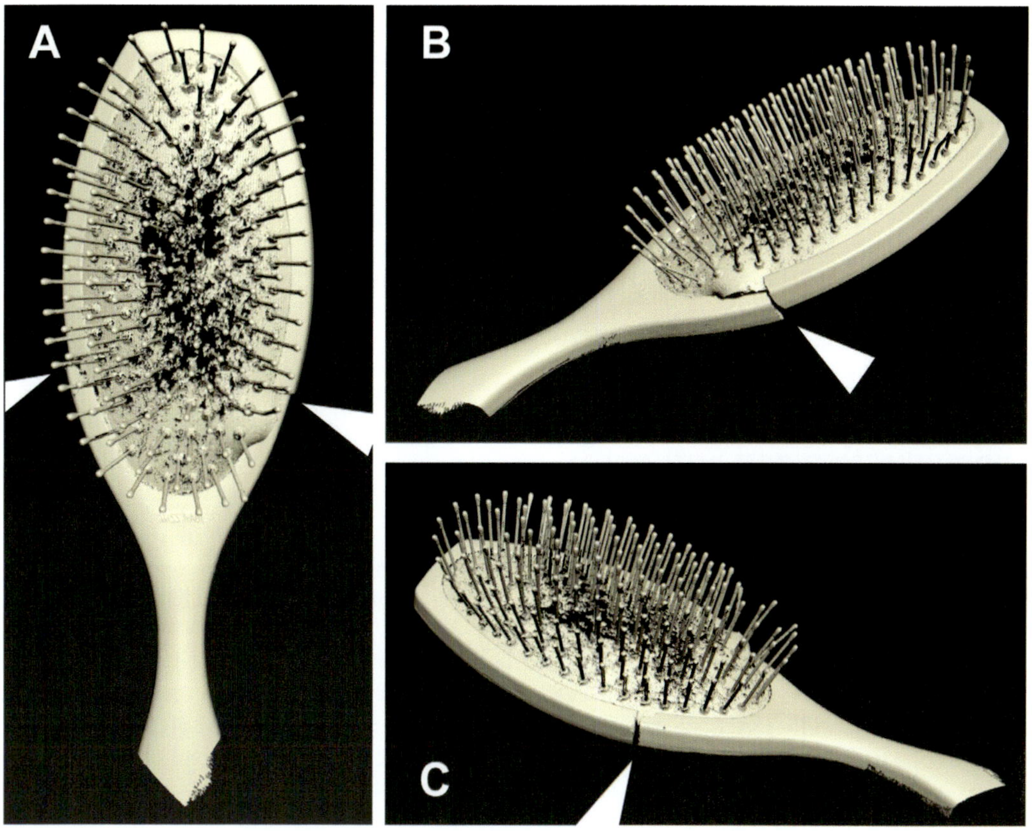

Figure 12.14 Polygon surface data visualization. Hair brush–3D-model from surface scan. Arrows: brush frame fracture lines. A: view from bottom. B: view from bottom left. C: view from bottom right. Figure re-published from publication [22], reuse licensed as 6040211304915.

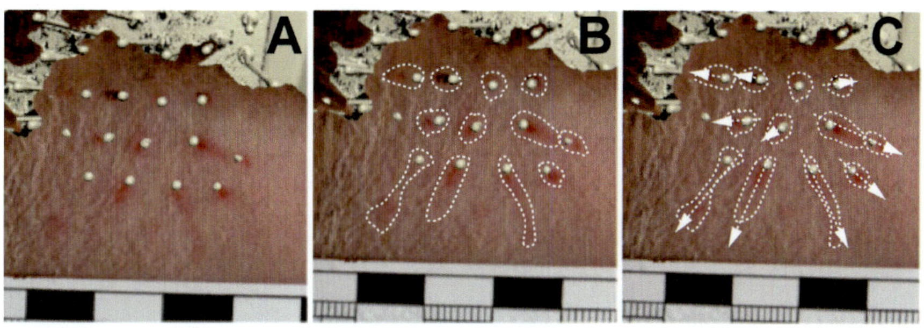

Figure 12.15 Polygon surface data visualization. Visually superimposed 3D models of skin injury (compare Figure 12.13) and brush (compare Figure 12.14). A: Superimposed bristles (beige) match the locations of the deeper and darker punctate abrasions, some of which contain dark dried-up exudate. It is easier to now recognize the striate red abrasions that are *not* matched by the super-imposed bristle tips. B: Dashed outlines highlight the skin abrasions. This allows the viewer to put the bristle locations (beige punctate shapes; see also A) that superimpose the punctate lesions in context of the overall injury constellation here. C: Clearly, a centrifugal pattern emerges which has been additionally highlighted and illustrated by added arrows. This describes the injury as one of centrally located deeper punctate abrasions that have added more superficial striate abrasions in centrifugal orientation; in context of the impact of the brush, an initially orthogonal and, with bristle deformation, increasingly tangential bristle tip impact explains the injury. Figure re-published from publication [22], reuse licensed as 6040211304915.

entrance defect of the skull in conjunction with the distribution of presumed blood and the exit defect of the skull. What is assumed to be blood, based on CT density, is not only present along the wound track but also in the ventricles.

The visualization of polygon data for medicolegal purposes may have to be guided by the forensic medicine specialist. A typical case is shown in Figures 12.13, 12.14, and 12.15 [22], where this forensic case study reports an application of 3D surface documentation in the evaluation of patterned skin injuries caused by a blunt object—a hairbrush—in a domestic assault. We identified limitations of static, rigid forensic shape comparisons and thus justify an integrative approach that combines technical modeling with medicolegal interpretation. There, a manual 3D superimposition revealed a partial but highly informative match: while punctate abrasions aligned well with bristle ends, accompanying striate abrasions did not. The mismatch between some injury features and the rigid brush model led us to interpret a dynamic deformation of the brush bristles upon impact. The centrifugal extension of abrasions suggested that the brush bristles, after an initially orthogonal impact, deformed outward, creating striate injuries not strictly matching the static geometry of the brush. This dynamic interpretation was further supported by the material properties of the brush and the presence of fractures in the brush frame. These observations underscore that injury morphology can be shaped not only by object geometry but also by interaction dynamics and deformation behavior—factors that static modeling alone cannot capture. Forensic shape matching, while potentially self-explanatory in static cases, often fails to account for deformation, velocity, and tangential forces. Dynamic injuries, especially those involving soft tissue, movement, or pliable materials, appear to demand expert interpretation beyond what automated or visual 3D match tools tend to routinely provide. In such cases, the integration of 3D surface documentation with medicolegal expertise may be important to arrive at reconstructively meaningful conclusions. The start of any further interpretation is, however, a solid visualization that is informed by medicolegal principles [39].

12.3.3 Communication and Software

For the communication of forensic medical findings in their respective anatomical context such as fracture patterns or the position of foreign bodies, 3D visualizations appear to be the tool of choice [1]. When multiple modalities, such as surface documentation, CT, and others, are combined, complex visualizations of trajectories, body postures, and sequences of events may be possible. However, postmortem imaging in forensic medicine always occurs within technical but also conceptual limitations [22, 40, 41].

A visualization or data processing pipeline may be regarded as an individual choice. It may depend on many factors such as the forensic question, the type of data that is available, the target audience, cost and available funding, and software and hardware aspects as well as user constraints or external requirements [42]. The software offered by CT manufacturers nowadays appears to be fairly sophisticated (e.g. [43, 44]). There is a whole range of freeware or open source software as well as various independent commercial software options [24]. In our forensic imaging team, we so far have often used syngo.via (Siemens, Erlangen, Germany), Amira (Thermo Fisher Scientific, Waltham, Massachusetts, USA) [45], Blender (Stichting Blender Foundation, Amsterdam, Netherlands [46]) [47], and 3D Slicer [48] as well as resorting to own programming on occasion [29, 49, 50].

Generally and for any application, the right reconstruction and visualization methods may have to be selected for the specific task [24].

ACKNOWLEDGMENTS

The authors express their gratitude to Emma Louise Kessler, MD, for her generous donation to the Zürich Institute of Forensic Medicine, University of Zürich, Switzerland. Spelling and grammar checked using AI tools. Image material: Figures 12.13, 12.14 and 12.15 re-published from publication [22], reuse licensed as 6040211304915. Released under CC BY 4.0. While the authors have made efforts to ensure clarity and structure, the development of this manuscript occurred under the constraints of available time and timelines. The retrospective use of anonymized data was submitted to and evaluated by the Ethics Committee of the Canton of Zürich (reference number 15–0686).

References

[1] Ampanozi G, Zimmermann D, Hatch GM, Ruder TD, Ross S, Flach PM, et al. Format preferences of district attorneys for postmortem medical imaging reports: Understandability, cost effectiveness, and suitability for the courtroom: A questionnaire based study. Legal Medicine. 2012;14(3):116–120.

[2] Schweitzer W, Bartsch C, Ruder TD, Thali MJ. Virtopsy approach: Structured reporting versus free reporting for PMCT findings. Journal of Forensic Radiology and Imaging. 2014;2(1):28–33.

[3] Thali MJ, Viner MD, Brogdon BG. Brogdon's forensic radiology. CRC Press; 2010.

[4] Grabherr S, Egger C, Vilarino R, Campana L, Jotterand M, Dedouit F. Modern postmortem imaging: An update on recent developments. Forensic Sciences Research. 2017;2(2):52–64.

[5] Christe A, Flach P, Ross S, Spendlove D, Bolliger S, Vock P, et al. Clinical radiology and postmortem imaging (Virtopsy) are not the same: Specific and unspecific postmortem signs. Legal Medicine. 2010;12(5):215–222.

[6] Schweitzer W, Ampanozi G, Ebert L, Thali M, Knaute DF. Possible XOR fallacy: Case report of combined foramen sternale with an osseous sternal knife stab injury. Journal of Forensic Radiology and Imaging. 2018;12:76–79.

[7] Foley JD, Van FD, Van Dam A, Feiner SK, Hughes JF, Hughes J. Computer graphics: Principles and practice. vol. 12110. Addison-Wesley Professional; 1996.

[8] Whitcher B, Schmid VJ, Thorton A. Working with the DICOM and NIf TI Data Standards in R. Journal of Statistical Software. 2011;44:1–29.

[9] Steffen C, Schaepman M, Hardmeier T, Schweitzer W. Digital 3D reconstruction of two parahissian accessory bundles in a case of Wolff-Parkinson-White syndrome. International Journal of Legal Medicine. 2004;118:101–105.

[10] Bryant J, Drage NA, Richmond S. CT number definition. Radiation Physics and Chemistry. 2012;81(4):358–361.

[11] Michienzi R, Meier S, Ebert LC, Martinez RM, Sieberth T. Comparison of forensic photo-documentation to a photogrammetric solution using the multi-camera system "Botscan". Forensic Science International. 2018;288:46–52.

[12] Schweitzer W, Fukuda H, Thali M, Bolliger S, Ebert L. Mobile forensic photogrammetry in the field: Conservative approach to non-collaborative surfaces. Forensic Imaging. 2024;38:2005910.

[13] Fukuda H, Decker SJ, Ford JM, Schweitzer W, Ebert LC. Neural radiance fields as a complementary method to photogrammetry for forensic 3D documentation: Initial comparative insights. Forensic Imaging. 2024;39:200605.

[14] Klein EC, Feddema JC, Chiu LZ. Closed iris, long exposure photography improves three-dimensional photogrammetric bone reconstructions. Forensic Imaging. 2024;39:200618.

[15] Kottner S, Ebert LC, Ampanozi G, Braun M, Thali MJ, Gascho D. VirtoScan-a mobile, low-cost photogrammetry setup for fast postmortem 3D full-body documentations in x-ray computed tomography and autopsy suites. Forensic Science, Medicine, and Pathology. 2017;13:34–43.

[16] Lozano M. Patience and Paranoia: The Necessary Art of Observation in a Modern World. Knihovna, Akademie múzicky`ch umění v Praze.; 2024. Bachelor of Arts Thesis. Praha, Czechia: Akademie múzicky`ch umění v Praze: Filmová a televizní fakulta.

[17] Locklair G. Digital data demise: Obsolete digital data formats. In: The Routledge companion to media technology and obsolescence. Routledge; 2018. p. 322–336.

[18] Graham RN, Perriss RW, Scarsbrook AF. DICOM demystified: A review of digital file formats and their use in radiological practice. Clinical Radiology. 2005;60(11):1133–1140.

[19] Bidgood Jr W, Horii SC. Introduction to the ACR-NEMA DICOM standard. Radiographics. 1992;12(2):345–355.

[20] Larobina M, Murino L. Medical image file formats. Journal of Digital Imaging. 2014;27:200–206.

[21] McHenry K, Bajcsy P. An overview of 3d data content, file formats and viewers. National Center for Supercomputing Applications. 2008;1205:22.

[22] Kreul D, Thali M, Schweitzer W. Case report: Forensic 3D-match of hair brush and scalp abrasions revealing dynamic brush deformation. Journal of Forensic Radiology and Imaging. 2019;16:34–310.

[23] Brodlie KW, Carpenter LA, Earnshaw RA, Gallop JR, Hubbold RJ, Mumford AM, et al. Scientific visualization: Techniques and applications. Springer Science & Business Media; 2012.

[24] Ebert LC, Franckenberg S, Sieberth T, Schweitzer W, Thali M, Ford J, et al. A review of visualization techniques of postmortem computed tomography data for forensic death investigations. International Journal of Legal Medicine. 2021:1–13.

[25] Schweitzer W, Schaepman M, Bär W. Geometrie von virtuellen Oberflächen: Darstellung der Passgenauigkeit durch Berechnung einer farbkodierten Fehleroberfläche. Rechtsmedizin. 2003;13(4):259–259.

[26] Röhrich E, Thali M, Schweitzer W. Skin injury model classification based on shape vector analysis. BMC Medical Imaging. 2012;12:1–12.
[27] Schweitzer W, Thali MJ, Flach PM. Plausibility of assisted suicide by ingestion in a case of limb-girdle muscular dystrophy. Journal of Forensic Radiology and Imaging. 2013;1(4):207–211.
[28] Schweitzer W, Ruder T, Thali M, Ringl H. Skull fractures in postmortem CT: VRT, flat and skin surface projections in comparison. Journal of Forensic Radiology and Imaging. 2015;3(4):214–220.
[29] Schweitzer W, Thali M. Computed tomography of scalp embedded gravel: Differentiation between falling and kicking. Journal of Forensic Radiology and Imaging. 2015;3(4):200–209.
[30] Schweitzer W, Thali M. Fatal obstructive asphyxia: Trans-pulmonary density gradient characteristic as relevant identifier in postmortem CT. Journal of Forensic Radiology and Imaging. 2019;19:1003310.
[31] Dobay A, Ford J, Decker S, Ampanozi G, Franckenberg S, Affolter R, et al. Potential use of deep learning techniques for postmortem imaging. Forensic Science, Medicine, and Pathology. 2020;16(4):671–679.
[32] Ebert LC, Rahbani D, Lüthi M, Thali MJ, Christensen AM, Fliss B. Reconstruction of full femora from partial bone fragments for anthropological analyses using statistical shape modeling. Forensic Science International. 2022;332:111196.
[33] Schweitzer W, Thali M, Aldomar E, Ebert L. Overview of the use of 3D printing in forensic medicine. Rechtsmedizin. 2020:1–8.
[34] Ampanozi G, Hatch GM, Ruder TD, Flach PM, Germerott T, Thali MJ, et al. Postmortem virtual estimation of free abdominal blood volume. European Journal of Radiology. 2012;81(9):2133–2136.
[35] Ebert LC, Ampanozi G, Ruder TD, Hatch G, Thali MJ, Germerott T. CT based volume measurement and estimation in cases of pericardial effusion. Journal of Forensic and Legal Medicine. 2012;19(3):126–131.
[36] Ebert LC, Heimer J, Schweitzer W, Sieberth T, Leipner A, Thali M, et al. Automatic detection of hemorrhagic pericardial effusion on PMCT using deep learning-a feasibility study. Forensic Science, Medicine and Pathology. 2017;13(4):426–431.
[37] Schweitzer W, Thali M, Aldomar E. Virtopsy visualisation: Mixed data gradient model for more accurate thin bone visualization in 3D rendering. Forensic Imaging. 2023;33:200529.
[38] Martinez RM, Ptacek W, Schweitzer W, Kronreif G, Fürst M, Thali MJ, et al. CT-guided, minimally invasive, postmortem needle biopsy using the B-R ob II needle-positioning robot. Journal of Forensic Sciences. 2014;59(2):517–521.
[39] Errickson D, Fawcett H, Thompson T, Campbell A. The effect of different imaging techniques for the visualisation of evidence in court on jury comprehension. International Journal of Legal Medicine. 2020;134:1451–1455.
[40] Schweitzer W, Röhrich E, Schaepman M, Thali MJ, Ebert L. Aspects of 3D surface scanner performance for postmortem skin documentation in forensic medicine using rigid benchmark objects. Journal of Forensic Radiology and Imaging. 2013;1(4):167–175.
[41] Schweitzer W, Thali MJ, Ampanozi G. Planned complex suicide combining pistol head shot and train suicide and Virtopsy examination. Forensic Imaging. 2022;28:200485.
[42] Ford JM, Decker SJ. Computed tomography slice thickness and its effects on three-dimensional reconstruction of anatomical structures. Journal of Forensic Radiology and Imaging. 2016;4:43–46.
[43] Jensen HW. Global illumination using photon maps. In: Eurographics workshop on rendering techniques. Springer; 1996. p. 21–30.
[44] Glemser PA, Engel K, Simons D, Steffens J, Schlemmer HP, Orakcioglu B. A new approach for photorealistic visualization of rendered computed tomography images. World Neuro-surgery. 2018;114:e283–e292.
[45] Ebert LC, Thali MJ, Ross S. Getting in touch—3D printing in Forensic Imaging. Forensic Science International. 2011;211(1–3):e1–e6.
[46] Community BO. Blender—a 3D modelling and rendering package. Stichting Blender Foundation, Amsterdam; 2018.
[47] Schweitzer W, Enders M, Thali M. Very affordable post mortem CT angiography kit: Feasibility study using immersion pump and 3D printed parts. Journal of Forensic Radiology and Imaging. 2019;16:11–18.
[48] Fedorov A, Beichel R, Kalpathy-Cramer J, Finet J, Fillion-Robin JC, Pujol S, et al. 3D Slicer as an image computing platform for the Quantitative Imaging Network. Magnetic Resonance Imaging. 2012;30(9):1323–1341.
[49] Ebert LC, Ptacek W, Naether S, Fürst M, Ross S, Buck U, et al. Virtobot—a multi-functional robotic system for 3D surface scanning and automatic post mortem biopsy. The International Journal of Medical Robotics and Computer Assisted Surgery. 2010;6(1):18–210.
[50] Ebert LC, Ptacek W, Breitbeck R, Fürst M, Kronreif G, Martinez RM, et al. Virtobot 2.0: The future of automated surface documentation and CT-guided needle placement in forensic medicine. Forensic Science, Medicine, and Pathology. 2014;10(2):179–186.

CHAPTER 13

MACHINE LEARNING TECHNIQUES

Akos Dobay and Raffael Golomingi

DOI: 10.4324/9781003413998-14

13.1 INTRODUCTION

Machine learning is foremost a statistical approach in which the algorithms and the mathematical equations defining the models rely on data [1]. Statistical learning plays a key role in many areas of science and in industry. It is especially useful in life sciences such as biology, biomedicine, or bioengineering due to the large amount of data, and the large number of possibilities displayed by living systems. But statistical learning is not limited to those fields. There are several domains where statistical learning appears as a powerful tool to create mathematical models.

The same observation can be made in medicine where each person is unique from its constitution and the same pathology might have slightly different etiologies. Another reason why statistical learning in medicine is needed comes from the amount of data that modern medical devices generate. This is the case for medical X-ray scanners able to produce three-dimensional (3D) images at a high resolution such that a single X-ray scan produces several thousand images [2]. Hence, modern technologies constitute a real challenge when it comes to analyze the data and there is a need to develop algorithms able to make useful prediction based on such a large and statistically diverse amount of data. The goal of machine learning is to devise an algorithm able to automatically adjust the parameters of a model used for making predictions.

Machine learning has two main components. First, the data. What makes a useful dataset for a machine learning approach is (1) the quality of the data and (2) the diversity of the samples. By quality we mean that the signal-to-noise ratio (SNR) should be such that it is possible for the model to extract features that are necessary for the model. The diversity in the dataset is instrumental for a machine learning algorithm to achieve a reliable predictability. If the samples in the dataset lack diversity or do not represent the problem, there is a risk of having a model that either learned the wrong features, or it has learned the right features but lacks generality. The lack of generality means that the model is constrained to recognize only specific aspects of a problem and new aspect that were not part of the training set will be falsely classified. The general term for this type of error in establishing a model based on machine learning is called *overfitting*. This can be compared to a human being learning how to recognize fruits. If the person is trained to recognize fruits by only seeing apples and pears, it is very likely that any other fruit will be recognized either as an apple or as a pear. On the other hand, if the person is presented only with red apples and green pears, it is likely that the person will decode color as the main feature characterizing the fruit and not necessarily the shape. Hence, a green apple will be classified as a pear, and a more yellow-type pear will be seen as an apple. Through this simple example, one can capture the essence of statistical learning and the role of machine learning techniques in devising computer algorithms to identify features in dataset. The quality and diversity of the data are therefore instrumental when developing machine learning algorithms.

How to improve the quality of the data and reduce the SNR in medical data constitute an active research field, and by now there are many techniques that can be used to clean data. When the data are images, the technique often involves filters, also called a *kernel* or *mask*, that are applied on the images at the pixel level. A filter can be seen as a matrix operation where each coefficient of the matrix is an integer carefully chosen to extract features. The mathematical operation for applying a filter to an image is called *convolution*. By doing a convolution on an image, we obtain a new image in which some elements have been highlighted. By applying multiple filters on the same image—sometimes hundreds of them—the result is a pile of images, also called a *tensor* in machine learning. This operation simplifies the complexity of an image to highlight characteristics that are necessary to train a model. Figure 13.1 provides a visually compelling example how the complexity of a problem such as defining the topology of a knot can be solved by having a projection in two dimensions of the same know. The projection includes all the characteristics of the original but in a lower dimension.

How to improve the diversity of the dataset when it comes to medical data also constitutes an active field of research. This technique is called *data augmentation*. Data augmentation consists of making small modification to already existing data, creating a pool of synthetic data whose variations are in the realm of what is possibly existing. Data augmentation can be seen as a useful technique to prevent overfitting [3]. Once the dataset has been prepared, the next step is usually to decide what type of model should be applied to the dataset. At this stage two scenario are possible. The dataset has been *labeled*, and each image in the dataset corresponds to an already-known pattern. Or the dataset is *unlabeled*, and the authors of the dataset do not know in advance what kind of features they are looking for. When the data have previously been labelled, we call this type of machine learning *supervised*. In a supervised setting, the model can be trained by splitting the dataset into three subsets: training, test, and validation. These three subsets will be used to optimize and validate a model. When the data are not labelled, we call this type of machine learning *unsupervised* or *non-supervised*. Some of the methods used in supervised machine learning can also be used in an unsupervised setting.

The optimization process is the second main aspect that characterizes any machine learning technique. In other words, machine learning is by itself an optimization process by which the computer runs an optimization algorithm based on what we call an *objective function* [1]. An objective function compares the actual prediction on the data made by the model with the expected prediction. Any discrepancy between these two outcomes means that the model is not

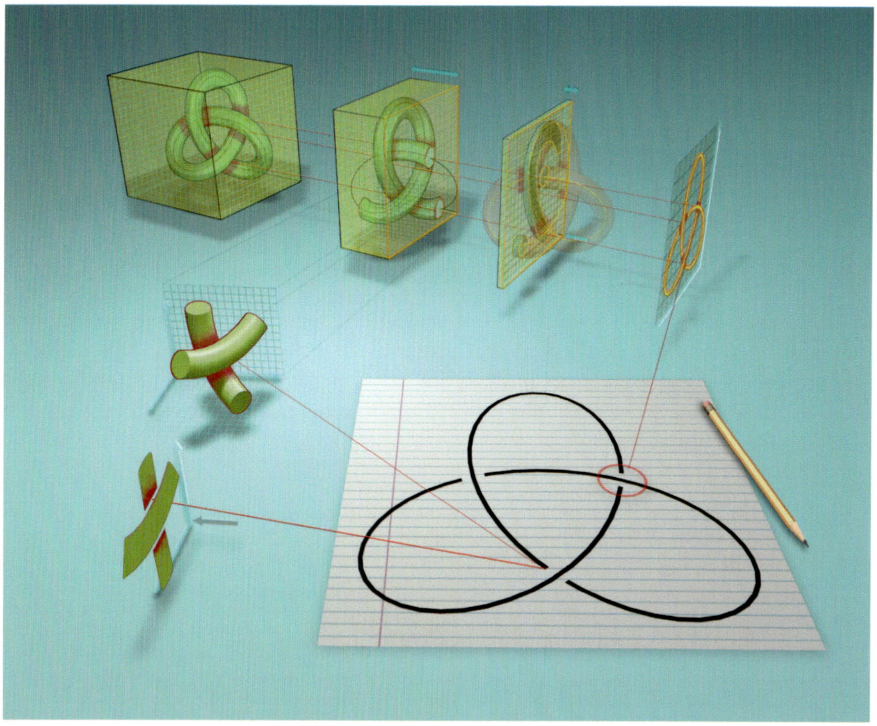

Figure 13.1 A visually compelling example of data compression or complexity reduction used in mathematical knot theory. A physical knot in three dimensions can be represented in two dimensions using a projection technique called regular projection that ensures a complexity reduction without loss of information. The knot drawn on the paper sheet contains all the characteristics of its three-dimensional counterpart such that the information on the sheet is enough to rebuild the knot in three dimensions.

yet optimized. Therefore, an optimization algorithm knows how to optimize an objective function by automatically reducing this discrepancy below a given threshold. Most of the optimization algorithms are based on a *gradient descent* principle. The gradient corresponds to the first derivative of objective function. The gradient indicates the direction in which the loss function increases. This information allows the computer to automatically choose a value that will decrease the loss function—the exact opposite direction—and reduce the discrepancy. The gradient descent stops when the threshold has been reached or the user has defined how many attempts should be made to reach that threshold. The role of the objective function is to represent the features as accurately as possible. For instance, a simple linear regression becomes accurate when we increase the number of data points over the whole range of possible points. Indeed, having two values or points corresponding to the same situation is not providing more information about the whole range of validity of a system. Instead, the more the data points we obtain that are representative of a system, the more the model will be also representative of the system.

As we mentioned earlier, the key aspect in machine learning is the correct sampling of the data. If the sampling is biased towards a feature—such as having only red apples—the model will overfit the data, and, no matter how many samples we collect, the model will not represent the system properly. Hence, ensuring a proper sampling is instrumental in any machine learning approach. The second aspect is how to choose the right model. Choosing an appropriate model is a combination of knowledge and experience. Even before training a model on a given dataset, it is useful and almost necessary to have an educated guess of the kind of model that can reach the best prediction level. A machine learning algorithm can learn from the data, and it can optimize its parameters, but it cannot modify its own architecture. This is an important remark, as it highlights the limitations of any machine learning techniques: a machine learning algorithm can learn from the data, but it cannot modify its own architecture. The predictive power of a model and its domain of validity is predefined by humans, and it is independent of the quality of the data. When choosing a model, one makes assumptions about the type of data and the distribution of the samples.

Finally, the choice of the objective function is also instrumental to achieve the best optimization possible. In the field of statistical learning and machine learning, the objective function is called the *loss function*, and several loss functions are available for machine learning techniques. The same is true of the gradient descent: several gradient descent algorithms can be implemented.

In this chapter, we will describe the most popular machine learning models used in medical image analysis. We will also describe some of the basic loss function, and its derivation, used to optimize these models. Among the typical choice of machine learning models, we can distinguish:

1. Linear Regressions
2. Logistic Regressions
3. Decision Trees
4. Support Vector Machines (SVMs)
5. Naïve Bayes Classifiers
6. Nearest-Neighbors (NN) and k-Nearest-Neighbors (kNNs)
7. K-Means
8. Random Forests (RFs)
9. Neural Networks (NNs) and Convolutional Neural Networks (CNNs)

Depending on the field, the variables may have different names. In machine learning we often use the names *input* and *output* to describe what is commonly defined in mathematics as *independent* and *dependent* variables or *functions*. Similarly, in statistical methods, the independent variables are called *predictors*. In the pattern recognition literature, the independent variables are called *features*. Readers who want to familiarize themselves with the topic should keep in mind that these terms may occur interchangeably.

13.2 LINEAR REGRESSION

Linear regression is one of the most common approaches to search for a correlation between a predictor and a response variable in form of a linear relationship [3]. For instance, we want to find out how the price of a real estate depends on its location (city versus countryside) and its surface area (in m²). The surface area is usually a reliable predictor for the price of a real estate if we compare real estates within the same location. However, we can also predict the price of a real estate based on its location from the center of a city or a town. Ultimately, we can derive a model that combines both predictors into one expression. To find a predictor such that

$$y \approx h(x) \tag{1}$$

we need a representation of our function $h(x)$. A simple representation is called a *linear representation* where y is a linear combination of some parameters, each with a given weight. Referring to our example, experience tells us that the price of a real estate not only depends on its location and its surface area, but the design and the quality of the materials will also play a role. Hence, each criterium can be included in the model and each of them will be weighted based on their level of importance. Such models are called *general linear models*, and the response variable y is a linear combination of multiple predictors. We can write this relationship as a linear combination of x_n variables:

$$h(x) = \beta_0 + \sum_{n=1}^{m} \beta_n x_n . \tag{2}$$

Other representations using more elaborate functions are possible. In a general linear model, each predictor x_n is weighted accordingly using a proportionality constant β_n. For a given price y, there is an infinite number of solutions that would satisfy equation (2). However, by fixing the value of all β_n in equation (2), we will soon discover that the prediction power of our model is useless. Therefore, the more information we collect on the values for x_n and y, the more likely it is to generalize our model by having an estimate of all the β_n. For the ease of calculation with a computer, equation (2) can be compressed using a vector representation. Nowadays, most modern programming languages have optimized standard libraries to perform matrix operations. In addition, the constant β_0 can be included in the summation by considering a term $x_0 = 1$:

$$h(x) = \sum_{n=0}^{m} \beta_n x_n = \vec{\beta}^{\mathsf{T}} \cdot \vec{x} = \begin{pmatrix} \beta_0 & \cdots & \beta_n \end{pmatrix} \cdot \begin{pmatrix} x_0 \\ \vdots \\ x_n \end{pmatrix}. \tag{3}$$

Standard linear regression tries to minimize the distance between the observed data and a model using the *least squares* approach: the squares of the difference between the predicted and the observed value are minimized. In other terms, the least squares method tries to minimize the residual sum of squares $\left(y^{(i)} - \sum_j \beta_j x_j^{(i)} \right)^2$, where $y^{(i)}$ is the observed value and $\sum_j \beta_j x_j^{(i)}$ the predicted one by adjusting each parameter β_j. The residual sum of squares corresponds to the loss function used in a linear regression model:

$$J(\beta) = \frac{1}{2} \sum_{i=1}^{N} \left(y^{(i)} - \sum_j \beta_j x_j^{(i)} \right)^2, \tag{4}$$

while the least squares method is the optimization algorithm by which the residual sum of squares is minimized. The directional derivative of $J(\beta)$ is given by the derivative of $J(\beta)$ along each parameter β_j:

$$\frac{\partial J(\beta)}{\partial \beta_j} = \frac{\partial}{\partial \beta_j} \left(\frac{1}{2} \sum_{i=1}^{N} \left(y^{(i)} - \sum_j \beta_j x_j^{(i)} \right)^2 \right) = \sum_{i=1}^{N} \left(y^{(i)} - \sum_j \beta_j x_j^{(i)} \right) x_j^{(i)} . \tag{5}$$

The quantity derived in equation (5) is the gradient of $J(\beta)$ along the direction of the parameter β_j. This quantity is either positive, negative, or equal to zero. It tells us how much discrepancy we have along the parameter β_j and in which direction we need to correct the value of β_j to get closer to the observations.

13.3 LOGISTIC REGRESSION

Linear regression works when the predictors are continuous-valued quantities such as the distance to the center of a city or the surface area in our example. However, in some cases the prediction falls into a category. A simple model of this kind is the *binary classifier,* where the prediction will fall in one of the two possible categories. To further evaluate the price of real estate, the question can be whether it has a garage or a swimming pool. The answer to this question is either a "yes" or a "no". Logistic regression is another type of regression where the function can perform a binary classification on an observable.

Like linear regression, logistic regression uses the same principle, but the relationship between the predictor and the response variable is based on a sigmoid function such as

$$h_\beta(x) = \frac{1}{1+e^{\left(-\sum_j \beta_j x_j\right)}}. \tag{6}$$

Depending on which combinations of $h_\beta(x)$ values will attain ½, the logistic function will convert a continuous predictor into a discrete observation. Hence, the logistic function can be used to turn a continuous predictor into a binary response variable. Using the same strategy as in the case of a linear regression model, one can build a cost function to optimize each parameter β_j in the model. The expression for the cost function uses the definition of the *cross-entropy* between the two classes:

$$J(\beta) = -\sum_{i=1}^{N}\left(y^{(i)}\log h_\beta\left(x^{(i)}\right) + \left(1-y^{(i)}\right)\log\left(1-h_\beta\left(x^{(i)}\right)\right)\right). \tag{7}$$

The directional derivative of $J(\beta)$ is given by the derivative of $J(\beta)$ along each parameter β_j:

$$\frac{\partial J(\beta)}{\partial \beta_j} = \frac{\partial}{\partial \beta_j}\left(-\sum_{i=1}^{N}\left(y^{(i)}\log h_\beta\left(x^{(i)}\right) + \left(1-y^{(i)}\right)\log\left(1-h_\beta\left(x^{(i)}\right)\right)\right)\right) \tag{8}$$

By redistributing the terms in equation (8), we can write

$$\frac{\partial J(\beta)}{\partial \beta_j} = -\sum_{i=1}^{N}\frac{y^{(i)}}{h_\beta\left(x^{(i)}\right)}x_j^{(i)} + \frac{\left(1-y^{(i)}\right)}{\left(1-h_\beta\left(x^{(i)}\right)\right)}x_j^{(i)}. \tag{9}$$

13.4 THE SOFTMAX REGRESSION

In the previous section we used the logistic function to turn a continuous-valued function into a binary classifier. The SoftMax regression is a generalization of the logistic regression procedure to multiple classes. In the case of a logistic regression the observables have labels that are either zero or one, i.e., $y^{(i)} \in \{0,1\}$. A SoftMax regression can handle K distinct labels. In other words, $y^{(i)} \in \{1,2,\ldots,K\}$. The generalization can be achieved by producing a probability distribution in form of a K-dimensional vector such that each entry will correspond the probability to belong to a class. A representation of the activation function $h(x)$ is given by a K-dimensional vector:

$$h_\beta(x) = \frac{1}{\sum_{k=1}^{K} e^{\left(\sum_j \beta_j^i x_j\right)}} \begin{bmatrix} e^{\left(\sum_j \beta_j^1 x_j\right)} \\ \vdots \\ e^{\left(\sum_j \beta_j^K x_j\right)} \end{bmatrix}. \tag{9}$$

The corresponding cost function $J(\beta)$ is given by a logarithmic function in the form of

$$J(\beta) = -\sum_{i=1}^{N}\sum_{k=1}^{K} \delta^{ik} \log \frac{e^{\left(\sum_j \beta_j^k x_j^{(i)}\right)}}{\sum_{p=1}^{K} e^{\left(\sum_j \beta_j^p x_j^{(i)}\right)}} \tag{10}$$

where the symbol δ^{ik} is called the delta Kronecker, and it is defined as

$$\delta^{ik} = \begin{cases} 1, & i = k \\ 0, & i \neq k \end{cases} \tag{11}$$

Minimizing equation (10) allows us to fit observables that have multiple labels.

13.5 NEAREST NEIGHBOR

The nearest-neighbor (NN) and the k-nearest neighbor algorithm (k-NN) is a non-parametric classification method [4]. By non-parametric we mean that the model is computed using a distance metric and no constant parameters are incorporated in the mathematical expression. During the training the observable values are groups into k different classes. The distance to the nearest neighbor is computed using a distance metric such as the Euclidean distance in case of continuous variables and a Hamming distance in case of discrete variables. Depending on the distance from the test sample to the nearest neighbors, one counts the number and type of points that are within a certain radius. The test sample will be assigned to the type represented by most of the points inside the radius. The k-NN algorithm is used when $k > 1$, in other words when the number of data points in the radius is at minimum two. In the case when $k = 1$, the k-NN method reduces to the NN method, also described as the 1-NN method. The K-means method is related to the k-NN algorithm. It is a clustering method that computes the centroids for a given set of points. First, the centers are placed randomly, and each data point is assigned to the closest center. A new center is computed using the points that have been previously assigned to the same center. This operation creates a new center, and the points are again reassigned to a center based on their proximity to a center. This procedure is repeated until the points are indefinitely assigned to the same center. All the points have been clustered around a center creating a k-partitioning of the space. Unlike the k-NN algorithm, the K-means method can be unsupervised.

The advantage of this method is that it does not require any assumptions prior to the classification. However, this method achieves good results if the data form some clusters. If the data are mixed, it might be hard to draw a boundary between the groups.

13.6 DECISION TREES

Decision trees represent a classification approach [5]. A dataset is composed of two types of outcomes. Each outcome can be characterized by a series of variables. A decision tree is like a test. Going through the test should predict with certainty a given outcome. For instance, we want to predict how reliable a car can be. To create such a test based on a decision tree, one would need to collect features that are common to all cars such as the number of doors, the engine size, the type of traction, the carmaker, etc. The question could be how what type of car reached 100,000 km without a repair. The variables in that case will be "number of doors", "engine size", "propulsion", etc. Each variable will have nodes corresponding to the actual possibilities such two or four doors. Engine size can vary between 1 and 6 liters. Propulsion can be diesel, gasoline, or electric. First, each of these features will constitute a root for a tree, and the tree with the highest score will be selected. The score is given based on whether a tree has a homogeneous branch, a branch that contains only one type of outcome. The depth of a tree is given by the number of nodes.

Once a tree has been selected, then we reapply the same features on the outcome of that branch and select for the highest score. This process is repeated until there is no homogeneous branch. The series of labels along the nodes from the root to the last node is a series of features that leads to the answer whether a car reached 100,000 km without a repair.

Each observation will therefore end up in a single class and the depth of the tree (number of branches) will depend on how many features can be extracted from the observation. Note that this technique works with small dataset. As soon as the dataset becomes very large, the likelihood to find a homogeneous branch decreases rapidly as one can expect.

13.7 THE RANDOM FOREST

The random forest (RF) algorithm relies on multiple decision trees [6]. The RF algorithm is similar to bootstrap aggregation or bagging (**b**ootstrap **agg**regat**ing**). Except that the RF algorithm allows for the control of the number of features used to define the decision trees as well as the number of trees in the forest. Figure 13.2 illustrates the main idea behind the RF algorithm using fruits in a basket. A random (decision) tree is a small tree based on a small subset of features. For each tree in the forest the subset is drawn randomly. In our example, for each tree we have chosen three features randomly. Because the number of features constituting the trees is smaller than a full decision tree based on the entire set of features, and the features has been drawn randomly, a tree in a RF algorithm is less effective than the whole tree. However, by generating many smaller trees, the RF algorithm can become more effective and less time consuming by looking at the proportion of successful trees. This method can be used to classify data.

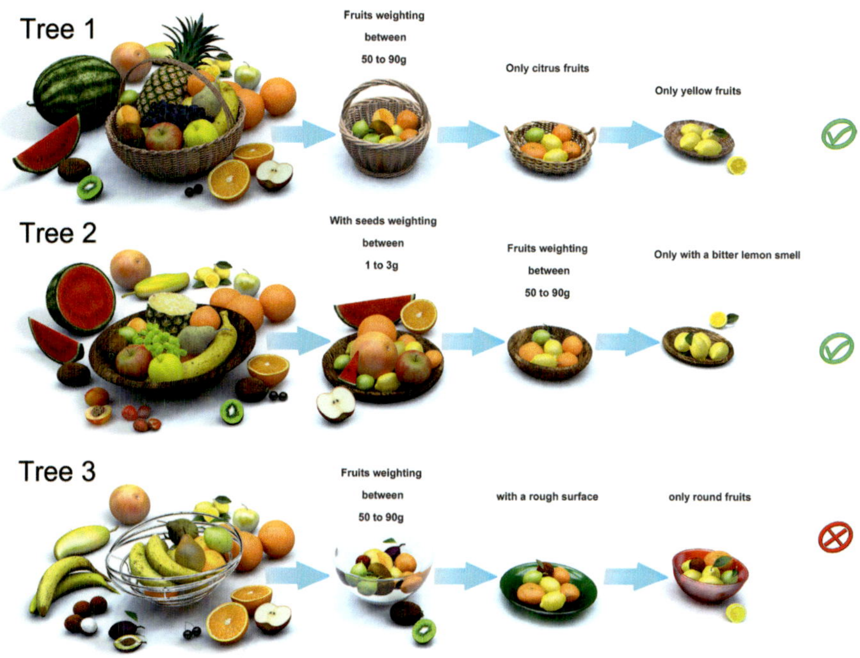

Figure 13.2 A random forest depicting three trees each with three nodes. At each node the tree is branching into two, but only one branch leading to a correct prediction is shown. A successful random decision tree contains at least one branch leading to a correct prediction. For each tree, the criteria are drawn randomly from a larger set and mixed to form a tree. Trees 1 and 2 can successfully classify lemons, while tree 3 would require additional nodes.

13.8 SUPPORT VECTOR MACHINE

Support vector machine (SVM) algorithms are commonly used to highlight pattern in non-supervised learning [7]. SVM models do not need to know the nature of the observables. By computing the boundaries between groups or clusters of observables, SVM tries to maximize the margin separating the groups. The key element of a support vector machine is the function defining the hyperplane separating the groups.

13.9 NAÏVE BAYES CLASSIFIER

Naïve Bayes classifier is another method that tries to separate observables based on their likelihood to belong to one of the prior distributions defining their class [8]. The prior distribution gives the probability to belong to a class, depending how likely it is that an observable has been drawn from a given probability distribution (prior). The term *naïve* is used to indicate that only a few assumptions have been made on the observables, and the features defining the dataset are independent of each other. Hence, using a similar example as earlier, if we take a pear that is green and is not round, but rather elongated, the model will treat these parameters as separate.

13.10 NEURAL NETWORK

Neural networks are one the most popular type of models. Neural networks can be traced back to the original idea of Warren McCullock and Walter Pitts (MPC), who published their description of a neuronal cell in 1943 [9]. A few years later, Frank Rosenblatt devised the concept of the perceptron learning rule based on the MCP neuronal model in 1957 [10]. The perceptron learning rule is a mechanistic rule that take several input values and turn it into a single output value. Each value has its own weight, and the weighted sum is compared to a threshold value. If the weighted sum is above the threshold, the perceptron will be activated, and a signal is transmitted. Figure 13.3 shows a schematic representation of a neural network model using the readout of an electroencephalogram (EGG).

The signal is not proportional to the sum but rather uses a discontinuous function that returns a number, usually −1 or 1 depending on whether the sum is below or above the threshold.

$$h_\beta(x) = \beta_0 + \sum_{n=1}^{m} \beta_n x_n = \begin{cases} -1, & h_\beta(x) < 0 \\ 1, & h_\beta(x) > 0 \end{cases}$$

Such computing units can be linked together to form a network. These types of networks are called feedforward neural networks [11]. In the 1980s, the success of convolutional neural networks gave a new impetus to the research field of computer vision in the sense that neural networks were presented as a solution for a various range of problems. The convolution process consists of applying different filters to an image to extract features. These features are grouped into maps and finally into a tensor, a series of matrices, each containing some features. The matrices are flattened to produce a single one-dimensional array containing all the features. This method is used to treat medical images. Since then, convolutional neural networks (CNNs) are widely used in many research areas in medicine such as genomics and other classification problems. Figure 13.4 shows the architecture of a neural network with one hidden layer between the input and the output layers.

Convolutional neural networks are the preferred type of neural networks when it comes to analyzing and classifying medical images, although new techniques such as *vision transformers* are gaining in efficiency [12–14]. The main components of convolutional neural networks are the backpropagation methods that correct for discrepancies and the convolutional part. The confusion matrix is used to report the results. The confusion matrix contains information about the precision, the recall, and the F_1 score. These values give information about the quality of the fit and the quality of the model.

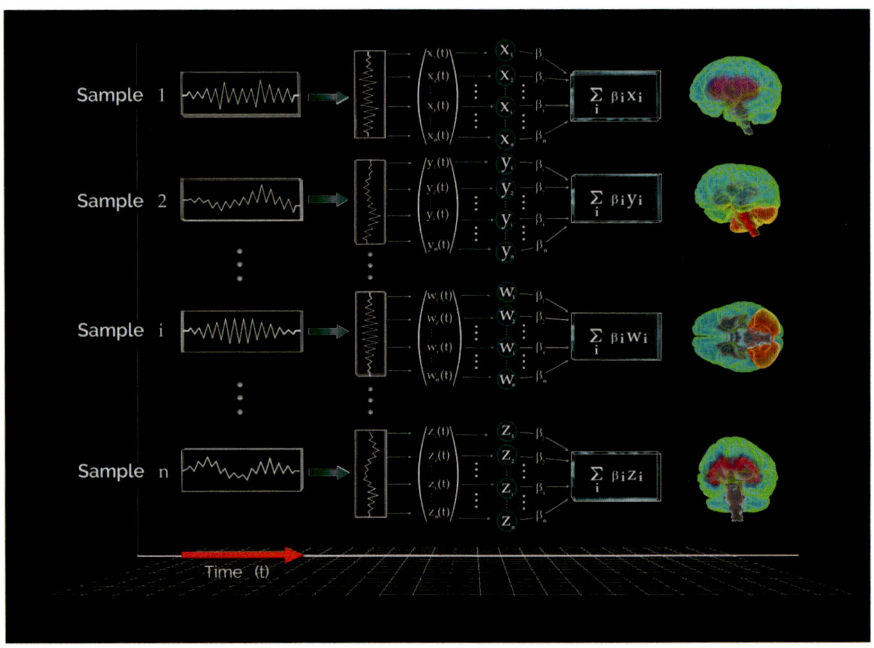

Figure 13.3 Schematic representation of a series of perceptrons. A time series is converted into a one-dimensional array of values extracted from the curve. Each time point translates into a node. The nodes form the input layer of the perceptron backbone. A weighted sum is then computed, and the result is compared to a threshold to create a binary classification.

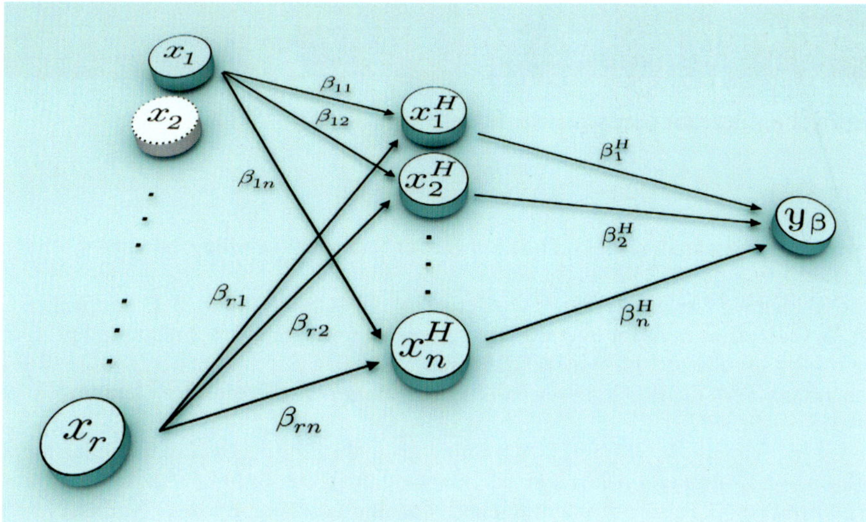

Figure 13.4 A fully connected neural network with one hidden layer (H) between the input (x_n) and the output layers (y_β). In a fully connected neural network, each node of the previous layer is connected to all nodes of the next layer.

The precision is defined as the fraction of true positives over the total number of positives. Optimizing the precision means reducing the number of false positives. False positives are samples that are predicted as containing a given feature when it is not the case. False positives are false alarms, and sometimes it is better to reduce the precision in order to prevent false negatives, samples that do contain the feature but are not detected. Recall is the fraction of true positives over the total of the actual number of positive samples. Recall is the ability of a model to detect a feature when presented with only samples that have the feature. Increasing the recall means decreasing the number of false negatives. Missing a prediction may have more tragic consequences than a false alarm. However, false alarm in medical science may have psychological side effects, or it can lead to administering a treatment for no purpose or to unwanted expenses.

In a recent review, we detailed how deep learning techniques can be used for postmortem image analysis [15].

ACKNOWLEDGMENTS

The authors thank Michael Bolliger for preparing the figures.

References

1. Hastie, T., Tibshirani, R., Friedman, J. (2009). The Elements of Statistical Learning, Springer Series in Statistics, Springer, New York, 745 pages.
2. Flach, P.M., Gascho, D., Schweitzer, W., Ruder, T.D., Berger, N., Ross, S.G., Thali, M.J., Ampanozi, G. (2014). Imaging in forensic radiology: An illustrated guide for postmortem computed tomography technique and protocols. Forensic Sci Med Pathol. December;10(4):583–606. https://doi.org/10.1007/s12024-014-9555-6. Epub 2014 April 11. PMID: 24723662.
3. Shorten, C., Khoshgoftaar, T.M. (2019). A survey on image data augmentation for deep learning. J Big Data. 6:60. https://doi.org/10.1186/s40537-019-0197-0
4. Boser, B.E., Guyon, I.M., Vapnik, V.N. (1992). A training algorithm for optimal margin classifiers. Proceedings of the Fifth Annual Workshop on Computational Learning Theory—COLT '92. p. 144.
5. Christopher, M. (2006). Bishop. Pattern Recognition and Machine Learning. Springer-Verlag, New York. 738 pages.
6. Piryonesi, S.M., El-Diraby, T.E. (2020). Role of data analytics in infrastructure asset management: Overcoming data size and quality problems. J Transp Eng, Part B: Pavements. 146(2):04020022.
7. Freedman, D.A. (2009). Statistical Models: Theory and Practice. Cambridge University Press.
8. Kamiński, B., Jakubczyk, M., Szufel, P. (2017). A framework for sensitivity analysis of decision trees. Cent Eur J Oper Res. 26(1):135–159.
9. Fix, E., Hodges, J.L. (1951). Discriminatory Analysis. Nonparametric Discrimination: Consistency Properties (Report). USAF School of Aviation Medicine, Randolph Field, Texas.
10. Ho, T.K. (1995). Random decision forests. Proceedings of the 3rd International Conference on Document Analysis and Recognition, Montreal, QC, 14–16 August. pp. 278–282.
11. Schmidhuber, J. (2015). Deep learning in neural networks: An overview. Neural Netw. January;61:85–1110. https://doi.org/10.1016/j.neunet.2014.09.003. Epub 2014 October 13. PMID: 254626310.
12. McCulloch, W.S., Pitts, W. (1943). A logical calculus of the ideas immanent in nervous activity. Bull Math Biophys. 5:115–133.
13. Rosenblatt, F. (1957). The Perceptron—a Perceiving and Recognizing Automaton. Report 85-460-1. Cornell Aeronautical Laboratory.
14. Dosovitskiy, A., Beyer, L., Kolesnikov, A., Weissenborn, D., Zhai, X., Unterthiner, T., Dehghani, M., Minderer, M., Heigold, G., Gelly, S., Uszkoreit, J., Houlsby, N. (2021). An Image is Worth 16x16 Words: Transformers for Image Recognition at Scale. arXiv Preprint.
15. Dobay, A., Ford, J., Decker, S. et al. (2020). Potential use of deep learning techniques for postmortem imaging. Forensic Sci Med Pathol. 16:671–679.

PART 2

CURRENT DEVELOPMENTS AND FUTURE PERSPECTIVES

CHAPTER 14

THE VIRTOBOT SYSTEM

Lars Ebert and Till Sieberth

DOI: 10.4324/9781003413998-16

14.1 INTRODUCTION

Forensic imaging encompasses a large variety of different image modalities, many of which require either high accuracy or are very repetitive [1]. From the beginning of the Virtopsy project around the year 2000 on, it was part of the vision to integrate and automate as many steps of the imaging procedure as possible. This not only included the automation of CT-guided needle placement, which had been shown to be able to successfully extract tissue samples of areas of interest visible on CT scans, but also the automation of the surface documentation and the analysis of the data generated. This vision of an imaging system was first presented at a meeting at the National Institutes of Health (USA) in 2003. The proposed system included a combined CT and MRI scanner, dedicated to postmortem investigations, that was integrated with a robotic arm for automated image-guided tissue sampling and a surface scanning device to get high-resolution colored scans of the body surface (Figure 14.1). Besides this main scanner, it was envisioned to have equipment to analyze the extracted tissue samples, such as a micro-CT scanner and/or micro-MRI scanner. On top of this, a sophisticated computer system was imagined to be able to automatically analyze the image data and present a preliminary report to the forensic pathologist.

Despite many technological uncertainties at the time, the project was pushed forward. By 2006, all components that were required for such a system were ready. A CT scanner specifically dedicated to postmortem imaging was installed at the Institute of Forensic Medicine in Bern. This system was combined with a heart-lung machine to allow for postmortem CT angiography. A tripod mounted surface scanner for high resolution postmortem surface documentation was adapted for routine use. While the data that was generated by this scanner was useful for creating 3D reconstructions and visualizations, the entire procedure, including preparation, CT scan, digital photos for texturing and the surface scan itself, could take 6 hours and more for supine and prone positions of the body. For image-guided postmortem tissue sampling, a surgical navigation system developed by the Institute of Surgical Technologies and Biomechanics was modified. This system combined a hybrid (active/passive) optical tracking system with a modified biopsy gun equipped with optical markers. In dedicated software, a point within the CT scan could be marked and the system would visually guide the user to the marked location by taking the position and angle of the biopsy gun into account [2]. All these components together demonstrated that an automation of the different components would indeed be beneficial and could help to increase accuracy and decrease investigation times. This led to the development of the Virtobot system, a robotic arm that is mounted to a ceiling rail above the CT scanner and which is able to pick up a variety of different tools to perform different tasks. As of today, two main prototypes have been built, both allowing for automated CT guided needle placement as well as postmortem surface documentation.

Figure 14.1 3D model of a deceased person created by the Virtobot system. Different positions for the photographic images (colored planes) and the surface scans (black-and-white scans) show the positions of the robot system during scanning.

14.1.1 First Prototype

The first prototype of a multi-functional robotic system for use in forensic imaging was dubbed the Virtual Autopsy Robot, or "Virtobot" [3]. The idea was to build a robotic system that could change its end effector and thus be able to conduct different types of tasks. This system had to overcome a variety of obstacles. Needle placement and surface scanning have contradictory requirements when it comes to relative distance to the body and angles. For a surface documentation using a structured light scanner, the distance to the body should be around 70 cm, whereas a biopsy requires the robot top to operate very close to the body, in the scanning field of the CT scanner. Surface scanners and biopsies require the robot to cover a large field of operation, including the entire scanning field of the CT scanner. Being able to access the CT scanning volume is relevant for both biopsy and surface scanning modules. For the biopsy, trajectories are planned using a CT scan, and for surface scanning, it is required that the surface information be fused with the documentation of the inside of the body, i.e. to be able to follow gunshot trajectories through the body. The robot hat to be able to carry the weight of the tools and operate with sufficient accuracy. In addition, one of the requirements was that it should be possible to move the robot out of the way if not needed. To fulfil all the requirements, it was finally decided to go for a ceiling mounted six-axis industrial robot. As a model that would satisfy the reach, accuracy, and load requirements, the Stäubli Tx-90L was selected. It was mounted on a ceiling mounted linear axis, which was aligned with the CT table and reached all the way to the opposing wall, where a tool rack with biopsy module, surface scanner, and needle magazine were attached to the wall. Because of different connection requirements for both tools (pneumatic and electric), it was decided to employ a semi-automatic tool changing mechanism; there the robot would place itself in close distance to the desired module, and the user had to manually use a mechanical coupling and connect cables between robot and end effector, which was a clear limitation of the first prototype.

For surface scanning, an ATOS III (Gesellschaft für optische Messtechnik mbH, Braunschweig, Germany) was selected. The surface documentation procedure consisted of four parts: A full-body CT, placement of coded and uncoded reference markers on the body, manual photography for texturing, and a 3D high-resolution surface scan (Figure 14.2). Due to the automation, the surface scanning procedure could be sped up to 3.5 hours instead of 6, including the required CT scan.

For tissue sampling, a biopsy module is attached to the robot. Dedicated planning software allows for planning of trajectories on CT datasets, defining target points inside the body, entry points on the skin, and the orientation of the end effector. The robot picks up a co-axial introducer needle from a needle rack and places it under the supervision of an active optical tracker. Using a manual approach, it is then possible to take either liquid samples by connecting a syringe or tissue samples by using a biopsy gun and needle, using the cannula of the trocar as a guide. Up to ten positions can be defined, as this is the limit of needles that can be placed in the needle magazine.

In addition, safety equipment in the form of light fences was integrated into the scanning room, stopping any robot movement as soon as someone enters the room while the robot is performing a task. Five different software packages were combined to operate the system. Robot control software provided by the manufacturer allowed for moving the robot axes and calculating kinematics. To incorporate the optical tracker, software called a tool server was used, which allowed reading out processed tracking data through a network interface. The surface scanner was operated with the Atos software provided by the manufacturer. Custom-made software for trajectory planning was developed, which enabled the user to receive a DICOM dataset pushed by the CT scanner workstation and plan trajectories by defining

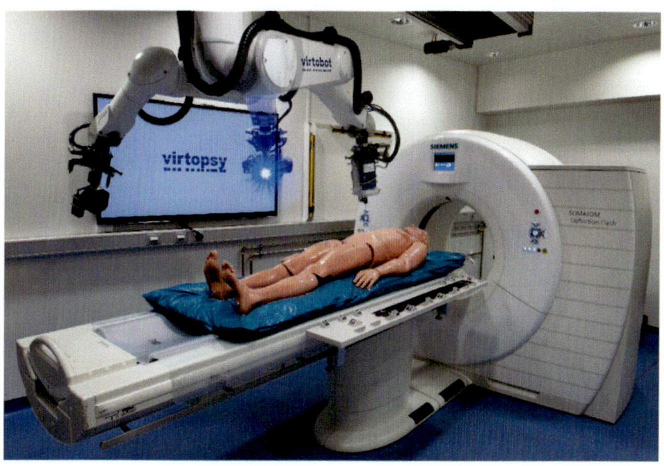

Figure 14.2 Virtobot 2.0 with surface scanner, camera, and biopsy needle introducer tool. The ceiling rail enables the system to access the entire volume that can be scanned by the CT scanner [2, Ebert et al. 2014 Virtobot 2.0].

target points, entry points, and the end effector rotation around the needle axis. Finally, custom-made software was developed to streamline the workflow and combine all the other software packages to work in conjunction, called the Virtopsy Control Center (VCC).

14.1.2 Second Prototype

The second prototype, which is currently in use at the IRM-UZH, was developed from scratch based on the experiences of the first prototype and installed in 2012 (Figure 14.2) [4]. In addition to redevelopment of the software with usability in mind as well as optimization of workflows such as registration, the second prototype features a new end-effector tool for photogrammetry, marker shield boards that can be mounted to the CT table rails, and a fully automatic tool change [5]. In addition, it features additional modules for tissue and liquid sampling (Figure 14.3).

The surface scanning tools are a digital camera and a structured light scanner. Currently we are using a Nikon D800 with a 20-mm fixed focal length lens and a GOM ATOS 5 as surface scanner. Both tools work in conjunction to perform surface scans. In the first step the camera documents 25 predefined positions of target markers applied to the body of the deceased, as well as the surface texture and color. In the second step, the pattern light scanner is also moved over 25 predefined positions and documents the surface structure and 3D shape. The use of photogrammetric dot targets is necessary for high-precision registration between the photographs and surface scans and supports the automation approach.

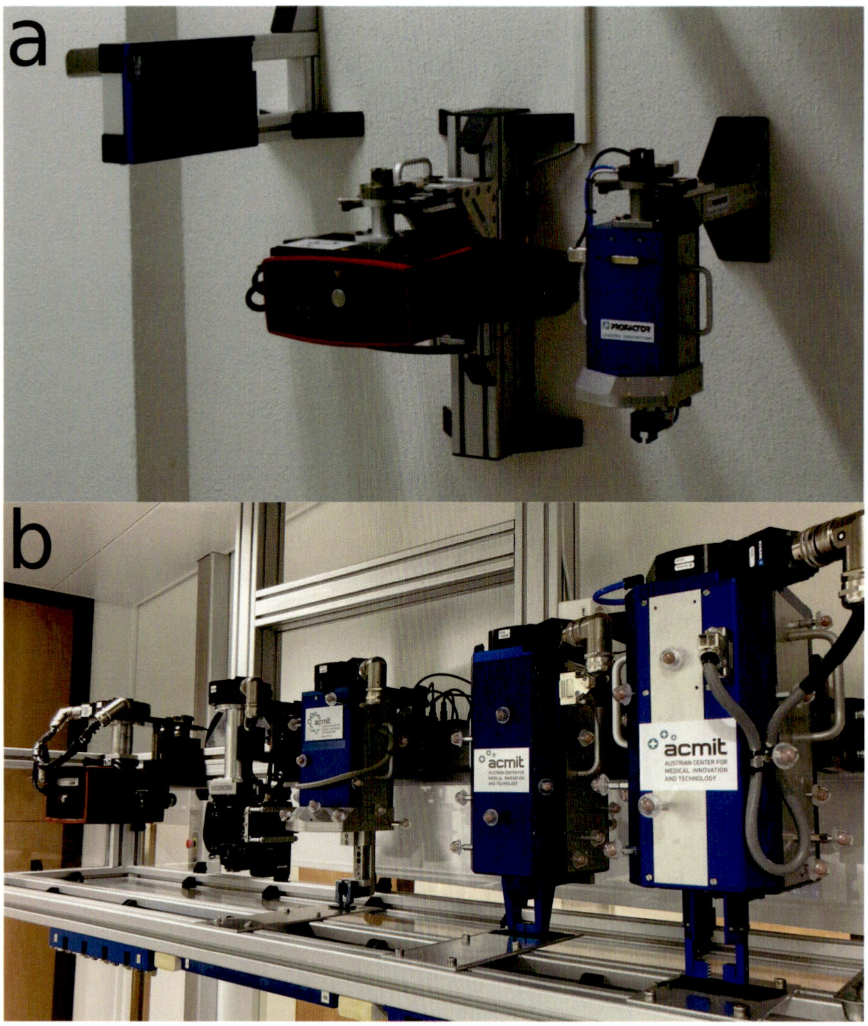

Figure 14.3 a: Toolstand for the first Virtobot prototype. Left to right: Needle magazine, surface scanner, module for placement of introducer needles. b: Toolstand for the second prototype of the Virtobot system. Left to right: surface scanner, digital photocamera, module for placement of introducer needles, module for automated liquid sampling, module for automated tissue sampling. The robot moves into place, and the tool has to be locked manually, including the connection of cables. Under the toolstand, needle magazines for coaxial introducer needles, disposable biopsy guns, and liquid sampling needles are visible. Tools are automatically picked up by the system, including connection of all electric and pneumatic systems.

This could also be solved by an increased number of scanning positions, which in return would increase the time of the scanning procedure, as about a third more scanning positions would be necessary. The drawback is that currently surface markers have to be applied, which is also a time-consuming procedure.

The biopsy tool of the second Virtobot prototype is able to place introducer needles based on previously defined trajectories in CT scans using a newly developed software. Like the first prototype, the biopsy tool can automatically pick up introducer needles from a needle magazine, movie to the planned position and then push the needle to defined position. With the upgraded software, the optimal entry point is automatically selected for any given target point, avoiding collisions with bones or other previously placed needles [6] (Figure 14.4).

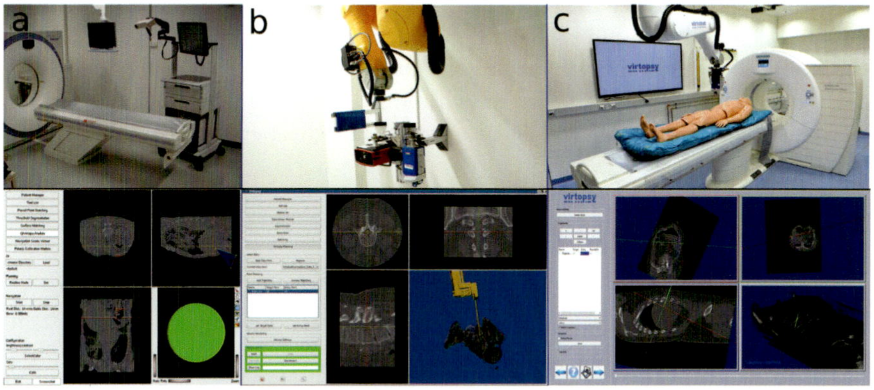

Figure 14.4 Software and hardware evolution. a: Modified surgical navigation system for placement of needles (top). The user has to orient the biopsy gun in such a way that the green circle is placed in the yellow one to hit the target (bottom). b: First prototype of the Virtobot system with surface scanner and biopsy module (top). The software allows for a manual registration and planning of needle trajectories (bottom). c: The second prototype of the Virtobot with biopsy module mounted (top). The software does not require manual registration and allows for semi-automated planning of trajectories (bottom).

14.2 ROUTINE AND RESEARCH

Surface documentation is conducted on a routine basis at the IRM-UZH as an essential procedure for evidence collection in a variety of cases. These include homicides, especially involving knives or firearms, where the surface injury in combination with a CT scan can be used to reconstruct the trajectory of the piercing objects for subsequent incident reconstructions. Furthermore, victims of traffic collisions should be documented, as these often have injuries which can be used to match body and vehicle parts and reconstruct the incident. The last category are pattern injuries that can be potentially matched to injury-causing objects in 3D. Surface documentation procedures are integrated into the standard workflow of the IRM-UZH [7], forming the basis for forensic 3D reconstructions.

In research, the Virtobot system has been used for various studies in forensic toxicology, because it allows for the collection of body fluids and tissue samples prior to autopsy in a minimally invasive way. Just recently it was shown that it is possible to use the Virtobot to automatically sample cerebrospinal fluid from a deceased [8]. Another use for the Virtobot is targeted angiography by injecting contrast fluid directly into the targeted vessels rather than filling the entire vascular system. This technique has the potential to be quicker and cheaper than full PMCT angiography while still being minimally invasive.

14.2.1 Benefits and Limitations of Automation

Using automation can speed up the process of a postmortem surface documentation, and this can also be true for postmortem biopsy, especially for needle placements out of the scanning plane and when multiple needles have to be placed with high accuracy. Being able to extract tissue and liquid samples without physical contact to the body reduces contamination risks. In theory, sampling can be performed by puncturing the needle through the body bag using a sticker to close the hole.

Like all optical surface scanning methods, structured light scanning can have problems with reflective surfaces and hair. This is solved by preparing the body, which means cleaning, drying, and shaving hair that covers up places of interest.

On the downside of automation are the costs of installing and maintaining a complex piece of equipment.

14.3 FUTURE IDEAS

For surface documentation, an upgrade to a GOM Atos Q is currently in planning which will include more scan positions and, due to the faster scanning procedure, will speed up the overall scanning time. In addition to this, the camera will be upgraded, and the data transmission rate between camera and computer will be improved. We expect a scanning time reduction of several minutes per side, further speeding up the documentation process.

Current developments in tissue and liquid sampling introduce two additional tools. Both tools aim at further automating the sampling procedure and thus additionally minimizing contamination. The first will be able to sample liquids fully automatically and place the sample in a rack. The second sampling tool will be able to automatically retrieve tissue samples and, after sampling, deposit them in a sample tube containing either formalin or other sampling solution where the sample is ejected from a needle using ultrasound.

With recent developments in machine learning techniques and the increased application of deep learning approaches on CT datasets, we aim to perform fully automated biopsies of predefined organs and tissues. While the deep learning algorithm provides knowledge about the location and volume of organs, tissues, and liquids, the Virtobot can automatically calculate the best trajectory to sample each tissue necessary for further examination. Neither an engineer nor physician would be necessary to control the Virtobot nor to define the sampling area in the CT scans. In the future, this could be used to automatically collect evidence of everybody, even if the body does not undergo an autopsy.

The automation of tasks such as photogrammetry, surface documentation, biopsy, liquid sampling, or injection of contrast agents can be beneficial to forensic medicine. Automation can reduce contamination risks while increasing accuracy and speed of the intervention, which has been acknowledged by other authors as well [9].

References

[1] Ebert LC, Franckenberg S, Sieberth T, Schweitzer W, Thali M, Ford J, et al. A review of visualization techniques of postmortem computed tomography data for forensic death investigations. Int J Legal Med 2021. https://doi.org/10.1007/s00414-021-02581-4.

[2] Aghayev E, Ebert LC, Christe A, Jackowski C, Rudolph T, Kowal J, et al. CT data-based navigation for postmortem biopsy: A feasibility study. J Forensic Leg Med 2008;15:382–7. https://doi.org/10.1016/j.jflm.2008.02.007.

[3] Ebert LC, Ptacek W, Naether S, Fürst M, Ross S, Buck U, et al. Virtobot: A multi-functional robotic system for 3D surface scanning and automatic post mortem biopsy. Int J Med Robot Comput Assist Surg MRCAS 2010;6:18–27. https://doi.org/10.1002/rcs.285.

[4] Ebert LC, Ptacek W, Breitbeck R, Fürst M, Kronreif G, Martinez RM, et al. Virtobot 2.0: The future of automated surface documentation and CT-guided needle placement in forensic medicine. Forensic Sci Med Pathol 2014. https://doi.org/10.1007/s12024-013-9520-9.

[5] Sekula AD, Thali MJ, Breitbeck R. Improving postmortem surface documentation with a CT mounted marker board. J Forensic Radiol Imaging 2014;2:213–6. https://doi.org/10.1016/j.jofri.2014.06.005.

[6] Ebert LC, Fürst M, Ptacek W, Ruder TD, Gascho D, Schweitzer W, et al. Automatic entry point planning for robotic postmortem CT-based needle placement. Forensic Sci Med Pathol 2016:1–7. https://doi.org/10.1007/s12024-016-9798-5.

[7] Ebert LC, Flach P, Schweitzer W, Leipner A, Kottner S, Gascho D, et al. Forensic 3D surface documentation at the Institute of Forensic Medicine in Zürich—Workflow and communication pipeline. J Forensic Radiol Imaging 2016;5:1–7. https://doi.org/10.1016/j.jofri.2015.11.007.

[8] Franckenberg S, Sieberth T, Frauenfelder T, Thali MJ, Ebert LC. Semiautomated robotic, CT-guided needle placement for postmortem CSF sampling: A Novel application of the Virtobot. Life 2021;14:75–9. https://doi.org/10.1080/26895293.2021.1875056.

[9] Gromniak M, Neidhardt M, Heinemann A, Püschel K, Schlaefer A. Needle placement accuracy in CT-guided robotic post mortem biopsy. Curr Dir Biomed Eng 2020;6. https://doi.org/10.1515/cdbme-2020-0031.

CHAPTER 15

FORENSIC HOLODECK

Till Sieberth and Lars Ebert

DOI: 10.4324/9781003413998-17

15.1 INTRODUCTION

The Forensic Holodeck is a virtual reality (VR) head-mounted display (HMD) setup that has been in development since 2013, first published in 2014 (Figure 15.1) (Ebert et al., 2014). The aim of the Forensic Holodeck is to present forensic data acquired in 3D, such as postmortem computer tomography (PMCT) scans; surface scans of scenes, objects and persons; and 3D reconstructions as they are performed by the 3D Zentrum Zürich. Traditionally, 3D data is provided to state attorneys and judges as 2D renderings, usually printed on paper or displayed on 2D screens. This was identified as a major bottleneck for the understanding of 3D scenes or the spatial relationship between different objects in the scene.

After the first prototype, we continued to develop the Forensic Holodeck system further and had the first real case application in early 2017. Since then, three main applications for VR in forensics with varying complexity could be identified. The first application is the case-related use of virtual reality by 3D experts, and the second is scene walks with forensic experts during the investigation, where a 3D expert and a forensic expert work together on a scene. The third application is the scene walk, which can, for example, happen as part of the court hearings where a witness explains their statement in VR (Figure 15.2).

Figure 15.1 Person wearing VR equipment. Headset and one controller per hand are standard. This can be accompanied by devices that ensure precise tracking of the user's headset and controllers.

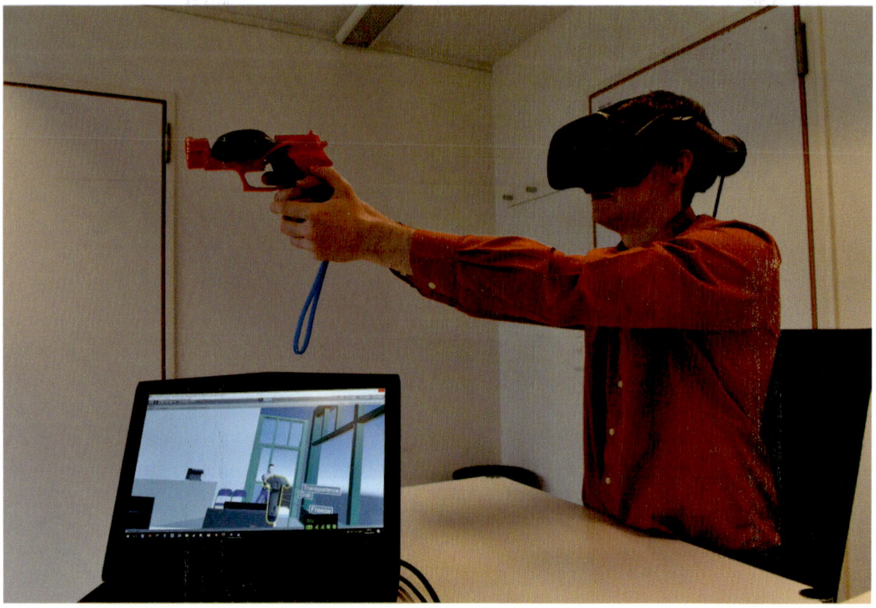

Figure 15.2 Recording of a VR visit. The person is documented from the outside, while the inset shows the view visible to this person.

15.2 EQUIPMENT

There is a variety of equipment for virtual reality visualizations. CAVE systems (Cruz-Neira et al., 1993, 1992) allow for visualization of 3D data through multiple 3D displays around an operator wearing tracked shutter glasses. By tracking the glasses, and therefore the position and orientation of the operator's head, the visualized content on the screen is adjusted to the viewpoint of the operator, allowing for 3D perception of the scene. This system, however, has some drawbacks, like acquisition cost of the hardware, difficulty of transport and size, which limit its use in forensic investigations (Havig et al., 2011). Therefore, the most-used VR systems in forensics are head-mounted displays. HMDs consist of a headset that displays a separate image for each eye and is able to measure its position in space by using gyroscopes, accelerometers and optical tracking. Because these types of HMDs are often used for gaming, they are available off-the-shelf and usually quite affordable and usually come with controllers, that allow for interaction with the virtual scene. Due to the small form factor, they are lightweight and mobile, which makes them feasible for use cases such as courtroom visits. Due to the low costs, it is more viable to purchase multiple HMDs, which can be connected via network, to allow for multi-user applications.

In addition to the VR equipment and the dedicated computer hardware, it is also recommended to add additional documentation hardware such as video cameras, capture devices and audio recorders to the setup (Figure 15.3). This hardware should be employed whenever the VR equipment is used in forensic applications like witness and expert testimony, as this can be relevant for subsequent court proceedings (Sieberth et al., 2021).

In addition, it is possible to provide other modifications to the VR equipment in the form of props. Those could be, for example, attachments to a controller to make it feel more like a certain object. A typical forensic example could be a 3D print of a weapon, which can be attached to the VR controller and has holes for all important buttons, e.g. the trigger (Figure 15.2). However, this is not visible to the VR user but just viewers, although it helps the user to feel not just the controller and its shape but the object they are holding.

Another addition could also be additional tracking equipment. This could be, for example, system-integrated trackers that can be attached to both real and virtual objects, so that the user has a better feeling for objects, their location and orientation. However, many of these trackers can only be used to track an object but do not allow for other interaction such as button pushes. These interactions require dedicated programming and integration in external hardware objects. Other trackers, for example, for fingers or eyes, often come integrated in VR equipment but are often provided by external companies e.g. Leap Motion or Tobii Eye Tracking. These usually also provide programs, plugins and software to integrate the measurements made into the VR environment.

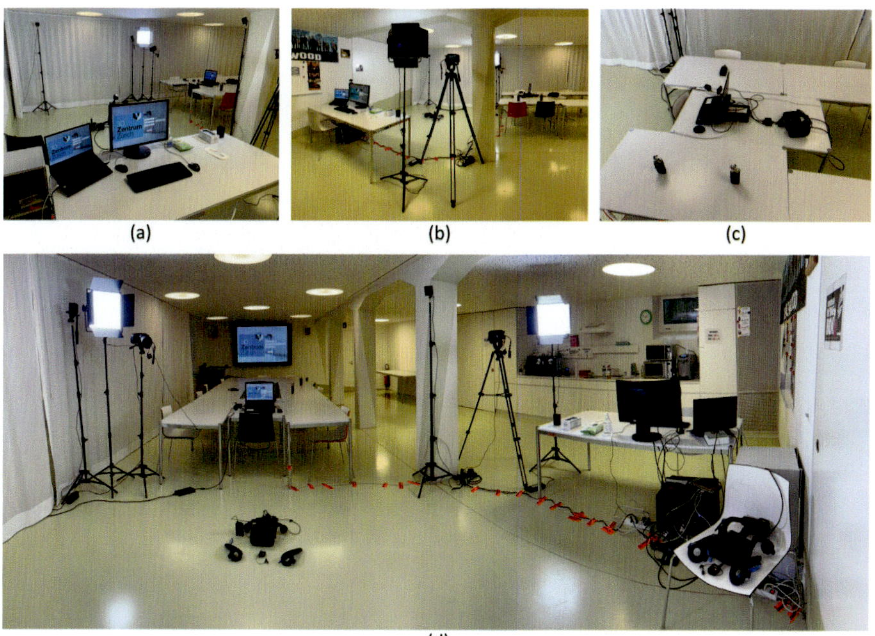

(a) (b) (c) (d)

Figure 15.3 Setup with a variety of additional recording equipment (Sieberth et al., 2021).

15.3 3D EXPERT USE OF VR

One field of application is the use of VR by forensic experts. For this application, VR is used by 3D experts to perform forensic investigations and discuss the case. There are two major examples where this can be applied, the first being weapon to injury matching, which is currently being performed on 2D computer displays; therefore, the third dimension is only perceivable indirectly by moving the viewpoint. This causes the problem that the matching of object and injury usually requires frequent rotation of the viewport and object. Each iteration of the object can be brought closer to the injury until a final match or no-match conclusion can be made. While this can already be a tedious process, it can become more complex with more complex objects or injuries, such as double contours by multiple impacts of the object or rounded surfaces such as imprints after rollovers. Furthermore, it can be necessary to evaluate the object-to-injury position to analyze the position of the injured person and/or attacking person had. Using virtual reality can speed up this process, as the injury-causing object can be moved while perceiving the scene in 3D. This is also the case for more complex scenarios (Figure 15.4). Reconstruction in 3D, however, also has some drawbacks. The fine positioning of objects can be especially challenging due to hand jitter, noise in the tracking signal and absence of support. While it is possible to identify a correct match, fine matching as necessary for court documents is especially difficult. Hence we suggest that VR should currently only be used as a first step and the match later fine-tuned using dedicated 3D software or specialized software that suppresses hand jitter (T. Sieberth et al., 2019).

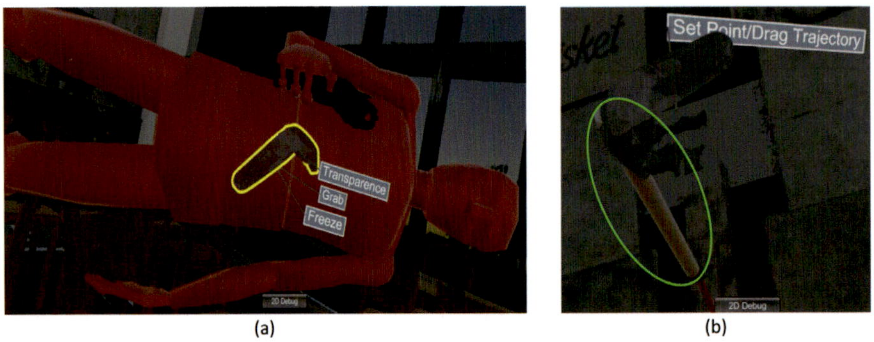

Figure 15.4 Example of Forensic VR use. (a) Interacting with an object. (b) Drawing a shot trajectory (Till Sieberth et al., 2019).

15.4 FORENSIC EXPERT USE OF VR

The next level of complexity is the use of VR for forensic experts. While 3D experts use the VR environment for 3D plausibility checks, other experts can use it to analyze and evaluate the forensic plausibility of events, statements or their own hypotheses. For this, the VR scene can be prepared by a 3D expert and tools to be used by the forensic expert can be introduced. In contrast to those of 3D experts, it should be ensured that the tools are reasonably easy to use to allow the users to focus on their expertise and the forensic facts and interpretation. Also, forensic experts might need different tools in virtual reality, which should be provided. This should be discussed with the experts ahead of a VR visit or based on previous experience for more efficient time use.

Typical examples for this kind of use are analysis of shot trajectories, connecting a victim and hole in a wall and discussing the probable position of a weapon. Another scenario of the use of virtual reality by forensic experts is medical examinations on 3D scans of persons (Leipner et al., 2019, 2016; Michienzi et al., 2018; Sieberth et al., 2020). It has been shown that 3D documentation of persons with injuries is possible and also that these 3D models can be used in virtual reality to perform medical examinations (Koller et al., 2019; Massini et al., 2021).

This use is usually part of the investigation and supports evidence interpretation for the forensic expert. The explanations can be included in a 3D reconstruction and reports. However, this can also lead to a case walk in court or other hearings.

15.5 USE OF VR IN HEARINGS AND COURT

The usage of VR in court and hearings can be performed with forensic experts but also with witnesses or perpetrators (Figure 15.5). They can be questioned and asked to explain their statement in detail, which allows for a more thorough examination and maybe to find contradictions (Norman et al., 2020; T. Sieberth et al., 2019). In Switzerland, these types of scene visits are common and usually conducted at the incident scene. However, it might have been altered, no longer exist anymore or be otherwise difficult to access. These issues can be circumvented by using 3D data and performing the visit in VR. It is important that prior to the event a plan of procedures be developed, which includes, for example, that the scene be presented to the judges prior to the scene visit to ensure that their questions can be answered.

Another possible application in court is to allow the judges or jury to visit the scene and provide them with a better spatial understanding by guiding them through the scene (Süncksen et al., 2019). This might allow for better judgment of the situation and a fair evaluation of the circumstances (Dath, 2017; Reichherzer et al., 2021).

In difference to the previously presented application, use in court requires sound equipment setup and scene programming. While programming mistakes during a use of VR "behind closed doors" can be inconvenient, the negative impact is limited. If they manifest during a hearing, they can significantly delay the proceedings, can be an embarrassment for all involved parties and can ultimately render a technology unpopular and less likely to be used in future hearings.

For this reason, it is important to perform proper debugging in advance of the hearing, to prepare the scene in collaboration with the requesting party and to train VR users in the use of the technology. Also the tools which are to be used in VR should be limited to the essentials to prevent overburdening the user. The area for movement in the VR scene should also be limited to the required area to prevent users from losing orientation.

Figure 15.5 Illustration by Christoph Fischer, VR im Gericht, UZH Magazin, 2, 2018.

15.6 OTHER USES RELATED TO FORENSICS AND FUTURE APPLICATIONS

There is potential for other uses of virtual reality related to forensics. These are using VR for teaching forensic crime scene work or performing forensic autopsies (Balzli et al.). The advantage of these approaches is that specific no-fail situations can be trained without any risk. This helps future forensic experts in their education and can improve forensic work.

There are several other applications in forensics that can be imagined, such as the implementation of eye tracking that could help to create heat maps during witness interrogations, which could help investigators to identify points of high interest and search for missing objects. Furthermore, the implementation of finger tracking could help to improve the use of tools in VR by making it more intuitive and closer to the natural behavior of the user.

There are also other potential applications of VR in a forensic setting. One would be to use VR for confrontation ID situations, where a witness identifies a suspect within a group of similar-looking people. This is currently performed using two separate rooms separated by a one-way mirror and requires significant effort and preparation. This could be made easier by using virtual reality where a 3D scan of the suspect and other people could be presented to the witness for identification purposes.

Another use could be the combination with body tracking. By using this, it would be possible to track and visualize the complete body of a person in VR. This could be used so that persons involved in an incident could be tracked in their complete posture, so that statements about body posture could be checked for plausibility. Another use would be by an actor who during an interrogation plays the victim and takes on the postures required by the witness. It could also be used during the 3D reconstruction phase to better understand positioning or movements necessary to perform certain actions.

15.7 CONCLUDING REMARKS

VR is a very useful technology that has not yet reached its full potential. It can be used in a variety of ways and in different stages of a forensic investigation. With a wide variety of applications, VR can help procedures and allow for better explanations of incidents and better understanding of a scene. However, it should be kept in mind that VR is only an additional tool to improve already existing procedures, as it cannot create any new evidence.

References

Cruz-Neira, C., Sandin, D.J., DeFanti, T.A., 1993. Surround-screen projection-based virtual reality: The design and implementation of the CAVE. Proc ACM SIGGRAPH 93 Conf Comput Graphics 135–142.

Cruz-Neira, C., Sandin, D.J., DeFanti, T.A., Kenyon, R.V., Hart, J.C., 1992. The CAVE: Audio visual experience automatic virtual environment. Communications of the ACM 35, 64–72. https://doi.org/10.1145/129888.129892

Dath, C., 2017. Crime scenes in virtual reality: A user centered study. Independent thesis, Advanced level. KTH, School of Computer Science and Communication (CSC), Media Technology and Interaction Design (MID).

Ebert, L.C., Nguyen, T.T., Breitbeck, R., Braun, M., Thali, M.J., Ross, S., 2014. The forensic holodeck: An immersive display for forensic crime scene reconstructions. Forensic Science, Medicine, and Pathology 10, 623–626. https://doi.org/10.1007/s12024-014-9605-0

Havig, P., McIntire, J., Geiselman, E., 2011. Virtual reality in a cave: Limitations and the need for HMDs? Head- and Helmet-Mounted Displays XVI: Design and Applications 8041, 804107. https://doi.org/10.1117/12.883855

Koller, S., Ebert, L.C., Maria, R., Sieberth, T., 2019. Using virtual reality for forensic examinations of injuries. Forensic Science International 295, 30–35. https://doi.org/10.1016/j.forsciint.2018.11.006

Leipner, A., Baumeister, R., Thali, M.J., Braun, M., Dobler, E., Ebert, L.C., 2016. Multi-camera system for 3D forensic documentation. Forensic Science International 261, 123–128. https://doi.org/10.1016/j.forsciint.2016.02.003

Leipner, A., Obertová, Z., Wermuth, M., Thali, M., Ottiker, T., Sieberth, T., 2019. 3D mug shot—3D head models from photogrammetry for forensic identification. Forensic Science International 300, 6–12. https://doi.org/10.1016/j.forsciint.2019.04.015

Massini, F., Ebert, L., Ampanozi, G., Franckenberg, S., Benz, L., Sieberth, T., 2021. Comparison of superficial wound documentation using 2D forensic photography, 3D photogrammetry, Botscan© and VR with real-life examination. Forensic Science, Medicine, and Pathology. https://doi.org/10.1007/s12024-021-00393-x

Michienzi, R., Meier, S., Ebert, L.C., Martinez, R.M., Sieberth, T., 2018. Comparison of forensic photo-documentation to a photogrammetric solution using the multi-camera system "Botscan". Forensic Science International 288, 46–52. https://doi.org/10.1016/j.forsciint.2018.04.012

Norman, D.G., Wade, K.A., Williams, M.A., Watson, D.G., 2020. Caught virtually lying—crime scenes in virtual reality help to expose suspects' concealed recognition. Journal of Applied Research in Memory and Cognition 9, 118–127. https://doi.org/10.1016/j.jarmac.2019.12.008

Reichherzer, C., Cunningham, A., Coleman, T., 2021. Bringing the jury to the scene of the crime: Memory and decision-making in a simulated crime scene. Conference on Human Factors in Computing Systems—Proceedings. https://doi.org/10.1145/3411764.3445464

Sieberth, T., Dobay, A., Affolter, R., Ebert, L., 2019. A toolbox for the rapid prototyping of crime scene reconstructions in virtual reality. Forensic Science International 305, 110006. https://doi.org/10.1016/j.forsciint.2019.110006

Sieberth, T., Dobay, A., Affolter, R., Ebert, L.C., 2019. Applying virtual reality in forensics: A virtual scene walkthrough. Forensic Science, Medicine, and Pathology 15. https://doi.org/10.1007/s12024-018-0058-8

Sieberth, T., Ebert, L.C., Gentile, S., Fliss, B., 2020. Clinical forensic height measurements on injured people using a multi camera device for 3D documentation. Forensic Science, Medicine, and Pathology 16, 586–594. https://doi.org/10.1007/s12024-020-00282-9

Sieberth, T., Seckiner, D., Dobay, A., Dobler, E., Golomingi, R., Ebert, L., 2021. The forensic holodeck: Recommendations after 8 years of experience for additional equipment to document VR applications. Forensic Science International 329, 111092. https://doi.org/10.1016/j.forsciint.2021.111092

Süncksen, M., Teistler, M., Hamester, F., Ebert, L.C., 2019. Preparing and guiding forensic crime scene inspections in virtual reality. ACM International Conference Proceeding Series 755–758. https://doi.org/10.1145/3340764.3344903

CHAPTER 16

AUGMENTED REALITY

Raffael Golomingi

During an autopsy it can be important to access image data by the attending forensic pathologist. To access a picture archiving and communication system (PACS) in an autopsy room a common setup is a large screen to provide the attending pathologist a clear view and mouse and keyboard as input devices. However, this kind of setup introduces some disadvantages. To control the system the autopsy has to be interrupted and gloves removed so as not to contaminate the input device. Furthermore, the input devices need to be cleaned as well because they could be contaminated with bacteria and other germs, especially in a multiuser environment (Anderson and Palombo 2009). Moreover, findings on cross-sectional images are often difficult to correlate with the actual findings without radiological training. In addition, this setup is limiting in its ability to give an overview to the pathologist due to the distance of the screen and the displayed information from the body.

Augmented reality (AR) is a technique to enhance objects of the real world by digital information. In contrast to virtual reality (VR), which creates an illusion of a new environment with the use of special headsets, AR can be created with the use of different devices. In this chapter, the basic technology of AR and examples of its uses in forensics and forensic medicine will be presented.

In order to create AR, different technologies are used together: Tracking, image registration a display technology and a human interface (Chen et al. 2019). In order to determine the position and orientation of a device to create AR, tracking is needed. Usually this is done with the use of multiple sensors like accelerometers, gyroscopes and optical tracking (Lang et al. n.d.; Zhou, Duh, and Billinghurst 2008). There are two ways to perform optical tracking. It is done either inside-out or outside-in. Inside-out tracking uses integrated optical sensors like cameras to estimate the position and orientation by tracking external features. Outside-in tracking on the other hand, uses outside trackers with markers to measure the position and orientation of the device. These trackers can be active or passive. With the use of infrared projectors, inside-out tracking can be performed relatively easily by just projecting a known pattern into the scene and computing the topology from the changes in the pattern (structured light) (Bell, Li, and Zhang 2016). In order to interact with AR, a human interface is required. This can either be a controller, tracked tools or finger gestures. To track gestures specialized gloves or an infrared structured light scanner could be used (Zhu et al. 2020; Melax, Keselman, and Orsten 2013). (Figures 16.1–16.3)

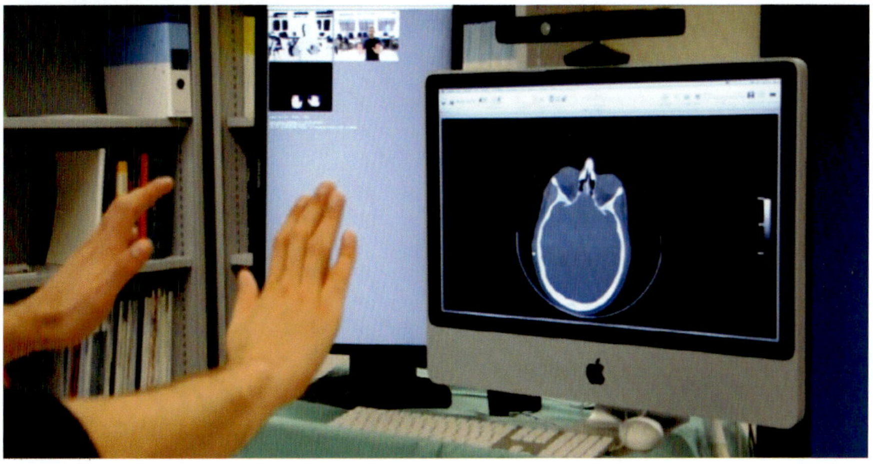

Figure 16.1 How the Kinect can be used.

DOI: 10.4324/9781003413998-18

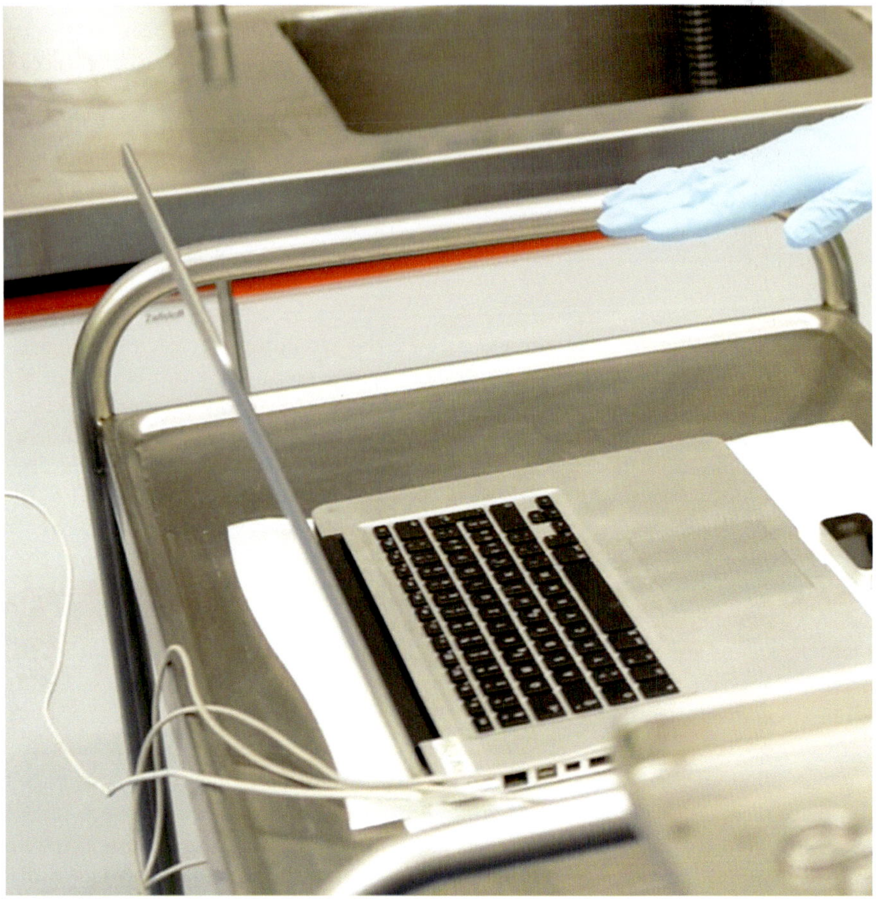

(a)

(b)

Figure 16.2 The leap controller (a) and (b) leap motion controller.

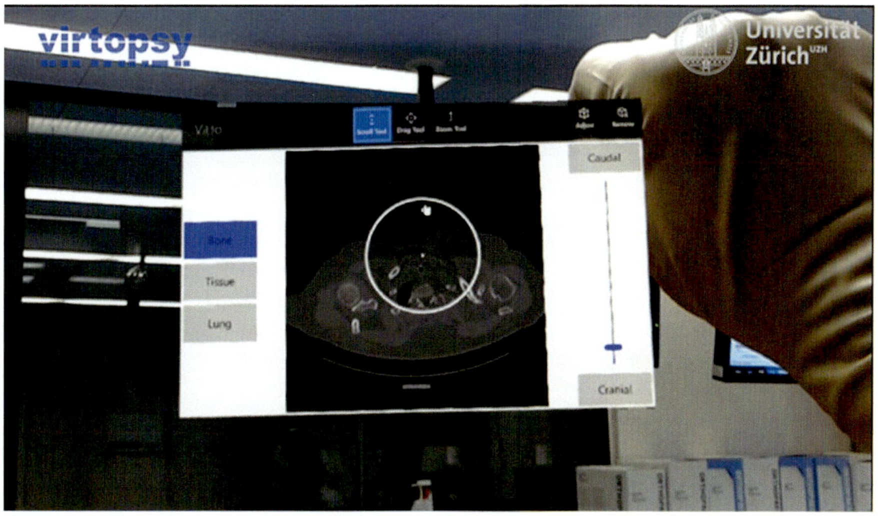

(a)

(b)

Figure 16.3 The HoloLens in action (a) and (b). (a) View from the HoloLens; (b) view from outside.

To render a 3D object with AR in the real world and place it at the right location, registration is required (Chen et al. 2019). Registration allows alignment of two coordinate systems. There are a variety of different algorithms depending on the type of data and the real-time requirements to perform registration (Chen et al. 2019).

In order to see the AR, different display technologies can be used (Figure 16.4). These can be divided into 3D and 2D display techniques. Mobile phones and tablet computers count as 2D displays and are the simplest and therefore most affordable ones. A disadvantage is that they lack depth perception and require at least one hand for holding them. However, with the use of shutter glasses, it is possible to create 3D visualizations (Teistler et al. 2016). Unfortunately this is not available for mobile devices, and therefore the screen stays stationary. More advanced is the head-mounted display (HMD) by Google, called Google Glass (Inc. 2013). This device projects a 2D image into the field of view of the wearer. A more modern HMD device is the Microsoft HoloLens (Figure 16.5). It uses a similar projection technique for both eyes and can therefore create a proper 3D overlay on the scene (Corporation 2015). With the use of external cameras and VR headsets, a similar goal can be achieved by displaying the real augmented scene through the headset displays. The advantage of HMDs over mobile displays is that the user has both hands free and has a more natural experience when interacting with the scene.

Figure 16.4 The Kinect motion controller from Microsoft. This device was initially developed as a motion controller for games on the XBOX 360 but could also be used as an input device for AR applications. Image by Evan-Amos.

Figure 16.5 The HoloLens from Microsoft.

There are already a couple of prototypes to use AR in a forensic setting. AR could be used by a forensic pathologist to access image data with anatomical landmarks in sight. Early prototypes used a 2D or 3D screen in combination with different input devices or gesture control (Teistler et al. 2016; Ebert et al. 2011, 2014, 2013). (Figure 16.3)

However, although that these systems allow for access of images during autopsy, they are not full AR devices. Ebert et al. presented a system with a combination of 2D images with optical tracking and a tracked surgical pointer that allowed defining a reslice plane. A laser line visualized the intersection between the image plane and the body on the real body itself (Ebert et al. 2015). The first and so far only system that uses a mobile display with the use of optical tracking to produce an 3D overlay in forensic medicine was the one by Kilgus et al. (2014). It was made with the use of a range camera mounted on a tracked tablet computer as a display. The images from the tablet were transmitted wirelessly to a dedicated computer, which made the calculations. Affolter et al. presented an AR device that utilized the Microsoft HoloLens as an HMD (2019). Despite not using 3D visualizations, it is able to show a medical image viewer in the field of view of the wearer. The digital image viewer can be placed wherever in the room one can interact with it using hand or finger gestures. This system was also tested for use during autopsy. Although there are some drawbacks for the wearer, like the weight, it appears to be useful in the first half hour of an autopsy (Bulliard et al. 2020).

AR has also been suggested for training in the field of forensics (Kommera, Kaleem, and Shah Harooni 2016; Leung and Blauw 2019). Furthermore, several systems for annotation of crime scenes have been implemented. Gee et al. constructed an AR system which uses a backpack computer with a camera and touch screen and global positioning system (GPS) as well as ultra-wideband (UWB) for tracking (2010). A similar but more lightweight system was presented by Streeferk et al. which uses only a tablet computer with an attached range camera (2013). With both systems, a forensic investigator is able to create a digital incident map of the scenery and annotate the findings for later review.

References

Affolter, Raffael, Sebastian Eggert, Till Sieberth, Michael Thali, and Lars Christian Ebert. 2019. "Applying Augmented Reality during a Forensic Autopsy—Microsoft HoloLens as a DICOM Viewer." *Journal of Forensic Radiology and Imaging* 16 (March). Elsevier BV: 5–8. https://doi.org/10.1016/j.jofri.2018.11.003.

Anderson, Glenn, and Enzo A. Palombo. 2009. "Microbial Contamination of Computer Keyboards in a University Setting." *American Journal of Infection Control* 37 (6). Elsevier BV: 507–9. https://doi.org/10.1016/j.ajic.2008.10.032.

Bell, Tyler, Beiwen Li, and Song Zhang. 2016. "Structured Light Techniques and Applications." *Wiley Encyclopedia of Electrical and Electronics Engineering*, February. John Wiley & Sons, Inc., 1–24. https://doi.org/10.1002/047134608x.w8298.

Bulliard, Jonas, Sebastian Eggert, Garyfalia Ampanozi, Raffael Affolter, Dominic Gascho, Till Sieberth, Michael J. Thali, and Lars C. Ebert. 2020. "Preliminary Testing of an Augmented Reality Headset as a DICOM Viewer during Autopsy." *Forensic Imaging* 23 (December). Elsevier BV: 200417. https://doi.org/10.1016/j.fri.2020.200417.

Chen, Yunqiang, Qing Wang, Hong Chen, Xiaoyu Song, Hui Tang, and Mengxiao Tian. 2019. "An Overview of Augmented Reality Technology." *Journal of Physics: Conference Series* 1237 (June). IOP Publishing: 022082. https://doi.org/10.1088/1742-6596/1237/2/022082.

Corporation, Microsoft. 2015. "Microsoft HoloLens." https://www.microsoft.com/en-us/hololens.

Ebert, Lars C., Gary Hatch, Garyfalia Ampanozi, Michael J. Thali, and Steffen Ross. 2011. "You Can't Touch This." *Surgical Innovation* 19 (3). SAGE Publications: 301–7. https://doi.org/10.1177/1553350611425508.

Ebert, Lars Christian, Thomas D. Ruder, Rosa Maria Martinez, Patricia M. Flach, Wolf Schweitzer, Michael J. Thali, and Garyfalia Ampanozi. 2015. "Computer-Assisted Virtual Autopsy Using Surgical Navigation Techniques." *American Journal of Roentgenology* 204 (1). American Roentgen Ray Society: W58–62. https://doi.org/10.2214/ajr.13.11957.

Ebert, L.C., P.M. Flach, M.J. Thali, and S. Ross. 2014. "Out of Touch—a Plugin for Controlling OsiriX with Gestures Using the Leap Controller." *Journal of Forensic Radiology and Imaging* 2 (3). Elsevier BV: 126–28. https://doi.org/10.1016/j.jofri.2014.05.006.

Ebert, L.C., G. Hatch, M.J. Thali, and S. Ross. 2013. "Invisible Touch—Control of a DICOM Viewer with Finger Gestures Using the Kinect Depth Camera." *Journal of Forensic Radiology and Imaging* 1 (1). Elsevier BV: 10–14. https://doi.org/10.1016/j.jofri.2012.11.006.

Gee, Andrew P., P.J. Escamilla-Ambrosio, Matthew Webb, Walterio Mayol-Cuevas, and Andrew Calway. 2010. "Augmented Crime Scenes." *Proceedings of the 2nd ACM Workshop on Multimedia in Forensics, Security and Intelligence—MiFor '10*. ACM Press. https://doi.org/10.1145/1877972.1877999.

Inc., Google. 2013. "Glass." https://www.google.com/glass/start/.

Kilgus, Thomas, Eric Heim, Sven Haase, Sabine Prüfer, Michael Müller, Alexander Seitel, Markus Fangerau, et al. 2014. "Mobile Markerless Augmented Reality and Its Application in Forensic Medicine." *International Journal of Computer Assisted Radiology and Surgery* 10 (5). Springer Science and Business Media LLC: 573–86. https://doi.org/10.1007/s11548-014-1106-9.

Kommera, Nikitha, Faisal Kaleem, and Syed Mubashir Shah Harooni. 2016. "Smart Augmented Reality Glasses in Cybersecurity and Forensic Education." *2016 IEEE Conference on Intelligence and Security Informatics (ISI)*, September. IEEE. https://doi.org/10.1109/isi.2016.7745489.

Lang, P., A. Kusej, A. Pinz, and G. Brasseur. n.d. "Inertial Tracking for Mobile Augmented Reality." *IMTC/2002. Proceedings of the 19th IEEE Instrumentation and Measurement Technology Conference (IEEE Cat. No.00CH37276)*. IEEE. https://doi.org/10.1109/imtc.2002.1007196.

Leung, Wai Sze, and Frans F. Blauw. 2019. "An Augmented Reality Approach to Delivering a Connected Digital Forensics Training Experience." *Information Science and Applications*, December. Springer Singapore, 353–61. https://doi.org/10.1007/978-981-15-1465-4_36.

Melax, Stan, Leonid Keselman, and Sterling Orsten. 2013. "Dynamics Based 3D Skeletal Hand Tracking." *Proceedings of the ACM SIGGRAPH Symposium on Interactive 3D Graphics and Games—I3D '13*. ACM Press. https://doi.org/10.1145/2448196.2448232.

Streefkerk, Jan Willem, Mark Houben, Pjotr van Amerongen, Frank ter Haar, and Judith Dijk. 2013. "The ART of CSI: An Augmented Reality Tool (ART) to Annotate Crime Scenes in Forensic Investigation." *Lecture Notes in Computer Science*. Springer Berlin Heidelberg: 330–9. https://doi.org/10.1007/978-3-642-39420-1_35.

Teistler, M., G. Ampanozi, W. Schweitzer, P. Flach, M.J. Thali, and L.C. Ebert. 2016. "Use of a Low-Cost Three-Dimensional Gaming Controller for Forensic Reconstruction of CT Images." *Journal of Forensic Radiology and Imaging* 7 (December). Elsevier BV: 10–13. https://doi.org/10.1016/j.jofri.2016.06.001.

Zhou, Feng, Henry Been-Lirn Duh, and Mark Billinghurst. 2008. "Trends in Augmented Reality Tracking, Interaction and Display: A Review of Ten Years of ISMAR." *2008 7th IEEE/ACM International Symposium on Mixed and Augmented Reality*, September. IEEE. https://doi.org/10.1109/ismar.2008.4637362.

Zhu, Minglu, Zhongda Sun, Zixuan Zhang, Qiongfeng Shi, Tianyiyi He, Huicong Liu, Tao Chen, and Chengkuo Lee. 2020. "Haptic-Feedback Smart Glove as a Creative Human-Machine Interface (HMI) for Virtual/Augmented Reality Applications." *Science Advances* 6 (19). American Association for the Advancement of Science (AAAS): eaaz8693. https://doi.org/10.1126/sciadv.aaz8693.

CHAPTER 17

THE VIRTOSCAN PROJECT

Sören Kottner and Dominic Gascho

DOI: 10.4324/9781003413998-19

17.1 INTRODUCTION

The VirtoScan project was started in 2015 at the Zürich Institute of Forensic Medicine (IRM) of the University of Zürich. The project is dedicated to whole-body spectral imaging and research and development of automated imaging systems. The aim of the project is to design imaging systems for facilities related to forensic medicine and to enhance and facilitate whole-body imaging for forensic documentation.

17.2 EXTERNAL BODY DOCUMENTATION & SURFACE DOCUMENTATION

A forensic documentation that exclusively focuses on the exterior of the body is generally referred to as "external body documentation" or "surface documentation". Its focus lies on the documentation of directly accessible and exposed parts of the human body. Generally, external body documentation aims to capture and preserve information regarding colors and textures on the surface of the body. The goal is to capture the current condition (i.e., as-is state) of the body in a two-dimensional (2D) or three-dimensional (3D) representation. Regions of interest during external body documentation are often associated with injuries on the surface of the body. In this context, findings usually address the integrity of the skin (e.g., open wounds, amputations, etc.), discolorations (e.g., hematomas, pattern injuries, livor mortis, etc.), foreign substances on the skin, and objects attached to the body (e.g., injury-causing objects, weapons, medical installations, etc.). However, findings may also address the condition of the skin. Depending on the circumstances of the case, scar tissue, dermatological conditions, state of decomposition, and body modifications (e.g., tattoos) are examples of information that could be relevant to the legal investigation.

Common methods for documentation are drawings and sketches in combination with voice recordings and written texts, photographs, and 3D scans. Photographs are nowadays arguably the most common and most available method to document and preserve information in this context. Especially with the increasing availability of digital cameras such as smartphone cameras, photo documentation of a body can be carried out almost everywhere and at any time. 3D documentation of a body usually requires additional software tools or even supplementary imaging hardware. 3D body documentations, therefore, generally demand a greater amount of work and might require additional equipment. This is why 3D documentations are not regularly performed for every case and may not yet be applicable to all forensic institutes around the world.

In order to perform an external body documentation that can be used in line with a legal investigation, certain documentation procedures and protocols need to be followed [1]. Overall, it is important that the documentation (i.e. the 2D and 3D optical imaging) is carried out in a technically correct manner and exhibits perfect image and data quality [2–8]. The objective of the VirtoScan project is to address these issues and to develop imaging solutions that allow for fast, cost-efficient, and easy-to-use 3D whole-body imaging.

17.3 A SHORT REVIEW OF OPTICAL 3D IMAGING IN FORENSIC MEDICINE

The use of optical 3D imaging, specifically the use of photogrammetry and its application for medico-legal investigations, started somewhere around the late 1990s. In 1997, Brüschweiler et al. [9] presented a methodology to match injuries with injury-causing objects using single-camera photogrammetry. Since then, the application of optical 3D imaging within the field of forensic medicine has been reported in various studies and case reports [10–28]. Over the years, different setups and methodologies were presented. Apart from a few studies, the most commonly used techniques for injury and whole-body documentation were generally based on either passive triangulation techniques such as close-range photogrammetry [18–21,23,25–27,29–31], or active triangulation techniques such as structured light scanning [11,12,15–17,22,24,32,33]. Apart from single- and multi-camera photogrammetry setups, tripod-based and handheld 3D scanners [12,15–17,22,24] as well as fully automated, robotic 3D scanning systems were introduced [34,35].

17.4 THE IDEA OF THE VIRTOSCAN PROJECT

At the IRM, optical 3D imaging is performed on selected cases. The indication for optical 3D imaging depends on the condition of the body and the background story of the case. Optical 3D imaging is usually carried out when the body exhibits visible defects and damage to the surface of the body that can contribute to further analysis such as 3D reconstructions. These cases are often related to violent crimes and accidents. The preselection of cases is necessary, due to the fact that the current workflow routine at the IRM takes about 3 hours for a complete 3D documentation of a body [35]. In general, the longer a specific examination procedure occupies the body, the more likely it is that this procedure is used less frequently for routine work. This is also true for optical 3D imaging. To address this issue and to reduce the duration of the 3D imaging process, the VirtoScan project was launched. One of the first goals was to develop an optical imaging system that can perform 3D whole-body imaging and reduce the overall examination time.

17.4.1 Optical 3D Imaging in Combination with CT Scanning

In 2015, the first VirtoScan prototype was built and tested at the IRM [20,21]. The prototype was built to perform optical 3D imaging in combination with X-ray computed tomography (CT). Optical 3D imaging was based on multicamera close-range photogrammetry. The overall setup was constructed using a wooden frame and a set of seven digital single-lens reflex (DSLR) cameras (EOS 100D, Canon Inc., Tokyo, Japan). Each camera was equipped with a 50-mm lens (Canon EF 50 mm 1:1.8 II, Canon Inc.) and a remote shutter release (Yongnou RF-603C II, YongNuo Photographic Equipment Co., Ltd., Shenzhen, China). For the imaging process, the setup was placed in front of the CT gantry. In conjunction with the CT scanner, consecutive images of the entire body were captured while the CT table was moved in a step-by-step motion through the CT gantry. An overview of the setup in front of the CT gantry is given in Figure 17.1. With the use of this prototype, photogrammetric datasets of one side of the body were acquired within 3 to 4 minutes. The 3D reconstruction of the photogrammetric dataset was carried out subsequently while the body was already free to undergo the following examinations. In addition to the fast image acquisition, the combination with the CT scanner allowed the alignment and merging of the optical 3D data with the corresponding CT data.

The resulting 3D model from a whole-body scan of a woman that suffered second- and third-degree scalding after a domestic accident is given in Figure 17.2. A detailed description of the setup, the imaging workflow, and the results can be found in Kottner et al. 2016 [20] and Kottner et al. 2017 [21].

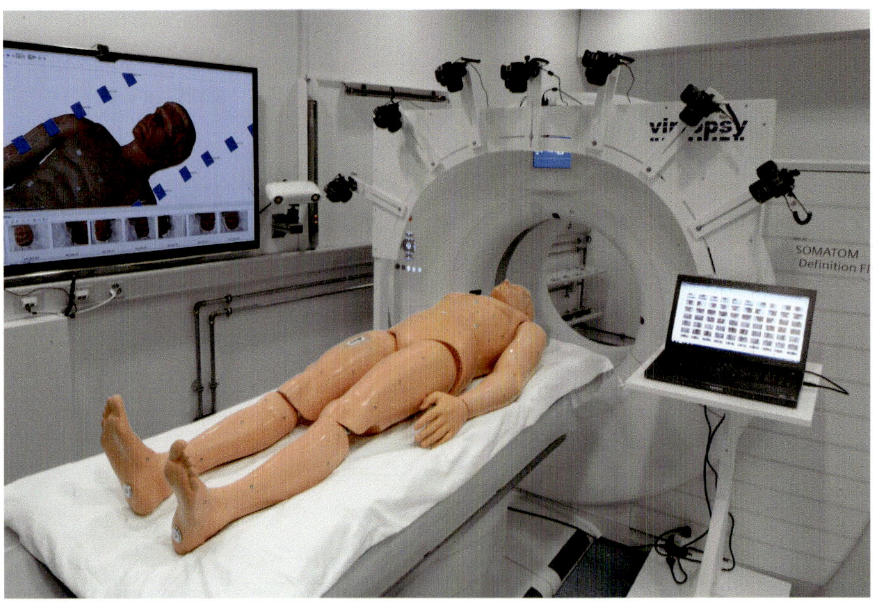

Figure 17.1 The first prototype of the VirtoScan project. Multicamera setup in conjunction with a medical X-ray computed tomography scanner at the Zürich Institute of Forensic Medicine at the University of Zürich.

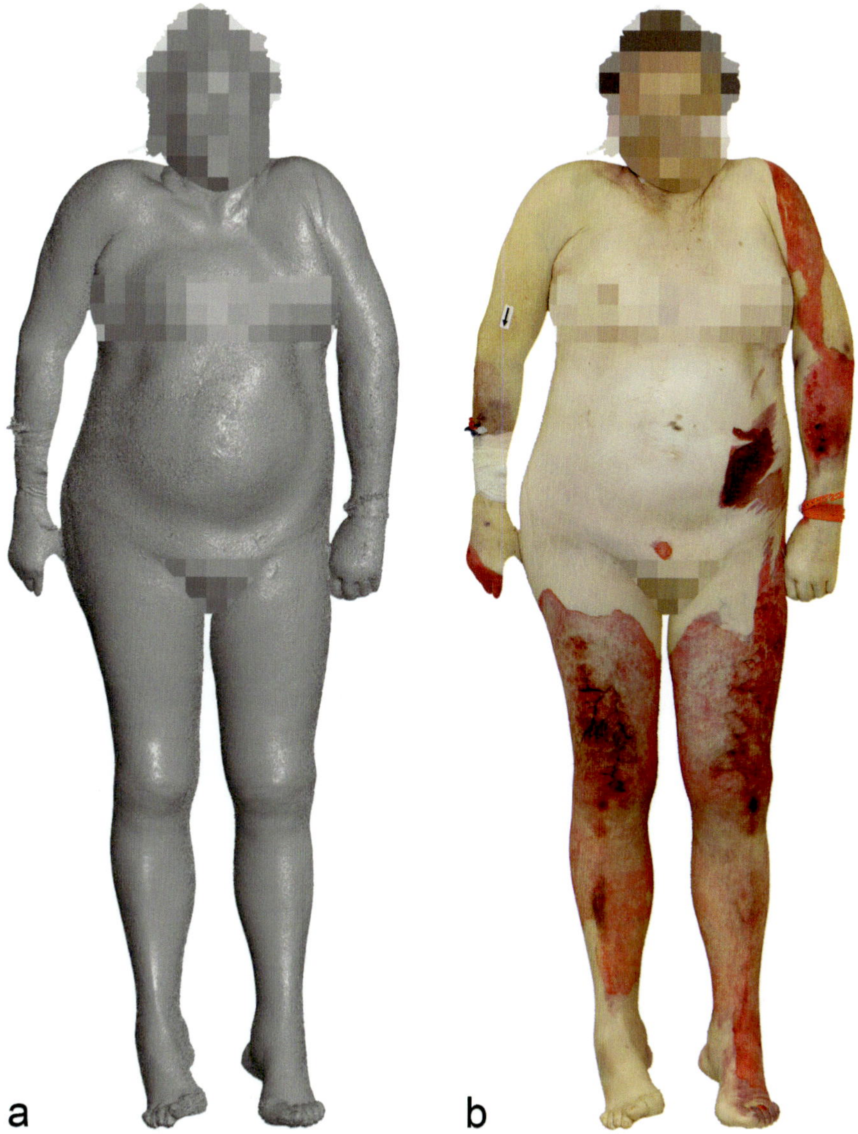

Figure 17.2 Reconstructed 3D model of a body displaying large area defects on the surface due to second- and third-degree scalding. Comparison of the 3D polygonal mesh representation (a) and its textured counterpart (b).

17.4.2 Automated Optical 3D Imaging in Autopsy Facilities

The combination of optical 3D data with radiological data (e.g. from CT scans) is commonly used for 3D reconstructions of crime and accident scenes [15–17,24]. It allows relating findings from the external body documentation with internal and anatomical information from the radiological data. Although CT scans at the IRM are performed on every body that is delivered to the institute, traditional autopsies are carried out as well. Autopsy examinations can confirm findings based on radiological data and allow obtaining additional information based on odor, color, and texture, which cannot be captured and visualized with radiological imaging modalities. Findings during autopsy examinations are commonly documented using photography.

To automate and extend the documentation procedure during autopsy examinations a new imaging setup in the frame of the VirtoScan project was developed. The setup called VirtoScan-on-Rails was implemented in 2018 at the autopsy facilities at the IRM. The construction and design of the setup were based on our first prototype. The hardware setup comprises three main components: a linear motion and lifting unit, an image-capturing unit, and an operating unit for user inputs [25]. The linear motion and lifting unit was mounted to the ceiling and carries the frame of the image-capturing unit. Furthermore, it enables to move the image-capturing unit on two parallel linear guide rails across the room. The image-capturing unit was equipped with 10 DSLR cameras (EOS 100D, Canon Inc., Tokyo, Japan) using 50-mm lenses (Canon EF 50 mm 1:1.8 II, Canon Inc.). The operating unit consists of a control panel and a hand-held enabling switch to select, start and stop predefined imaging protocols. The protocol for the image acquisition was adapted from our previous prototype. The photogrammetric dataset for a body is captured by moving the image-capturing unit in a step-by-step motion across the autopsy table. In between each step, the multicamera array is triggered automatically. This allows capturing a full set of images from head to toe in a prone or supine position. An overview of the setup is given in Figure 17.3.

Image acquisition for one side of the body takes about 2 minutes and can be carried out on two different autopsy tables. Optical 3D imaging within the autopsy facility enables documenting bodies at different stages during the external and internal body examination. This way, the state of a body can be recorded directly after the arrival at the institute, after cleaning, and during the internal examination (i.e., autopsy). An example of a 3D reconstruction before and during an autopsy procedure is given in Figure 17.4.

The rapid image acquisition of the VirtoScan-on-Rails enables the incorporation of whole-body 3D imaging into routine work at autopsy facilities. Furthermore, the automated workflow enables forensic pathologists and morgue technicians with a basic knowledge of photography to perform whole-body 3D imaging in an objective and reproducible manner. A detailed description of the setup, the imaging workflow, and the results can be found in Kottner et al., 2019 [25].

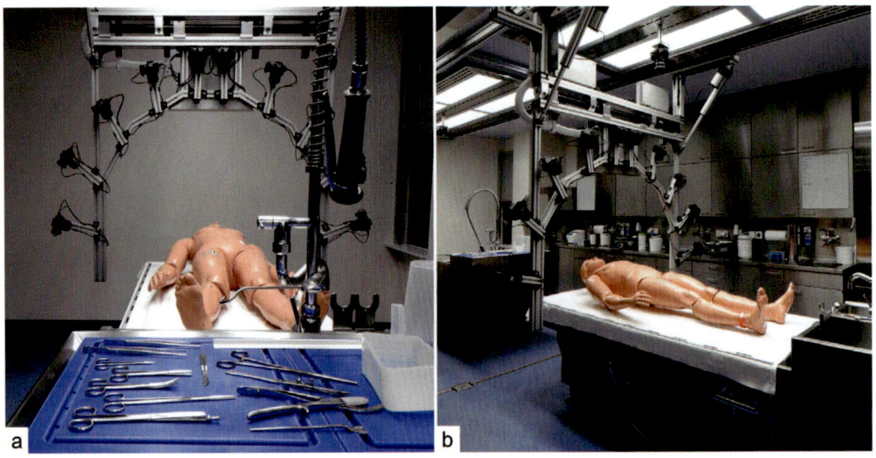

Figure 17.3 VirtoScan-on-Rails at the autopsy facility of the Zürich Institute of Forensic Medicine at the University of Zürich.

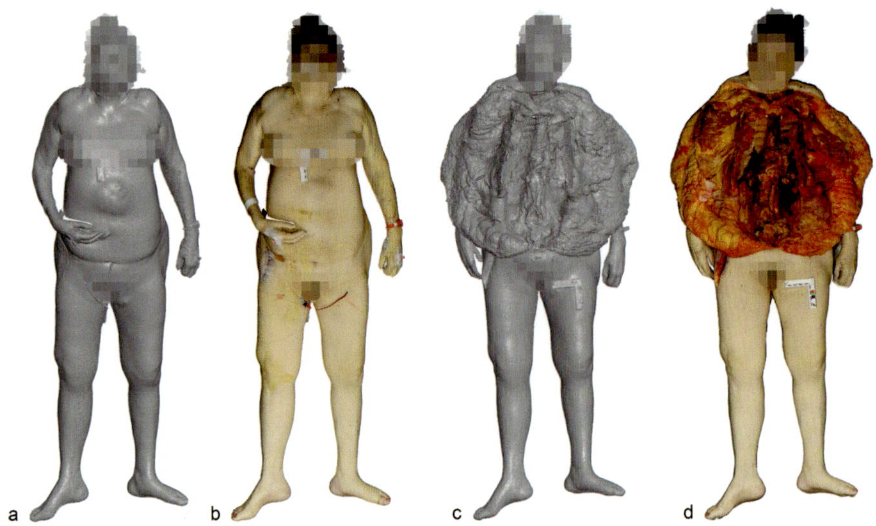

Figure 17.4 Reconstructed 3D model of a body before and during an autopsy procedure. Images a and c illustrate the 3D polygonal mesh of the reconstructed data, whereas images b and d refer to the corresponding view of the textured 3D model.

17.4.3 Rapid and Color Accurate 3D Imaging in Combination with CT Scanning

From the first references of optical 3D imaging applied to forensic medicine until today, optical 3D imaging is almost always used to analyze pattern injuries and related wound morphologies based on the color information from the reconstructed 3D data. Although there are approaches, which assessed wound morphologies and spatial relationships solely based on 3D polygonal data [36] the quality of the color information on the reconstructed and textured 3D models is usually of greater importance to the following reconstructive analysis. The quality of the color representation in the reconstructed 3D data among the commercially available 3D scanners however varies. Additionally, the lighting conditions under which optical 3D imaging is performed have a significant influence on the color representation of the captured data. Certain lighting conditions may positively contribute to and enhance the perceptibility of injuries while others may impair it. To address this issue a new imaging setup in the frame of the VirtoScan project, was developed. The design of the imaging setup was adapted and improved from our two previous systems. The frame was constructed of aluminum profiles and was equipped with custom-made mounting elements for the camera and light modules. Overall, eight industrial R-G-B area scan cameras (Apex AP-3200T-PGE, JAI A/S, Copenhagen, Denmark) were used in combination with eight 5600 K LED light sources (Lykos Daylight MLL1500-D, Manfrotto, The Vitec Group plc, Richmond, United Kingdom). The imaging setup was built to acquire images in conjunction with a medical CT scanner. The added light sources allowed reducing the exposure times of the cameras. This way, image acquisition was carried out at the scanning speed of the CT scanner, and allowed to perform image acquisition during CT scanning. An overview of the setup is given in Figure 17.5.

This setup allows capturing external and internal information of a body in 72 seconds. The prism-based cameras record the color information for red, green, and blue on separate complementary metal–oxide–semiconductor (CMOS) sensors and allow for a higher color representation in the images. The added LED light sources feature a high color rendering index and guarantee a consistent light distribution on the body. This setup combines scanning speed with a high-quality color representation and enables to align and merge the 3D surface information with the radiological data from the CT scan. An example of a reconstructed 3D model of a female with several stab wounds is given in Figure 17.6.

17.4.4 Multispectral 3D Imaging in Combination with CT Scanning

The use of optical 3D imaging has been mainly focused on the documentation of color information based on the visible part of the electromagnetic spectrum (i.e., visible light). Visible light corresponds to the spectral range of the electromagnetic spectrum that is visible to the human eye. Although two-dimensional multispectral imaging such as ultraviolet (UV) photography and near-infrared (NIR) photography are established tools to

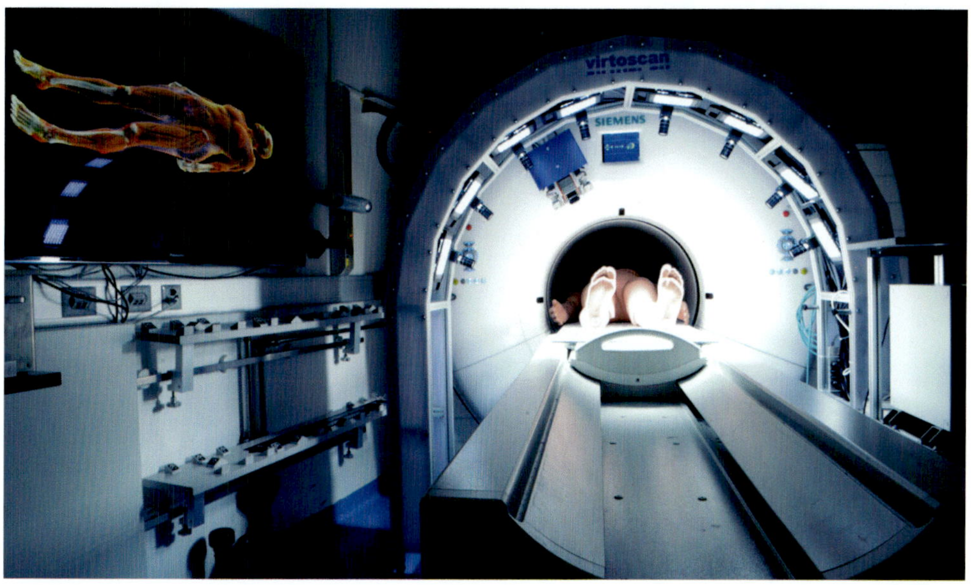

Figure 17.5 The latest VirtoScan prototype in conjunction with the medical X-ray computed tomography scanner at the Zürich Institute of Forensic Medicine at the University of Zürich.

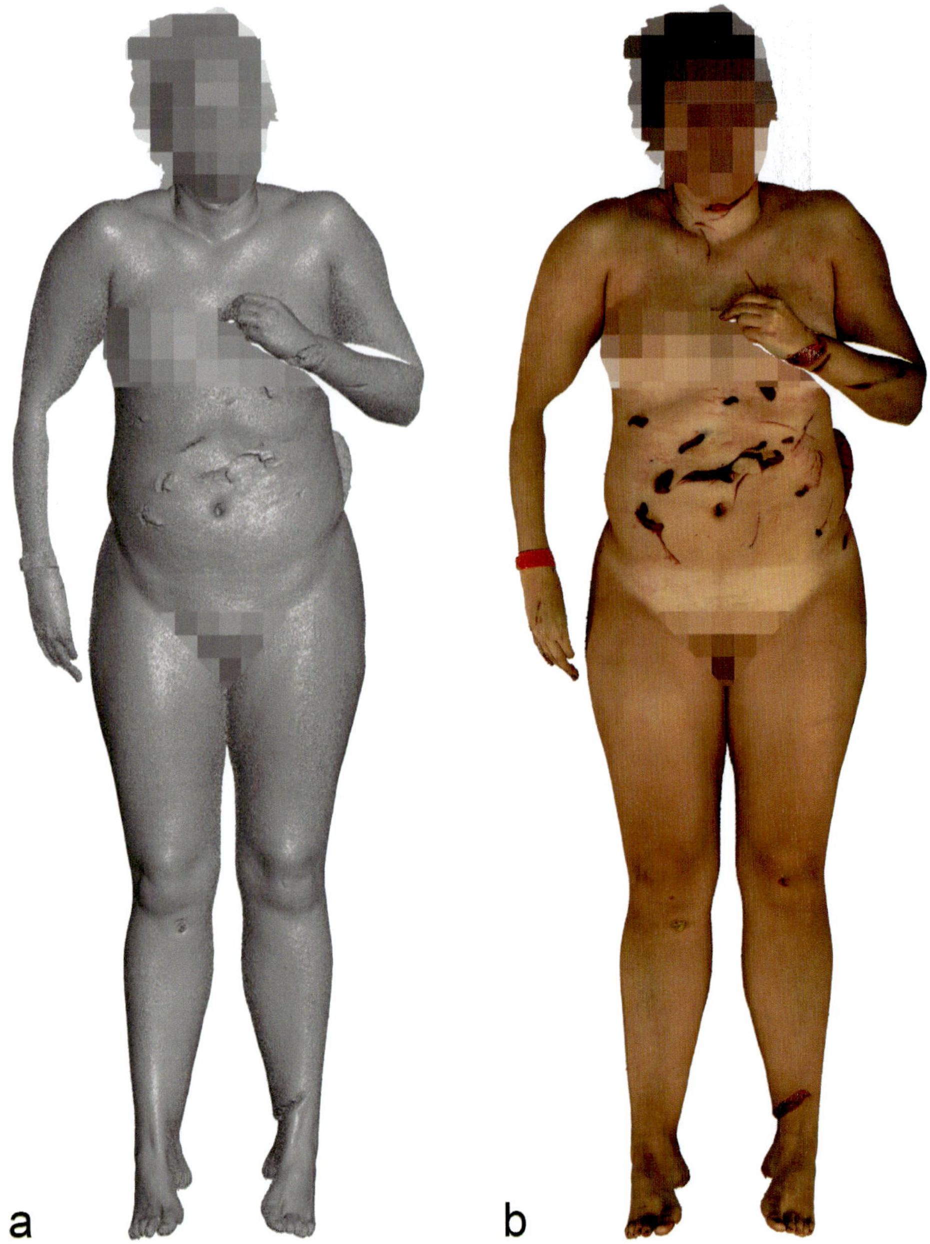

Figure 17.6 Reconstructed 3D model of a body with several stab wounds. 3D polygonal mesh of the reconstructed data (a) and the corresponding view of the textured 3D model (b).

detect, visualize, and document evidence that is otherwise invisible to the naked eye [26,27], it seems that 3D multispectral imaging is neither well known nor commonly applied in the field of forensic medicine. UV and NIR radiation is adjacent to the visible part of the electromagnetic spectrum. UV radiation is characterized by shorter wavelengths (100–400 nm), while NIR radiation is characterized by longer wavelengths (780–1400 nm) [37]. An application that is most commonly associated with multispectral photography within the field of forensic medicine is the detection and visualization of latent bruises and injuries on the body [26,27]. Additionally, multispectral photography is commonly used for the detection and documentation of gunshot residue and bodily fluids on textiles and skin [26,27]. To address the absence of studies regarding multispectral 3D imaging of injuries and whole bodies, a prototype called VirtoScan Multispectral was developed. The setup was based on the

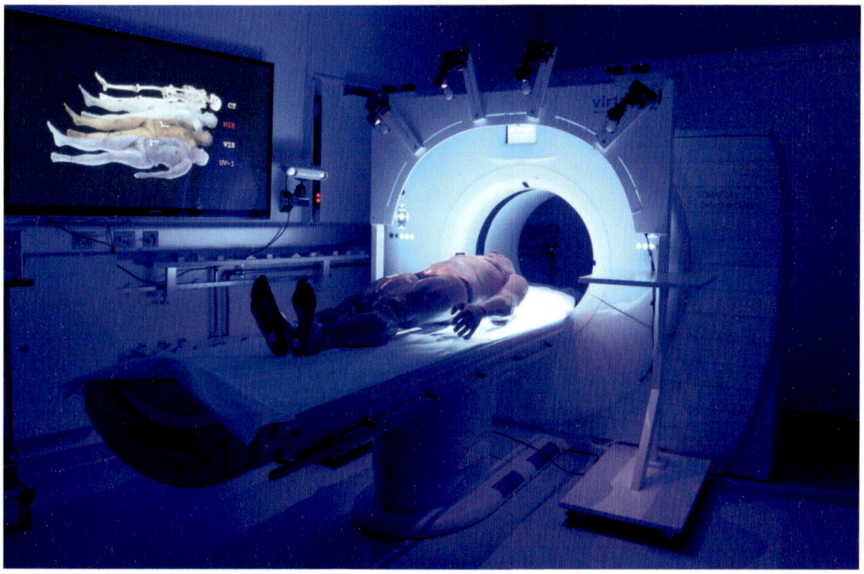

Figure 17.7 3D multispectral imaging setup in conjunction with the medical X-ray computed tomography scanner at the Zürich Institute of Forensic Medicine at the University of Zürich.

Adapted from Kottner et al. 2021 [26]

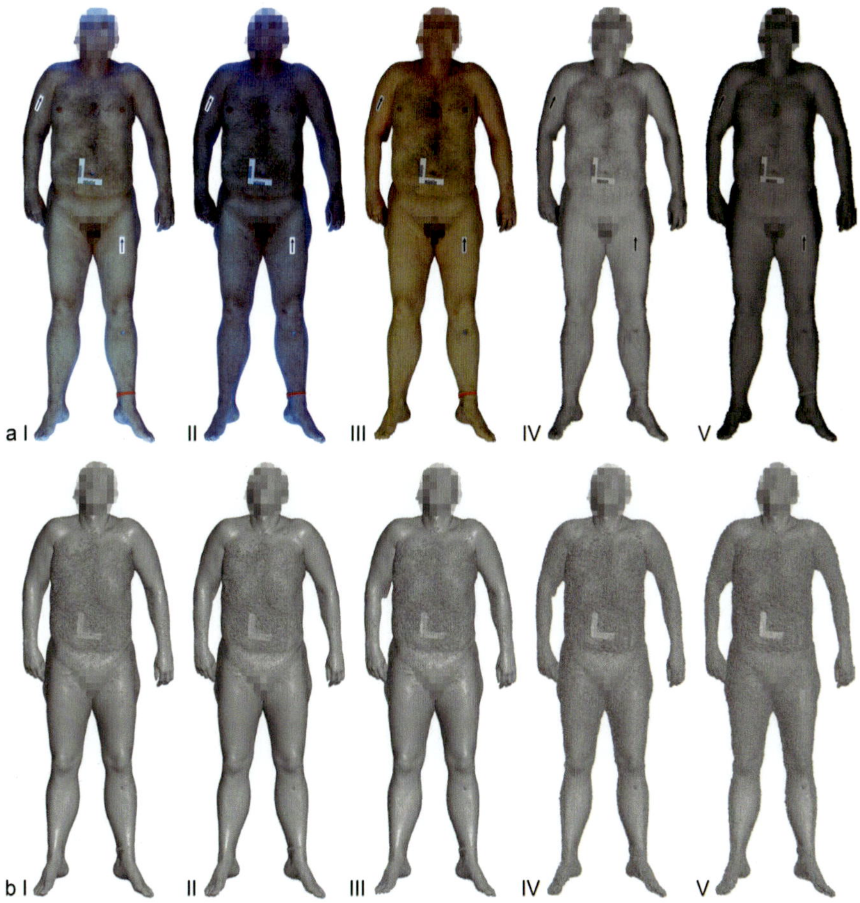

Figure 17.8 Reconstructed textured 3D models (a I-V) and their corresponding 3D polygonal meshes (b I-V) from a multispectral dataset of one body. Image a I refers to the photogrammetric dataset under UV illumination at 365 nm in combination with a yellow filter with 50% transmission at 460 nm. Image a II refers to the photogrammetric dataset under UV illumination at 400 nm in combination with a yellow filter with 50% transmission at 460 nm. The photogrammetric dataset captured during visible light using a IR-neutralization filter (i.e., hot mirror) is represented by image a III. The remaining images of a IV and a V refer to the photogrammetric datasets under NIR illumination at 860 nm and 960 nm, respectively. NIR images were acquired in combination with a longpass filter with 50% transmission at 850 nm. Images b I to b V refer to the corresponding 3D polygonal meshes.

Adapted from Kottner et al. 2021 [26].

multicamera frame from our first prototype. The setup was equipped with four modified DSLR cameras (EOS 200D, Canon Inc., Tokyo, Japan), 40-mm lenses (EF 40 mm f/2.8 STM, Canon Inc), and remote shutter controls (Yongnou RF-603C II, YongNuo Photographic Equipment Co., Ltd.). Camera modification involved the removal of the original camera filters in front of the sensor and the replacement by a multispectral glass filter (transmission range: 190–4000 nm). A photo of the 3D multispectral imaging setup in combination with the CT scanner is given in Figure 17.7.

Multiple sets of whole-body photogrammetry, using different combinations of light sources and lens filters, were used to acquire a 3D multispectral dataset of a body. The imaging procedure was adapted from our first prototype and was based on step-by-step image acquisition. An example of a 3D reconstruction based on multispectral photogrammetry of a body is given in Figure 17.8. 3D multispectral imaging extends standard optical 3D imaging and allows one to detect, visualize and document forensic evidence that lies beyond the visible spectrum and cannot be perceived by the human eye. The multicamera setup based on four modified DSLR cameras presents a simple and inexpensive multispectral imaging setup that can be used semiautomatically in conjunction with a medical CT scanner to perform 3D whole-body documentation of external and internal findings in human bodies. A detailed description of the setup, the imaging workflow, and the results can be found in Kottner et al. 2021 [26] and Kottner et al. 2021 [27].

References

[1] S. Kottner, P.M. Flach, D. Gascho, G. Ampanozi, M. Thali, L.C. Ebert, Communicating 3D data—interactive 3D PDF documents for expert reports and scientific publications in the field of forensic medicine, International Journal of Legal Medicine. 134 (2020) 1175–1183. https://doi.org/10.1007/s00414-019-02156-4.

[2] D. Dolinak, E.W. Matshes, E.O. Lew, Forensic pathology: Principles and practice, Elsevier/Academic Press, Amsterdam; Boston, 2005.

[3] R.J. Riviello, ed., Manual of forensic emergency medicine: A guide for clinicians, Jones and Bartlett Publishers, Sudbury, MA, 2010.

[4] O. Ozkalipci, M. Volpellier, Photographic documentation, a practical guide for non professional forensic photography, Torture. 20 (2010) 45–52.

[5] M. Grassberger, K. Yen, E. Türk, eds., Klinisch-forensische Medizin: interdisziplinärer Praxisleitfaden für Ärzte, Pflegekräfte, Juristen und Betreuer von Gewaltopfern, Springer, Wien, 2013.

[6] M.A. Verhoff, A. Gehl, M. Kettner, K. Kreutz, F. Ramsthaler, Digitale forensische Fotodokumentation, Rechtsmedizin. 19 (2009) 369–381. https://doi.org/10.1007/s00194-009-0626-z.

[7] M.A. Verhoff, M. Kettner, A. Lászik, F. Ramsthaler, Digital photo documentation of forensically relevant injuries as part of the clinical first response protocol, Deutsches Ärzteblatt International. (2012). https://doi.org/10.3238/arztebl.2012.0638.

[8] S. Evans, S. Baylis, R. Carabott, M. Jones, Z. Kelson, N. Marsh, J. Payne-James, J. Ramadani, P. Vanezis, A. Kemp, Guidelines for photography of cutaneous marks and injuries: A multi-professional perspective, Journal of Visual Communication in Medicine. 37 (2014) 3–12. https://doi.org/10.3109/17453054.2014.911152.

[9] W. Brüschweiler, M. Braun, H.J. Fuchser, R. Dirnhofer, Photogrammetrische Auswertung von Haut- und Weichteilwunden sowie Knochenverletzungen zur Bestimmung des Tatwerkzeuges—grundlegende Aspekte, Rechtsmedizin. 7 (n.d.) 76–83. https://doi.org/10.1007/BF03042360.

[10] M.J. Thali, M. Braun, W. Brüschweiler, R. Dirnhofer, Matching tire tracks on the head using forensic photogrammetry, Forensic Science International. 113 (2000) 281–287. https://doi.org/10.1016/S0379-0738(00)00234-6.

[11] J. Subke, H.-D. Wehner, F. Wehner, S. Szczepaniak, Streifenlichttopometrie (SLT): A new method for the three-dimensional photorealistic forensic documentation in colour, Forensic Science International. 113 (2000) 289–295. https://doi.org/10.1016/S0379-0738(00)00236-X.

[12] M.J. Thali, M. Braun, R. Dirnhofer, Optical 3D surface digitizing in forensic medicine: 3D documentation of skin and bone injuries, Forensic Science International. 137 (2003) 203–208. https://doi.org/10.1016/j.forsciint.2003.07.009.

[13] M.J. Thali, M. Braun, W. Brueschweiler, R. Dirnhofer, 'Morphological imprint': Determination of the injury-causing weapon from the wound morphology using forensic 3D/CAD-supported photogrammetry, Forensic Science International. 132 (2003) 177–181. https://doi.org/10.1016/S0379-0738(03)00021-5.

[14] M.J. Thali, M. Braun, Th.H. Markwalder, W. Brueschweiler, U. Zollinger, N.J. Malik, K. Yen, R. Dirnhofer, Bite mark documentation and analysis: The forensic 3D/CAD supported photogrammetry approach, Forensic Science International. 135 (2003) 115–121. https://doi.org/10.1016/S0379-0738(03)00205-6.

[15] U. Buck, S. Naether, M. Braun, S. Bolliger, H. Friederich, C. Jackowski, E. Aghayev, A. Christe, P. Vock, R. Dirnhofer, M.J. Thali, Application of 3D documentation and geometric reconstruction methods in traffic accident analysis: With

high resolution surface scanning, radiological MSCT/MRI scanning and real data based animation, Forensic Science International. 170 (2007) 20–28. https://doi.org/10.1016/j.forsciint.2006.08.024.

[16] U. Buck, S. Naether, B. Räss, C. Jackowski, M.J. Thali, Accident or homicide—Virtual crime scene reconstruction using 3D methods, Forensic Science International. 225 (2013) 75–84. https://doi.org/10.1016/j.forsciint.2012.05.015.

[17] J. Tschui, N. Feddern, N. Schwendener, L. Campana, S. Utz, M. Schweizer, C. Jackowski, W.D. Zech, When the prey gets too big: An uncommon road accident involving a motorcyclist, a car and a bird, International Journal of Legal Medicine. 130 (2015) 463–467. https://doi.org/10.1007/s00414-015-1188-0.

[18] P. Urbanová, P. Hejna, M. Jurda, Testing photogrammetry-based techniques for three-dimensional surface documentation in forensic pathology, Forensic Science International. 250 (2015) 77–86. https://doi.org/10.1016/j.forsciint.2015.03.005.

[19] M.M. de Sainte Croix, D. Gauld, A.H. Forgie, R. Lowe, Three-dimensional imaging of human cutaneous forearm bite marks in human volunteers over a 4 day period, Journal of Forensic and Legal Medicine. 40 (2016) 34–39. https://doi.org/10.1016/j.jflm.2016.02.003.

[20] S. Kottner, L.C. Ebert, G. Ampanozi, M. Braun, M.J. Thali, D. Gascho, A Mobile, Multi camera setup for 3D full body imaging in combination with postmortem computed tomography procedures, in: Proceedings of the 7th International Conference on 3D Body Scanning Technologies, Lugano, Switzerland, 30 Nov.–1 Dec. 2016, Hometrica Consulting—Dr. Nicola D'Apuzzo, Lugano, Switzerland, 2016: pp. 53–60. https://doi.org/10.15221/16.053.

[21] S. Kottner, L.C. Ebert, G. Ampanozi, M. Braun, M.J. Thali, D. Gascho, VirtoScan—a mobile, low-cost photogrammetry setup for fast postmortem 3D full-body documentations in x-ray computed tomography and autopsy suites, Forensic Science, Medicine and Pathology. 13 (2017) 34–43. https://doi.org/10.1007/s12024-016-9837-2.

[22] J. Sivanandan, E. Liscio, Assessing structured light 3D scanning using Artec Eva for injury documentation during autopsy, Journal of the Association for Crime Scene Reconstruction. 21 (2017) 5–13.

[23] G.J. Edelman, M.C. Aalders, Photogrammetry using visible, infrared, hyperspectral and thermal imaging of crime scenes, Forensic Science International. 292 (2018) 181–189. https://doi.org/10.1016/j.forsciint.2018.09.025.

[24] U. Buck, K. Buße, L. Campana, F. Gummel, C. Schyma, C. Jackowski, What happened before the run over? Morphometric 3D reconstruction, Forensic Science International. 306 (2020) 110059. https://doi.org/10.1016/j.forsciint.2019.110059.

[25] S. Kottner, S. Schaerli, M. Fürst, W. Ptacek, M. Thali, D. Gascho, VirtoScan-on-Rails—an automated 3D imaging system for fast postmortem whole-body surface documentation at autopsy tables, Forensic Sci Med Pathol. 15 (2019) 198–212. https://doi.org/10.1007/s12024-019-00095-5.

[26] S. Kottner, M.M. Schulz, F. Berger, M. Thali, D. Gascho, Beyond the visible spectrum—applying 3D multispectral full-body imaging to the VirtoScan system, Forensic Sci Med Pathol. 17 (2021) 565–576. https://doi.org/10.1007/s12024-021-00420-x.

[27] S. Kottner, M.M. Schulz, F. Berger, M.J. Thali, D. Gascho, Multispectral 3D whole-body imaging of dressed and undressed bodies in combination with postmortem X-ray computed tomography, in: Proceedings of 3DBODY.TECH 2021—12th International Conference and Exhibition on 3D Body Scanning and Processing Technologies, Lugano, Switzerland, 19-20 October 2021, Hometrica Consulting—Dr. Nicola D'Apuzzo, Lugano, Switzerland, 2021. https://doi.org/10.15221/21.24.

[28] J. Heimer, D. Gascho, R. Odermatt, N. Zoelch, S. Kottner, M.J. Thali, V. Chatzaraki, Full virtual autopsy in a case of a suicidal transthoracic gunshot injury, Forensic Imaging. 21 (2020) 200368. https://doi.org/10.1016/j.fri.2020.200368.

[29] A. Leipner, R. Baumeister, M.J. Thali, M. Braun, E. Dobler, L.C. Ebert, Multi-camera system for 3D forensic documentation, Forensic Science International. 261 (2016) 123–128. https://doi.org/10.1016/j.forsciint.2016.02.003.

[30] C. Villa, Forensic 3D documentation of skin injuries, International Journal of Legal Medicine. 131 (2017) 751–759. https://doi.org/10.1007/s00414-016-1499-9.

[31] C. Villa, M.J. Flies, C. Jacobsen, Forensic 3D documentation of bodies: Simple and fast procedure for combining CT scanning with external photogrammetry data, Journal of Forensic Radiology and Imaging. 12 (2018) e2–e7. https://doi.org/10.1016/j.jofri.2017.11.003.

[32] A. Shamata, T. Thompson, Using structured light three-dimensional surface scanning on living individuals: Key considerations and best practice for forensic medicine, Journal of Forensic and Legal Medicine. 55 (2018) 58–64. https://doi.org/10.1016/j.jflm.2018.02.017.

[33] A. Shamata, T. Thompson, Documentation and analysis of traumatic injuries in clinical forensic medicine involving structured light three-dimensional surface scanning versus photography, Journal of Forensic and Legal Medicine. 58 (2018) 93–100. https://doi.org/10.1016/j.jflm.2018.05.004.

[34] L.C. Ebert, W. Ptacek, S. Naether, M. Fürst, S. Ross, U. Buck, S. Weber, M. Thali, Virtobot—a multi-functional robotic system for 3D surface scanning and automatic post mortem biopsy, International Journal of Medical Robotics and Computer Assisted Surgery. 6 (2010) 18–27. https://doi.org/10.1002/rcs.285.

[35] L.C. Ebert, W. Ptacek, R. Breitbeck, M. Fürst, G. Kronreif, R.M. Martinez, M. Thali, P.M. Flach, Virtobot 2.0: The future of automated surface documentation and CT-guided needle placement in forensic medicine, Forensic Science, Medicine and Pathology. 10 (2014) 179–186. https://doi.org/10.1007/s12024-013-9520-9.

[36] E. Röhrich, M. Thali, W. Schweitzer, Skin injury model classification based on shape vector analysis, BMC Med Imaging. 12 (2012) 32. https://doi.org/10.1186/1471-2342-12-32.

[37] CIE JTC 8, CIE S 017/E:2020 ILV: International Lighting Vocabulary, 2nd Edition, International Commission on Illumination (CIE), n.d. https://doi.org/10.25039/S017.2020.

CHAPTER 18

IMAGING-RING FORENSICS—A MOBILE CT SCANNER WITH INTEGRATED CAMERAS FOR SIMULTANEOUS 3D SURFACE DOCUMENTATION

Dominic Gascho

DOI: 10.4324/9781003413998-20

18.1 INTRODUCTION

At the heart of the "virtual autopsy" is computed tomography (CT), a technology that makes it possible to look inside the body without a single incision. Whether as an extension of an external examination or as a supplement to a conventional autopsy, CT offers a special forensic role that is becoming increasingly important [1–4]. The ability to visualize the CT data set in any oblique view with various three-dimensional (3D) reconstruction options and to preserve this detailed reconstruction for future analyses has established CT scanning in forensic medicine [5,6]. For forensic investigations, particularly in cases of homicide or pedestrian accidents, the ideal approach combines CT scans with detailed surface documentation of the body [7]. Special scanning systems such as the VirtoScan were developed for the purpose of forensic surface documentation (see the chapter on the VirtoScan Project in this book). For the CT examinations, standard clinical CT scanners are used and scan and reconstruction parameters are adapted for optimized postmortem imaging [6]. Such clinical CT scanners, while effective, are often prohibitively expensive for many institutions. Purchase costs, service contract costs, costs for the technical equipment of the examination room (power supply, water cooling) and the radiation protection add up. Some institutions transport the deceased to a clinical radiology facility outside normal operating hours and scan there. Other institutions purchase whole-body X-ray scanners as an alternative, but these only provide 2D X-ray projections of the deceased, so the diagnostic value of this form of postmortem imaging is considerably less than that of 3D CT. Other alternatives are therefore being sought that offer 3D CT data of the entire body within an affordable financial framework.

This chapter introduces the ImagingRing, a cost-effective mobile imaging robot capable of acquiring both conventional 2D X-rays and cone-beam CT data, all while boasting built-in cameras for computer vision. Through a research collaboration with the manufacturer, it was feasible to develop a specialized protocol for postmortem forensic whole-body CT examinations. This, combined with an innovative approach utilizing the scanner's cameras for photogrammetric documentation of the entire body surface, opens up exciting possibilities for forensic investigations.

18.2 WHERE DID THE IMAGINGRING SYSTEM COME FROM?

The ImagingRing is an imaging system that was originally developed for image-guided radiation therapy and received its CE clearance in January 2015. In December 2016, the ImagingRing system received its FDA 510(k) clearance. It was developed by medPhoton (www.medphoton.at, Salzburg, Austria), a company that develops modern, high-end technologies for the application in radiotherapy [8]. The ImagingRing is available in various designs for movement on the floor and on the ceiling as well as a mobile unit. This chapter refers exclusively to the mobile version of the ImagingRing. This versatile imaging system also found application in image-guided brachytherapy and is currently being marketed for this application by Elekta (Stockholm, Sweden). Its potential soon caught the attention of the intraoperative surgery field, leading to its acquisition by Brainlab (Munich, Germany) and a rebranding as Loop-X for the application in surgery.

18.3 WHAT EXACTLY IS THIS VERSATILE IMAGING SYSTEM?

The ImagingRing is a medical imaging system that enables both 2D X-ray and 3D cone-beam CT imaging and can be easily controlled via a tablet (Figure 18.1). The independently movable X-ray source and detector arms are mounted on an ultra-large gantry with an internal diameter of around 120 cm. The gantry can be tilted, and in combination with slight horizontal rotations during data acquisition, so-called wobble scans can be performed, which allows one to reduce image artifacts. In the case of metallic implants or foreign bodies, reconstructions can be performed using a metal artifact reduction (MAR) algorithm. The gantry is also equipped with an infrared tracking system for surgical navigation. Furthermore, there are camera systems in the gantry and the detector. These camera systems are prepared for stereoscopic computer vision applications and can also be used for monitoring during the scan.

In the following, the X-ray source, the detector system, and the stereoscopic optical camera systems are described in more detail, although it should be noted that the components and functions are constantly being further developed or may even be replaced by other, improved components in the future.

The X-ray source (IAE RTM 780H) provides a tube voltage range of 40 to 120 kV and a CT scan can be generated in continuous mode (tube current range: 0.2–8 mA) or, for higher dose scans, in pulsed mode (tube current range with small focal spot: 5–30 mA, with large focal spot: 40–150 mA). A dual-energy scan is possible in pulsed mode. The imaging system also offers a fluoroscopy mode. In addition, dose modulation can be performed during acquisition. The ImagingRing operates with input voltages of 230 V (+/– 10%), 50–60 Hz, 2 kVA (4 kVA Peak), main fuse 13 A or 10 A (configurable) or 120 V (+/– 10%), 50–60 Hz, 2 kVA (4 kVA Peak), main fuse 25 A or 20 A (configurable). But it is also battery powered maneuverable. The cooling system consists of a mono-tank filled with dielectric oil and a fan-assisted cooling control system.

The detector (Varex XRD 4343RF) consists of an amorphous silicon flat panel. Although the detector size offers only a maximum 2D field of view (FOV) of 25.4 × 25.4 cm^2 and a maximum in-plane diameter of 25.1 cm for the 3D CT scan, the system is not limited to these FOV sizes because the independent rotation and swiveling capability of both the detector and X-ray source arm enables non-isocentric imaging. This offers two key advantages: (1) positioning a desired anatomical region outside the gantry's isocenter into the scan's isocenter for optimal image quality and (2) combining two scans with different isocenters into an extended field-of-view through image stitching. This means that an in-plane diameter of 80 cm (ultra-large FOV) is feasible, even if only an in-plane diameter of around 50 cm (large FOV) is currently applicable. The voxel size reaches 0.5 mm for this large FOV but even an isotropic spacing 0.15 mm for a very small FOV with an in-plane diameter of <6 cm.

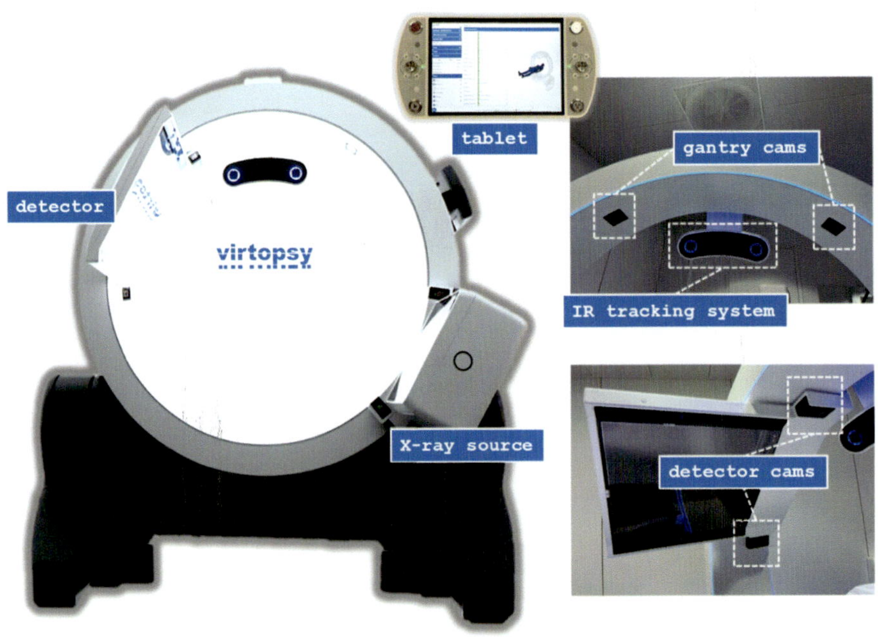

Figure 18.1 The ImagingRing consists of a mobile, ultra-large gantry with independently movable detector and X-ray arms. It can be controlled wirelessly with a tablet. In addition, the ImagingRing has an infrared (IR) tracking system and two optical cameras that are integrated into the gantry and the detector.

The gantry contains two CMOS-sensor cameras model ELP-SUSB1080P01-LC1100 with a resolution of 1080p (50 fps), and the detector contains two CMOS-sensor cameras model ELP-USBFHD01M-L36 with a resolution of 1080p (30 fps). The cameras are connected to a Raspberry Pi 4 model B (64 bit quad-core, dual display support, max. 2160p [60 fps]). The gantry cameras are positioned at a horizontal distance of 712.77 mm. They are inclined at a horizontal angle of 55° and a vertical angle of 43.64° towards the central axis of the gantry. The distance between the detector cameras measures 589 mm. They are inclined at an angle of 57.02° to the center axis of the detector, but are also inclined at an angle of 35.52° away from the detector surface and thus directed towards the scan center line.

18.3.1 What Does ImagingRing Forensics Mean?

The research collaboration between the manufacturing company and the authors of this book chapter is referred to as ImagingRing Forensics. The name comes from the aim of adapting this system for use in postmortem forensic imaging with a focus on whole-body CT with simultaneous photogrammetric surface documentation by the integrated cameras.

The ImagingRing already offers examination protocols for various anatomical regions. Although these were programmed for image-guided radiotherapy and surgical use, they can also be adapted for targeted postmortem examinations or for the radiologic identification of decedents (RadID [9,10]) (Figure 18.2).

Previously, the system lacked a protocol for scanning entire bodies. Our research collaboration developed and implemented such a protocol for postmortem whole-body CT examinations. For this purpose, an examination table was constructed that allows the mobile ImagingRing to scan over a body length of about 220 cm. In principle, the scan length is only limited by the distance between the table supports. The table consisted of a hard foam board (L: 2400 mm, W: 500 mm, H: 30 mm) mounted on two glass-fiber reinforced plastic tubes (L: 3500 mm, W: 50 mm, THK: 5 mm), which were mounted on two wooden trestles. The walls of the examination room, which did not have the required concrete thickness, were covered with lead foil in accordance with radiation protection regulations. Owing to the reduced radiation dosage emitted by the ImagingRing compared to a standard clinical computed tomography (CT) scanner, a thinner and consequently less expensive lead shielding could be employed.

For the planning of the CT scan, a 2D topogram (scout view) protocol was implemented for which the user defines the start and end positions in order to cover the entire body. While the imaging ring moves caudo-cranially, an anterior-posterior topogram is produced and while the imaging ring moves back to its starting position, a lateral topogram is produced. The CT scan length and the size of the in-plane FOV can be planned on the basis of these two orthogonal whole-body topograms. The whole-body CT datasets are generated by performing overlapping sequential scans (Figure 18.3).

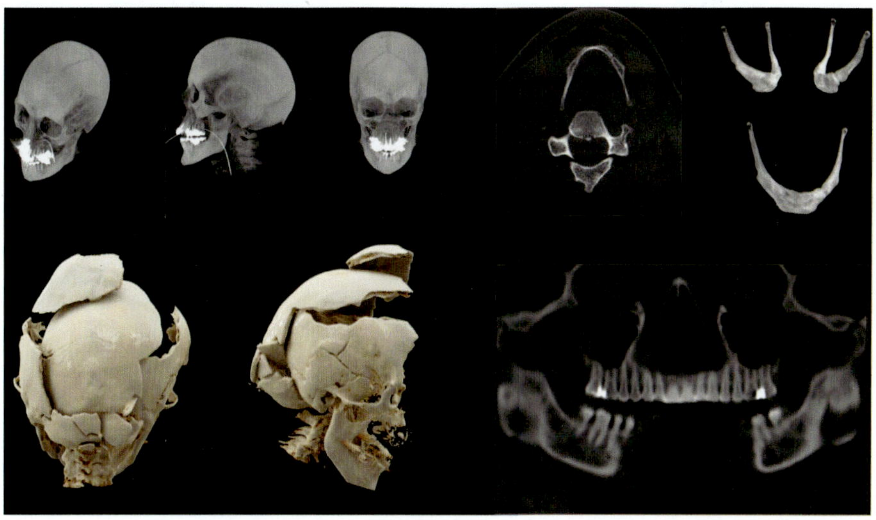

Figure 18.2 The CT protocols provided by the system for individual anatomical regions over a limited scan length can be adapted for forensic postmortem examinations. As examples, maximum intensity projections of a postmortem CT scan of the skull are shown here at the top left, highlighting the metallic dental implants and the intubation tube. An in situ representation of severe skull fractures using cinematic renderings is illustrated at the bottom left. A postmortem CT scan of the larynx with 3D volume renderings of the hyoid bone reconstructed from this CT data set are depicted at the top right. A panoramic reconstruction (curved multi-planar reformation) of the upper row of teeth with the implants shown brightly is presented at the bottom right. CT reconstructions such as this one are performed for comparison with antemortem dental images for radiological identification.

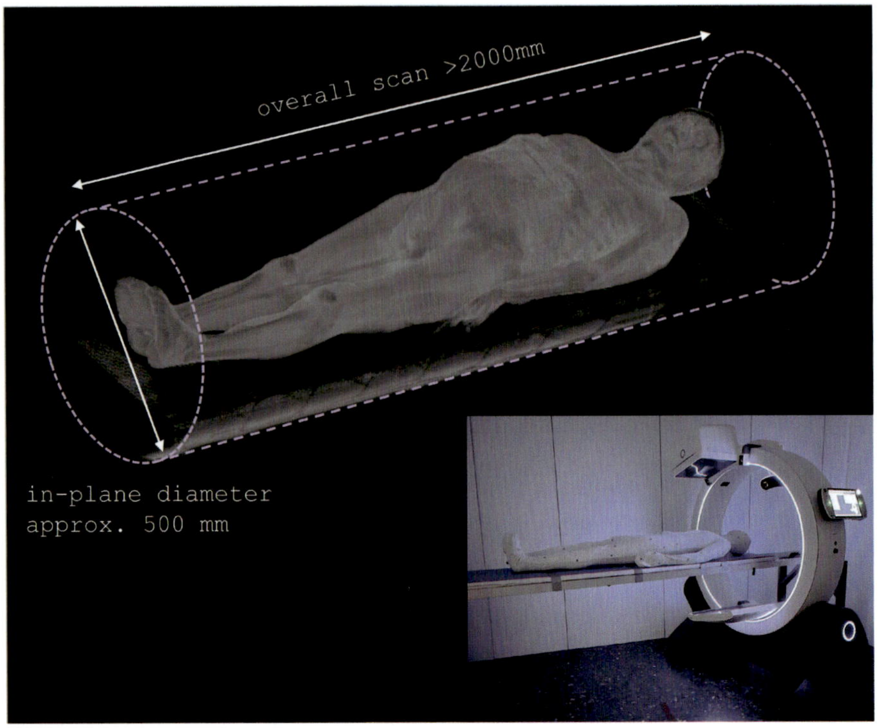

Figure 18.3 The large field-of-view has a maximum in-plane diameter of approx. 500 mm. The maximum scan length is limited to approx. 2200 mm due to the distance between the trestles of the examination table. The provisional examination table for postmortem whole-body examinations can be seen in the small image.

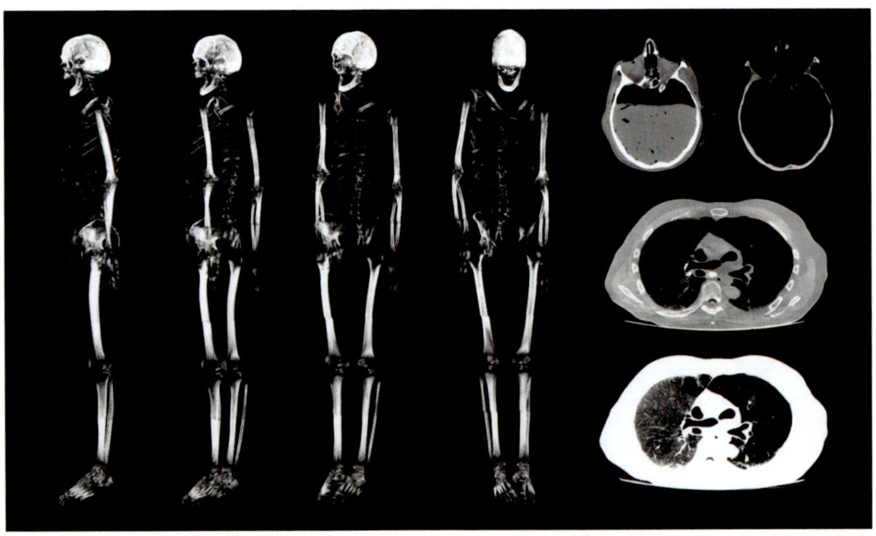

Figure 18.4 Maximum intensity projections in four different views of a whole-body CT scan virtually rotated around its longitudinal axis. Note the step artifact in the middle of the thigh, tibia and fibula. The merging of the sequentially acquired data sets was not yet optimized at the time this book chapter was written. Sectional images through the skull in the soft tissue and lung window are shown at the top right, and sectional images through the thorax in the soft tissue and lung window are depicted at the lower right.

During each sequential scan, X-ray images are taken over a 360° rotation, resulting in a whole-body examination that takes 10–15 minutes, depending on the length of the body. The infrared tracking system enables the data from the individual scans along the longitudinal axis of the body to be combined more precisely for subsequent reconstruction of the whole-body data set. For this purpose, spherical markers are placed on the head side so that they can be seen and tracked by the infrared tracking system during each sequential scan. The merging of the individual sequential data sets into a whole-body CT data set is performed automatically after all scans have been performed (Figure 18.4). In addition to the usual soft and hard reconstruction kernels, the system offers further kernels for individual reconstruction, which are continuously being further developed.

While the detector rotates around the body when acquiring CT data, photographic images can be taken simultaneously with the detector cameras. Adaptations in the ImagingRing software allowed modifying the image capturing settings and granted access to the acquired images. These modifications allowed capturing an optical dataset with sufficient overlap to carry out photogrammetric reconstructions. Ultimately, Agisoft Metashape Professional (Version 2.0.1 build 16069, Agisoft LLC, St. Petersburg) was used for the multiview 3D reconstruction of the captured images (Figure 18.5). The detector cameras thus enable a 3D surface documentation simultaneously with the whole-body CT scan. When using a transparent table top, the color information of the entire body surface can be detected, as the detector rotates 360° around the body. In this way, the back of the body lying on the table can also be optically captured and serve as a rough color supplement to the CT. In order to adequately assess and measure injuries to the back of the body, it is essential to scan the body in the prone position simultaneously with the CT and the detector camera. At the moment, however, the resolution of the reconstructed 3D mesh and the quality of the texture can only help to provide a rough overview of the body's condition in 3D. The camera settings are still being improved in order to achieve a higher quality of the photographic images and thus ultimately also of the 3D model, which might then meet the requirements of a forensic medical examination.

The ImagingRing presents a viable solution for three-dimensional radiologic and optical postmortem forensic documentation encompassing the entire body, aligning with the Virtopsy approach. However, its cost-effectiveness hinges on a trade-off, where the simplified power supply and cooling system compromise image quality and prolong scan times compared to conventional clinical CT scanners. Furthermore, careful consideration must be given to the inherent differences between cone-beam and fan-beam CT technologies within this context [11]. A noteworthy feature of the ImagingRing, particularly advantageous for postmortem forensic imaging, is its concurrent acquisition of photographic images. This capability offers potential for streamlined documentation and enhanced visualization of external findings. Furthermore, its straightforward power connection from a standard electrical outlet, self-contained cooling system, and portability render it potentially suitable for disaster victim identification (DVI) occurring in remote and resource-constrained environments.

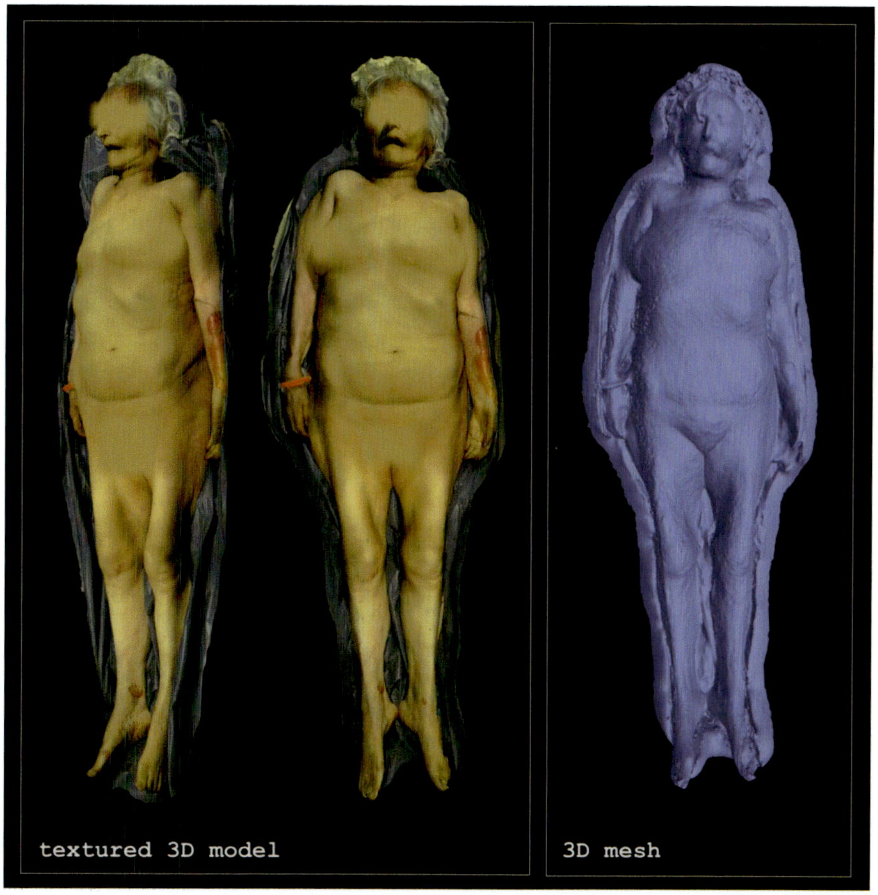

Figure 18.5 Textured 3D model and 3D mesh of a deceased person. The detector cameras enable photographic images to be captured during a CT scan. However, the camera settings need to be improved in order to achieve a higher quality of the photographic images and thus ultimately also of the 3D model, which could then meet the requirements of a forensic medical examination.

There are currently several automated tasks that are to be further optimized for forensic applications as part of the research cooperation and in the course of the further development of the ImagingRing. Additionally, several functionalities require further investigation for forensic postmortem imaging. Notably, metal artifact reduction algorithms and wobble scan protocols lack evaluation and potential optimization for foreign bodies commonly encountered in forensic settings, such as projectiles. The potential advantages of utilizing the gantry tilt functionality for postmortem CT examinations remain unexplored. Furthermore, the quality of postmortem CT angiography and the optimal composition of contrast media necessitate systematic evaluation. The ImagingRing offers unique add-ons with forensic value, including the surgical intervention package and Criq robotic arm. Their suitability for minimally invasive postmortem biopsies and fluid removal requires thorough investigation. For surface documentation, incorporating additional lights attached to the detector to ensure uniform illumination warrants exploration. Additionally, the extensive testing of gantry cameras is essential, particularly considering their current near-infrared recording capabilities, which present intriguing possibilities for surface documentation. In conclusion, the continued development and implementation of novel imaging solutions with the ImagingRing system hold significant promise for postmortem forensic applications, and future advancements in this field remain eagerly anticipated.

ACKNOWLEDGMENTS

The author thanks Heinz Deutschmann, Phil Steininger, Anastasiia Parii, Michael Hubauer-Brenner, Felix Ginziger, Sabrina Brückner, Julia Coste-Marin, and Herbert Biber from MedPhoton (part of the Brainlab company) for their research collaboration and Sören Kottner from the Zürich Institute of Forensic Medicine.

References

[1] R. Dirnhofer, C. Jackowski, P. Vock, K. Potter, M.J. Thali, VIRTOPSY: Minimally invasive, imaging-guided virtual autopsy, Radiographics 26 (2006) 1305–1333. https://doi.org/10.1148/rg.265065001.

[2] V. Chatzaraki, J. Heimer, M. Thali, A. Dally, W. Schweitzer, Role of PMCT as a triage tool between external inspection and full autopsy–case series and review, Journal of Forensic Radiology and Imaging 15 (2018) 26–38.

[3] D. Gascho, S. Schaerli, L. Tuchtan-Torrents, M.J. Thali, G. Gorincour, Use of postmortem computed tomography to detect bowel obstruction and its relationship to the cause of death, The American Journal of Forensic Medicine and Pathology 39 (2018) 30. https://doi.org/10.1097/PAF.0000000000000365.

[4] L. Filograna, L. Pugliese, M. Muto, D. Tatulli, G. Guglielmi, M.J. Thali, R. Floris, A practical guide to virtual autopsy: Why, when and how, Seminars in Ultrasound, CT and MRI 40 (2019) 56–66. https://doi.org/10.1053/j.sult.2018.10.011.

[5] P.M. Flach, D. Gascho, W. Schweitzer, T.D. Ruder, N. Berger, S.G. Ross, M.J. Thali, G. Ampanozi, Imaging in forensic radiology: An illustrated guide for postmortem computed tomography technique and protocols, Forensic Science, Medicine and Pathology 10 (2014) 583–606. https://doi.org/10.1007/s12024-014-9555-6.

[6] D. Gascho, M.J. Thali, T. Niemann, Postmortem computed tomography: Technical principles and recommended parameter settings for high-resolution imaging, Medicine, Science and the Law 58 (2018) 70–82.

[7] M.J. Bolliger, U. Buck, M.J. Thali, S.A. Bolliger, Reconstruction and 3D visualisation based on objective real 3D based documentation, Forensic Science, Medicine and Pathology 8 (2012) 208–217. https://doi.org/10.1007/s12024-011-9288-8.

[8] M. Stock, D. Georg, A. Ableitinger, A. Zechner, A. Utz, M. Mumot, G. Kragl, J. Hopfgartner, J. Gora, T. Böhlen, The technological basis for adaptive ion beam therapy at MedAustron: Status and outlook, Zeitschrift Für Medizinische Physik 28 (2018) 196–210. https://doi.org/10.1016/j.zemedi.2017.09.007.

[9] D. Gascho, H. Philipp, P.M. Flach, M.J. Thali, S. Kottner, Standardized medical image registration for radiological identification of decedents based on paranasal sinuses, Journal of Forensic and Legal Medicine 54 (2018) 96–101. https://doi.org/10.1016/j.jflm.2017.12.003.

[10] D. Gascho, P.M. Flach, S. Schaerli, M.J. Thali, S. Kottner, Application of 3D image fusion for radiological identification of decedents, Journal of Forensic Radiology and Imaging 13 (2018) 12–16.

[11] L. Lechuga, G.A. Weidlich, L. Lechuga, G.A. Weidlich, Cone Beam CT vs. Fan Beam CT: A comparison of image quality and dose delivered between two differing CT imaging modalities, Cureus 8 (2016). https://doi.org/10.7759/cureus.778.

CHAPTER 19

7 TESLA MRI AND MRS—BASICS AND FORENSIC POSTMORTEM APPLICATIONS

Dominic Gascho and Niklaus Zoelch

DOI: 10.4324/9781003413998-21

19.1 INTRODUCTION

Magnetic resonance (MR) scanners with magnetic field strengths of 7 Tesla are generally referred to as ultra-high-field (UHF) scanners in comparison to the high-field (HF) scanners with a magnetic field strength of 3 Tesla, which are currently more commonly used in the clinic. The first UHF scanners for neuroscience purposes were already installed in the late 1990s and early 2000s [1,2]. Nevertheless, it was not until 2017 that the first UHF system, a 7 Tesla MR scanner, was approved by the FDA (U.S. Food and Drug Administration) for clinical use as a medical device (Figure 19.1). Due to the stronger magnetic field, such a UHF scanner has different conditions in terms of signal strength and susceptibility sensitivity compared to HF scanners, which is beneficial for magnetic resonance imaging (MRI) and spectroscopy (MRS). The benefits resulting from these changing conditions are described in the following sections before discussing the use or potential use of postmortem 7 Tesla MRI and MRS for forensic purposes.

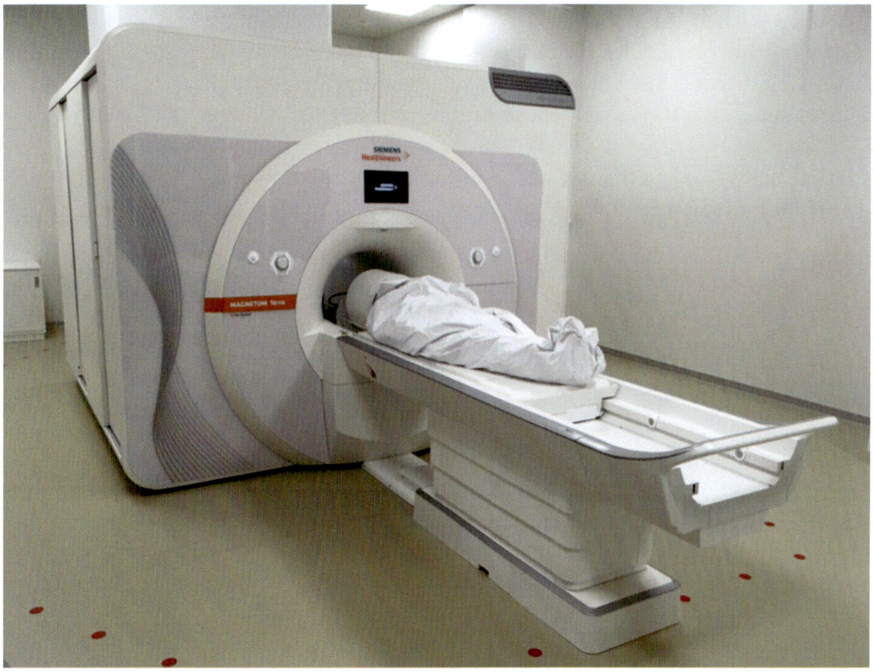

Figure 19.1 FDA-approved 7 Tesla MRI scanner (Magnetom Terra, Siemens Healthineers) at the Swiss Center for Musculoskeletal Imaging (SCMI) for examination of deceased persons in research collaboration with the Zürich Institute of Forensic Medicine.

19.2 WHY IS THE SIGNAL STRONGER WITH 7 TESLA THAN WITH 3 TESLA?

In the chapter about the MR fundamentals, it is mentioned that the initial magnetization M_0, the "source" of the MR signal, depends on the strength of the external field and the temperature. M_0 can be described as follows:

$$M_0 = \frac{n \cdot \gamma^2 \cdot h^2 \cdot B_0}{16\pi^2 \cdot k_b \cdot T}$$

where n is the number of spins with spin nuclear number ½, γ the gyromagnetic ratio, h is the Planck constant, k_b is Boltzmann's constant, T is the absolute temperature, and B_0 is the polarizing static magnetic field.

M_0 is tilted upon excitation and is therefore proportional to the signal amplitude that can be observed during an MR experiment. Consequently, the measurable signal from hydrogen protons can be amplified mainly either by low temperatures, which is used in nuclear magnetic resonance (NMR) examinations of small samples, or by increasing the B_0 field. The signal detection itself using MR coils is also B_0 dependent. The tilted M_0 induces a voltage in the coil based on Faraday's law of induction, which states that the induced voltage (U_{signal}) is proportional to the time rate of change of magnetic flux (ϕ). This change in magnetic flux is proportional to the Larmor frequency (ω), which is determined by the gyromagnetic ratio times the B_0 field, and to the net magnetization vector (M_0), which, as already mentioned, is proportional to the B_0 field. Consequently, the signal is theoretically proportional to B_0 squared [3]:

$$U_{signal} \propto \frac{\delta\phi}{\delta t} \propto \omega \cdot M_0 \propto \gamma \cdot B_0 \cdot M_0 \propto B_0^2$$

The noise, however, also increases linearly with the field strength, and therefore a rather linear increase of the signal-to-noise ratio (SNR) can be expected with the increase of the field strength [4]. The variation of SNR with field strengths above 1.5 Tesla is quite a complex function depending on the size, shape, and composition of the object under examination and due to differences in radiofrequency coils and other hardware aspects, it is difficult to predict the exact SNR gain for a given experiment at higher field strength. [5]. In general, 7 Tesla MRI offers a substantial higher SNR than 3 Tesla MRI. This increase in SNR can be used for higher spatial resolution or shorter scan times. In addition, the higher field strength can also lead to stronger contrasts, which will be exemplary shown in the following sections.

19.3 WHY IS 7 TESLA MRI MORE SENSITIVE TO SUSCEPTIBILITY EFFECTS THAN 3 TESLA MRI?

T_2* describes the observable or effective decay of the transverse magnetization together with the inhomogeneity in the magnetic field itself and the susceptibility-induced field distortions generated by the tissue. T_2* decreases approximately linearly with increasing magnetic field strength for all tissue types. Therefore, the generated signals decay faster at 7 Tesla than at 3 Tesla [6]. More importantly, a stronger magnetic field leads to a larger dispersion of T_2* values, resulting in accentuated signal differences at locations where the susceptibilities of a structure differ strongly from those of the surrounding tissue, and thus, for example, very small hemorrhages (microbleeds) with paramagnetic components are highlighted more at 7 Tesla than at 3 Tesla.

Susceptibility-weighted imaging (SWI) uses the magnitude and phase information of the complex-valued signal, and therefore, not only benefits from higher SNR but also from larger phase effects at 7 Tesla. Phase changes ($\Delta\varphi$) that originate from susceptibility changes are proportional to the B_0 field [5]:

$$\Delta\varphi = -\gamma \cdot \chi \cdot B_0 \cdot TE$$

where χ is the magnetic susceptibility.

Since 7 Tesla MRI also allows for higher resolution at the same scan time, 7 Tesla SWI is much more sensitive than 3 Tesla SWI for the detection of microbleeds indicative of pathological or traumatic influences (Figure 19.2) [7].

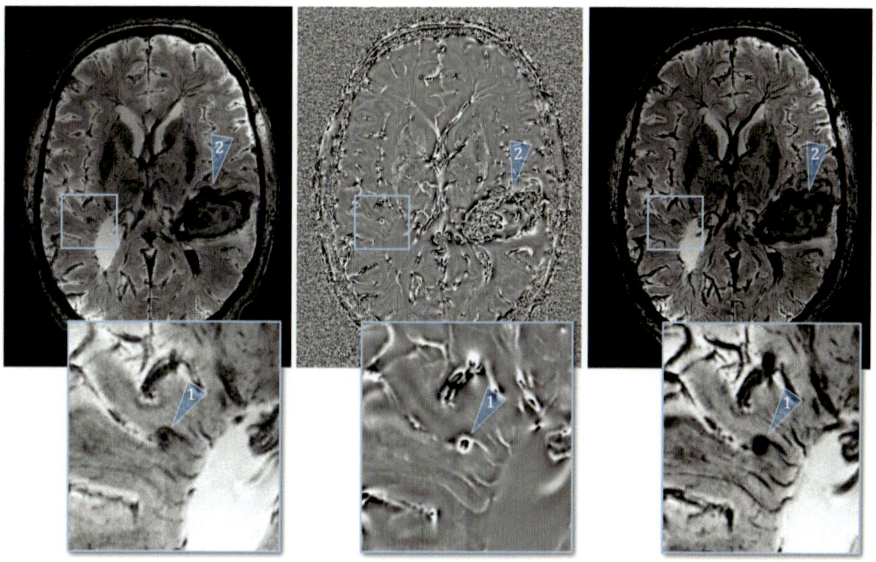

Figure 19.2 Postmortem 7 Tesla SWI: magnitude image (left), phase image (middle), and SWI image (right). Microbleeding in the right lobe (arrowhead 1) highlighted due to susceptibility effects in a natural death of a 68-year-old man who died from intracerebral hemorrhage in the left temporal lobe (arrowhead 2). For more details on this case and information on the direct comparison with a 3 Tesla study, the reader is referred to the article "7-T MRI for Brain Virtual Autopsy: A Proof of Concept in Comparison to 3-T MRI and CT" [7].

19.4 WHAT IS THE ADVANTAGE OF 7 TESLA OVER 3 TESLA FOR MRS?

As for the free water signal used in magnetic resonance imaging, the higher field strength also results in a direct improvement of the SNR for the metabolite signals. And thus, the mathematical description of M_0 is also directly applicable to metabolite signals. In addition, the signals from the different metabolites in the spectrum may overlap less at 7 Tesla compared to 3 Tesla (for an illustrated presentation, the reader is referred to the chapter "MRS in Forensic Medicine: Current Developments and Future Perspectives"). This is because the splitting of the J-coupled protons becomes smaller in proportion to the differences due to different chemical shifts of the metabolites. For the metabolites detectable by MRS, the shielding is field independent and as expressed by the shielding constant σ (see the chapter on the fundamentals of MRS in this book). The electrons shield the same amount of the external field at 7 Tesla and 3 Tesla. This is why, for example, creatine always appears at about 3.02 ppm and N-acetylaspartate (NAA) at 2.02 ppm, regardless of field strength. However, the frequency differences between creatine and NAA ($\upsilon_{Cre\text{-}NAA}$) expressed in hertz (Hz) increase with the strength of the external field (B_0):

$$\upsilon_{Cre-NAA} = \frac{\gamma}{2\pi} \cdot B_0 \cdot (\sigma_{NAA} - \sigma_{Creatine})$$

J-coupling causes field-independent splittings on the order of a few Hz, typically ≈ 7 Hz. In spectra with the x-axis given in parts per million (ppm), this leads to the fact that these splittings appear smaller at higher field strength as the peaks move closer together.

The potential reduction in the overlap of the signals from the different metabolites in the spectrum may be compromised by an increased linewidth due to shorter T_2 (and T_2*) relaxation times and a potential increase in field inhomogeneity within the measured volume.

19.5 HOW MIGHT FORENSIC POSTMORTEM IMAGING BENEFIT FROM STUDIES ON A 7 TESLA MR SCANNER?

Due to the fact that 7 Tesla MR scanners are not yet widely available and therefore access to these ultra-high-field scanners is limited, the literature on postmortem examinations in a forensic context is correspondingly sparse. The potential advantages and limitations of using 7 Tesla MRI for postmortem examination of fetuses have been outlined in the literature, with the authors concluding that it is an safe alternative approach to stereo-microscopic autopsy for the assessment of fetal structural abnormalities at the end of the first trimester of pregnancy [8]. The MR scanner used for this purpose was a preclinical scanner that does not allow the examination of larger bodies due to the gantry diameter of a few centimeters. A clinical 7 Tesla MR scanner with a gantry diameter of 60 cm allows not only postmortem examinations of children (Figure 19.3) but also of adults. However, there is the problem of B_0 and B_1 inhomogeneity on the 7 Tesla MR scanner, which complicates studies over the torso area of the adult body, and suitable coils for studies of the thorax and abdomen are not yet available.

Studies of head penetrations on 7 Tesla MRI compared to 3 Tesla MRI showed primarily increased sensitivity for injuries attributable to the temporal wound cavity, as bleeding into injured tissue is highlighted due to the higher sensitivity to susceptibility effects at 7 Tesla [8]. In a 92-year-old man, postmortem 7 Tesla MRI examination showed coup and contrecoup injuries in high resolution compared with a 3 Tesla MRI examination [7]. The detailed presentation of coup and contrecoup injury may allow to obtain new information on the origin of the contrecoup injuries. The high resolution of 7 Tesla MRI makes it possible to move beyond the macroscopic range to a smaller scale between the macro and micro scales, referred to here as the "mesoscopic scale", which also allows for detailed studies of brain structures in a non-invasive way [9]. This could also be useful for the study of minute traumatic injuries to determine the origin of these injuries, whether from impact trauma or projectile trauma.

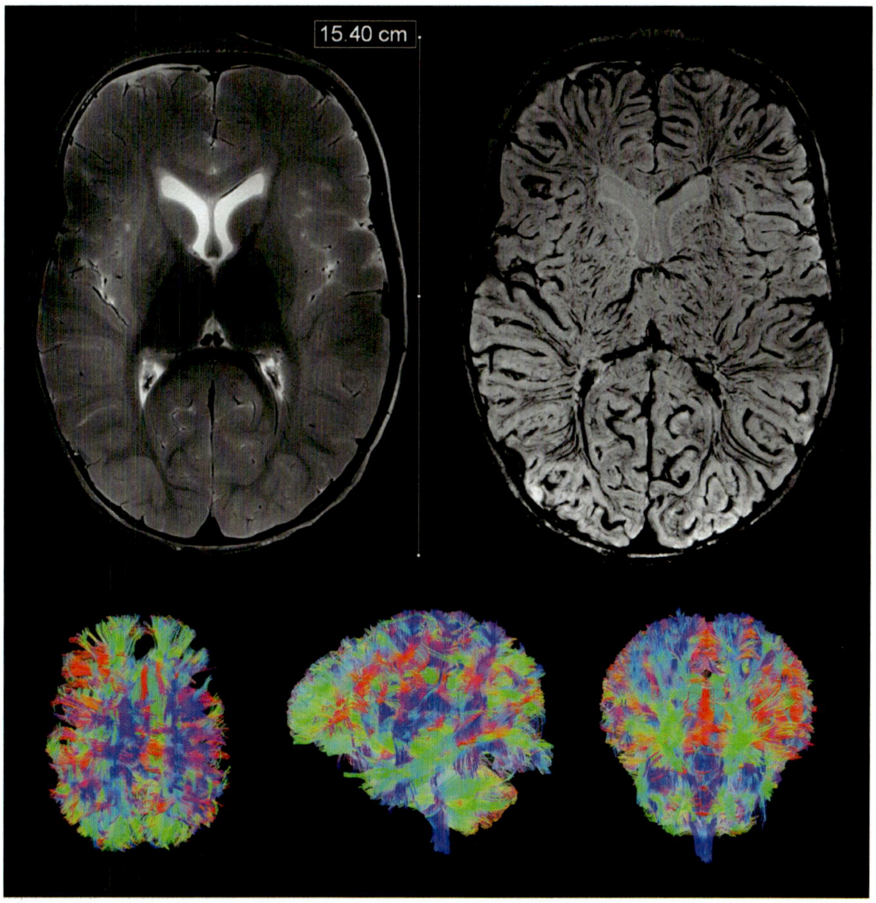

Figure 19.3 Postmortem 7 Tesla MRI of a 6-month-old child showing a T_2-weighted sequence with a transverse slice thickness of 2 mm (upper left image), a susceptibility-weighted imaging (SWI) sequence with a transverse slice thickness of 1.5 mm (upper right image), and diffusion tensor imaging (DTI) reconstructions (second row). The 7 Tesla MRI examination convinces with its high-resolution images. The 7 Tesla MRI examination was conducted in research collaboration between the Zürich Institute of Forensic Medicine and the Swiss Center for Musculoskeletal Imaging (SCMI).

ACKNOWLEDGMENTS

The authors thank Daniel Nanz, Stefan Sommer, and Natalie Hinterholzer of the Swiss Center for Musculoskeletal Imaging (SCMI) for their collaboration in the performance and assessment of the 7 Tesla MRI examinations.

References

[1] P.-M.L. Robitaille, R. Warner, J. Jagadeesh, A.M. Abduljalil, A. Kangarlu, R.E. Burgess, Y. Yu, L. Yang, H. Zhu, Z. Jiang, R.E. Bailey, W. Chung, Y. Somawiharja, P. Feynan, D.L. Rayner, Design and assembly of an 8 Tesla whole-body MR scanner, Journal of Computer Assisted Tomography. 23 (1999) 808–820.

[2] E. Yacoub, A. Shmuel, J. Pfeuffer, P.-F. Van De Moortele, G. Adriany, P. Andersen, J.T. Vaughan, H. Merkle, K. Ugurbil, X. Hu, Imaging brain function in humans at 7 Tesla, Magnetic Resonance in Medicine. 45 (2001) 588–594. https://doi.org/10.1002/mrm.1080.

[3] N.B. Smith, A. Webb, Introduction to medical imaging: Physics, engineering and clinical applications—Chapter 5: Magnetic resonance imaging (MRI), Cambridge University Press, 2010.

[4] P.M. Cunningham, M. Law, M.E. Schweitzer, High-field MRI, Orthopedic Clinics. 37 (2006) 321–329. https://doi.org/10.1016/j.ocl.2006.05.002.

[5] M.E. Ladd, P. Bachert, M. Meyerspeer, E. Moser, A.M. Nagel, D.G. Norris, S. Schmitter, O. Speck, S. Straub, M. Zaiss, Pros and cons of ultra-high-field MRI/MRS for human application, Progress in Nuclear Magnetic Resonance Spectroscopy. 109 (2018) 1–50.

[6] A.M. Peters, M.J. Brookes, F.G. Hoogenraad, P.A. Gowland, S.T. Francis, P.G. Morris, R. Bowtell, T2* measurements in human brain at 1.5, 3 and 7 T, Magnetic Resonance Imaging. 25 (2007) 748–753. https://doi.org/10.1016/j.mri.2007.02.014.

[7] D. Gascho, N. Zoelch, S. Sommer, C. Tappero, M.J. Thali, E. Deininger-Czermak, 7-T MRI for brain virtual autopsy: A proof of concept in comparison to 3-T MRI and CT, European Radiology Experimental. 5 (2021) 3. https://doi.org/10.1186/s41747-020-00198-7.

[8] D. Gascho, E. Deininger-Czermak, N. Zoelch, C. Tappero, S. Sommer, N. Hinterholzer, M.J. Thali, Noninvasive 7 tesla MRI of fatal craniocerebral gunshots—a glance into the future of radiologic wound ballistics, Forensic Science, Medical and Pathology. (2020). https://doi.org/10.1007/s12024-020-00300-w.

[9] L.J. Edwards, E. Kirilina, S. Mohammadi, N. Weiskopf, Microstructural imaging of human neocortex in vivo, NeuroImage. 182 (2018) 184–206. https://doi.org/10.1016/j.neuroimage.2018.02.055.

CHAPTER 20

MRS IN FORENSIC MEDICINE

Current Developments and Future Perspectives

Niklaus Zoelch and Dominic Gascho

DOI: 10.4324/9781003413998-22

20.1 INTRODUCTION

Among the noninvasive diagnostic procedures applied in forensic pathology, the application of MRS for inference in forensic pathology has been envisioned for two decades[1]. MRS can be a powerful adjunct to postmortem imaging, as it primarily augments imaging with information on metabolism. This way, MRS contributes to one of the primary goals of noninvasive postmortem procedures: gaining information on the cause (and manner) of death before the onset of the autopsy, which potentially enables one to waive it entirely[2],[3]. In comparison to conventional biochemistry, which demands an indication, MRS is untargeted, as it will simultaneously estimate all attainable metabolite concentrations within the target volume. Furthermore, MRS allows for repeated measurements of the volume without any alteration of the subject, which is of great benefit to longitudinal studies in forensic science research.

Initial work using in situ MRS in forensic medicine focused on postmortem metabolic changes as a function of PMI, thus taking full advantage of the ability of MRS to perform longitudinal studies[4–6]. The primary goal of these studies was to improve time-of-death estimates in the late postmortem phase, where PMI estimates based on temperature measurements are becoming increasingly difficult. These findings have had little impact on forensic practice, due in part to the availability of MR scanners at forensic institutes. However, there are still some uncertainties regarding the applicability of the results to humans, since the fundamentals rely on studies in severed animal heads.

In this section, we want to show further possibilities of MRS in forensic medicine beyond PMI determination. The versatility of MRS, which allows us to collect not only metabolic information but also other parameters of the measured body, should also be emphasized in this context. These additional parameters might potentially be case relevant, but more importantly could improve postmortem imaging itself. We do this based on various current publications of the Virtopsy Group and would like to create an outlook on what could be possible in the future.

20.2 DETECTION OF METABOLITES RELEVANT FOR FORENSIC MEDICINE

The use of MRS in forensic medicine for the direct detection of metabolites or substances relevant to the forensic issue is a relatively new approach in which our group has pioneered. At best, this approach could replace conventional biochemical testing (toxicological or traditional biochemical testing) or at least help to improve the subsequent investigations by being able to narrow down the search.

The primary advantage of the MRS approach is that the information is obtained non-invasively. With non-invasiveness, the expense of sample collection, preparation and storage disappears and the consumption of a sample, and thus the destruction of the evidence, is avoided. The elimination of collection and preparation can mean a significant time gain, so that results are available prior to any invasive procedure. It can also increase employee safety in cases of infectious disease or the like. In combination with MRI, MRS offers the possibility to obtain metabolic information relatively quickly at, in principle, any location in the body (that even may be difficult to access). This allows rapid comparative measurements, such as between healthy and non-healthy tissue.

Another advantage is the fact that the metabolite signals ideally also contain information about the immediate environment of the metabolites. Such information can be tapped e.g. by relaxation times, visibility of the metabolite (NMR visibility) or certain interactions (residual dipolar coupling in muscle) and could provide further insights especially in the field of postmortem studies.

A crucial question here is whether reliable estimates of substance quantities are possible in the field of forensic medicine using MRS. Two main difficulties usually arise from the peculiarities of MRS: the overlap of the signals from different metabolites and the conversion of the signals into reliable quantities that are comparable across different measurements and modalities (data calibration, see technical chapters). The first problem limits the detectability of certain compounds whereas the second limits the interpretability of the signals. The same problems also arise in the application of MRS in vivo. However, postmortem measurements have the advantage that the subject can be measured over a longer time without dealing with patient comfort or motion artifacts and therefore allow additional measurements potentially necessary for a proper data calibration or to simply improve SNR. In addition, in general only larger deviations from normal values are of interest in postmortem investigations. In the following, we would like to provide more insight into the use of MRS for the detection of relevant metabolites in forensic medicine based on some recent studies.

20.3 POSTMORTEM ETHANOL DETECTION WITH MRS

In a first work in this direction by Zoelch et al[7], a total of 15 cadavers with suspected antemortal alcohol consumption were investigated to determine whether MRS can be used to reliably detect ethanol and how accurately the measured concentrations agree with conventionally determined alcohol concentrations.

MRS measurements were performed in situ in CSF of the ventricles and compared with alcohol concentrations determined biochemically (headspace gas chromatography coupled to flame ionization detection) in blood and when available in CSF sampled in the spinal canal. CSF is well suited for MRS measurements, because due to the long T2 times of the contained substances and the associated narrow linewidth of the corresponding peaks, the overlapping of the signals of lactate and ethanol is less pronounced than for example in brain tissue. At the same time, the uncertainty of how large the fraction of bound alcohol (and thus probably not measurable by MRS) is in CSF is not present, which simplifies comparisons to biochemical/toxicological studies[8].

In 8 of the 15 cadavers, alcohol was measured in the blood; in all these cases, MRS in the CSF also detected alcohol. However, the measured concentrations differed between 4% and 45% percent from the toxicological values measured in the CSF, with MRS mostly overestimating the alcohol concentration. The differences in concentrations can be attributed in large part to inaccuracies in the fit and an inaccurate correction of the relaxation attenuation. The inaccuracies in the fit were particularly large when the measured volume contained portions of brain tissue. According to the literature, the T_1 of ethanol at body temperature is 4400 ms (measured in monkey brains). At a repetition time of 3000 ms, as used in this work, this leads to a significant T_1 weighting of the ethanol signal, since the magnetization cannot fully relax in a series of measurements. For a proper concentration estimation, this T_1 influence must be corrected. Zoelch et al. used the literature value mentioned before, but this value is likely higher than the effective value in the individual case. A lower temperature or long PMI might lead to shorter T_1 and thus leads to an overestimation of the ethanol concentration.

This work clearly shows the need to adapt protocols optimized for in vivo measurements to postmortem measurements. Ideally, postmortem protocols focusing on accurate concentration measurements should be run with long TR and short TE. Even more than in vivo, the relaxation time plays a major role, as it can differ significantly between different cadavers due to large temperature differences but also due to different PMI. A correction based on a literature value is therefore not sufficient here, unlike perhaps with in vivo measurements. The above-mentioned influence of T_1 on the ethanol measurement could be avoided, for example, by using a longer TR. A longer TR > 10 s would indeed prolong the measurement significantly, but this might be possible in the postmortem setting.

In addition to the determination of ethanol concentration, MRS could also provide information about alcohol tolerance or chronic alcohol consumption by evaluating other substances or comparing different regions. For example, clinical studies have shown differences in brain scyllo-inositol concentrations between non-detoxified chronic alcoholics and healthy controls[9,10]. Or, increased MR visibility of ethanol in brain tissue has been shown in alcohol tolerant individuals[11]. This difference in visibility could potentially be assessed by measurements in CSF and brain tissue (see Figure 20.1).

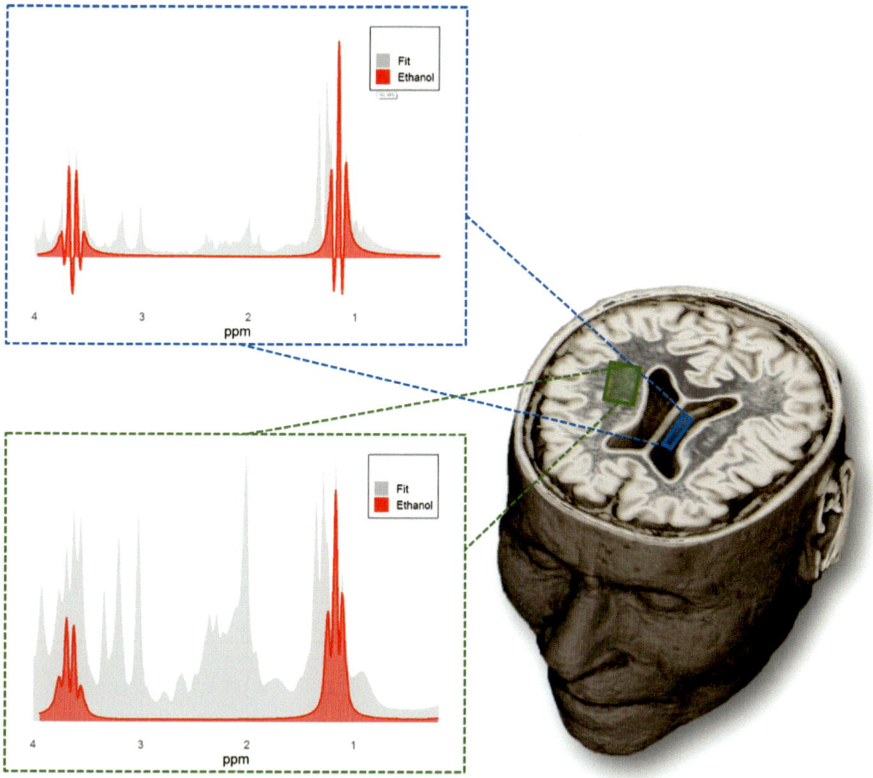

Figure 20.1 Example fitted spectra recorded in CSF (above, outlined in blue) and WM tissue (below, outlined in green) in a cadaver. Both fits show the signal components of ethanol colored in red (triplet at 1.17 ppm and the quartet at ~3.64 ppm). The ethanol signal is partially overlapped by the lactate signal at 1.17 ppm. The WM spectrum also shows a clear scyllo-inositol peak.

20.4 ^{1}H-MRS FOR THE DIAGNOSIS OF FATAL METABOLIC DISORDERS

A similar picture to the ethanol measurements is seen in the work of Heimer et al.[12–14] on the diagnosis of fatal metabolic disorders using MRS. In the most extensive work to date, a total of 20 bodies with suspected fatal metabolic disorders were included to compare the concentrations of beta-hydroxybutyrate (BHB), glucose (GLC), and lactate (LAC) derived from both biochemical analyses (BCA) and ^{1}H-MRS for the diagnosis of fatal metabolic disorders[14]. For this purpose, the concentrations in CSF and vitreous humor were compared. Primarily, it was investigated whether the same diagnosis was made based on given threshold values using the different approaches (BCA and MRS). Spectra from this study are illustrated in Figure 20.2.

Except for a single case, the MRS values gave the same diagnosis as based on the BCA values. The MRS was just as good in detecting pathological concentrations of GLU and BHB. This clearly shows the suitability of MRS for such investigation. In the case of the single disagreement, the spectral quality of the MRS measurement was not good enough to determine the concentration of GLC or BHB (see Figure 20.3). The poor quality in this case is clearly visible when compared to other measurements from the same study. Poor signal quality can be detected by checking metrics such as linewidth, SNR, quality of water suppression and Cramér-Rao lower bounds (CRLB), and thus spectra can be rejected for further analysis. A good list of rejection criteria can be found for example in the work from de Barros et al.[15]. Similarly, the reliability of quantification can be estimated by metrics such as CRLB or fit quality number[15]. However, poor quality is not always easy to detect. For example, we may have poor localization or signal extinction for individual metabolites without the spectrum looking badly or being conspicuous in the metrics mentioned above. To avoid this, a meaningful quality management is necessary[15]. For the application of MRS in forensic medicine, this is of great importance and further studies in this area are necessary.

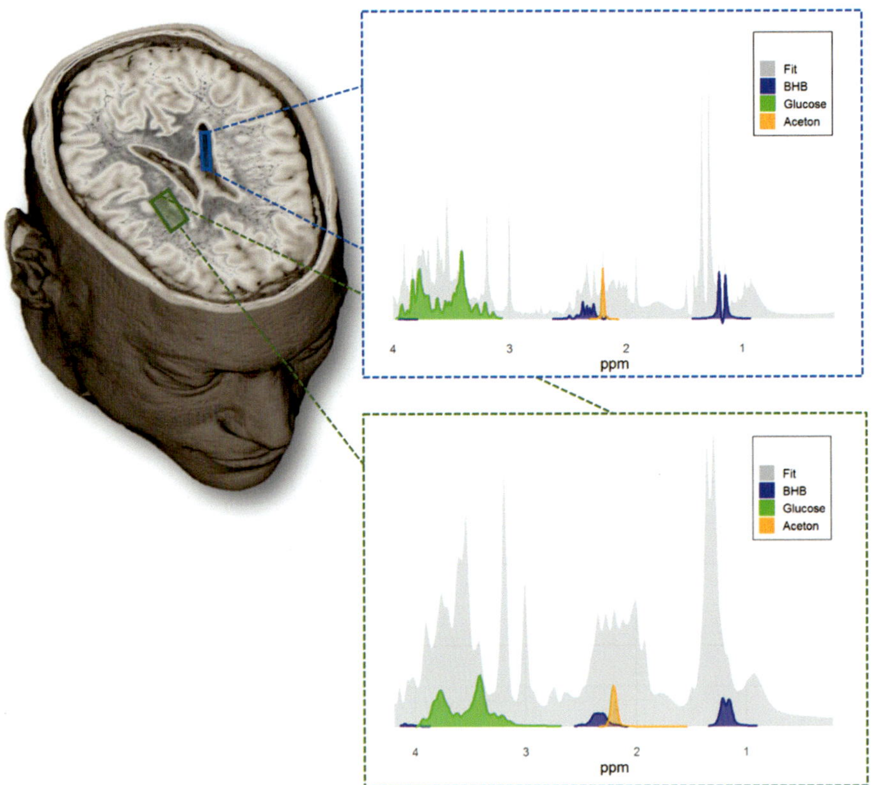

Figure 20.2 Example fitted spectra in a case of diabetic ketoacidosis recorded in the CSF above (outlined in blue) and white matter tissue (outlined in green) in one cadaver. The contributions of BHB, glucose and acetone to the fit are colored accordingly. While in WM BHB and glucose are more difficult to delineate, in CSF this succeeds very well thanks to the narrow linewidth. The small tip of acetone is also clearly visible in WM.

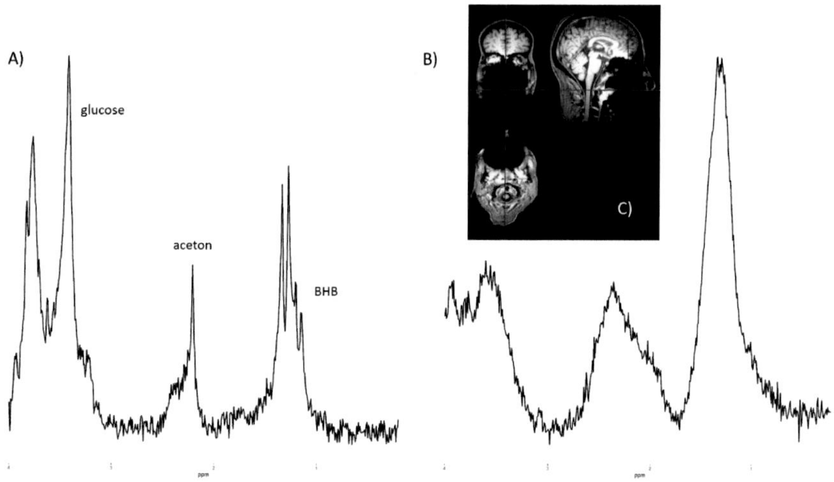

Figure 20.3 Example of good (A) and bad (B) spectral quality in vitreous humor. While the narrow linewidth and straight baseline allows the detection and quantification of the relevant metabolites to detect ketoacidosis in A), no clean separation of the metabolites signals is possible in spectrum B). The increased glucose and BHB concentrations that would be expected from biochemical analysis and CSF measurements can only be guessed at. Not even the duplet of the lactate peak is visible or recognizable as such. The poor spectral quality, manifested in wide linewidth and unstable baseline are due to due to a dental brace and the associated field disturbances. The effect of these field disturbances can also be seen in the corresponding MRI images (C).

Apart from the perfect binary agreement, the differences in the absolute concentrations of both methods show high SDs, which are a compulsory consequence of the low accuracy (systematic errors) and precision (random errors) of both methods.

Three main problems limiting the accuracy of the ^{1}H-MRS results could be identified which correspond well with the problems found in the ethanol study. These problems include the non-consideration of the relaxation effects of the metabolite signals, the overlap of the signals of interest with other metabolite signals, and a mixed tissue composition within the voxel. Different tissue types are primarily a problem when MRS measurements are to be compared with measurements in collected samples, which usually contain just a specific tissue type, such as CSF. This problem can be to some degree solved by using smaller MRS voxel sizes. Sometimes, however, it is not possible to avoid mixing tissue types due to anatomical conditions or high intracranial pressure. A reduction of the voxel size also results in a significant reduction of the SNR, which can be compensated with a longer measurement time. SNR is directly proportional to the voxel size and to the square root of the number of averages. This means that, for example, a reduction in volume by a factor of three must be compensated for by an increase in the number of averages by a factor of nine, which means that the scanning time is nine times longer. The influence of T_1 relaxation on metabolite concentrations can be reduced by choosing longer TR, as mentioned above. For a $TR > 4 \cdot T_1$, the influence is below 2%. While the extension of the TR for metabolite measurements leads to a considerable extension of the measurement time (TR · needed number of averages), the measurement of the water signal for referencing with a long TR is always worthwhile. In the Heimer et al. study, water was measured with a TR of 15 s for CSF and vitreous humor and 7 s for WM to be used as reference for conversion of metabolite signals into concentrations. In addition, also a series of echo times was measured to minimize the influence of T_2 on the water reference signal. In principle this could be done for the metabolites too, but since these signal are much lower, a long extra measurement time would be required. Another approach is to use localization sequences which allow for shorter possible echo times such as STEAM or Image-selected in Vivo spectroscopy (ISIS)[16]. ISIS is a multi-shot technique, where the spatial selection of the signals is calculated using data from eight different scans. This technique is highly susceptible to artifacts related to subject motion and therefore has been replaced in ^{1}H-MRS by single-shot techniques like PRESS or sLASER. For postmortem applications, however, this is definitely an interesting alternative that has not yet been exploited. The problem that signals from low-concentrated metabolites are obscured by signals from higher concentrated metabolites is generally not easily overcome. Sometimes, reducing the size of the voxel can lead to a better shim and thus to an improved linewidth and reduce the spectral overlap. Or for example, with smaller voxels, excitation of lipid resonances can be avoided, reducing overlap due to lipid signals. In general, however, special sequences are needed to improve the spectral separation of metabolites[17]. One way to achieve this is spectral editing, where the acquisition of the spectrum is altered to reduce the overlap of certain metabolites. J-difference editing is probably the most common way, how spectral editing is achieved today in in vivo measurements[17]. Signals from metabolites with J-coupling can be separated from metabolites without J-coupling by influencing the J-coupling evolution. Alternatively, two-dimensional spectra can be acquired as shown in Figure 20.4, where a 2D J-resolved spectroscopy (JPRESS) spectrum[18],[19] is shown. In this case, the J-coupling evolution is used to shift only signals from metabolites with J-coupling into the second dimension of the final spectrum. This generally improves spectral separation.

Unlike using advanced sequences, spectral separation can also be improved by measuring at field strengths higher than 3T[20]. Figure 20.5 demonstrates improved spectral separation at 7T compared to a measurement at 3T.

Excursus: The relative weakness of common MRS sequences in the determination of small concentrations is even more evident in the study by Heimer at al. than in the ethanol study mentioned above. Due to the weak sensitivity of MR in general and MRS in particular, concentrations below 1 mM/L are hardly detectable. Lower concentrations are not sufficient for a sufficient SNR at common voxel sizes and acquisition duration. However, even at sufficiently high concentrations, a metabolite may be overlapped by another substance in the spectrum, leading to uncertainty in the fit. On the one hand, this can cause low-concentrated metabolites to go undetected when they are actually there. For example, if the BHB signal at 1.19 ppm is overlapped by a high ethanol signal, which is resonating right next to it at 1.17 ppm. On the other hand, if there is a very high ethanol signal, some of the ethanol signal may be assigned to BHB in the fitting process, although there is little or no BHB present. The latter may partially be due to inaccuracies in the simulation of the metabolite signals, insufficiently modeled contributions to the spectrum or due to measurement artifacts[21].

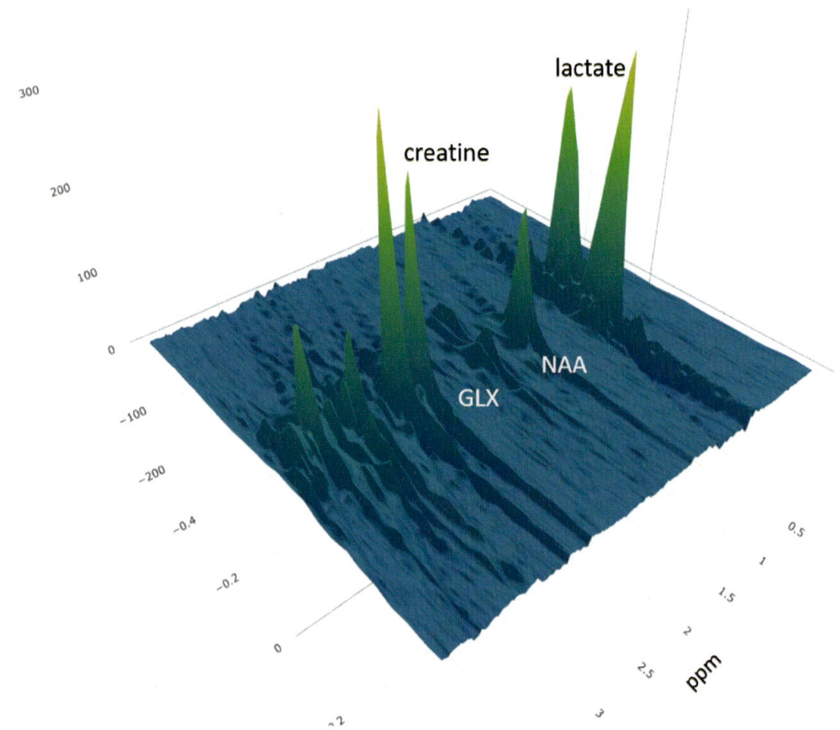

Figure 20.4 Experimental postmortem 2D JPRESS spectrum. JPRESS consists of a series of measurements with PRESS localization and different echo times. After 2D Fourier transform, this leads to a 2D spectrum where the evolution due to J-coupling (only observed for J-coupled protons) leads to a shift of the peaks into the second dimension. This can be observed for example for the peaks from lactate or glutamate and glutamine (GLX) while the signals from uncoupled protons (e.g. from NAA or Creatine) remain on the zero line. The available information is spread over a larger space, which can lead to improvements in the separation of the signals from different metabolites.

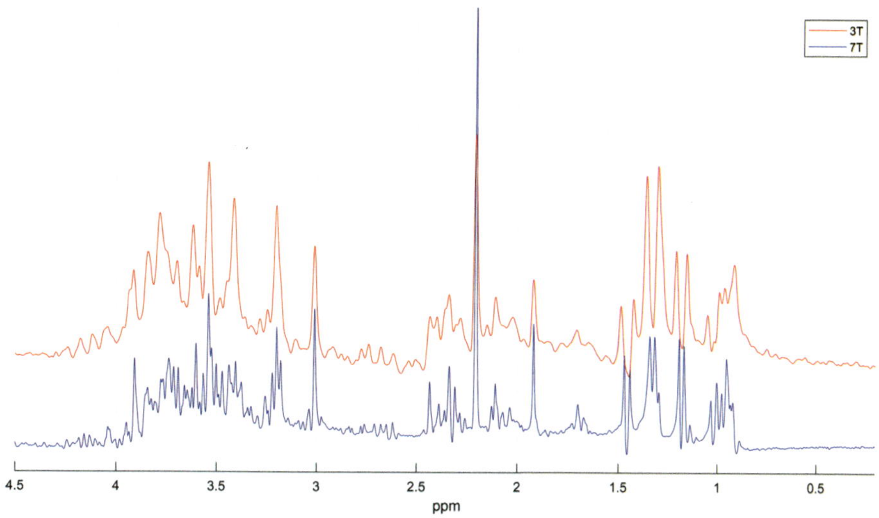

Figure 20.5 Postmortem spectra acquired at 3T and 7T in situ in the CSF of a deceased patient with fatal diabetic ketoacidosis. The spectra were acquired with different sequences and settings, therefore peak heights are not directly comparable. However, the 7T spectrum shows significantly improved spectral separation. Compared to the frequency differences associated with the chemical shifts, which increase with field strength, the splitting into multiple peaks due to J-coupling, which is independent of B0, becomes smaller. As a result, the signal pattern from J-coupled protons appear in a smaller region of the spectrum, reducing spectral overlap. This can be observed at several positions in the spectrum, but especially in the 1 to 1.5 ppm region where we observe the doublets of alanine, lactate and BHB. Compared to 3T, these doublets are well separated at 7T, which ultimately facilitates quantification. In brain tissue, the improvements in spectral separation are normally less obvious. There, the increased chemical shift dispersion is partly offset by an increase in linewidth. Details of the acquisitions parameters: 3T: (PRESS localization, TE 28ms, TR = 2300 ms, voxel size 31 × 9 × 9 mm^3, NSA = 5 × 64) vs 7T: (sLASER localization, TE 35 ms, TR = 10,000ms, voxel size 31 × 7 × 8 mm^3, NSA = 64).

20.5 METABOLOMIC CLASSIFICATION OF CAUSE OF DEATH

Metabolomic classification of cause of death is a future idea for postmortem MRS, which so far has not been exploited. This idea is based on the notion that finely tuned concentrations of endogenous metabolites depend on the cause of death, as described by Maeda et al[22]. Taking into account the preceding "deterioration of vital functions" during the agonal period[22], which influences metabolite concentrations, the identification of the cause of death might become tangible. According to Maeda et al.[23], despite convincing results, such approaches have hardly found their way into routine casework, perhaps because of the expected interference of the agonal period and the postmortem interval and the lack of confirmed data. This lack of data is a consequence of the high variance in postmortem metabolite concentrations as a function of PMI, sample temperature, sample tissue or fluid, pre- and post-sampling metabolite stability, and assay procedures and instrumentation that are often optimized for clinical biochemistry[24]. MRS can potentially solve these problems: Because MRS always estimates the concentrations of several compounds (>15) for each measurement, no a priori assumptions and no specific indication are required. If the composition of the metabolites derived from the MRS spectrum changes as a function of the cause of death, a conclusion can be drawn. A cause-of-death prediction model can be built based on the metabolomic pattern, given a sufficient number of human postmortem MRS measurements.

An inherent obstacle in forensic pathology is the uncertainty of the "cause of death" classification, which makes unsupervised learning clustering, such as k-means or principal component analysis, an attractive choice[25]. An example for a metabolic classification of the cause of death is given by Hirakawa et al.[26], who studied samples from 72 rat muscles using NMR-spectroscopy. The rats were sacrificed in three different ways: Asphyxiation, cocaine overdose, and pancuronium-induced asphyxia. The authors used principal component analysis for dimension reduction and found that the different causes of death were separable 15 min postmortem, but the principal component values became more similar with increasing PMI. They also predicted PMI (<15 minutes, 1–4 hours, >8 hours) using a support vector machine, which was able to correctly classify 96.8% of the training data.

20.6 INTRACORPOREAL FLUIDS

Another possible application of MRS in forensic medicine is shown in the work of Heimer et al. on a fatal traffic accident[27]. CT images had initially shown a peculiar lesion in the left thigh of a 12-year-old boy, which subsequent MRS and MRI examination identified as fluid collection formed due to a misplaced intraosseous infusion. In this case, the fluid surrounded the uninjured muscle fibers. The exact identification of the fluid was of little relevance in this specific case, but nevertheless the strengths of MRS are clearly demonstrated here. MRS allows measurements at the desired location in fluid that otherwise could have been examined only by sampling and that might be difficult to collect or easily overlooked during an autopsy. MRS measurements in this volume, which contained a mixture of muscle tissue[28] and fluid, showed significant changes compared with measurement in the right thigh at the same location as shown in Figure 20.6. Thus, the high peak at 1.9 ppm, which can be assigned to acetate, led to the identification of the fluid. Based on the different linewidths of the peaks, which can be traced back to the strongly different T_2 relaxation times, it was possible to distinguish whether the substances are present in the muscle tissue or in the liquid itself.

This example shows the potential of 1H-MRS to investigate and distinguish intracorporeal fluids. This potential should also be applicable for other intracorporeal fluids e.g. hematoma, effusions, abscesses, ascites, putrefaction fluids and provide useful information for forensic investigations.

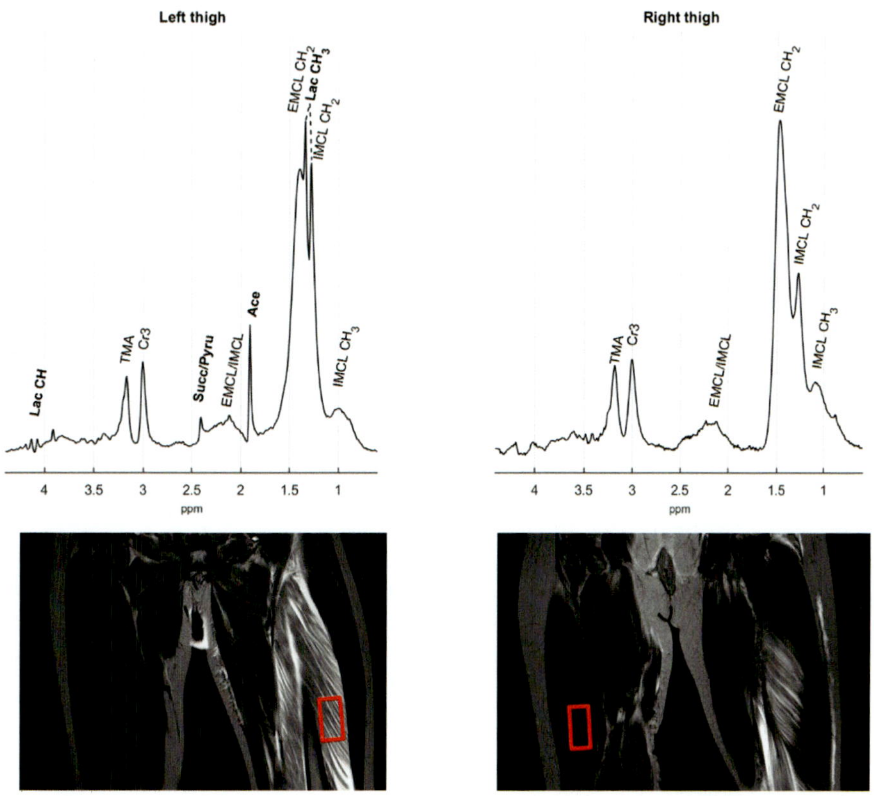

Figure 20.6 Obtained 1H-MRS spectra from the left intermediate vastus saturated in the unknown fluid (left image) and from the unaffected intermediate vastus of the right thigh (right image). The common peaks observed in muscle spectra are denoted according to Boesch et al.[27]: TMA: trimethylammonium-containing compounds, Cr3: the methyl group of creatine, EMCL: extramyocellular, and IMCL: intramyocellular lipids. Additionally observed metabolites in the left thigh: Ace: acetate, Lac: lactate and Succ/Pyru: succinate and/or pyruvate.

20.7 REPEATED MEASUREMENTS

Due to its non-invasiveness, MRS offers the possibility to detect metabolic changes in an object over a longer period without causing changes to the object itself. This feature makes MRS excellent for studying postmortem changes under real-world conditions as demonstrated by the first applications of MRS in legal medicine[4–6]. A more recent work that has exploited this peculiarity of MRS is the work of Gascho et al.[29]. They investigated whether the storage at low temperatures leaves traces in muscle tissue or bone marrow that can be distinguished from normal postmortem changes using MRS and MRI. These studies in intact sheep legs make the best use of the power of MRS examination. In addition to the fact that a sheep leg can be measured multiple times without alteration/intervention, several pieces of information are utilized that could be lost if samples were taken and the subsequent necessary processing was performed. Thus, in this work not only a decrease in certain lipid signals (decrease in the bulk methylene signal) was detected, but also changes in the relaxation behavior (decay pattern) of two peaks were observed. Relaxation times are particularly sensitive to the motion of the molecules (overall tumbling time) and thus also on the environment of the molecules. Thus, MRS allows in principle to detect subtle changes in the environment of the measurable molecules and not only of the water as with imaging.

MRS could be used to study a variety of postmortem changes. For example, based on our experience, changes in glucose concentration in different tissue types would be a worthwhile future investigation. Or the postmortem changes in muscle tissue, which has only been studied in muscle samples so far. The only (known) influence of the MR measurement on the measured object is a possible temperature increase. Temperature rises are determined by the specific absorption rate of the sequences used, which is usually lower for MRS measurements than for imaging sequences and can lead to heating in the range of a few degrees. As explained in the following section, the temperature can also be measured by MRS during the measurement and thus this influence can be controlled.

20.8 MRS THERMOMETRY

This section illustrates the versatility of MRS, which goes far beyond the collection of the metabolic information. As explained in the technical chapter on MRS, the field at the location of the proton is strongly influenced by the surrounding electrons, which leads to an attenuation of the applied B_0 field and thus a decrease in the measured resonance frequency, i.e. peak position. In the case of water, this attenuation and hence the resonance frequency is temperature dependent[30]. The temperature dependence is governed by hydrogen bonds that the water molecules form. At low temperatures, these hydrogen bonds are strong and the oxygen atoms pull the electrons away from the protons. As the temperature rises, the water molecules move more strongly, resulting in weaker hydrogen bonds. As a result, the electrons are more likely to be at the site of the nucleus than at low temperatures, meaning that the attenuation is stronger at higher temperatures and thus the resonance of water protons is lower at high temperatures than at cold temperatures[31]. However, since the resonance frequency also depends on B_0 and is therefore affected by field inhomogeneities or field drifts, the temperature cannot be directly determined from the position of the water peak. To do this, temperature

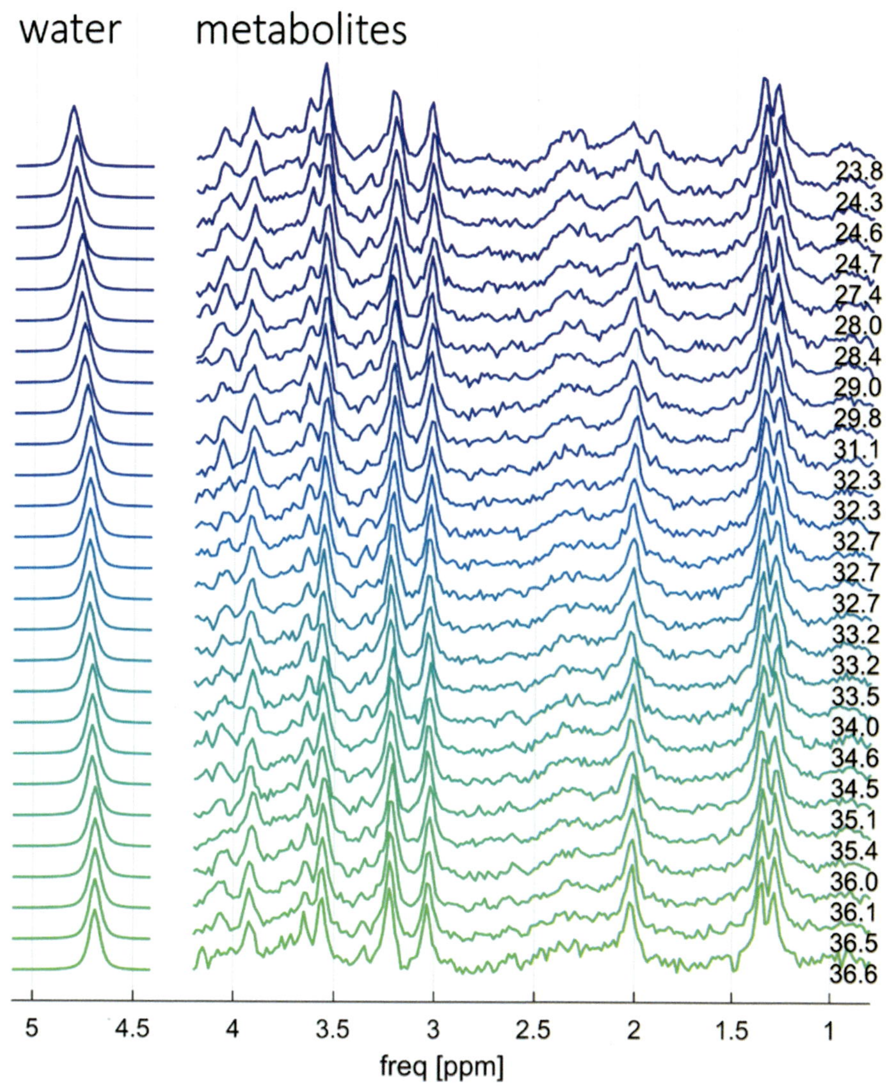

Figure 20.7 MRS measurements at different temperatures in brain tissue of a severed sheep head. Measurements were started shortly after death and continued in irregular time intervals until the head reached room temperature. The first spectrum is shown at the bottom drawn in light green. Both the spectrum of the metabolites and the water spectrum, which was recorded in a separate scan without water suppression, are shown. On the right side, the measured temperature is given for each spectrum in. The signal from lactate was used as temperature stable reference, that is, each spectrum was shifted so that the methyl signal from lactate appeared at 1.31 ppm. While only minor temperature effects are visible for metabolites (creatine getting closer and closer to 3 ppm) and rather PMI effects dominate (decrease of NAA 2.02 ppm, increase of acetate 1.91 ppm) water shows a clear increase of resonance frequency with decreasing temperature.

stable reference peaks are needed. This is where MRS comes into play. Peaks of N-acetylaspartate (NAA), creatine and lactate[30,32] are known to be relatively temperature stable over the range of about 5 to 40°, which is the relevant range for postmortem studies. NAA and creatine are metabolites that are readily measurable in vivo. If a temperature stable reference peak is available, the temperature can be read from the position difference of water and the reference peak with the help of so-called calibration curves. Calibration curves are obtained by simultaneously measuring the temperature and the position difference[32].

Using calibration curves obtained in living pigs[33], we have recently gained some initial experience with this technique in the postmortem muscle tissue of the human thigh in 24 intact cadavers with a PMI up to 50 h. We compared the MRS based temperature measurements (T_{MRS}) with the rectally measured body temperature. T_{MRS} values were in general slightly lower than the rectal temperature (median difference −0.9°C) as would be expected based on the slightly different cooling rates. The MRS temperatures showed a very good agreement between the left and right thigh (median difference −0.4°C). Primarily, there is no systematic influence of the PMI on the MRS temperature in the measured cases.

For postmortem measurements in brain tissue, creatine or lactate, which is easily detectable within few minutes after death, can be used as a reference peak. NAA has been described to decrease with increasing PMI and therefore seems to be an option only for short PMIs. How well the calibration curves obtained in vivo or in phantoms work also for postmortem MRS temperature measurements depends on the stability of the reference signals against pH and ionic strength[34]. We could show that calibrations determined in vivo or in phantoms and adapted for lactate, which is hardly measurable in vivo, are well suited for postmortem MRS thermometry[35]. Even in severed sheep heads with pronounced postmortem effects observed in the spectrum, MRS thermometry allowed reliable temperature measurements up to 50 hours after death (see Figure 20.7) with no influence of the PMI on the precision. After that, the spectral quality might be degraded due to gas formation or other postmortem effects, making the determination of the lactate peak unreliable. In intact cadavers, temperature measurement should be possible even with PMIs twice or three times as long. Based on our initial experience, it seems that the precision of postmortem MRS thermometry should be at least good enough for corrections for relaxation times or concentration measurements, where a precision in the order ±2°C is sufficient. In addition, parameter adjustments for MRI protocols based on MRS thermometry could also be an interesting application. Or the recording of the temperature distribution within an object, which is otherwise not possible to access and could possibly play a role in the analysis of the position of the corpse or PMI estimation.

References

1. Thali MJ, Dirnhofer R, Vock P. *The virtopsy approach: 3D optical and radiological scanning and reconstruction in forensic medicine*. CRC Press; 2009.
2. Chatzaraki V, Heimer J, Thali M, Dally A, Schweitzer W. Role of PMCT as a triage tool between external inspection and full autopsy—case series and review. *Journal of Forensic Radiology and Imaging*. 2018;15:26–38.
3. Tappero C, Fliss B, Gentile S, Thali MJ, Gascho D. Postmortem computed tomography and postmortem magnetic resonance imaging as triage tools for autopsy—a case report. *Journal of Forensic Radiology and Imaging*. 2018;14:65–67.
4. Ith M, Bigler P, Scheurer E, et al. Observation and identification of metabolites emerging during postmortem decomposition of brain tissue by means of in situ 1H-magnetic resonance spectroscopy. *Magnetic Resonance in Medicine*. 2002;48(5):915–920.
5. Scheurer E, Ith M, Dietrich D, et al. Statistical evaluation of time-dependent metabolite concentrations: Estimation of postmortem intervals based on in situ1H-MRS of the brain. *NMR in Biomedicine*. 2005;18(3):163–172.
6. Ith M, Scheurer E, Kreis R, Thali M, Dirnhofer R, Boesch C. Estimation of the postmortem interval by means of 1H MRS of decomposing brain tissue: Influence of ambient temperature. *NMR in Biomedicine*. 2011;24(7):791–798.
7. Zoelch N, Hock A, Steuer AE, et al. In situ postmortem ethanol quantification in the cerebrospinal fluid by non-water-suppressed proton MRS. *NMR in Biomedicine*. 2019;32(5):e4081.
8. Meyerhoff DJ. Brain proton magnetic resonance spectroscopy of alcohol use disorders. In: *Handbook of Clinical Neurology*. Vol 125. Elsevier; 2014:313–337.
9. Viola A, Nicoli F, Denis B, et al. High cerebral scyllo-inositol: A new marker of brain metabolism disturbances induced by chronic alcoholism. *MAGMA*. 2004;17(1):47–61.
10. Deelchand DK, Eberly LE, McCarten JR, Hemmy LS, Auerbach EJ, Marjanska M. Scyllo-inositol: Transverse relaxation time constant at 3 T and concentration changes associated with aging and alcohol use. *NMR in Biomedicine*. 2023:e4929.
11. Chiu T-M, Mendelson JH, Woods BT, Teoh SK, Levisohn L, Mello NK. In vivo proton magnetic resonance spectroscopy detection of human alcohol tolerance. *Magnetic Resonance in Medicine*. 1994;32(4):511–516.
12. Heimer J, Gascho D, Chatzaraki V, et al. Postmortem ^{1}H-MRS—detection of ketone bodies and glucose in diabetic ketoacidosis. *International Journal of Legal Medicine*. 2018;132(2):593–598.
13. Heimer J, Gascho D, Fliss B, Martinez RM, Zoelch N. Detection of elevated ketone bodies by postmortem 1H-MRS in a case of fetal ketoacidosis. *Journal of Forensic and Legal Medicine*. 2018;59:16–19.

14. Heimer J, Gascho D, Madea B, et al. Comparison of the beta-hydroxybutyrate, glucose, and lactate concentrations derived from postmortem proton magnetic resonance spectroscopy and biochemical analysis for the diagnosis of fatal metabolic disorders. *International Journal of Legal Medicine.* 2020:1–10.
15. de Barros NP, Slotboom J. Quality management in in vivo proton MRS. *Analytical Biochemistry.* 2017;529:98–116.
16. Ordidge RJ, Connelly A, Lohman JA. Image-selected in vivo spectroscopy (ISIS): A new technique for spatially selective NMR spectroscopy. *Journal of Magnetic Resonance (1969).* 1986;66(2):283–294.
17. Choi IY, Andronesi OC, Barker P, et al. Spectral editing in (1) H magnetic resonance spectroscopy: Experts' consensus recommendations. *NMR in Biomedicine.* 2020:e4411.
18. Fuchs A, Boesiger P, Schulte RF, Henning A. ProFit revisited. *Magnetic Resonance in Medicine.* 2014;71(2):458–468.
19. Schulte RF, Lange T, Beck J, Meier D, Boesiger P. Improved two-dimensional J-resolved spectroscopy. *NMR in Biomedicine.* 2006;19(2):264–270.
20. Xin L, Tkac I. A practical guide to in vivo proton magnetic resonance spectroscopy at high magnetic fields. *Analytical Biochemistry.* 2017;529:30–39.
21. Kreis R. Issues of spectral quality in clinical 1H-magnetic resonance spectroscopy and a gallery of artifacts. *NMR in Biomedicine.* 2004;17(6):361–381.
22. Maeda H, Ishikawa T, Michiue T. Forensic biochemistry for functional investigation of death: Concept and practical application. *Legal Medicine.* 2011;13(2):55–67.
23. Maeda H, Zhu B-L, Ishikawa T, Quan L, Michiue T. Significance of postmortem biochemistry in determining the cause of death. *Legal Medicine.* 2009;11:S46–S49.
24. Madea B, Musshoff F. Postmortem biochemistry. *Forensic Science International.* 2007;165(2–3):165–171.
25. Celebi ME, Aydin K. *Unsupervised learning algorithms.* Springer; 2016.
26. Hirakawa K, Koike K, Uekusa K, Nihira M, Yuta K, Ohno Y. Experimental estimation of postmortem interval using multivariate analysis of proton NMR metabolomic data. *Legal Medicine.* 2009;11:S282–S285.
27. Heimer J, Gascho D, Tappero C, Thali MJ, Zoelch N. Noninvasive analysis and identification of an intramuscular fluid collection by postmortem 1H-MRS in a case of a fatal motor vehicle accident. *International Journal of Legal Medicine.* 2019.
28. Boesch C. Musculoskeletal spectroscopy. *Journal of Magnetic Resonance Imaging: An Official Journal of the International Society for Magnetic Resonance in Medicine.* 2007;25(2):321–338.
29. Gascho D, Richter H, Karampinos DC, et al. Noninvasive in situ proton MRS in muscle tissue and bone marrow as a novel approach to identify previous freezing in a completely thawed cadaver. *NMR in Biomedicine.* 2020;33(2):e4220.
30. Lutz N, Kuesel A, Hull W. A 1H-NMR method for determining temperature in cell culture perfusion systems. *Magnetic Resonance in Medicine.* 1993;29(1):113–118.
31. Kuroda K. Non-invasive MR thermography using the water proton chemical shift. *International Journal of Hyperthermia.* 2005;21(6):547–560.
32. Zhu M, Bashir A, Ackerman JJ, Yablonskiy DA. Improved calibration technique for in vivo proton MRS thermometry for brain temperature measurement. *Magnetic Resonance in Medicine.* 2008;60(3):536–541.
33. Yoshioka Y, Oikawa H, Ehara S, et al. Noninvasive measurement of temperature and fractional dissociation of imidazole in human lower leg muscles using 1H-nuclear magnetic resonance spectroscopy. *Journal of Applied Physiology (1985).* 2005;98(1):282–287.
34. Babourina-Brooks B, Simpson R, Arvanitis TN, Machin G, Peet AC, Davies NP. MRS thermometry calibration at 3 T: Effects of protein, ionic concentration and magnetic field strength. *NMR in Biomedicine.* 2015;28(7):792–800.
35. Zoelch N, Heimer J, Richter H, Thali MJ, Gascho D. Finding the optimal temperature calibration for postmortem MRS thermometry in forensic medicine. ISMRM Annual Meeting; 2022; London, England, UK.

CHAPTER 21

BONE MRI AND SYNTHETIC CT—BASICS AND FORENSIC POSTMORTEM APPLICATIONS

Dominic Gascho and Niklaus Zoelch

DOI: 10.4324/9781003413998-23

21.1 INTRODUCTION

For the three-dimensional radiological examination of deceased persons in the context of "Virtopsy" or "virtual autopsy", computed tomography (CT) is usually mentioned as the examination method of choice and magnetic resonance imaging (MRI) as a potential supplementary examination to it. While a postmortem CT scan radiologically documents the whole body in a few minutes, an additional MRI examination is used for specific postmortem examinations, for example of the head and neck region after trauma. The strengths of a CT scan are visualization of bone and lung tissue and highlighting of metallic foreign bodies, such as lodged projectiles. The limited contrast of soft tissue in the X-ray absorption images of a CT scan is considered a weakness compared to MRI, which overcomes this limitation. That is why it is often read that a complementary MRI provides better visualization of soft tissue anatomy and soft tissue injuries than CT. Since soft tissue injuries are relevant for the radiological assessment of traumatic brain injuries in addition to fractures and bone defects, complementary MRI is important for proper postmortem forensic neuroimaging examinations. Finally, tomography data from both imaging modalities can then be visualized side by side in the same anatomical orientation on a split screen using image registration or overlaid using image fusion (Figure 21.1) [1–3].

Although the image registration and image fusion in the head area works well, the positioning of the head in a head coil for the MRI examination results in differences in the degree of flexion and extension of the neck spine and also the position of the jaw can change. Altogether, this leads to difficulties for the registration and fusion. Image fusion offers the advantage of displaying the bony anatomy and the soft tissue anatomy in one image stack, but the transparency of the superimposed data set required for this leads to losses in image quality. It would be advantageous to have a single imaging modality in a traumatic injury to show both soft tissue injuries and adjacent fractures in good image quality. And indeed, this is possible in the field of CT as well as MRI. In CT imaging, the implementation of phase-contrast and dark-field imaging in medical CT scanners in the coming years will open up new possibilities for three-dimensional imaging of soft tissue with X-rays [4–6]. In MRI, on the one hand, special sequences already offer the possibility of displaying bony structures in good image quality, and on the other hand, atlas-based or deep learning algorithms are used to generate synthetic CT images from MRI data sets. This chapter presents bone MRI using special sequences and describes the creation of a synthetic CT dataset with a focus on postmortem forensic examinations. For the basic principles of MRI and for the general description of deep learning, the reader is referred to the corresponding chapters in this book.

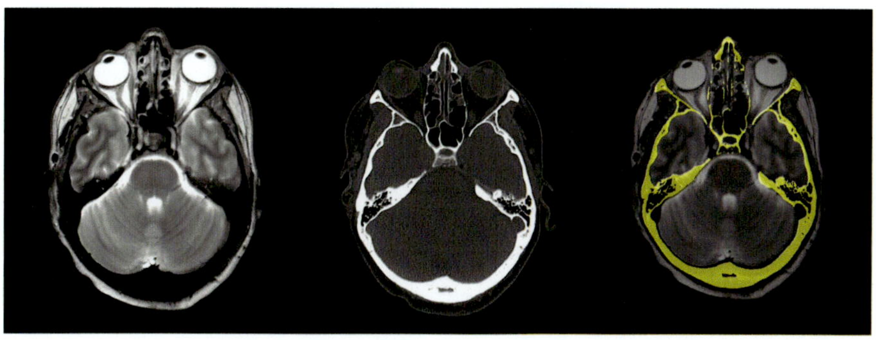

Figure 21.1 Axial cross-sectional images of the head. The MRI (image on the left) shows good contrast of the soft tissue but the bony structures are poorly visible. In contrast, CT (image in the middle) shows the bones in detail but the soft tissue contrast is poor. One way to take advantage of both imaging modalities is to fuse the data. Image fusion (right image) allows the good soft tissue contrast provided by MRI and the detailed representation of the bones provided by CT to be reproduced together in one image, simplifying the assessment of bony defects and soft tissue injuries associated with the same traumatic impact, as is the case, for example, with gunshot injuries [3]. However, the transparency of the superimposed images also leads to a reduction of the original image contrast of the respective imaging modality.

21.2 WHAT IS BONE COMPOSED OF AND WHAT DOES THIS MEAN FOR MRI?

In the human body there are more than 200 bones, which can be divided into long bones, short bones, flat bones, and irregular bones. For example, the occipital, parietal, and frontal bones are flat bones, while the temporal bone and vertebrae are irregular bones. The outer dense part of the bone, which consists of a solid bone matrix, is called compact bone (cortical bone), while the interior of the bone, which consists of a cancellous structure with numerous holes and passages through which nerves and blood vessels pass, is called spongy bone (trabecular bone) (Figure 21.2).

The main components of cortical bone are the bone mineral and the organic matrix. The bone mineral consists of calcium and phosphate in the form of hydroxyapatite (HA), which accounts for about 40% of the cortical bone volume. The MR signal of the protons bound in the chemical structure of the HA crystals cannot be detected by a conventional clinical MR unit. The organic matrix consists of collagen fibrils and water. The (type I) collagen fibrils account for about 30% of the cortical bone volume. The T_2* decay of collagen backbone protons is extremely fast (20 µs) [7]. The water accounts for about 20% of the cortical bone volume. However, a large part of the water is bound to HA crystals and collagen (bound water) and only a smaller part of the total water is in "free" form in pores such as Haversian channels (pore water). The T_2* decay of bound water is very fast (300 to 400 µs) [7]. The molecular tumbling of pore water is restricted to the tiny space in the pores, and thus, these hydrogen proton pools show very short T_2* values in the order of a few milliseconds [7,8]. Like compact cortical bone, the trabeculae in trabecular bone are mainly composed of hydroxyapatite, collagen, and bound water. Accordingly, the MR signal of the trabeculae is also too short to be detected by conventional MR sequences and therefore appears dark on conventional MRI. However, in trabecular bone, the total water accounts for more than 80% of the trabecular bone volume and is present mainly as compounds with fat and red bone marrow in the interstitial spaces of the trabeculae [7]. The fat and water in the interstitial spaces of the trabeculae provide a sufficiently long MR signal to be displayed on conventional T_1-weighted imaging. This is the reason why in bone the inner trabecular part is shown bright on T_1 and only the outer cortical bone layer is shown dark.

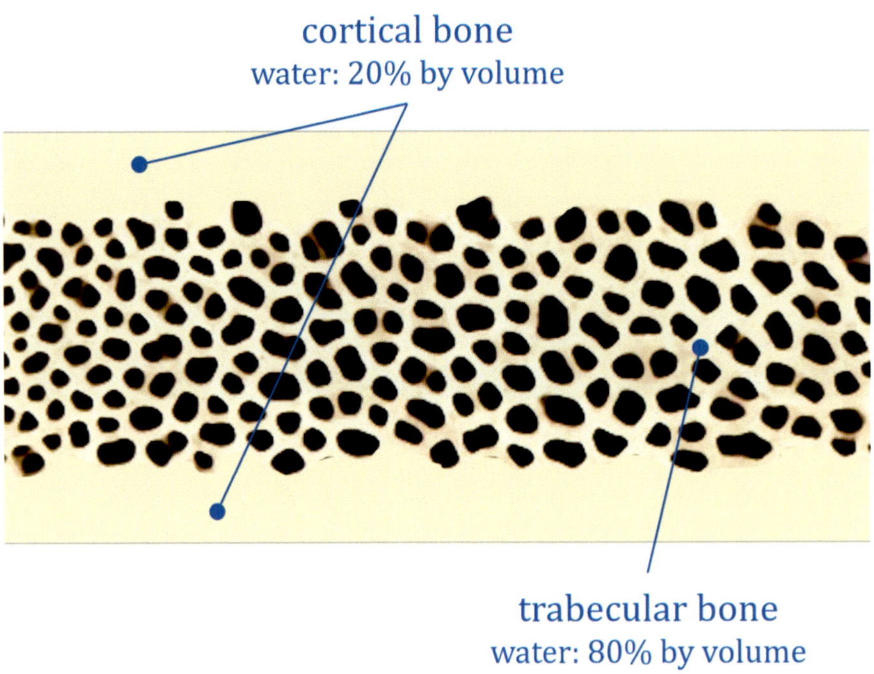

Figure 21.2 The solid part of the bone is called cortical bone, while the cancellous part of the bone is called *diploë* or trabecular bone. Small pores in the cortical bone, such as Haversian canals, allow tiny blood vessels and nerves to pass through them and supply the osteocytes. Water in the cortical bone accounts for 20% of the volume, much of which is bound water. The water in the trabecular bone accounts for 80% of the volume and is present in the form of red bone marrow.

21.3 WHAT IS MEANT BY "BLACK BONE" MRI AND FRACTURE MRI?

For imaging biological structures with rapid T_2* decay, a short echo time (TE) is set. However, with standard MR scanning equipment and conventional MR sequences, the TE can only be reduced to a few milliseconds. Accordingly, it is not possible to detect the MR signal of the aforementioned components of the bones with very fast T2* decay. Thus, even at a TE of a few milliseconds (ms), the cortical bone is still dark and only a weak signal of fat and water in the interstitial spaces of the trabeculae can be visualized. For example, the skull plate bones with few trabecular components are largely dark or "black". Hence the name "black bone" MRI, in which, in addition to a short TE, the flip angle is also chosen to be small, so that the soft tissue is shown in low-contrast grayscale and the bony structures are shown in black due to their low signal [9,10]. Based on images obtained from a standard 1.5 Tesla MR unit, where the TE of a gradient echo sequence could be reduced to 4.2 ms, a flip angle of 5° was found the ideal flip angle to effectively suppress signal from both fat and water [9]. These parameters were later adopted for studies on 3 Tesla MR units. This method of "black bone" imaging was presented as an alternative to CT scanning in pediatric examinations, such as in the preoperative evaluation of craniosynostosis or in the evaluation of skull fractures, although the assessment of bones in areas where they are adjacent to air, such as in the craniofacial sinuses, is challenging because the cortical bone and air are black in the images [11–13].

Another "black bone" imaging approach is FRACTURE, an abbreviation for "FFE resembling a CT using restricted echo-spacing" [14,15], where FFE (fast field echo) is a manufacturer-specific designation for the gradient echo. As described in the literature [14], this three-dimensional gradient echo sequence uses a three times larger flip angle (15°) compared to "black bone" MRI, but acquires four echoes starting at a TE of 4.6 ms and using a constant echo spacing of 4.6 ms. For example, in an MRI of the head, the contrast between the different types of brain tissue increases over the four echo times, and the signal can also be enhanced when the four echoes are combined in a postprocessing step. Inverting the grayscale, which can be performed in an automated postprocessing step, results in images that resemble a CT image [14,15]. While the first image at a TE of 4.6 ms corresponds to the "black bone" image, the cumulated image allows the visualization of those regions from which no signal could be detected, such as the bone components with very fast T2* decay, together with soft tissue regions from which signal was detected (Figure 21.3). The cumulated FRACTURE image thus allows the visualization of those regions from which no signal could be detected, such as the bone components with very fast T2* decay, together with soft tissue regions from which signal was detected. The acquisition of multiple echoes with in-phase temporal resolution promises a better delineation and localization of the bone with less edge blurring, and a reduction in signal loss at the bone tissue interfaces [15].

Accordingly, both "Black Bone" and FRACTURE MRI play with the contrast difference between the tissue from which MR signals are detected and bone components with very rapid T2* decay from which MR signals are not detected. However, special sequences have also been developed that actually allow MR signals to be detected from these bone components, which are described in the following sections.

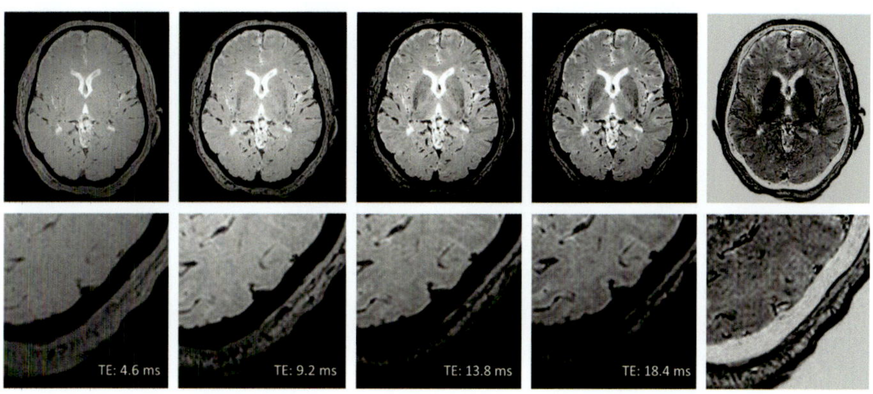

Figure 21.3 Postmortem FRACTURE MRI of the skull in transverse view using a 3 Tesla MR unit. The images with the four individual echo times are shown in ascending order from left to right and the cumulated and inverted image calculated from them is shown on the far right. At a very short TE of 4.6 ms, the black bone area is well delineated from the gray soft tissue, and the soft tissue shows little contrast. This representation corresponds to "black bone" MRI. While the soft tissue contrast increases with increasing TE, the black bone region shows more and more blurring. The cumulative image shows a good compromise between the sharp delineation of the bone region and the soft tissue contrast. The cumulated image was additionally inverted to show the bone region bright, as one is used to from CT images.

21.4 WHAT ARE UTE AND ZTE SEQUENCES?

Special sequences with an extremely short echo time (TE) allow the detection of rapidly decaying bone signals, which is why these sequences are called ultra-short TE (UTE) sequences.

However, if the duration of the RF pulses is similar in length or longer than the T_2 of the component to be excited, then the relaxation during the pulse cannot be neglected. The relaxation during the RF pulse leads to a decrease of the transverse magnetization without a compensatory increase of the available longitudinal magnetization as there is a competition between the RF pulse, which rotates the magnetization into the xy-plane, and the dephasing that occurs in this plane [16]. For the UTE sequence, therefore, very short radiofrequency (RF) pulses are required. To save further time special k-space trajectories are used for non-Cartesian readout. An example for non-Cartesian readout is center-out spiral encoding, in which no prephaser gradient is used, i.e. the signal readout starts directly in the center of k-space and then spirals outwards. A very commonly used readout method for 3D UTE is radial center-out encoding (Figure 21.4). An advantage of radial center-out encoding is that the center of the k-space is oversampled and thus finally the low spatial frequencies of the image have a better signal-to-noise ratio (SNR) than high frequencies [16] (Figure 21.5).

In summary, UTE sequences allow the detection of the MR signal from bound water in cortical bone (in addition to the MR signal from free water in the interstitial spaces of the trabeculae), and the bone itself is accordingly represented in the image. For the optimization of the contrast between the bone components with short T_2 values and the tissue types with long T_2 values, the latter are additionally suppressed efficiently, for which there are different methods mostly based on saturation or inversion [17]. The protons on the native collagen molecule, however, do not contribute directly to the image with UTE sequences. Even faster methods are needed for this purpose.

An even faster method for encoding the rapid decaying MR signal of bone is a method termed zero TE (ZTE), in which the readout gradient is ramped up before the excitation with a short, hard RF pulse is performed, so that data acquisition immediately follows excitation [18]. As early as the 1990s, such ZTE methods, in which the RF excitation is performed with the gradient switched on, were introduced as back-projection low angle shot (BLAST) and rotating ultra-fast imaging sequence (RUFIS) in nuclear magnetic resonance (NMR) imaging [19,20]. Due to the fact that the

Figure 21.4 Cranial images of a deceased person obtained with a UTE sequence using a 3 Tesla MR unit in the form in which they were acquired (left row of images) and, for better illustration of the cranial bones, in inverted form (right row of images). The repetition time (TR) was 10.6 ms and the echo time (TE) was 0.2 ms. The flip angle was 10°. The flat cranial bones are clearly delineated from the soft tissues and significantly more MR signal could be detected from the bone region than is possible with "black bone" and FRACTURE MRI. Note also the low-contrast of the soft tissue.

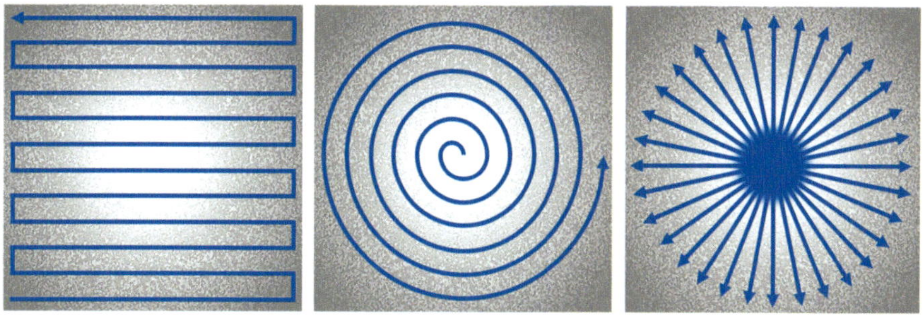

Figure 21.5 Schematic representation of fast k-space trajectories for signal encoding. The rapid Cartesian readout (left image), shown here for comparison, is used for echo-planar imaging (EPI) sequences. For UTE imaging the non-Cartesian center-out spiral encoding (middle image) or more frequently center-out radial encoding is employed. To cover the entire 360° with radial encoding, the read-out must be repeated accordingly many times. The sampled data points are then gridded onto a Cartesian grid, a rectangular matrix, prior to Fourier transformation to create the final image.

readout gradient is switched on before the excitation, the ZTE is an inherent 3D sequence, also referred to as free induction decay (FID) projection imaging, because the simultaneous application of a slice selection gradient is not possible. However, the application of ZTE MRI in humans is constrained by the limited RF switching speed of clinical scanners, and in addition, the hard RF pulse must be sufficiently short, while at the same time its amplitude must be large enough to allow a suitable flip angle [18]. To overcome these limitations, custom-made coils and special insert gradient systems have been developed for high-bandwidth ZTE imaging of the human head and lower extremities of humans [21,22]. Special pulse sequence techniques to circumvent the need for a short, hard pulse with very high amplitude have also been developed, such as sweep imaging with Fourier transform (SWIFT). This method uses a frequency-modulated pulse sequence, i.e., the pulse is divided into N segments, and signal acquisition is performed after each segment, which initially results in frequency-encoded projections that are then reconstructed using a 3D back-projection algorithm or gridding [23]. On the one hand, the SWIFT method allows a higher SNR with equivalent T_2 sensitivity compared to a hard pulse excitation, since larger flip angles are possible, but on the other hand, the SWIFT method is accompanied by repeated dead time gaps that take up significant SNR credit, ultimately favoring hard-pulse excitation when the RF amplitude is sufficient [24].

In addition to the excitation pulse required for ZTE imaging, there is another problem related to the change of the coil from transmit to receive mode, which means that the TE is only theoretically zero because the acquisition starts with a delay (dead time) caused by the transmit/receive change, This acquisition gap, also referred to as dead time gap, is usually filled by oversampling the acquisition using a linear algebra-based image reconstruction technique [23]. Other methods to obtain additional data as a substitute for the missing signal samples at longer dead time gaps are water- and fat-suppressed proton projection MRI (WASPI), PETRA (point-wise encoding time reduction with radial acquisition), and hybrid filling (HYFI). WASPI reduces the number of missing samples and by performing additionally a small number of radial projections with reduced gradient strength to sample points near the k-space center [25]. PETRA performs radial half-projections to fill the k-space, but the k-space center is measured individually point-wise on a Cartesian trajectory [26]. By using the PETRA method for ZTE imaging, the k-space center can be filled by an exact value obtained from the single point acquisition resulting in higher SNR as more readout points are acquired before the signal decays [23]. HYFI fills the inner k-space with a small single-point-imaging core surrounded by a stack of shells acquired on radial readouts in an onion-like fashion, and based on simulations, the SNR efficiency is expected to be higher for HYFI than for PETRA while maintaining image quality, so the advantage of HYFI increases with the size of the dead-time gap [27].

In summary, ZTE sequences not only enable direct imaging of cortical bone collagen, but generally offer applications for T2 mapping of ultrafast relaxation signals [27]. In addition to bone imaging, ZTE-based MRI is also of interest in the field of neuroimaging for direct imaging of the myelin bilayer in the brain. The interest in direct imaging of ultrashort T_2 components using ZTE imaging gives hope that this method will become more widely available in the future.

21.5 WHAT IS MEANT BY SYNTHETIC CT?

Synthetic CT, pseudo CT, or substitute CT refers to an absorption image calculated from MRI data. The motivation for this approach comes mainly from radiation therapy, where soft tissue information from MRI and absorption information from CT, i.e. CT numbers in Hounsfield units, are used to plan radiation treatment. For such MRI-only radiation therapy planning, so-called atlas-based techniques with a single standard MRI sequence or voxel-based techniques with T_1/T_2^*-weighted MRI, Dixon imaging, or UTE imaging are used [28] (Figure 21.6).

The atlas-based approaches use image registration to align a target MR image to an atlas MR and the correspondence information can then be used to align a corresponding atlas CT image to the target MR image and generate an sCT, while the voxel-based technique classifies MRI voxels into a discrete set of tissue types, such as air, fat, soft tissue, and bone, and then assigns a different CT number for each tissue type based on bulk density assignments [29]. Special MRI sequences must be used for good discrimination between tissue types. Since atlas-based conversion is highly dependent on accurate registration and has difficulty with atypical anatomy, and voxel-based conversions require special sequences that increase examination time, deep learning-based methods have recently been considered [29–31].

Deep learning (also deep structured learning) is a subfield of machine learning that involves enabling computers to perform tasks based on artificial intelligence without explicit programming. The former learning-based approaches use statistical learning or model-fitting techniques to create a mapping function that then associates MRI voxel intensities or intensity patterns with the corresponding CT numbers [29]. As with the voxel-based technique, however, this also requires the special sequences such as the UTE sequence. Recent developments in deep learning have led to the use of deep convolutional neural network (DCNN) methods and U-net architectures for the generation of synthetic CT data [29,30]. The use of convolution and max-pooling layers for encoding, where the resolution decreases and the number of channels increases, and un-pooling and convolution layers for decoding, where the resolution again increases and the number of channels decreases, describes one of the first U-shaped DCNN models that uses standard T_1-weighted 3D gradient echo sequences as input to generate synthetic CT images as output [29] (Figure 21.7).

Deep learning is a rapidly evolving area of research, and it is expected that future models will be able to generate synthetic CT images from MRI data even more accurately in terms of representing the bony structures and CT numbers.

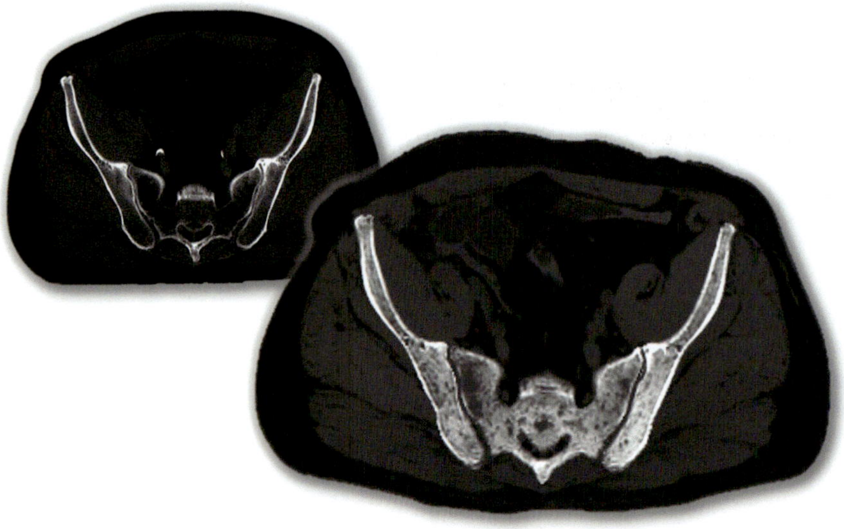

Figure 21.6 Cross-section of the pelvis reconstructed from a postmortem CT scan (background, left image) and the reconstructed synthetic CT dataset from a postmortem MRI examination (foreground, right image). The synthetic CT dataset is based on a Dixon MRI sequence and was calculated using MRCAT (MR for Calculating Attenuation, Philips Healthcare), which is an abbreviation for magnetic resonance for calculating attenuation. MRCAT provides excellent soft tissue contrast for target and organs at risk (OAR) declination, as well as continuous Hounsfield units for dose calculations.

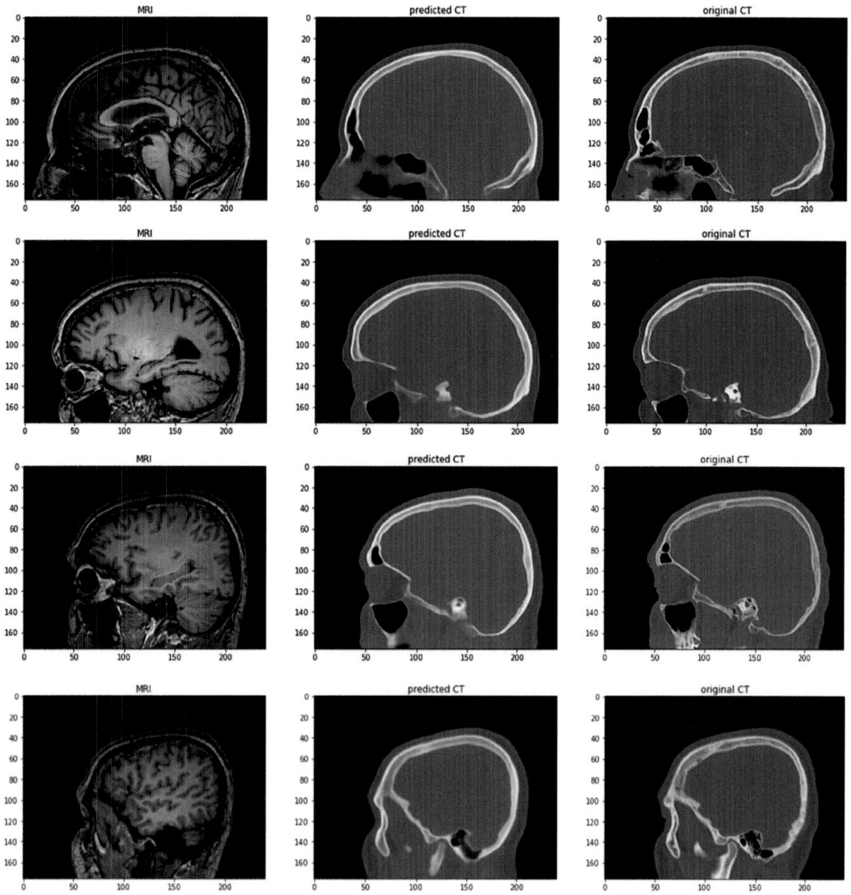

Figure 21.7 Sagittal view of postmortem MRI showing a T1-weighted 3D FFE sequence (left column: MRI), the DCNN model generated from the MRI data (middle column: predicted CT), and the postmortem CT scan (right column: original CT). The DCNN model is based on a model presented in the literature [29]. However, strided convolutional transpose layers were used instead of unpooling layers. The advantage of strided transpose convolutions is that the upsampling is learned and fixed equations are not used as in unpooling layers. Also, the resolutions within the network differ since input and output resolution is 176 × 240 instead of 256 × 256. Therefore, the model applied here uses resolutions of 176 × 240, 88 × 120, 44 × 60, 22 × 30, and 11 × 15 instead of 256 × 256, 128 × 128, 64 × 64, 32 × 32, and 16 × 16. The DCNN model was trained for 10,057 steps (113 epochs). The DCNN model was developed as part of a bachelor thesis entitled "Cracking the Bone" in collaboration with the FHNW University of Applied Sciences and Arts Northwestern Switzerland—School of Engineering.

21.6 WHY ARE BONE MRI AND SYNTHETIC CT OF INTEREST FOR POSTMORTEM EXAMINATIONS?

MR-based imaging of bone basically offers three potential advantages for postmortem examinations. First, the additional cost of a CT scan could be eliminated, which is an organizational advantage. Second, the assignment of bone defects to soft-tissue injuries can be made directly and without positional changes, which is a diagnostic advantage. Third, there are new opportunities to study the bone postmortem because the image is essentially based on the MR signal rather than the CT signal, which is a scientific advantage.

Based on the results of fetal examinations using the "black bone" MRI method, it has been noted that this method of "bone imaging" could adequately represent the fetal skeleton in postmortem examinations of fetuses and eliminate the need for an additional CT examination [32]. FRACTURE MRI has actually been used for postmortem examinations in a forensic context, where this multi-echo "bone imaging" sequence has been evaluated for the forensic assessment of cranio-cerebral gunshot injuries and the resulting damage to the plate bones, such as beveled fractures, radiating fracture lines, intracranial bone fragments or gunshot debris, which provide characteristic findings for identifying the direction of the firing [14]. However, this evaluation has also shown that it can be difficult to distinguish between small bone fragments and air or gas resulting from trauma or decomposition. Consequently, these "bone imaging" sequences are less suitable for the examination of bone defects with adjacent air and deceased persons with a long postmortem interval and corresponding gas formation. In these cases, UTE sequence or ZTE sequences would probably be much more suitable, which can also detect the signal of bound water or the signal of bound water and collagen, respectively. These bone imaging sequences are applicable not only in case of gas formation in the process of decomposition but also in case of dehydration in the process of mummification. The visualization of dehydrated tissues using ZTE imaging was impressively demonstrated for the investigation of mummified human body parts from ancient Egypt [33]. Therefore, such bone imaging techniques are expected to become increasingly common in postmortem MRI examinations (Figure 21.8).

While these MR sequences are intended to depict bone in an MR image, the goal of synthetic CT imaging is to generate a familiar CT image from the MR data. Both approaches offer the possibility to adequately image and assess not only the soft tissues but also the bones with only one imaging modality, namely MRI. While for clinical imaging the avoidance of radiation exposure is in the foreground, the presented bone MRI methods also provided diagnostic advantages, which are also of interest postmortem. The limitations of MRI in assessing bony structures can be substantially reduced compared to CT imaging, and postmortem MRI in particular also has fewer limitations in cases of decay and desiccation. Therefore, it is expected that these methods will find increased application in postmortem forensic medicine.

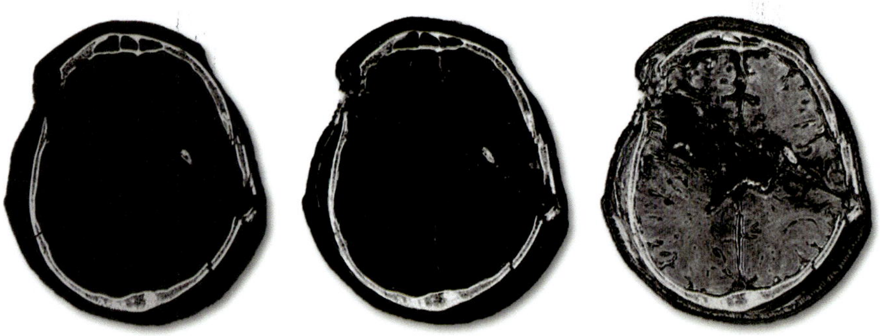

Figure 21.8 Noninvasive examination of a deceased person after a gunshot wound to the head. CT reconstruction with a hard kernel for bone visualizations (left image), the inverted FRACTURE image with the shortest TE of 4.6 ms (middle image), and the inverted cumulated FRACTURE image. Inwardly beveled fractures on the right side correspond to the entry wound, while outwardly beveled fractures on the left side correspond to the exit wound. These two characteristic findings are clearly visible not only on CT but also on FRACTURE MRI. The cumulated FRACTURE image allows the assessment of bone defects along with the associated soft tissue injuries, which is a particular advantage of this bone imaging method.

ACKNOWLEDGMENTS

The authors thank Michael Graber, Florian Schärer, Wai Tsun Chen, and Marco Willi of the FHNW University of Applied Sciences and Arts Northwestern Switzerland—School of Engineering for their work on the DCNN model.

References

[1] D. Gascho, H. Philipp, P.M. Flach, M.J. Thali, S. Kottner, Standardized medical image registration for radiological identification of decedents based on paranasal sinuses, Journal of Forensic and Legal Medicine. 54 (2018) 96–101. https://doi.org/10.1016/j.jflm.2017.12.003.

[2] D. Gascho, P.M. Flach, S. Schaerli, M.J. Thali, S. Kottner, Application of 3D image fusion for radiological identification of decedents, Journal of Forensic Radiology and Imaging. 13 (2018) 12–16. https://doi.org/10.1016/j.jofri.2018.04.002.

[3] D. Gascho, C. Tappero, N. Zoelch, E. Deininger-Czermak, H. Richter, M.J. Thali, S. Schaerli, Synergy of CT and MRI in detecting trajectories of lodged bullets in decedents and potential hazards concerning the heating and movement of bullets during MRI, Forensic Science, Medicine and Pathology. 16 (2020) 20–31. https://doi.org/10.1007/s12024-019-00199-y.

[4] T. Thüring, R. Guggenberger, H. Alkadhi, J. Hodler, M. Vich, Z. Wang, C. David, M. Stampanoni, Human hand radiography using X-ray differential phase contrast combined with dark-field imaging, Skeletal Radiology. 42 (2013) 827–835. https://doi.org/10.1007/s00256-013-1606-7.

[5] K. Willer, A.A. Fingerle, L.B. Gromann, F.D. Marco, J. Herzen, K. Achterhold, B. Gleich, D. Muenzel, K. Scherer, M. Renz, B. Renger, F. Kopp, F. Kriner, F. Fischer, C. Braun, S. Auweter, K. Hellbach, M.F. Reiser, T. Schroeter, J. Mohr, A. Yaroshenko, H.-I. Maack, T. Pralow, H. van der Heijden, R. Proksa, T. Koehler, N. Wieberneit, K. Rindt, E.J. Rummeny, F. Pfeiffer, P.B. Noël, X-ray dark-field imaging of the human lung: A feasibility study on a deceased body, PLoS One. 13 (2018) e0204565. https://doi.org/10.1371/journal.pone.0204565.

[6] L. Birnbacher, E.-M. Braig, D. Pfeiffer, F. Pfeiffer, J. Herzen, Quantitative X-ray phase contrast computed tomography with grating interferometry, European Journal of Nuclear Medcine and Molecular Imaging. 48 (2021) 4171–4188. https://doi.org/10.1007/s00259-021-05259-6.

[7] S. Jerban, Y. Ma, E.W. Dorthe, L. Kakos, N. Le, S. Alenezi, R.L. Sah, E.Y. Chang, D. D'Lima, J. Du, Assessing cortical bone mechanical properties using collagen proton fraction from ultrashort echo time magnetization transfer (UTE-MT) MRI modeling, Bone Reports. 11 (2019) 100220. https://doi.org/10.1016/j.bonr.2019.100220.

[8] S. Abbasi-Rad, H. Saligheh Rad, Quantification of human cortical bone bound and free water in vivo with ultrashort echo time MR imaging: A model-based approach, Radiology. 283 (2017) 862–872. https://doi.org/10.1148/radiol.2016160780.

[9] K.A. Eley, A.G. Mcintyre, S.R. Watt-Smith, S.J. Golding, "Black bone" MRI: A partial flip angle technique for radiation reduction in craniofacial imaging, BJR. 85 (2012) 272–278. https://doi.org/10.1259/bjr/95110289.

[10] K.A. Eley, S.R. Watt-Smith, S.J. Golding, "Black bone" MRI: A potential alternative to CT when imaging the head and neck: Report of eight clinical cases and review of the Oxford experience, BJR. 85 (2012) 1457–1464. https://doi.org/10.1259/bjr/16830245.

[11] K.A. Eley, S.R. Watt-Smith, F. Sheerin, S.J. Golding, "Black Bone" MRI: A potential alternative to CT with three-dimensional reconstruction of the craniofacial skeleton in the diagnosis of craniosynostosis, European Radiology. 24 (2014) 2417–2426. https://doi.org/10.1007/s00330-014-3286-7.

[12] M.H.G. Dremmen, M.W. Wagner, T. Bosemani, A. Tekes, D. Agostino, E. Day, B.P. Soares, T.A.G.M. Huisman, Does the addition of a "black bone" sequence to a fast multisequence trauma MR protocol allow MRI to replace CT after traumatic brain injury in children?, American Journal of Neuroradiology. 38 (2017) 2187–2192. https://doi.org/10.3174/ajnr.A5405.

[13] A. Saarikko, E. Mellanen, L. Kuusela, J. Leikola, A. Karppinen, T. Autti, P. Virtanen, N. Brandstack, Comparison of black bone MRI and 3D-CT in the preoperative evaluation of patients with craniosynostosis, Journal of Plastic, Reconstructive & Aesthetic Surgery. 73 (2020) 723–731. https://doi.org/10.1016/j.bjps.2019.11.006.

[14] D. Gascho, N. Zoelch, C. Tappero, S. Kottner, E. Bruellmann, M.J. Thali, E. Deininger-Czermak, Fracture MRI: Optimized 3D multi-echo in-phase sequence for bone damage assessment in craniocerebral gunshot injuries, Diagnostic and Interventional Imaging. 101 (2020) 611–615. https://doi.org/10.1016/j.diii.2020.02.010.

[15] B. Johnson, H. Alizai, M. Dempsey, Fast field echo resembling a CT using restricted echo-spacing (Fracture): A novel MRI technique with superior bone contrast, Skeletal Radiology. 50 (2021) 1705–1713. https://doi.org/10.1007/s00256-020-03659-7.

[16] D.J. Tyler, M.D. Robson, R.M. Henkelman, I.R. Young, G.M. Bydder, Magnetic resonance imaging with ultrashort TE (UTE) PULSE sequences: Technical considerations, Journal of Magnetic Resonance Imaging. 25 (2007) 279–289. https://doi.org/10.1002/jmri.20851.

[17] A.M. Afsahi, Y. Ma, H. Jang, S. Jerban, C.B. Chung, E.Y. Chang, J. Du, Ultrashort echo time magnetic resonance imaging techniques: Met and unmet needs in musculoskeletal imaging, Journal of Magnetic Resonance Imaging. 55 (2022) 1597–1612. https://doi.org/10.1002/jmri.28032.

[18] M. Weiger, D.O. Brunner, B.E. Dietrich, C.F. Müller, K.P. Pruessmann, ZTE imaging in humans, Magnetic Resonance in Medicine. 70 (2013) 328–332. https://doi.org/10.1002/mrm.24816.

[19] S. Hafner, Fast imaging in liquids and solids with the Back-projection Low Angle ShoT (BLAST) technique, Magnetic Resonance Imaging. 12 (1994) 1047–1051.

[20] D.P. Madio, I.J. Lowe, Ultra-fast imaging using low flip angles and FIDs, Magnetic Resonance in Medicine. 34 (1995) 525–529.

[21] M. Weiger, D.O. Brunner, T. Schmid, R. Froidevaux, M.B. Rösler, S. Gross, K.P. Pruessmann, A virtually 1H-free birdcage coil for zero echo time MRI without background signal, Magnetic Resonance in Medicine. 78 (2017) 399–407. https://doi.org/10.1002/mrm.26368.

[22] M. Weiger, J. Overweg, M.B. Rösler, R. Froidevaux, F. Hennel, B.J. Wilm, A. Penn, U. Sturzenegger, W. Schuth, M. Mathlener, M. Borgo, P. Börnert, C. Leussler, R. Luechinger, B.E. Dietrich, J. Reber, D.O. Brunner, T. Schmid, L. Vionnet, K.P. Pruessmann, A high-performance gradient insert for rapid and short-T2 imaging at full duty cycle, Magnetic Resonance in Medicine. 79 (2018) 3256–3266. https://doi.org/10.1002/mrm.26954.

[23] S. Mastrogiacomo, W. Dou, J.A. Jansen, X.F. Walboomers, Magnetic resonance imaging of hard tissues and hard tissue engineered bio-substitutes, Molecular Imaging and Biology. 21 (2019) 1003–1019. https://doi.org/10.1007/s11307-019-01345-2.

[24] M. Weiger, K.P. Pruessmann, F. Hennel, MRI with zero echo time: Hard versus sweep pulse excitation, Magnetic Resonance in Medicine. 66 (2011) 379–389. https://doi.org/10.1002/mrm.22799.

[25] Y. Wu, G. Dai, J.L. Ackerman, M.I. Hrovat, M.J. Glimcher, B.D. Snyder, A. Nazarian, D.A. Chesler, Water- and fat-suppressed proton projection MRI (WASPI) of rat femur bone, Magnetic Resonance in Medicine. 57 (2007) 554–567. https://doi.org/10.1002/mrm.21174.

[26] D.M. Grodzki, P.M. Jakob, B. Heismann, Ultrashort echo time imaging using pointwise encoding time reduction with radial acquisition (PETRA), Magnetic Resonance in Medicine. 67 (2012) 510–518. https://doi.org/10.1002/mrm.23017.

[27] R. Froidevaux, M. Weiger, M.B. Rösler, D.O. Brunner, K.P. Pruessmann, HYFI: Hybrid filling of the dead-time gap for faster zero echo time imaging, NMR in Biomedicine. 34 (2021) e4493. https://doi.org/10.1002/nbm.4493.

[28] E. Johnstone, J.J. Wyatt, A.M. Henry, S.C. Short, D. Sebag-Montefiore, L. Murray, C.G. Kelly, H.M. McCallum, R. Speight, Systematic review of synthetic computed tomography generation methodologies for use in magnetic resonance imaging–only radiation therapy, International Journal of Radiation Oncology*Biology*Physics. 100 (2018) 199–217. https://doi.org/10.1016/j.ijrobp.2017.08.043.

[29] X. Han, MR-based synthetic CT generation using a deep convolutional neural network method, Medical Physics. 44 (2017) 1408–1419. https://doi.org/10.1002/mp.12155.

[30] D. Gupta, M. Kim, K.A. Vineberg, J.M. Balter, Generation of synthetic CT images from MRI for treatment planning and patient positioning using a 3-channel U-net trained on sagittal images, Frontiers in Oncology. 9 (2019) 964. https://doi.org/10.3389/fonc.2019.00964.

[31] E. Palmér, A. Karlsson, F. Nordström, K. Petruson, C. Siversson, M. Ljungberg, M. Sohlin, Synthetic computed tomography data allows for accurate absorbed dose calculations in a magnetic resonance imaging only workflow for head and neck radiotherapy, Physics and Imaging in Radiation Oncology. 17 (2021) 36–42. https://doi.org/10.1016/j.phro.2020.12.007.

[32] A.J. Robinson, S. Blaser, A. Vladimirov, D. Drossman, D. Chitayat, G. Ryan, Foetal "black bone" MRI: Utility in assessment of the foetal spine, BJR. 88 (2015) 20140496. https://doi.org/10.1259/bjr.20140496.

[33] E.L. Baadsvik, M. Weiger, R. Froidevaux, M.B. Rösler, D.O. Brunner, L. Öhrström, F.J. Rühli, P. Eppenberger, K.P. Pruessmann, High-resolution MRI of mummified tissues using advanced short-T2 methodology and hardware, Magnetic Resonance in Medicine. 85 (2021) 1481–1492. https://doi.org/10.1002/mrm.28530.

CHAPTER 22

PHOTON-COUNTING CT—BASICS AND FORENSIC POSTMORTEM APPLICATIONS

Dominic Gascho, Victor Mergen, and Patrik Wili

DOI: 10.4324/9781003413998-24

22.1 INTRODUCTION

In postmortem forensic imaging, image quality is crucial for the detection and interpretation of findings obtained non-invasively from inside the body. But what does the term "image quality" mean? This term encompasses both the subjective perception of image quality and objective image quality criteria. While subjective image quality depends primarily on individual preferences and past experience, objective image quality is measurable in terms of spatial resolution, contrast, and noise [1]. These objective image quality criteria limit diagnostic evaluability. For example, subtle findings such as non-displaced skull base fractures may be missed due to limited spatial resolution. In addition, excessive image noise, which is particularly prevalent in obese patients, can affect diagnostic accuracy. And due to the low inherent contrast between different types of soft tissue, the ability to reliably distinguish between altered and normal tissue using conventional CT is limited. A decisive limiting factor for these objective image quality criteria are the conventional energy integrating detectors (EIDs). Photon-counting detectors (PCDs) have the potential to outperform EIDs and impress with high spatial resolution, inherent spectral information, reduced image noise, and increased contrast-to-noise ratio [2–16].

In 2021, the first whole-body full field-of-view PCD-CT scanner (NAEOTOM Alpha, Siemens Healthineers) has received clearance from the U.S. Food and Drug Administration (FDA). Based on our experience with this CT system (Figure 22.1), this chapter explains how photon-counting detectors work and how this novel technology can affect spatial resolution, contrast, and noise. Ultimately, potential benefits of PCD-CT in postmortem forensic imaging will be discussed. For comparison, data acquired using a CT with EID (SOMATOM Definition Flash, Siemens Healthineers) are presented.

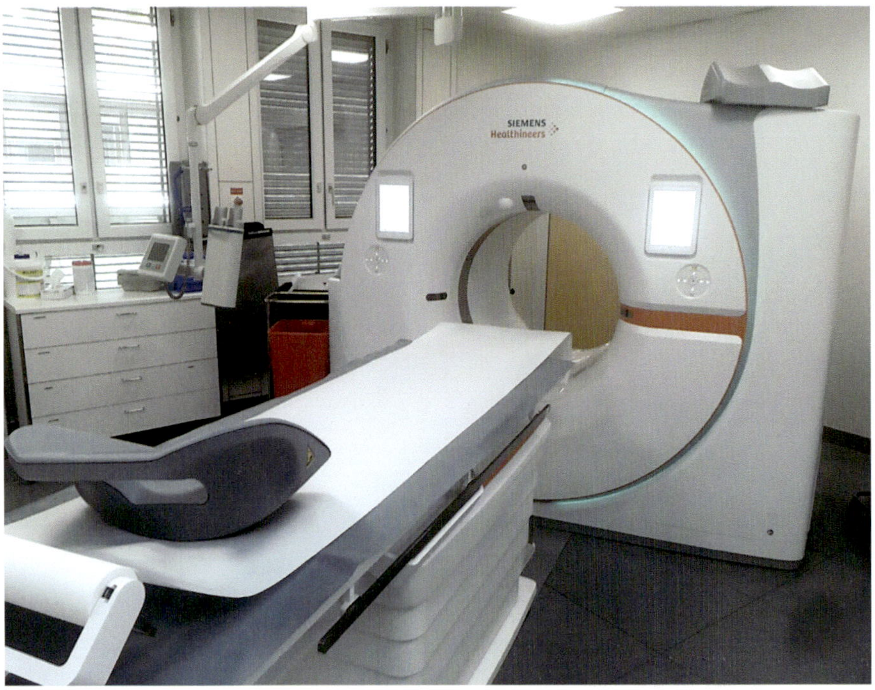

Figure 22.1. First-generation PCD-CT with quantum technology (NAEOTOM Alpha, Siemens Healthineers) installed at the Institute of Diagnostic and Interventional Radiology of the University Hospital Zürich, enabling a 500 mm scanning field-of-view and 57.6 mm longitudinal detector coverage.

22.2 HOW DO ENERGY-INTEGRATING DETECTORS WORK?

Before focusing on the technical and physical principles of PCD, a brief summary describes the technical and physical principles of conventional EID systems, particularly solid-state scintillation detectors (SSDs). SSDs use a scintillator layer (Figure 22.2), typically consisting of gadolinium-oxide or gadolinium-oxysulfide, and photodiodes attached to the bottom side [2,17]. When excited by ionizing radiation, such detector materials exhibit scintillation, i.e. visible flashes of typically blue light are generated as X-ray photons enter the scintillator layer. The intensity of such a visible flash is proportional to the energy of the X-ray photon. The scintillation light generated by the individual flashes is subsequently converted into an electrical signal by the connected photodiodes. This detector system integrates the energy of all detected X-ray photons and thus does not provide information about the energy of individual photons. Furthermore, the scintillator layer must be divided into individual cells by optically opaque septa to prevent optical crosstalk. This leads to "dead zones", i.e. zones where X-ray photons cannot be detected. These have a thickness of 0.1 mm and are located between the individual detector cells [2].

In SSDs, both the intensity of the scintillation light and the amplitude of the induced current pulse are proportional to the energy E of the absorbed X-ray photon. The detector responsivity (R_D) is proportional to the energy (E) of the converted X-ray photons. The scintillation signal of such SSDs (S_{SSD}) is given by the integral of all signal pulses that are registered during the measurement time of one reading, i.e. during a single projection [2]:

$$S_{SSD} = \int_0^{E_{max}} R_D(E) \cdot \Phi \, dE$$

In SDDs, R_D and E are approximately linearly related, thus the above equation can be simplified to:

$$S_{SSD} \approx \int_0^{E_{max}} E \cdot \Phi \, dE$$

where Φ is the absorbed X-ray flux during one reading, and E_{max} is the maximum energy of the X-ray beam. The equation above shows that low energy photons contribute less to the detector signal than high energy photons

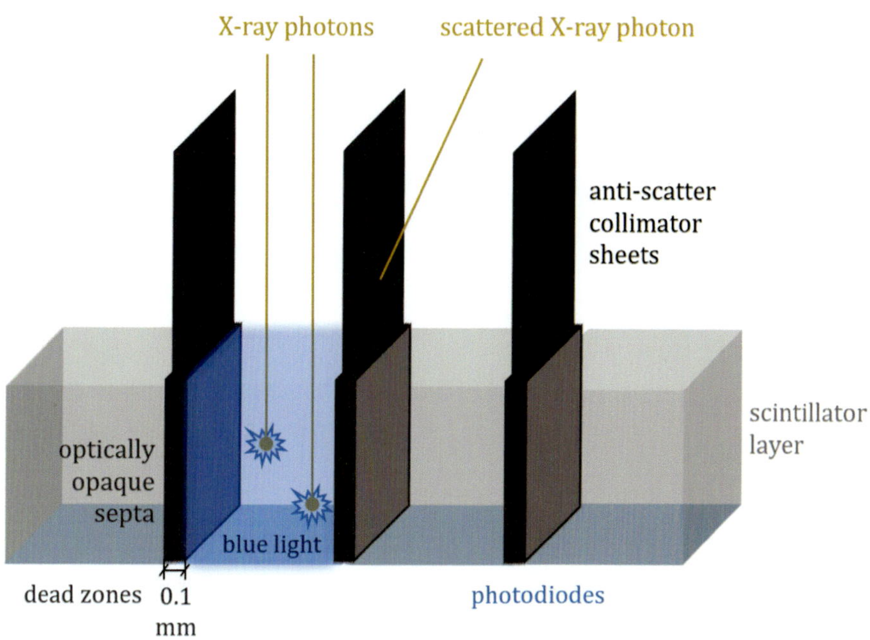

Figure 22.2 Schematic representation of a solid-state scintillation detector (SSD). This detector system consists of a scintillation layer divided into individual cells by optically opaque septa. When entering the scintillator layer, X-ray photons generate visible flashes of light. The typically blue light is subsequently converted into an electrical signal by the photodiodes attached at the bottom. The detected light intensity is proportional to the energies of all the detected X-ray photons. The optically opaque septa prevent optical crosstalk between the detector cells. These septa with a thickness of about 0.1 mm induce so-called "dead zones" in which no X-ray photons can be detected. The anti-scatter collimation sheets on top suppress scattered X-ray photons.

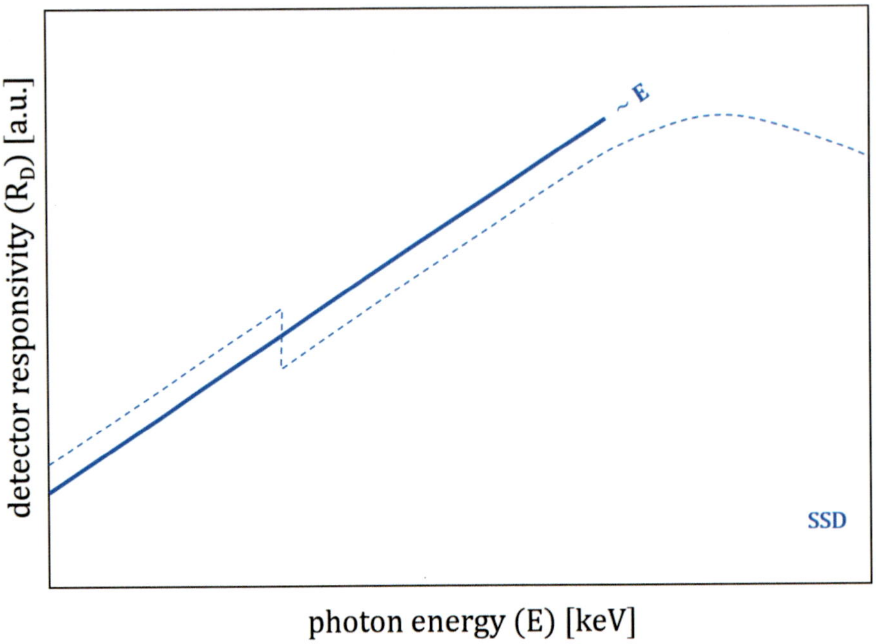

Figure 22.3 The detector responsivity as a function of the photon energy for a solid-state scintillation detector. The graph was derived from Flohr et al. [2]. The dashed line represents the real responsivity of SSDs, while the continuous line shows the approximated behavior. SSDs are less sensitive for low energy photons and more sensitive for high energy photons. The detector responsivity of SSDs increases linearly with increasing photon energy. Thus, detector responsivity is directly proportional (~) to photon energy and the scintillation signal depends on the photon energy.

(Figure 22.3). In other words, SSDs are less sensitive to low energy photons, so information in the low energy region is reduced or even lost.

22.3 HOW DO PHOTON-COUNTING DETECTORS WORK?

In contrast to the indirect conversion process of EIDs via a scintillator layer, PCDs use a semiconductor layer (Figure 22.4), commonly made of cadmium-telluride (CdTe) or cadmium-zinc-telluride (CdZnTe or CZT), and directly convert incident X-ray photons into electrical signals [18]. The thickness of such a semiconductor layer varies between 1.6 to 30 mm [19]. When a photon hits the semiconductor layer, an electron-hole pair is created. An electron-hole pair consists of an electron and an electron hole. The electron is a negatively charged particle, whereas the electron hole can be described as quasiparticle that behaves similarly to positively charged particles. When a high voltage ranging between 0.8 and 1.0 kV is applied between a negatively charged cathode on the top surface and positively charged pixelated anode at the bottom surface of the semiconductor layer, the electrons drift towards the anode and induce a short current pulse of about 10 nanoseconds [2]. A pulse shaping circuit converts these current pulses into voltage pulses that have a full width at half maximum of 10 to 15 nanoseconds, with the pulse height proportional to the energy of the absorbed X-ray photons. When the pulse height exceeds a predefined energy threshold above the electronic baseline noise, the pulse is counted and the photon count is increased by one.

The energy threshold (T_0) is typically set at 20 to 25 keV. The detector responsivity R_D is approximately constant within an X-ray energy range of 30 to 100 keV (Figure 22.5) and all photons contribute equally to the measurement signal regardless of their energy [2]. The PCD signal (S_{PCD}) is given by the integral of all signal pulses that exceed the energy threshold T_0 and that are registered during the measurement time of one reading, i.e. during a single projection [2]:

$$S_{PCD} = \int_{T_0}^{E_{max}} R_D(E) \cdot \Phi \, dE \approx \int_{T_0}^{E_{max}} \text{constant} \cdot \Phi \, dE$$

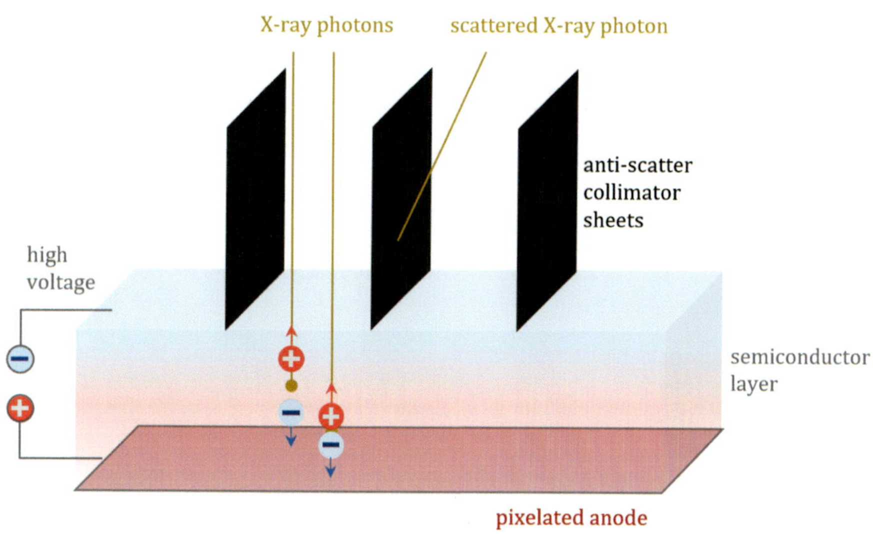

Figure 22.4 Schematic representation of a photon-counting detector. When an X-ray photon hits the semiconductor layer, an electron-hole pair is generated. By applying a high voltage of about 1 kV, the electron is attracted to the pixelated anode and is subsequently converted into an electrical pulse. When such an electrical pulse exceeds an energy threshold above the electronic baseline noise, it is counted which means that the photon count is increased by one.

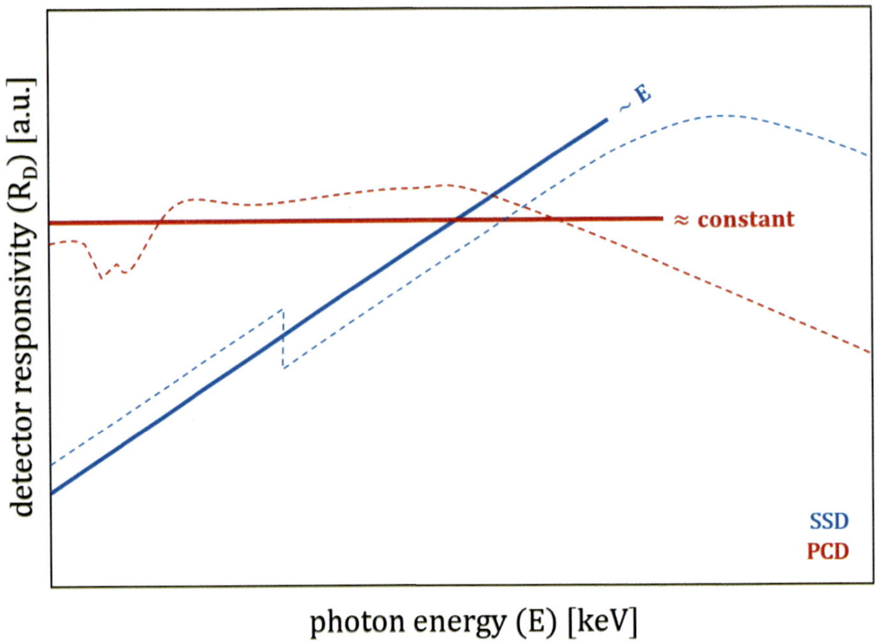

Figure 22.5 The detector responsivity as a function of the photon energy for a solid-state scintillation detector (blue lines) compared with a photon counting detector (red lines). The graph was derived from Flohr et al. [2]. The dashed lines represent the real responsivity and the continuous lines the approximated behavior of the SSD and the PCD, respectively. Note the approximately (≈) constant detector responsivity of the PCD over a broad range of photon energies. The PCD signal is independent of the photon energy, which is not the case for SSD. The responsivity of a SSD is directly proportional (~) to the photon energy, i.e., it is less sensitive to low energy photons and more sensitive to high energy photons.

22.4 WHY DO PHOTON-COUNTING DETECTORS ENABLE INCREASED SPATIAL RESOLUTION?

Since PCDs with the semiconductor layer and the pixelated anodes do not require any additional optically opaque septa, there are also no "dead zones". The pixels between the inevitable anti-scatter collimator sheets can be divided into small sub-pixels of less than 0.2 × 0.2 mm^2 (Figure 22.6), allowing excellent spatial resolution [5,12]. However, the z-coverage in the ultra-high resolution data acquisition mode is limited to 24 mm. In a "standard" data acquisition mode the PCD data can be binned to macro-pixels of 0.4 × 0.4 mm^2 for a z coverage of 57.6 mm [5].

By using sub-pixels, a high spatial resolution can thus be achieved (Figure 22.7). Postmortem forensic CT examinations may benefit from such high spatial resolution, e.g., for the detection of small fracture lines, as is often the case in laryngeal fractures or fractures of the facial skull, or for quantitative CT examinations to determine volumetric bone mineral density.

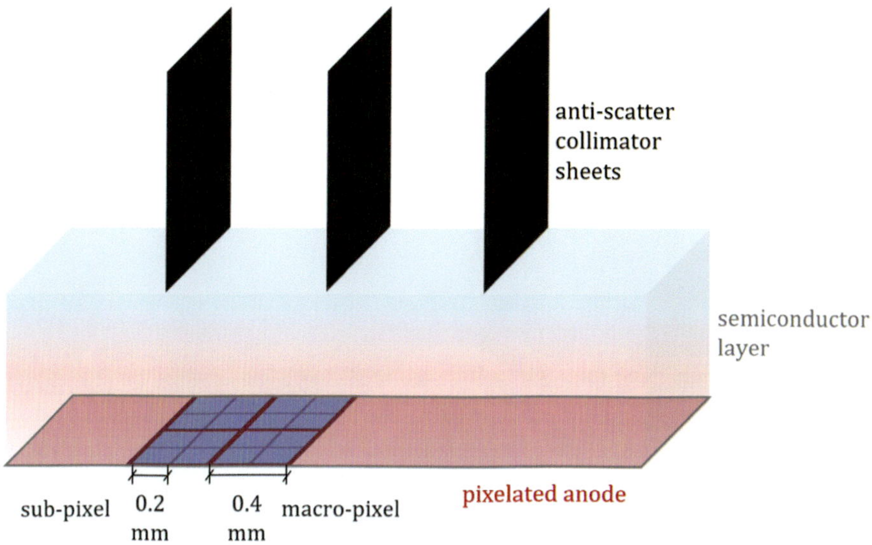

Figure 22.6 The pixels of the pixelated anode can be divided into small sub-pixels of 0.2 × 0.2 mm^2. In a "standard" data acquisition mode the PCD data can be binned to macro-pixels of 0.4 × 0.4 mm^2.

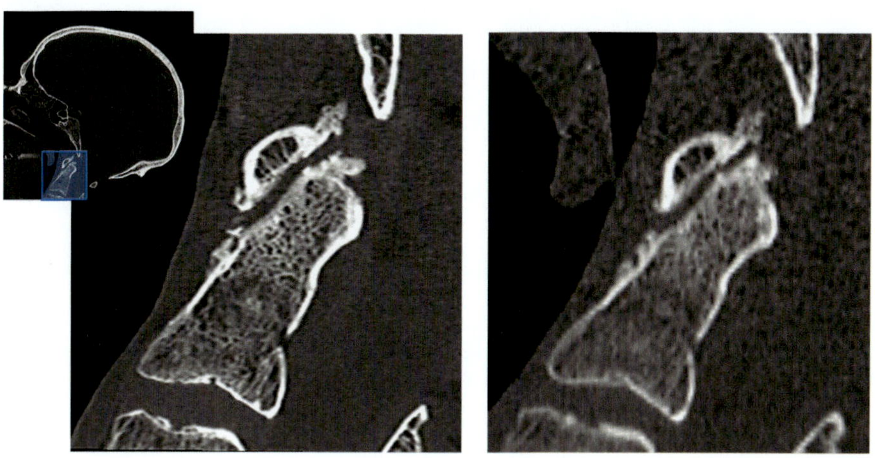

Figure 22.7 Zoomed-in images of the dens axis in sagittal view using multi-planar reconstruction. The left image shows the reconstructions of the PCD data (acquisition collimation: 120 × 0.2 mm) using a slice thickness of 0.2 mm, a reconstruction matrix of 1024 × 1024, and a field-of-view (FOV) of 200 × 200 mm^2 to create isotropic voxels. The image on the right shows the reconstruction of the EID data (acquisition collimation: 128 × 0.6 mm) with a slice thickness of 0.6 mm, a matrix of 512 × 512, and a FOV of 200 × 200 mm^2. A hard kernel (Hr68) was used for both reconstructions. The scan parameters were 120 kVp and 600 mAs for both scans. Note the improved resolution of the PCD-CT image compared with the EID-CT image, which is reflected by an excellent visualization of trabecular structures within the dens axis.

22.5 WHY IS THE IMAGE NOISE REDUCED WITH PHOTON-COUNTING DETECTORS?

In EIDs, the low-level analogue electrical signal from the photodiodes of the EID is distorted by electrical noise. At low X-ray flux, the electrical noise becomes larger than the quantum noise of the X-ray photons resulting in a disproportional increase in image noise and instability of the low CT numbers [2]. In PCDs, the baseline energy threshold for counting an energy equivalent signal pulse is set above the low amplitude baseline noise, so the baseline noise does not trigger counts. Therefore, compared to SSDs, image noise is reduced and CT numbers are more stable with PCD [2]. Even in low-dose CT, only quantum noise is present. Although postmortem imaging is performed at high radiation doses, diagnostic assessment of obese bodies may be limited due to excessive image noise. In these cases, PCD technology enables reduced image noise.

To reduce image noise in general, iterative reconstruction algorithms can be used. The so-called quantum iterative reconstruction (QIR) algorithm is optimized for the processing of PCD-CT data and can be set to four different strength levels [20]. The NAEOTOM Alpha does not use pure filtered back projection algorithms to reconstruct the spectral information; instead, "QIR-off" is selectable (Figure 22.8). This algorithm uses minimal statistical optimization compared to the standard weighted filtered back projection [20].

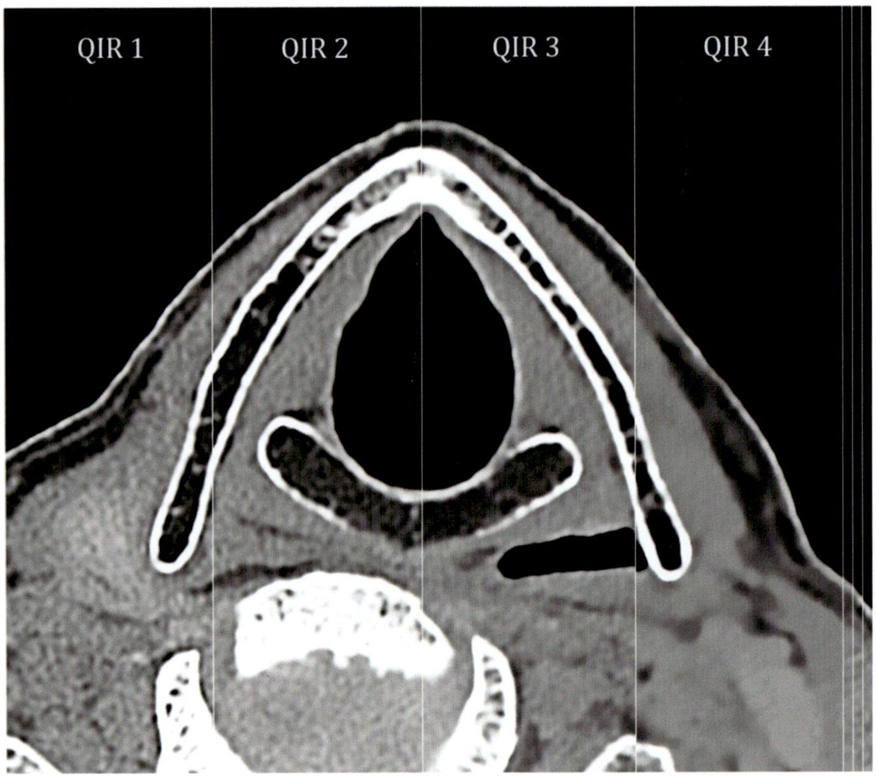

Figure 22.8 The composed image shows reconstructions of the larynx using quantitative iterative reconstruction (QIR) at four different strength levels. The strength level of QIR increases from left to right. Note the decreasing image noise with increasing QIR strength level best seen on the soft tissue. Depending on the desired image information, the corresponding QIR strength level can be selected.

22.6 WHAT ARE THE ADVANTAGES OF PHOTON-COUNTING DETECTORS IN TERMS OF CONTRAST?

In SSDs, X-ray photons with high energy contribute more to the signal than X-ray photons with low energy. This imbalance reduces the contrast-to-noise ratio (CNR) in reconstructed images, since the lower energy X-ray photons provide most of the contrast information.

PCDs, in turn, have no such imbalance, since all photons contribute equally to the measurement signal regardless of their energy. In addition, signal pulses of detected X-ray photons can be divided into "energy bins" according to their energy using different energy thresholds. Thus, not only are all signal pulses that exceed the base threshold (T_0) counted, but they can also be assigned to "energy bins" defined by higher thresholds (T_i) [2]:

$$S_{PCD} = \int_{T_i}^{E_{max}} R_D(E) \cdot \Phi \, dE \approx \int_{T_i}^{E_{max}} \text{constant} \, \Phi \, dE$$

During a single CT scan, several signals can be detected according to the respective threshold (Figure 22.9), i.e. for the energy thresholds T_0, T_1, T_2, and T_3 the signals S_0, S_1, S_2, and S_3 are detected. The difference of two signals, e.g. the difference between S_2 and S_1, gives the "energy bin" data, which in this case means that the "energy bin" contains all counted photons between the threshold T_1 and the threshold T_2 [2]. Thus, the PCD allows spectrally resolved measurements with a single energy scan. This makes so-called energy bin imaging feasible. Currently, reconstructed virtual monoenergetic images (VMI), as already known from dual-energy CT imaging, represent the new standard reconstruction with PCD-CT in clinical applications. Therefore, the PCD-CT data are often expressed in kilo-electron volts (keV) rather than kilo-voltage peaks (kVp).

However, energy separation with PCDs using semiconductor layers made of cadmium-telluride or cadmium-zinc-telluride is also limited by unavoidable physical effects [2]. One of such unintended effects is related to the K-edge energies of cadmium (26.7 keV) and telluride (31.8 keV), while zinc (9.7 keV) is less of a problem here due to its low K-edge energy. In the case of cadmium and telluride there is the possibility that an incident X-ray photon with the corresponding energy will kick out a K-electron. To maintain the stable configuration, the resulting "hole" in the K-shell is immediately refilled. Such refill processes generate characteristic X-ray photons with a certain fluorescence energy, which are finally absorbed and counted in the detector cell itself or in neighboring detector cells. The initial X-ray photon loses energy by kicking out the K-electron and is therefore incorrectly counted as a lower energy X-ray photon, which is called the "K-escape peak". Furthermore, "charge sharing" leads to incorrect counting. If the absorption of an X-ray photon occurs too close to pixel borders of the pixelated anode, the electrical pulse generated by the X-ray photons can be shared between neighboring detector cells. As a result, the incident X-ray photon is incorrectly counted as X-ray photons with lower energies.

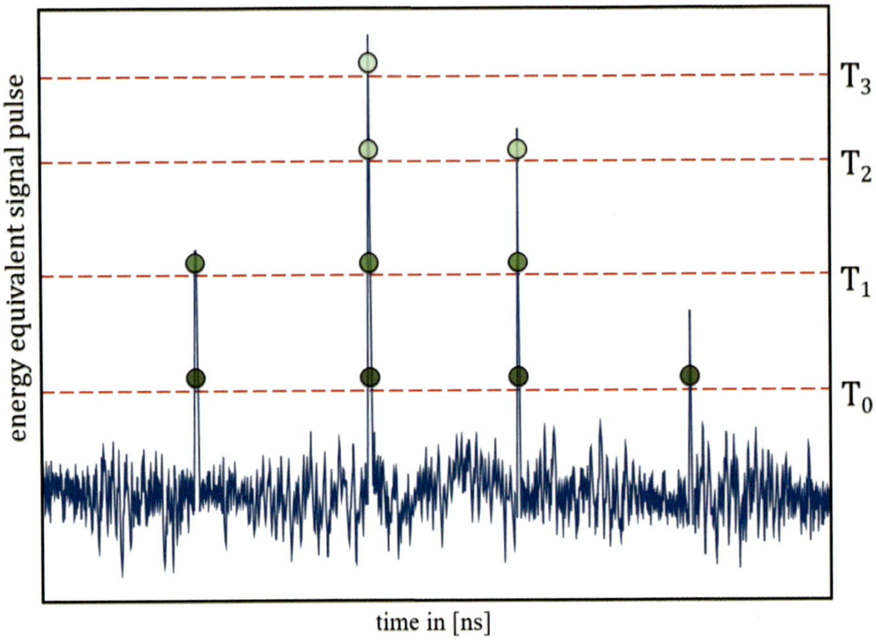

Figure 22.9 Schematic representation of the working principle for the signal detection of a PCD system. Each photon detected by the detector generates a pulse with a height proportional to its energy. If the pulse height reaches a certain threshold (T_0 to T_3) the photon will be accounted to this specific energy range and signals S_0-S_3 are generated. This enables to differentiate between photons of e.g. four different energy ranges and makes it possible to generate an approximated monoenergetic image for each energy threshold (T_0-T_3) or signal (S_0-S_3) respectively. Furthermore, this information can be used for iodine maps, virtual noncontrast reconstruction, spectral decomposition of materials/tissue, the measurement of the bone volume fraction in a specific region of interest or visualization of bone edema.

22.7 WHAT CAN BE EXPECTED FROM PCD-CT FOR POSTMORTEM APPLICATION?

The inherent spectral capabilities of PCD-CT enable new avenues of advanced data processing such as direct computation of virtual monoenergetic images, iodine maps, virtual noncontrast reconstructions, multi-material decomposition, bone volume fraction measurement or visualization of bone edema [3,4,6–8,11,16,21–24]. However, unlike clinical radiology, postmortem forensic radiology has not yet extensively evaluated the capabilities of current dual-energy and multi-energy CT for diagnostic purposes. Predominantly, dual-energy and multi-energy studies have been used in forensic medicine for non-invasive material differentiation of drugs or projectiles in the body [25–33]. However, there are only a few publications on the diagnostic capabilities of dual-energy and multi-energy CT in forensic medicine [34–36]. With the spread of new PCD-CT scanners, it is likely that spectral imaging will be increasingly used in the field of virtual autopsy.

ACKNOWLEDGMENTS

The authors thank Hatem Alkadhi (University Hospital Zürich) and Bernhard Schmidt (Siemens Healthineers) for their collaboration in the performance and assessment of the PCD-CT examinations.

References

[1] D. Gascho, M.J. Thali, T. Niemann, Postmortem computed tomography: Technical principles and recommended parameter settings for high-resolution imaging, Medicine Science and the Law. 58 (2018) 70–82. https://doi.org/10.1177/0025802417747167.

[2] T. Flohr, M. Petersilka, A. Henning, S. Ulzheimer, J. Ferda, B. Schmidt, Photon-counting CT review, Physica Medica. 79 (2020) 126–136. https://doi.org/10.1016/j.ejmp.2020.10.030.

[3] A. Euler, K. Higashigaito, V. Mergen, T. Sartoretti, B. Zanini, B. Schmidt, T.G. Flohr, S. Ulzheimer, M. Eberhard, H. Alkadhi, High-pitch photon-counting detector computed tomography angiography of the aorta: Intraindividual comparison to energy-integrating detector computed tomography at equal radiation dose, Investigative Radiology. 57 (2022) 115–121.

[4] T. Sartoretti, V. Mergen, K. Higashigaito, M. Eberhard, H. Alkadhi, A. Euler, Virtual noncontrast imaging of the liver using photon-counting detector computed tomography: A systematic phantom and patient study, Investigative Radiology. 57(7) (2022 July) 488–493. https://doi.org/10.1097/RLI.0000000000000860. PMID: 35136003.

[5] T. Sartoretti, D. Racine, V. Mergen, L. Jungblut, P. Monnin, T.G. Flohr, K. Martini, T. Frauenfelder, H. Alkadhi, A. Euler, Quantum iterative reconstruction for low-dose ultra-high-resolution photon-counting detector CT of the lung, Diagnostics. 12 (2022) 522.

[6] V. Mergen, T. Sartoretti, E. Klotz, B. Schmidt, L. Jungblut, K. Higashigaito, R. Manka, A. Euler, M. Kasel, M. Eberhard, Extracellular volume quantification with cardiac late enhancement scanning using dual-source photon-counting detector CT, Investigative Radiology. 57 (2022) 406–411.

[7] V. Mergen, E. Ried, T. Allmendinger, T. Sartoretti, K. Higashigaito, R. Manka, A. Euler, H. Alkadhi, M. Eberhard, Epicardial adipose tissue attenuation and fat attenuation index: Phantom study and in vivo measurements with photon-counting detector CT, American Journal of Roentgenology. 218 (2022) 822–829.

[8] T. Sartoretti, V. Mergen, L. Jungblut, H. Alkadhi, A. Euler, Liver iodine quantification with photon-counting detector CT: Accuracy in an abdominal phantom and feasibility in patients, Academic Radiology. 30(3) (2022) 461–469. DOI: https://10.1016/j.acra.2022.04.021. PMID: 35644755.

[9] M. Mannil, T. Hickethier, J. von Spiczak, M. Baer, A. Henning, M. Hertel, B. Schmidt, T. Flohr, D. Maintz, H. Alkadhi, Photon-counting CT: High-resolution imaging of coronary stents, Investigative Radiology. 53 (2018) 143–149.

[10] S.A. Si-Mohamed, S. Boccalini, H. Lacombe, A. Diaw, M. Varasteh, P.-A. Rodesch, R. Dessouky, M. Villien, V. Tatard-Leitman, T. Bochaton, Coronary CT angiography with photon-counting CT: First-in-human results, Radiology. 303 (2022) 303–313.

[11] L. Jungblut, T. Sartoretti, D. Kronenberg, V. Mergen, A. Euler, B. Schmidt, H. Alkadhi, T. Frauenfelder, K. Martini, Performance of virtual non-contrast images generated on clinical photon-counting detector CT for emphysema quantification: Proof of concept, The British Journal of Radiology. 95 (2022) 20211367.

[12] V. Mergen, T. Sartoretti, M. Baer-Beck, B. Schmidt, M. Petersilka, J.E. Wildberger, A. Euler, M. Eberhard, H. Alkadhi, Ultra-high-resolution coronary CT angiography with photon-counting detector CT: Feasibility and image characterization, Investigative Radiology. (n.d.) 10.1097.

[13] K. Rajendran, M. Petersilka, A. Henning, E. Shanblatt, J. Marsh, J. Thorne, B. Schmidt, T. Flohr, J. Fletcher, C. McCollough, Full field-of-view, high-resolution, photon-counting detector CT: Technical assessment and initial patient experience, Physics in Medicine & Biology. 66 (2021) 205019.

[14] K. Higashigaito, A. Euler, M. Eberhard, T.G. Flohr, B. Schmidt, H. Alkadhi, Contrast-enhanced abdominal CT with clinical photon-counting detector CT: Assessment of image quality and comparison with energy-integrating detector CT, Academic Radiology. 29 (2022) 689–697.

[15] L. Jungblut, A. Euler, J. von Spiczak, T. Sartoretti, V. Mergen, V. Englmaier, A. Landsmann, C.-M. Mihai, O. Distler, H. Alkadhi, Potential of photon-counting detector CT for radiation dose reduction for the assessment of interstitial lung disease in patients with systemic sclerosis, Investigative Radiology. (2022) 10.1097.

[16] V. Mergen, D. Racine, L. Jungblut, T. Sartoretti, S. Bickel, P. Monnin, K. Higashigaito, K. Martini, H. Alkadhi, A. Euler, Virtual noncontrast abdominal imaging with photon-counting detector CT, Radiology. (2022) 213260. https://doi.org/10.1148/radiol.213260.

[17] S.S. Hsieh, S. Leng, K. Rajendran, S. Tao, C.H. McCollough, Photon counting CT: Clinical applications and future developments, IEEE Transactions on Radiation and Plasma Medical Sciences. 5(4) (2020) 441–452. https://doi.org/10.1109/trpms.2020.302021. PMID: 34485784; PMID: PMC8409241.

[18] S. Leng, M. Bruesewitz, S. Tao, K. Rajendran, A.F. Halaweish, N.G. Campeau, J.G. Fletcher, C.H. McCollough, Photon-counting detector CT: System design and clinical applications of an emerging technology, Radiographics. 39 (2019) 729–743.

[19] M.J. Willemink, M. Persson, A. Pourmorteza, N.J. Pelc, D. Fleischmann, Photon-counting CT: Technical principles and clinical prospects, Radiology. 289 (2018) 293–312. https://doi.org/10.1148/radiol.2018172656.
[20] T. Sartoretti, A. Landsmann, D. Nakhostin, M. Eberhard, C. Roeren, V. Mergen, K. Higashigaito, R. Raupach, H. Alkadhi, A. Euler, Quantum iterative reconstruction for abdominal photon-counting detector CT improves image quality, Radiology. 303 (2022) 339–348.
[21] L. Jungblut, D. Kronenberg, V. Mergen, K. Higashigaito, B. Schmidt, A. Euler, H. Alkadhi, T. Frauenfelder, K. Martini, Impact of contrast enhancement and virtual monoenergetic image energy levels on emphysema quantification: Experience with photon-counting detector computed tomography, Investigative Radiology. 57 (2022) 359–365.
[22] J.A. Decker, S. Bette, C. Scheurig-Muenkler, B. Jehs, F. Risch, P. Woźnicki, F.M. Braun, M. Haerting, C. Wollny, T.J. Kroencke, Virtual non-contrast reconstructions of photon-counting detector CT angiography datasets as substitutes for true non-contrast acquisitions in patients after EVAR—performance of a novel calcium-preserving reconstruction algorithm, Diagnostics. 12 (2022) 558.
[23] T. Sartoretti, M. Eberhard, T. Nowak, R. Gutjahr, G. Jost, H. Pietsch, B. Schmidt, T. Flohr, H. Alkadhi, A. Euler, Photon-counting multienergy computed tomography with spectrally optimized contrast media for plaque removal and stenosis assessment, Investigative Radiology. 56 (2021) 563–570.
[24] T. Nowak, M. Eberhard, B. Schmidt, D. Frey, O. Distler, N. Saltybaeva, H. Alkadhi, A. Euler, Bone mineral density quantification from localizer radiographs: Accuracy and precision of energy-integrating detector CT and photon-counting detector CT, Radiology. 298 (2021) 147–152.
[25] S. Leschka, J. Fornaro, P. Laberke, S. Blum, A. Hatem, I. Niederer, C. Meile, D. Hibbeln, R. Hausmann, S. Wildermuth, D. Eisenhart, Differentiation of cocaine from heroine body packs by computed tomography: Impact of different tube voltages and the dual-energy index, Journal of Forensic Radiology and Imaging. 1 (2013) 46–50. https://doi.org/10.1016/j.jofri.2013.03.041.
[26] J. Grimm, R. Wudy, E. Ziegeler, S. Wirth, M. Uhl, M.F. Reiser, M. Scherr, Differentiation of heroin and cocaine using dual-energy CT—an experimental study, International Journal of Legal Medicine. 128 (2014) 475–482. https://doi.org/10.1007/s00414-014-0980-6.
[27] P.J. Laberke, J. Fornaro, S. Kim, S. Blum, M. Augsburger, H. Alkadhi, S. Wildermuth, R. Hausmann, S. Leschka, Dual-energy CT behavior of heroin, cocaine, and typical adulterants, Forensic Science, Medicine, and Pathology. 11 (2015) 20–28.
[28] A. Boizet, J. Ognard, O. Hmandi, C. Saccardy, D. Bourhis, D.B. Salem, Contribution of dual-energy computed tomography in the differentiation of illicit drugs, Archiwum Medycyny Sądowej i Kryminologii/Archives of Forensic Medicine and Criminology. 70 (2020).
[29] D. Gascho, N. Zoelch, H. Richter, A. Buehlmann, P. Wyss, S. Schaerli, Identification of bullets based on their metallic components and X-Ray attenuation characteristics at different energy levels on CT, American Journal of Roentgenology. (2019) 1–9. https://doi.org/10.2214/AJR.19.21229.
[30] J. Ognard, B. Dissaux, I. Diallo, L. Attar, C. Saccardy, D.S. Ben, Manual and fully automated segmentation to determine the ferromagnetic status of bullets using computed tomography dual-energy index: A phantom study, Journal of Computer Assisted Tomography. (2019). https://doi.org/10.1097/RCT.0000000000000899.
[31] D. Gascho, N. Zoelch, H. Richter, A. Buehlmann, P. Wyss, M.J. Thali, S. Schaerli, Heavy metal in radiology: How to reliably differentiate between lodged copper and lead bullets using CT numbers, European Radiology Experimental. 4 (2020) 43. https://doi.org/10.1186/s41747-020-00168-z.
[32] D. Gascho, N. Zoelch, E. Deininger-Czermak, C. Tappero, A. Buehlmann, P. Wyss, M.J. Thali, S. Schaerli, Visualization and material-based differentiation of lodged projectiles by extended CT scale and the dual-energy index, Journal of Forensic and Legal Medicine. 70 (2020) 101919. https://doi.org/10.1016/j.jflm.2020.101919.
[33] N. Norman, M.R. Dimmock, K. Lee, J. Graham, R. Bassed, The applicability of Dual-Energy Computed Tomography (DECT) in forensic odontology: A review, Journal of Forensic Radiology and Imaging. 10 (2017) 15–22.
[34] A. Persson, C. Jackowski, E. Engström, H. Zachrisson, Advances of dual source, dual-energy imaging in postmortem CT, European Journal of Radiology. 68 (2008) 446–455. https://doi.org/10.1016/j.ejrad.2008.05.008.
[35] L. Filograna, N. Magarelli, A. Leone, R. Guggenberger, S. Winklhofer, M.J. Thali, L. Bonomo, Value of monoenergetic dual-energy CT (DECT) for artefact reduction from metallic orthopedic implants in postmortem studies, Skeletal Radiology. 44 (2015) 1287–1294. https://doi.org/10.1007/s00256-015-2155-z.
[36] D. Kreul, D. Gascho, S. Franckenberg, S. Eggert, B. Fliss, R.A. Kubik-Huch, M.J. Thali, T. Niemann, Postmortem determination of hepatic steatosis: Comparing Rho/Z and fat fraction measurements on dual-energy CT for histological grading: A retrospective analysis, Forensic Imaging. 23 (2020) 200422. https://doi.org/10.1016/j.fri.2020.200422.

CHAPTER 23

BULLET IDENTIFICATION—VIRTUAL INVESTIGATION OF LODGED BULLETS

Dominic Gascho and Niklaus Zoelch

DOI: 10.4324/9781003413998-25

23.1 INTRODUCTION

Depending on the position of a bullet that is lodged in the body, the bullet is not necessarily removed from a patient with gunshot wounds. It is believed that there are actually only a few medical indications for the removal of a bullet. For the assessment of whether surgical intervention should take place, a radiological examination is usually performed beforehand to assess the gunshot wound and also to localize a lodged bullet. In criminal investigations where surgical intervention to recover a lodged bullet is not supported from a medical perspective, such radiological examinations can provide important information about the type of ammunition used. For example, a recently published forensic case report [1] describes how a computed tomography (CT) scan eventually exonerated a police officer, who was initially blamed for a shooting during a violent demonstration. The bullet was lodged directly behind the shooting victim's right eye and was therefore not removed to avoid the risk of surgical intervention. Visual assessments and caliber measurements on the CT data were able to rule out that the bullet was a 9-mm Luger, the ammunition used by police officers. Another forensic case report [2] describes the use of CT data for visual assessments and caliber measurements of a bullet that had lodged in close proximity to the thoracolumbar junction. The investigators were able to assign the projectile to .38 Special ammunition and ruled out a .40 S&W ammunition, both of which were involved in police shootings that resulted in the paraplegia of a fugitive. The felon died in prison 2 years later, and the bullet was removed at autopsy, allowing investigators to confirm the CT-based classification of the bullet. In addition to caliber measurements, three-dimensional CT data sets also allow in situ examination of a bullet in terms of its shape and design, fragment distribution, and metallic components [3].

Not only in living patients but also in fatal gunshot victims, the increasingly used CT can provide important information not only about the location of the bullet, the distribution of its fragments, and the injuries caused, but also information about the ammunition used. Although in the case of a violent crime, removal of the bullet for detailed forensic analysis is performed, making the virtual examination of the bullet seem less relevant. However, if a weapon is recovered, the CT data could quickly reveal whether it is a single type of ammunition that corresponds to the weapon, or whether another weapon or perpetrator may need to be searched for.

This chapter describes how bullets and their fragments can be virtually examined from CT data generated using a clinical CT scanner. At the end, the use of radiological techniques for ballistic examinations of soap and gelatin blocks is presented. At the beginning, however, the limitations and possibilities of virtual examinations of lodged bullets using two-dimensional X-ray projection images, so-called radiographs, is briefly discussed in order to reaffirm later the advantages of three-dimensional CT imaging.

23.2 HOW TO EXAMINE LODGED BULLETS VIRTUALLY ON RADIOGRAPHIC IMAGES

Bullets are typically made of metallic material and due to the high radiopacity of metallic materials, bullets and bullet fragments contrast well with biological tissues on radiographic images. In fact, shortly after the discovery of X-rays in the late 19th century, radiographs were taken in a forensic case to locate a lodged bullet in the lower leg of a gunshot victim after attempts to find the bullet with a probe failed [4]. Over the course of the last century, radiographs became a widely used imaging technique in both clinical patient care and in forensic medicine. In the latter, it is used as an adjunct to the autopsy to detect and locate bullets, bullet fragments, or shot pellets. Due to the sharp-edged representation of a bullet on conventional radiographs, this imaging technique allows classification of an intact bullet based on its shape and size and, depending on the degree of deformation, virtual caliber estimates. However, magnification factors must be taken into account, since the dimensions of the radiographic representation of a lodged bullet can differ from its actual dimensions depending on the distances from the X-ray tube focal spot to the X-ray film and the position of the bullet in between as shown in Figure 23.1 [5]. Thus, without taking into account the magnification factors, there may be gross errors in the estimation of the caliber based on radiographic images.

A radiographic image allows only a two-dimensional summation image, that is, a projection representing the degree of X-ray absorption of the bullet along the X-ray path. Therefore, at least three orthogonal radiographic images are required for accurate calibration measurements, since the spatial orientation of a bullet in the body is most likely not parallel to the imaged plane [6].

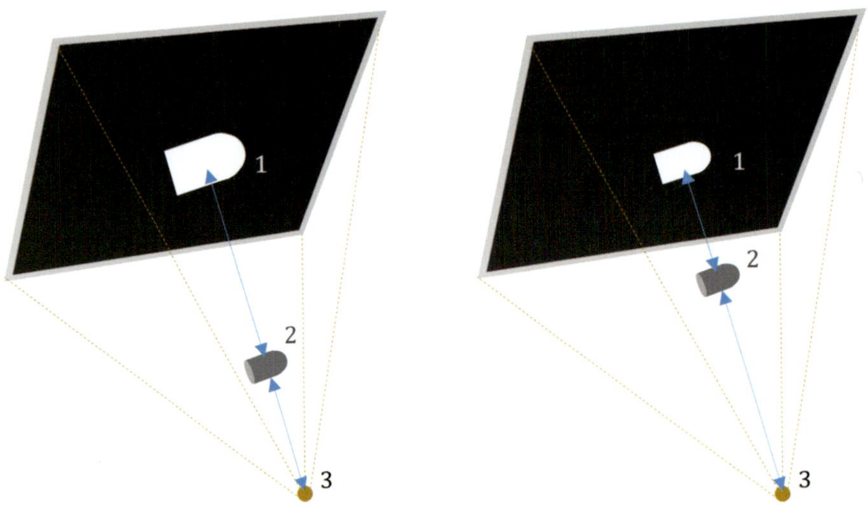

Figure 23.1 Schematic representation of the radiographic projection (1) of a bullet (2) at different distances from the X-ray tube focal spot (3). The smaller the distance between the X-ray tube focal spot and the bullet, the larger the bullet appears on the radiographic image. Consequently, the dimensions of the X-ray representation of an enclosed projectile on a radiographic image may differ from its actual dimensions, so that magnification factors must be taken into account for virtual caliber measurements.

23.3 HOW TO EXAMINE LODGED BULLETS VIRTUALLY ON CT

In contrast to planar radiographs, the problems of magnification and spatial orientation do not exist for X-ray computed tomography. Furthermore, the 3D CT data allow multi-planar reformations, whereby a bullet can be virtually rotated in all three spatial axes and thus aligned for accurate caliber measurements with software integrated measuring tools. However, metallic objects cause considerable artifacts on CT images. Such metal artifacts represent as light and dark streaks across the cross-sectional images. To reduce these streaks in the image, special metal artifact reduction algorithms based on iterative reconstruction or deep learning have been developed (Figure 23.2) [7].

Although such metal artifact reduction algorithms allow to reduce these light and dark streaks in the image, they hardly optimize the representation of the metallic object, in our case the lodged bullet. A more precise representation of a lodged bullet can be achieved by the so-called windowing. Usually CT reconstructions have a 12-bit depth and accordingly offer 4096 values (2^{12} values). In CT imaging, these 4096 values are called CT numbers and are given in Hounsfield units (HUs). This is why these values are often referred to as HU values, and this term will now be used in this chapter. The HU values are calibrated to distilled water, which takes the value 0 HU. Hence, the HU scale extends over a range from −1024 to 3071 HU. Medical monitors typically display a CT image as an 8-bit grayscale image with 256 tone options, making it impossible to display 4096 different gray values on the screen. Regardless of this, the human eye is not capable of perceiving 4096 different gray values. This is why a technique called windowing is used, i.e. a range, the window, is selected on the HU scale and all values below it are displayed in black and all above in white. In other words, windowing is used to select the region of the histogram, i.e. the number of voxels as a function of gray values, to be used for visualization (Figure 23.3).

By windowing in the high range of the HU scale, the metal artifacts displayed as light and dark streaks are no longer visible and only the shape of the metallic foreign body is displayed. Since anatomical structures are also rendered

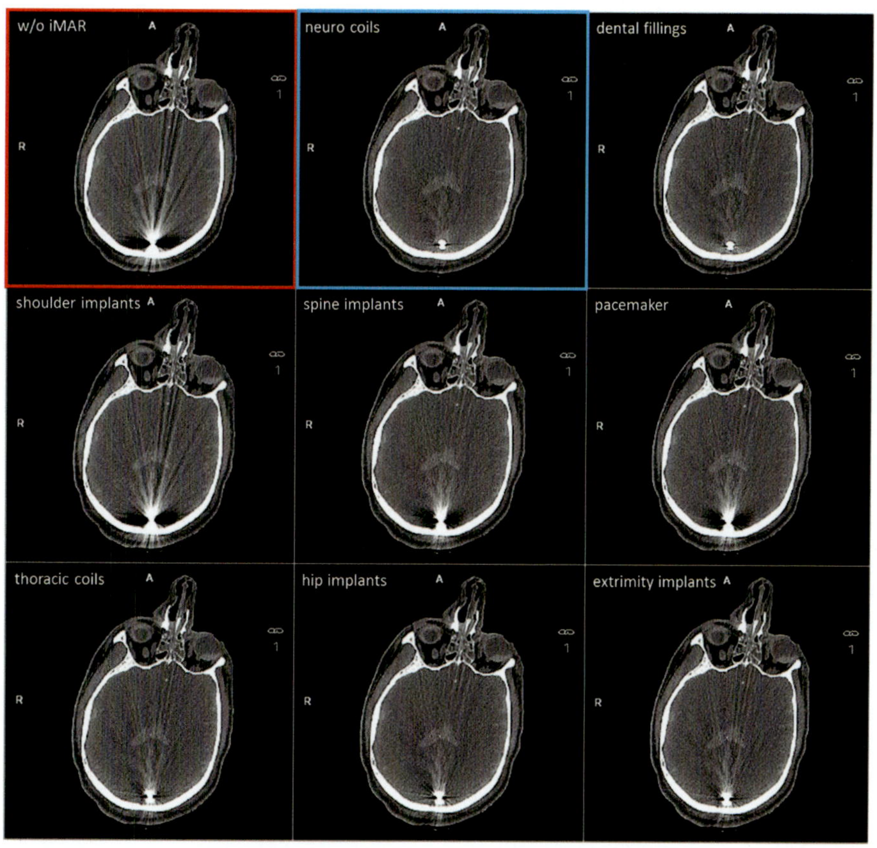

Figure 23.2 The standard CT reconstruction with the typical metal artifacts caused by a lodged bullet are demonstrated in the top left image (red frame). An iterative metal artifact reconstruction (iMAR) algorithm was used to reduce these metal artifacts. The images show the effect of different algorithms designed for common medical implants. A study on pig models has shown that the algorithm designed for neuro-coils is best suited to reduce metal artifacts in stuck projectiles [7], which is also shown in the pictures presented here (blue frame).

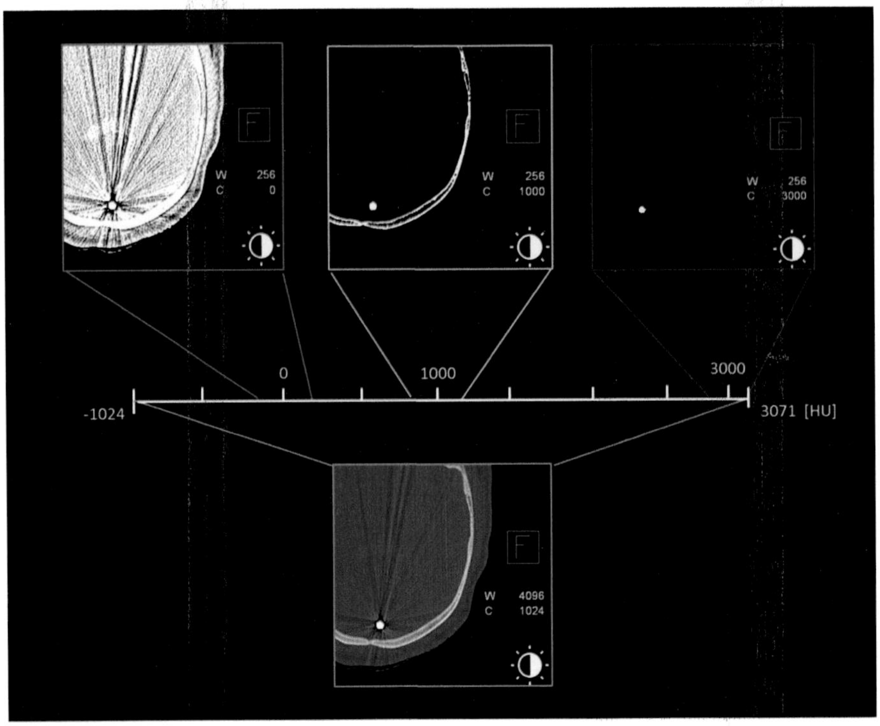

Figure 23.3 A window over the entire HU scale, i.e. in this case from −1024 to 3071 HU, is displayed by a screen with an 8-bit grayscale image in 256 tone options, where the gray value 0 is black and the value 255 is white, and thus one gray value on the screen corresponds to the range of 16 HU values (image at the bottom). The window can be narrowed, as in this case to 256 (images in the upper row), thus one gray value on the screen corresponds to one HU value. The window can be moved along the HU scale. By moving the window to higher HU values, the light and dark stripes caused by the lodged bullet can be made invisible with the disadvantage that the tissue and finally the bones are no longer visible because all values below the lower window limit are displayed in black. In the end, only the lodged bullet is visible in the CT image (red bordered image).

invisible (displayed in black) if their HU values are below the lower limit of window, this windowing method is not considered as true metal artifact reduction. Nevertheless, the shape and caliber of the bullet can now be examined more closely. However, for an accurate representation, the reconstruction field-of-view (FOV) must be taken into account, which has an influence on the in-plane pixel size [8]. The user has the possibility to select the in-plane field which will be reconstructed with the available matrix, usually a 512 × 512 matrix. With a small reconstruction FOV that is aligned to the size of the lodged bullet, the display of the bullet can be improved considerably in terms of resolution (Figure 23.4).

In this context it should also be noted that a lodged bullet may appear distorted on the CT images, if the bullet is located far from the iso-center of the CT gantry (Figure 23.5), i.e. at the edge of the maximum reconstruction field-of-view. This may be the case, for example, when the bullet is located in the upper arm.

Since CT scan data are three-dimensional (3D) data sets, a lodged bullet can be virtually rotated. At this point it should be remembered that the reconstruction FOV defines the in-plane resolution and the third dimension is determined by the slice thickness, which must be small to ensure high resolution. While multi-planar reconstructions (MPR) can be used to align the dataset to the cross section of the bullet for caliber measurements, the volume rendering technique (VRT) can be used to inspect the shape of a lodged bullet from all sides (Figure 23.6). The shape of a lodged bullet can provide additional information that allows further classification of the bullet. For instance, the characteristic mushroom shape of a deformed bullet is an indication that it is most likely a hollow-point or soft-point bullet. In a previous study [9], it was shown that CT reconstructions can provide accurate caliber information when measured at the rear end of the lodged bullet, and that classification based on the depicted shape or deformation of the bullet is very possible. In addition, the degree of mushrooming can provide information about the bullet, since, for example, special bullets that are used by the police are designed to mushroom only slightly in order to avoid "over-penetration" and to cause a smaller wound cavity than common hollow-point bullets [10].

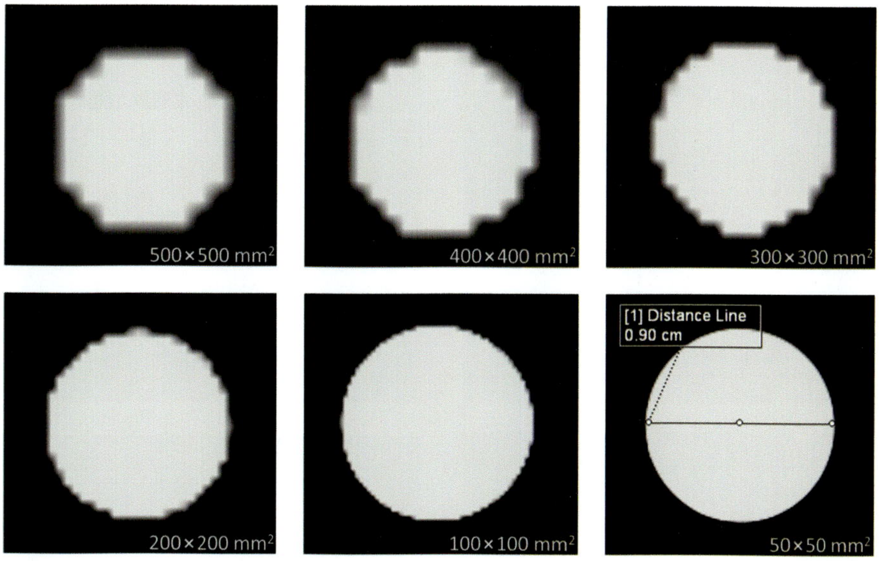

Figure 23.4 Depictions of a 9-mm Luger in the same magnified image section reconstructed with a progressively decreasing field of view (FOV) starting from a FOV of 500 × 500 mm². All reconstructions were generated from the same ex-vivo acquired scan data. It can be clearly seen how the resolution of the bullet improves when the reconstruction FOV is reduced.

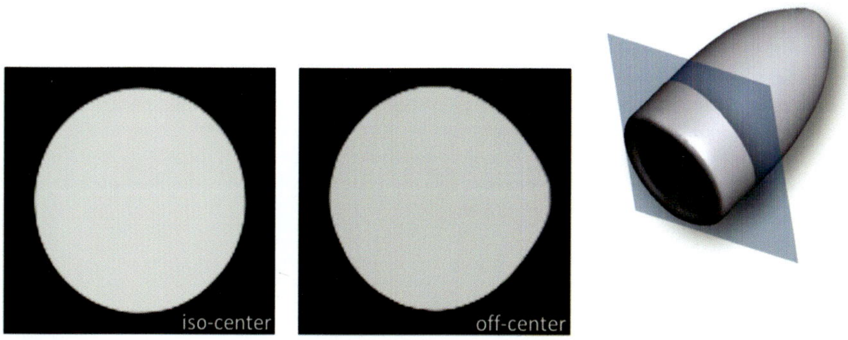

Figure 23.5 Ex vivo scans of a 9-mm Luger bullet positioned at the iso-center (left image) and 25 cm off-center (middle image). In both scans, the bullet was aligned parallel to the scan direction so that the cylindrical rear end of the bullet should appear as a circle in the cross-sectional image. The blue area in the right image shows the orientation and position of the cross-section planes. At the edge of the 500 × 500 mm² reconstruction FOV the cylindrical shape of the rear end of the bullet is distorted on the CT image due to the divergence of the X-ray beam and its rotation around the center.

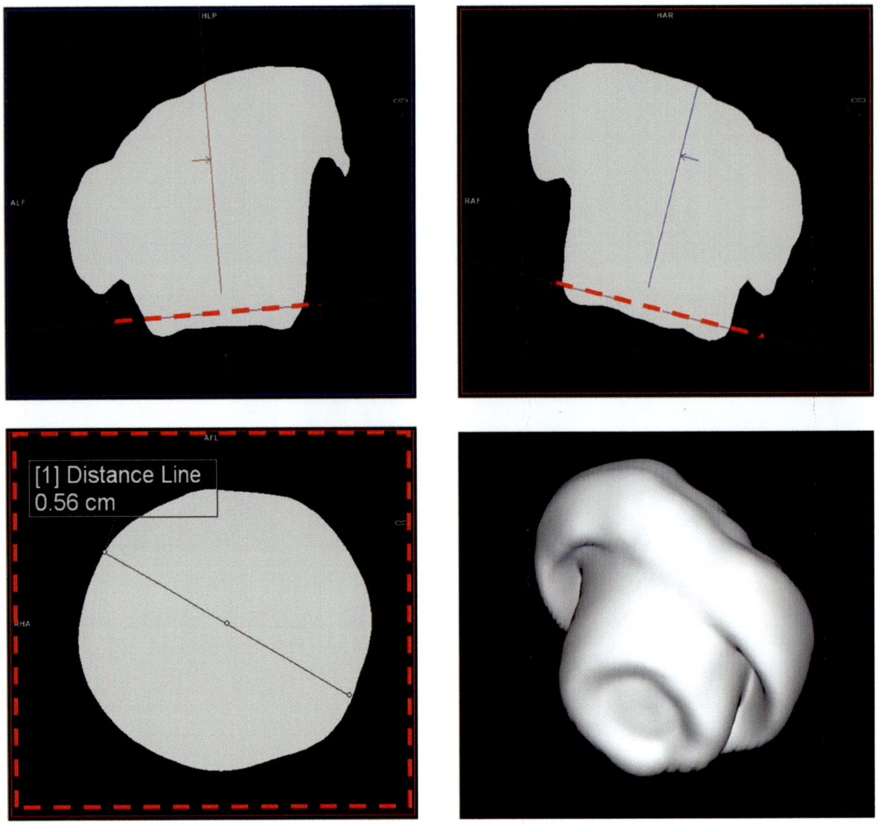

Figure 23.6 Multi-planar reconstructions allow the lodged bullet to be aligned and the section planes to be adjusted accordingly in order to perform caliper measurements. It is also important to pay attention to the location at which caliber measurements are performed. In the case of this .22 long rifle hollow-point bullet illustrated here, measurements were taken at the rear end (dashed lines). The bullet has deformed and thus caliber measurements at other locations would indicate an incorrect caliber. The volume rendering technique, on the other hand, allows a three-dimensional representation of the bullet, which clearly illustrates the characteristic mushroom shape, which is an indication for a hollow-point or soft-point bullet.

23.4 HOW TO EXAMINE THE INTERNAL STRUCTURE OF A LODGED BULLET

Not only is the external shape of a lodged bullet used for classification, but also the visualization of structures inside a bullet can help to classify the bullet more accurately. Special bullets designed for more controlled deformation and slight mushrooming can have internal cavities that are characteristic of that particular bullet. Examples of such a special ammunition are the Action 4, SECA (SEcurity CArtridge/Safe Environment Controlled Action) and QD-PEP (Quick Defense—Polizei Einsatz Patrone) bullets. These bullets can be identified on CT images by detecting internal cavities and can even be distinguished from each other by the characteristic shape of these cavities (Figure 23.7) [10].

However, to visualize the interior of such a bullet, the usual 12-bit image reconstruction is not sufficient, where the maximum HUvalue is 3071 HU. Rather, it requires a 16-bit image reconstruction, thus 2^{16} grayscale values, to allow a much higher maximum value than 3071 HU. Alternatively, the so-called extended CT scale (ECTS) algorithm is used for a tenfold increase of the maximum attenuation value from 3071 HU to 30710 HU (Figure 23.8). The ECTS algorithm is sometimes interpreted as a metal artifact reduction algorithm, although this is not actually the case. Rather, such reconstructions allow to window into a much higher range of HU values as the HU scale is extended.

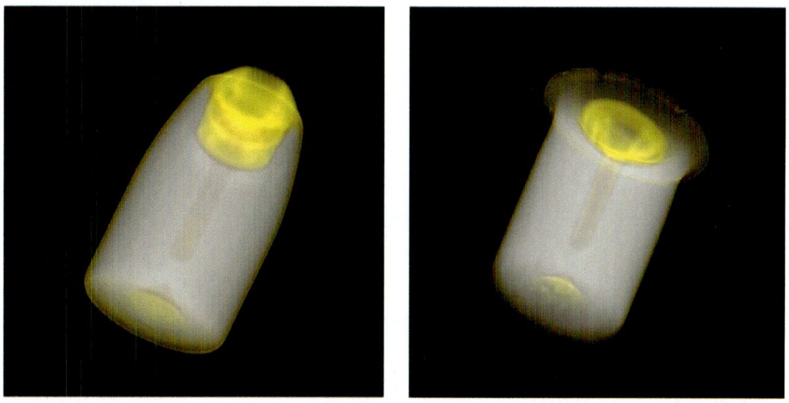

Figure 23.7 Volume rendering of an intact (left image) and a deformed (right image) Action 4 bullet reconstructed with an extended CT scale to show the interior of the bullet. When the bullet hits tissue, the plastic tip is pressed inwards and a slight mushroom shape is created at the front end of the bullet. It is intended that a slight mushrooming occurs, so that the cross-sectional area of the projectile is slightly increased, thus more energy is transferred (intended effectiveness) and as a result the penetration "power" decreases, since it is intended that the projectile lodges in the fugitive and does not hit bystanders.

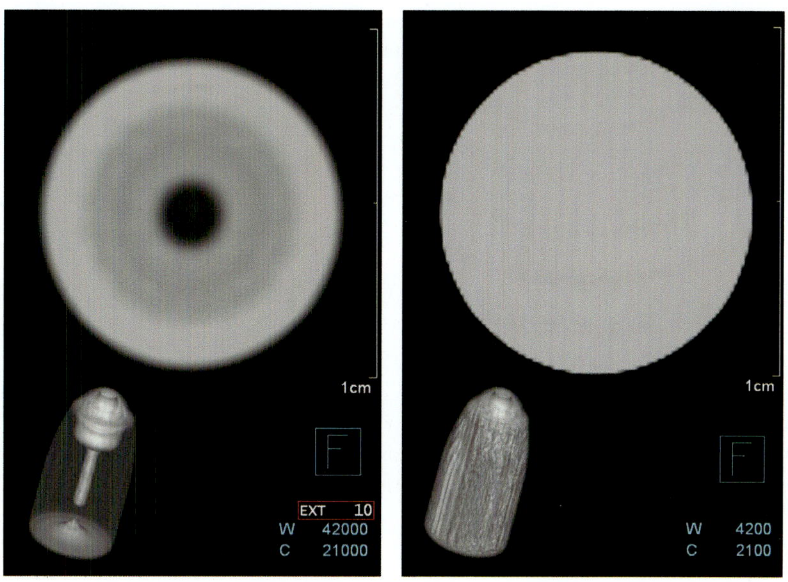

Figure 23.8 Multi-plane reconstructions of an Action 4 bullet reconstructed with the ECTS algorithm (left) and with the 12-bit grayscale range (right), shown in the transverse plane (large images), and as a volume rendering (lower left images). One can clearly see here that a 12-bit reconstruction is not sufficient to visualize the inner cavity of such a special bullet.

23.5 WHAT CAN BE CONCLUDED FROM THE FRAGMENTS?

When a projectile is in a highly fragmented and deformed state, the original shape can hardly be inferred. Nevertheless, it is worthwhile to examine the individual fragments in detail virtually. For example, the shape of individual bullet fragments can also indicate specific projectile types, such as the G2 Research Radical Invasive Projectile (RIP) [3,11–14]. This is a fragile bullet consisting of a solid base slug and—depending on the caliber—six or eight trocars (Figure 23.9). These trocars expand radially as the bullet penetrates tissue, while the base slug continues to penetrate the tissue along its trajectory. Identification of such trocars allows identification of that particular type of bullet. Virtual investigation of the base slug, in addition to the number of trocars detected, can also be useful for caliber measurements.

The fragmentation behavior of the bullet can also give clues about its metallic material. Furthermore, a closer look at the individual fragments could reveal the jacket of the bullet [15,16]. A jacket can be identified when it has separated from the core of the projectile, splitting in such a way that the jacket is finally star-shaped (Figure 23.10). Among fragments, such a detached jacket can be readily visualized on CT. In case a jacket is identified among the fragments, it can be excluded that it is a monobloc or monolithic bullet which has no jacket.

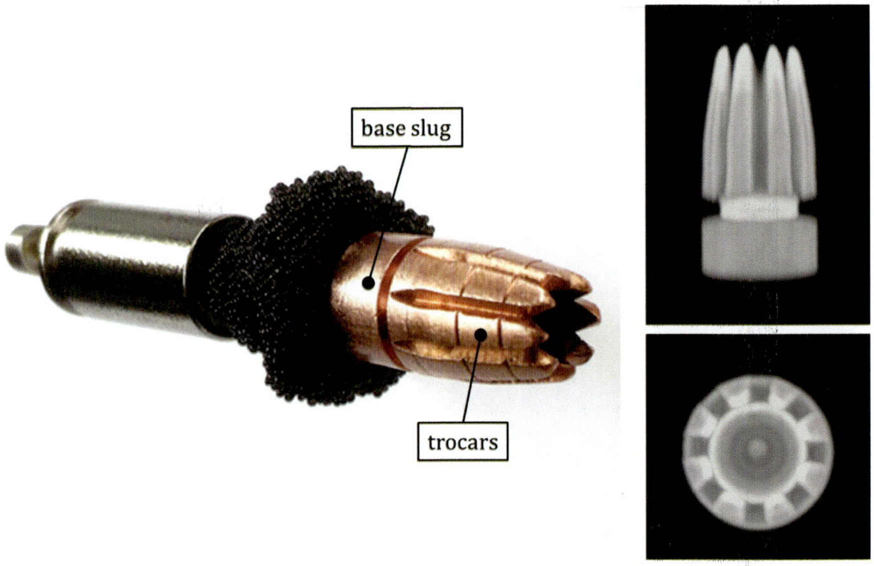

Figure 23.9 Photo of a handmade 3D expanded ammunition cartridge ammo display model of a 9 mm G2 Research Radically Invasive Projectile by Ballistic Concepts, and maximum intensity projections of the CT data in sagittal and transverse view. Note the eight trocars of the projectile, which detach from the base slug upon entering tissue and spread radially.

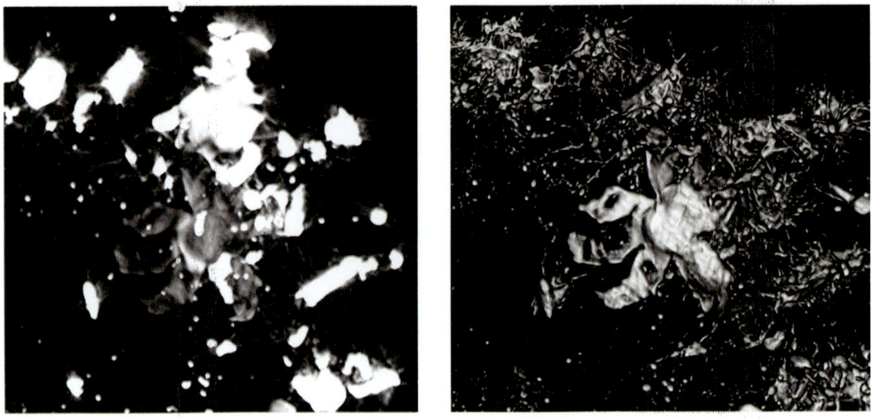

Figure 23.10 Maximum intensity projection (left image) and volume rendering (right image) of a fragmented flat point full metal jacket lead bullet with a ferromagnetic steel jacket retained in a ballistic glycerin soap block after impact with a steel plate. The burst steel jacket is visually distinguishable from the surrounding lead fragments.

23.6 HOW TO DISTINGUISH LODGED BULLETS BASED ON HU VALUES (CT NUMBERS)

Two types of bullets are most common. On the one hand those with a lead core and a jacket of iron in form of steel, copper, or a copper-zinc alloy such as brass or tombac. And on the other hand, those without jacket, which are made of iron, copper or zinc. Some bullets are additionally plated with a very thin layer of a more noble metal, especially as a rust protection, which is not considered here. The atomic number (Z) of lead (Z = 82) is much higher than that of iron (Z = 26), copper (Z = 29) or zinc (Z = 30). Lead should therefore have much higher HU values than iron, copper or zinc, since the photoelectric effect is strongly dependent on Z. Consequently, lead bullets should be easily distinguishable from, copper, and zinc using X-rays. However, this is not straightforward on conventional CT imaging because the X-ray beam is polychromatic and cupping and beam hardening effects occur (Figure 23.11). The cupping effect causes the edge of metallic objects to appear bright, while directly following this bright edge the HU values decrease more and more towards the center of the metallic object due to beam hardening. The bright edge of a bullet on CT does not represent the jacket of a bullet and can also be observed in bullets without a jacket. Depending on the metallic element or alloy, these two effects are more or less pronounced [17]. Here it should be mentioned that jackets made of steel, copper, brass, or tombac, have no appreciable effect on the beam hardening effect in the inner core made of lead, which means that ROI measurements in the lead core are not indicative of the materials in the jacket [18].

On the one hand, bone in the immediate vicinity can affect the HU values, and on the other hand, the material thickness, i.e. the distance the polychromatic X-ray beam has to travel through the bullet, plays a decisive role in the result of ROI measurement. In CT imaging, the location in the body, the caliber and the orientation of the bullet lead to difficulties in material-based discrimination of metallic objects in the body (Figure 23.12).

Therefore, ROI-based measurement of HU values with only single-energy CT data should not be considered reliable for discrimination of metallic objects. However, if CT data are collected at two different energies, the ratio of HU values at these different energies can provide a much more reliable discrimination between lead and copper bullets. For this purpose, the so-called dual-energy index (u) is calculated:

$$u = \frac{(x_{E1} + 1000) - (x_{E2} + 1000)}{(x_{E1} + 1000) + (x_{E2} + 1000)} = \frac{x_{E1} - x_{E2}}{x_{E1} + x_{E2} + 2000}$$

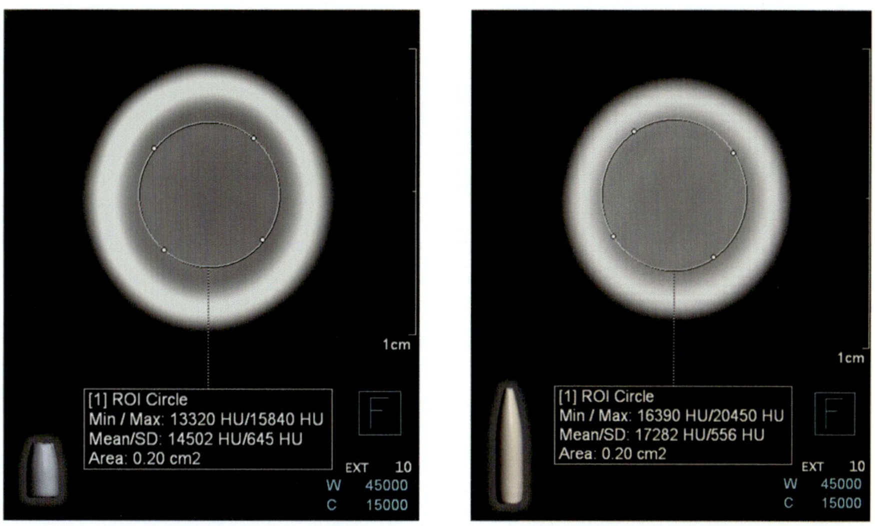

Figure 23.11 ROI measurement in the cross-sectional plane of a Hydra-Shok bullet (left image) and an elongated .30–06 Springfield bullet (right image). The Hydra-Shok bullet (caliber: 9 mm) is made of lead and is coated with an approx. 0.5 mm thick copper jacket while the .30–06 Springfield bullet (caliber: 7.62 mm) is made of copper and has no jacket. Both bullets are inserted into muscle tissue of severed sheep legs for the measurements and are aligned parallel to the scan direction. Note that the outer brighter edge of the bullet is not the jacket but a cupping effect. Although lead is more radiopaque than copper, here the ROI measurement in the lead core of the Hydra-Shok bullet shows lower HU values than the ROI measurement in the copper core of the .30–06 Springfield bullet. The reason for this is beam hardening in combination with the different caliber of the bullets.

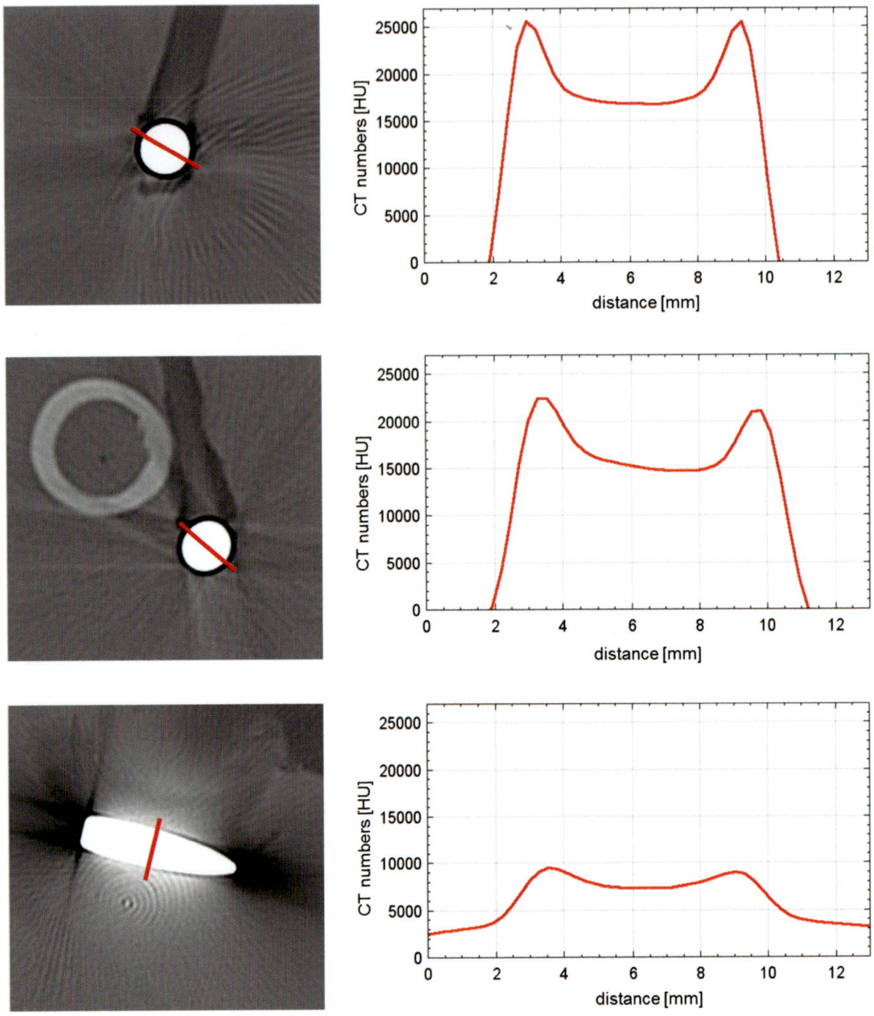

Figure 23.12 HU values as a function of distance measured through the cross-section of the .30–06 Springfield bullet, parallel to the scan direction (first row), parallel to the scan direction immediately adjacent to the bone (middle row), and across the scan direction (bottom row). The respective peaks show the cupping effect at the edge bullet. It can be seen that the bone in the immediate vicinity does not change the shape of the HU profile much, but reduces the HU values considerably overall. The HU profile almost collapses when the bullet is aligned perpendicular to the scan direction. Note: The bullets were inserted by hand into a severed sheep leg after incision with a scalpel.

where x_{E1} is the HU value at the lower energy and x_{E2} is the HU value at higher energy. The summand 1000 for each HU value, results from the attenuation difference between air and water. For the material-based differentiation, ROI measurements are performed at the same position within the bullet on the CT data set acquired with lower and that acquired with higher energy. Calculating this normalized difference in HU values considerably reduces the aforementioned influences on the HU values and allows a reliable differentiation between, for example, lead bullets and copper bullets [15,19]. The evaluation of the dual-energy index for the differentiation of copper and lead bullets showed that HU values measured further at the edge of the bullet are better suited for differentiation than HU values measured deeper inside the bullet. In addition, it was shown that the maximum values of the HU values are better suited for the differentiation than the mean values. This simplifies this method considerably in the application since the ROI can be drawn over the entire metallic object which is to be seen in the cross-section image, and from this measurement the maximum HU values can be taken for the calculation of the dual-energy index (Figure 23.13).

Studies have shown that of the selectable tube voltages of 80, 100, 120, and 140 kV, photon energies obtained from 120 and 140 kV are best suited for distinction of copper and lead bullets based on the dual-energy index [15,19]. When the HU values were obtained from these two photon energy levels, a negative value often results for lead bullets, which is related to the increase of HU values with increasing photon energy, which in turn is related to the k-edge energy of lead [19]. Lead has a k-edge absorption energy of 88.0045 kilo-electron-volt (keV), while, for comparison, the k-edge absorption energy for copper is only 8.9789 keV. When the energy of the X-ray photons approaches the energy of the

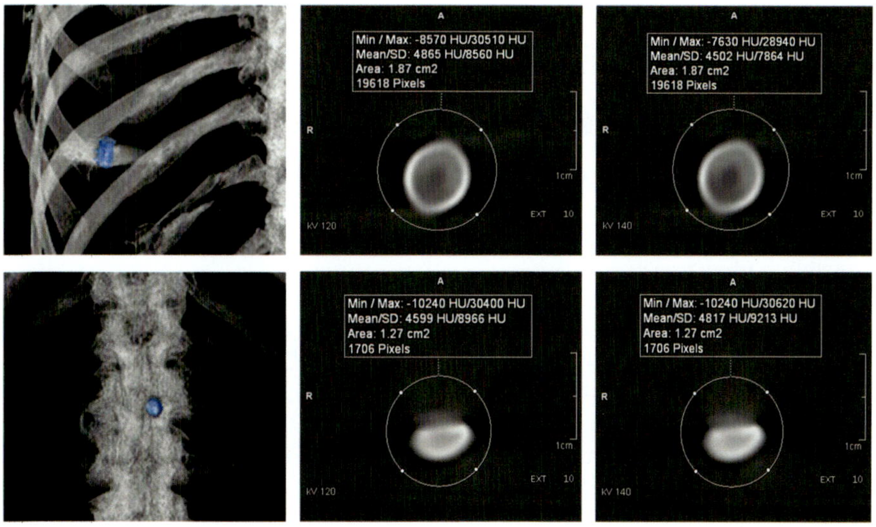

Figure 23.13 ROI measurements of a lodged copper bullet (top row) and a lodged lead bullet (bottom row) on CT datasets acquired at tube voltages of 120 kV and 140 kV. Substituting these values into the DEI equation results in a DEI of +0.025 for the copper bullet and an DEI of −0.003 for the lead bullet. These values are consistent with the values presented in the literature [15,19].

k-edge energy, the photoelectric effect becomes more dominant and the X-ray photons are absorbed more strongly. As the tube voltage increases from 80 to 140 kV, the mean energy of the X-ray spectrum increases and approaches the k-edge energy of lead, causing the HU values of lead to increase.

The dual-energy index approach for material differentiation is not limited to intact bullets, but can also be applied to deformed bullets and bullet fragments. Although this index is referred to as the dual-energy index, its use for projectile differentiation does not necessarily require a dual-energy CT scanner. Rather, dual-energy CT datasets usually cannot be reconstructed with 16-bit or ECTS, which means that metals with high HU values cannot be differentiated. For the differentiation of metals based on the dual-energy index, therefore, two single-energy scans are usually required or, in addition to the basic scan at 120 kV, an additional scan at 140 kV over the region where the metallic foreign body is located. For the ROI measurements, care should be taken that the reconstructions of the two CT scans have the same reconstruction kernel, the same in-plane FOV, the same slice thickness, and the same slice increment.

23.7 APPENDIX 1: THE USE OF CLINICAL CT SCANNERS IN TERMINAL BALLISTICS

As mentioned in the introduction to this chapter, a radiological examination is a primary examination for gunshot victims, but conventional radiographs are more and more superseded by CT examination. CT imaging is also used in forensic investigations of firearm deaths along with 3D optical imaging of the body surface to reconstruct the crime scene [20–23]. The possibility of virtual examination of an object using 3D CT data also found its way into experimental studies in terminal ballistics. For example, for a detailed study of the effectiveness of a bullet based on the cavitation retained in ballistic soap or ordnance gelatin [16,24–29]. CT data of a soap or gelatin block into which a bullet has penetrated allows for ballistic measurements and to study the behavior of a bullet in 3D.

For experiments on wound ballistics, velocity measurements of the projectile should always be carried out in advance at a distance between weapons and the intended position of the soft tissue simulant (soap or gelatin). The kinetic energy (E_{kin}) of the projectile can be calculated from the velocity (v) and the mass (m) of the projectile:

$$E_{kin} = \frac{1}{2} m \cdot v^2$$

For projectiles lodged in the simulant, E_{kin} of the projectile (immediately before entering the simulant) corresponds to the transferred energy (E_{ab}) in the simulant. If the projectile is expected to pass through the simulant, velocity measurements should also be made on the exit side of the simulant, because the E_{ab} then corresponds to the difference between E_{kin} before penetration and E_{kin} after exit. Other energy conversion processes such as energy emitted in the form of heat or energy required for the deformation of the projectile are often disregarded here.

The energy density (ED) or sectional density (SD) is usually specified for the penetration potential of a projectile into a simulant, i.e. E_{kin} is divided by the cross-sectional area of the projectile. For handgun projectiles, the caliber (cal) or caliber area (1/4 cal2 π) is commonly used as the cross-sectional area for estimating the penetration potential. For full metal jacket projectiles of handguns which generate a long narrow shot channel, this gives a good estimate of the penetration depth. In the case of hollow point and soft point projectiles, however, the mushrooming of the projectile must be taken into account when penetrating the simulant, since this results in a considerably larger cross-sectional area and thus releases more energy over a shorter distance, thus resulting in greater effectiveness and less penetration depth. In these cases, the transferred energy per distance (E'_{ab}) in joules per centimeter is of interest to describe the generated expansion of the cavity, as this allows the wound potential to be estimated. Under the assumption that the projectile does not fragment when penetrating the simulant, E'_{ab} can be approximated for soap blocks and gelatin blocks as described in the following.

A soap block preserves the expansion of the stretched cavity, the temporary cavity, which collapses within a very short time in biological tissue. Using volume rendering technology and adjusting the transfer function, the soap block can be made transparent [30]. Additionally, a clip-box tool can be used to virtually slice the soap block. Finally, the temporary cavity can be visualized and virtually examined in 3D (Figure 23.14).

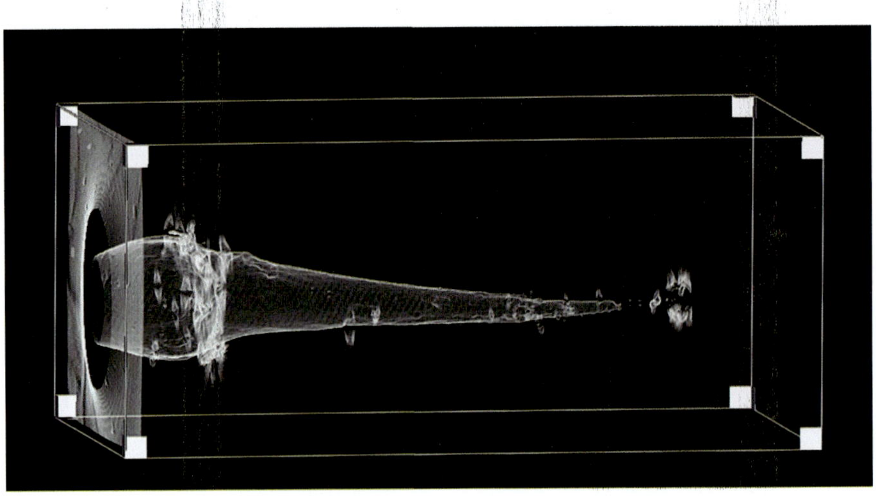

Figure 23.14 Visualization of the temporary cavity of a .357 MAG FTX (140 gr) with a special tip (Flex Tip Technology) using the volume rendering technique with a suitable transfer function and a clip-box to virtually punch the temporary cavity out of the soap block.

The volume (V) is proportional to the energy transferred to the soap block (E_{ab}):

$$E_{ab} \cdot \mu = V$$

where μ is the proportionality constant, which was experimentally determined to be 0.182 cm^3/J for soap [31].

The volume of the temporary cavity can be measured directly with segmentation methods such as region growing. This is necessary for example to calculate the energy transferred from the bullet to the soap block (Figure 23.15) [16]. The principle of region growing segmentation is to find regions of connected voxels with similar characteristics, such as HU values in our case. The algorithm starts from a voxel or region called seed, which is defined manually. The growing from the seed is governed by similarity and the region grows, for example, until a threshold is reached or a distance is exceeded. The cavity differs considerably in its HU values from the HU values of the soap block, and therefore the segmentation algorithm fills the cavity well. However, at the entrance hole and, if present, also at the exit hole, the cavity is connected to the background, and thus the region continues to grow at these points. The regions that extend beyond the cavity can be virtually cut off or, before the scan, the entrance and exit holes can be closed with plastic plates, for example. The seed should not be defined too close to a retained bullet because of the metal artifacts. For perforating shots, the seed should be defined at half the penetration depth.

Alternatively, the cross-sectional area of the temporary cavity can be measured on the individual cross-sectional CT images without having to cut open the soap block and the volume of the temporary cavity can be approximated from the sum of the individual cross-sectional areas (Figure 23.15) (Figure 23.16).

Thus a data set with a certain slice thickness (Δsl) can be reconstructed, and the partial volume (ΔV) can be calculated from the area of the cavity shown in the cross-sectional image multiplied by the slice thickness. From this, the transferred energy per path length (E'_{ab}), where in this case the selected slice thickness determines the path length, can now be calculated for individual volume segments or CT slices:

$$(E'_{ab})_i \cdot \Delta sl \cdot \mu = (\Delta V)_i$$

In contrast to the "frozen image" of the maximum cavity expansion in the soap block, the cavity expansions in a gelatin block are only temporary which, however, is more equivalent to human soft tissue. Thus, a gelatin block preserves only a smaller permanent cavity, while the destruction of the gelatin is indicated by cracks formed during the expansion of the temporary cavity. Since there is a proportionality between the transferred energy and the volume of the maximum cavity expansion, as described for soap block experiments, as well as a proportionality between such a volume and the resulting cracks in gelatin, a proportionality between the transferred energy and the cracks in the gelatin block is also assumed:

$$E_{ab} \cdot c = TCL$$

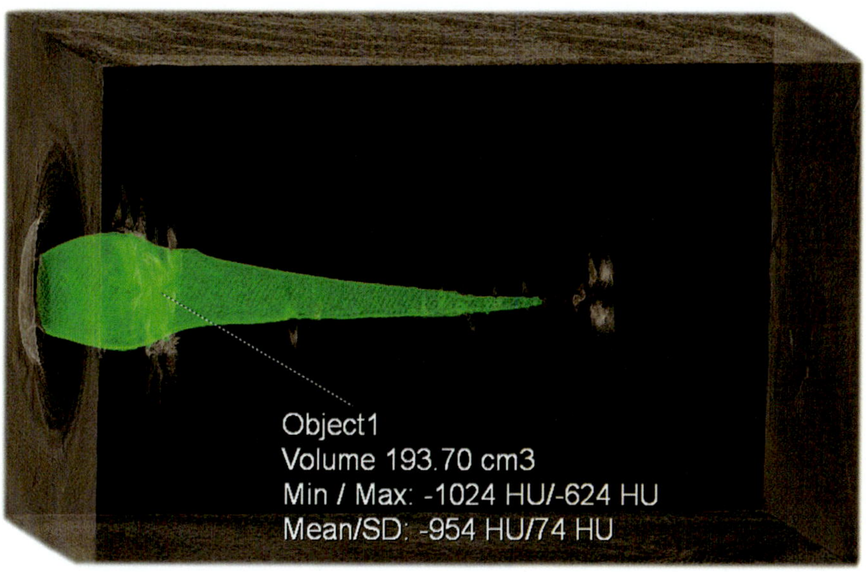

Figure 23.15 Segmentation using region growing to quickly and easily measure the volume of the temporary cavity in the soap block. https://creativecommons.org/licenses/by/4.0/.

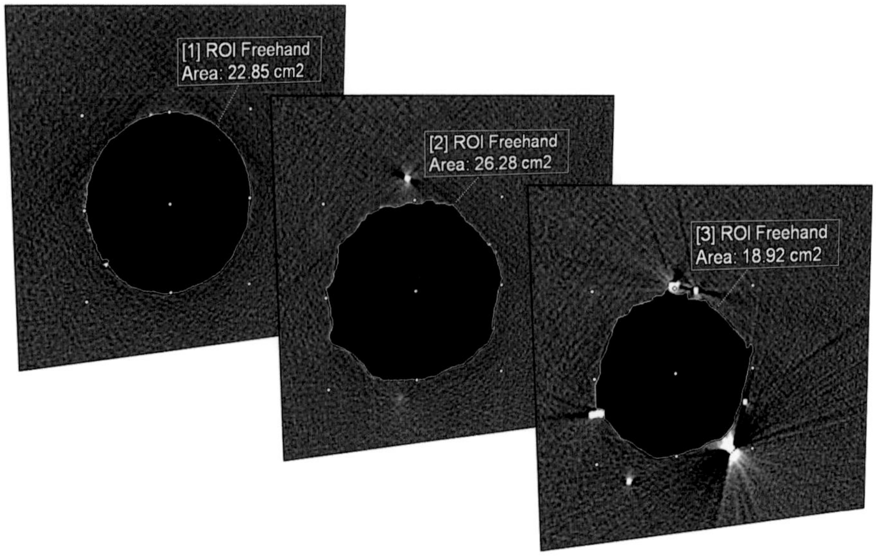

Figure 23.16 Freehand drawn region-of-interest (ROI) measurements to determine the cross-sectional area of the cavity performed on individual cross-sectional images. From the sum of the individual cross-sectional areas measured along the entire cavity, the volume of the cavity can finally be approximated.

where c is the proportionality constant and TCL is the total crack length. The proportionality constant c results from TCL, which can be calculated from CT measurements as described in the following, and E_{ab}, which can be determined from velocity measurements and the mass of the projectile as described earlier. TCL is the sum of the crack lengths (r_i) within sequential slices (j) along the bullet path through the gelatin block summed over all slices:

$$TCL = \sum_i \left(\sum_j r_{i,j} \right)$$

When cutting a gelatin block manually, it is difficult to maintain a constant slice thickness on the one hand, and on the other hand, the slice thickness cannot be chosen to be very thin or the gelatin slice would fall apart. The CT scan of a gelatin block allows it to be virtually sectioned into thin slices with a constant slice thickness. Thus, the energy transferred per cm (E'_{ab}) can be calculated for each slice (i) via the sum of the crack lengths in the slice:

$$\left(E'_{ab}\right)_i \cdot \Delta sl \cdot c = \sum_j r_{i,j}$$

As mentioned before, the constant slice thickness (Δsl) can be selected for CT reconstruction, whereby the thinner the slice, the more accurate the calculations (Figure 23.17).

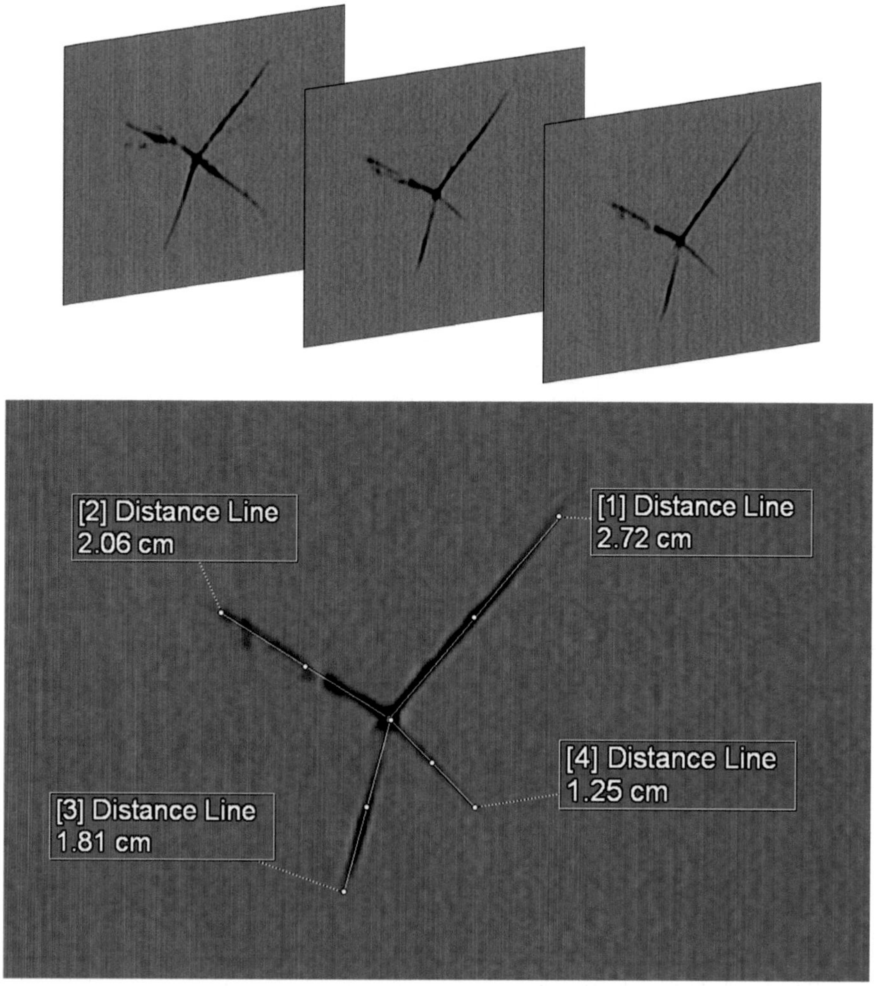

Figure 23.17 Cross-sectional images perpendicular to the shot channel of a GP11 bullet illustrate the radial cracks originating from the permanent cavity (upper row). The sum of the lengths of these cracks in a given cross-sectional area is proportional to the energy transferred to the gelatin per unit length.

23.8 APPENDIX 2: THE USE OF CLINICAL MRI SCANNERS IN TERMINAL BALLISTICS

Finally, the use of magnetic resonance imaging should also be mentioned briefly. By means of special sequences, such as the so-called FRACTURE MRI, which is described in more detail in the chapter on bone MRI, an impressive representation of the radial cracks originating from the permanent cavity is possible (Figure 23.18).

However, due to the strong magnetic field of an MRI scanner, this investigation is not suitable for ballistic experiments of bullets with ferromagnetic components such as steel. This is due to the fact that on the one hand ferromagnetic metal is strongly attracted by the static magnetic field of the MRI scanner and thus a ferromagnetic bullet can migrate through the gelatin and on the other hand ferromagnetic metal leads to a strong signal loss and thus causes significant metal artifacts in the images [32,33]. For non-ferromagnetic metals such as lead and copper, there is no risk of being attracted by the static magnetic field, nor are severe metal artifacts caused [33]. (Figure 23.19)

For the examination, the gelatin block is positioned in the center of the MRI scanner gantry. Due to the shape and size of bullets and bullet fragments, there is minimal risk of heating during the examination [33].

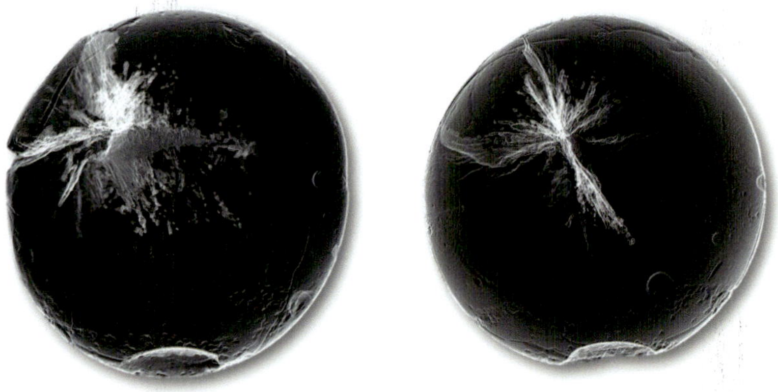

Figure 23.18 FRACTURE MRI volume rendering visualization of the permanent cavity and radiating cracks caused by a 5.56 NATO in a gelatin head model. The surrounding synthetic bone material is not visible on MRI which is advantageous for the 3D visualization of the permanent cavity and its radial cracks using the volume rendering technique.

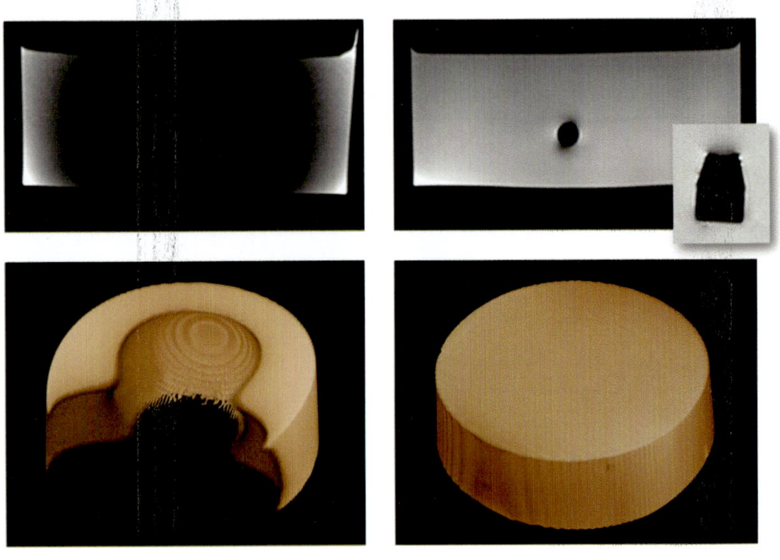

Figure 23.19 Cross-sectional images and 3D volume renderings of an MRI examination of bullets embedded in gelatin in a bucket. The plastic bucket is not visible in the images. In the left images the strong signal losses through the ferromagnetic steel jacket of a leaded 7.62 NATO rifle bullet can be seen. The right images show a non-ferromagnetic .44 Magnum bullet which causes only minimal metal artifacts. In the coronal view, even the shape of the .44 Magnum bullet can be seen clearly (small picture on the right).

References

[1] A.A.S. Marais, H.J. Dicks, Utilization of X-ray computed tomography for the exclusion of a specific caliber and bullet type in a living shooting victim, Journal of Forensic Sciences. 64 (2019) 264–269. https://doi.org/10.1111/1556-4029.13805.

[2] A.M. Alves, F.F. Picoli, R.J. Silveira, L.G. Rodrigues, S.L.R. Lenharo, R.F. Silva, A. Franco, When forensic radiology meets ballistics–in vivo bullet profiling with computed tomography and autopsy validation: A case report, Forensic Imaging. 20 (2020) 200357. https://doi.org/10.1016/j.fri.2020.200357.

[3] D. Gascho, Lodged bullets on computed tomography: Three classification procedures for the virtual investigation of bullets or their fragments that cannot be recovered from the living patient, Medicine, Science and the Law. 60 (2020) 245–248. https://doi.org/10.1177/0025802420962697.

[4] B.G. Brogdon, J.E. Lichtenstein, Brogdon's forensic radiology—Chapter 2. Forensic radiology in historical perspective. In: Thali, M.J., Viner, M.D., Brogdon, B.G., editors. Brogdon's Forensic Radiology, Second Edition. CRC Press—Taylor & Francis Group, 2011, pp. 9–24., n.d.

[5] G.D. Dodd 3rd, R.F. Budzik Jr, Identification of retained firearm projectiles on plain radiographs, AJR. American Journal of Roentgenology. 154 (1990) 471–475.

[6] R.P. Bixler, C.R. Ahrens, R.P. Rossi, D. Thickman, Bullet identification with radiography, Radiology. 178 (1991) 563–567. https://doi.org/10.1148/radiology.178.2.1987625.

[7] F. Berger, T. Niemann, R.A. Kubik-Huch, H. Richter, M.J. Thali, D. Gascho, Retained bullets in the head on computed tomography: Get the most out of iterative metal artifact reduction, European Journal of Radiology. 103 (2018) 124–130. https://doi.org/10.1016/j.ejrad.2018.04.019.

[8] D. Gascho, M.J. Thali, T. Niemann, Postmortem computed tomography: Technical principles and recommended parameter settings for high-resolution imaging, Medicine, Science and the Law. 58 (2018) 70–82. https://doi.org/10.1177/0025802417747167.

[9] D. Gascho, N. Zoelch, E. Deininger-Czermak, C. Tappero, A. Buehlmann, P. Wyss, M.J. Thali, S. Schaerli, Visualization and material-based differentiation of lodged projectiles by extended CT scale and the dual-energy index, Journal of Forensic and Legal Medicine. 70 (2020) 101919. https://doi.org/10.1016/j.jflm.2020.101919.

[10] D. Gascho, N. Zoelch, E. Deininger-Czermak, C. Tappero, H. Richter, M.J. Thali, S. Schaerli, In situ identification of Action 4, SECA and QD-PEP bullets from special police ammunitions by computed tomography, Medicine, Science and the Law. 60 (2020) 188–195. https://doi.org/10.1177/0025802420911555.

[11] K.R. Iverson, E. Curtis, I.E. Brown, E.S. Salcedo, J.T. Anderson, New fragmenting bullet leads to unexpected injury pattern: A case report involving the Radically Invasive Projectile, Trauma. 21 (2019) 73–76.

[12] N. Ditkofsky, H.S. Maresky, S. Steenburg, Radically invasive projectiles—first reports and imaging features of this new and dangerous bullet, Emergency Radiology. (2020) 1–5. https://doi.org/10.1007/s10140-020-01766-5.

[13] G.S. Stacy, S. Thomas, The G2 research RIP fragmenting bullet—radiographic features of a recently encountered projectile, Skeletal Radiology. 49(10) (2020) 1163–1168. https://doi.org/10.1007/s00256-020-03486-w. PMID: 32504097.

[14] X. Zhang, M.D. Cain, C.D. Williams, T.A. Spears, C.K. Poulos, G2 Research radically invasive projectile: The importance of recognizing its imaging and autopsy patterns, The American Journal of Forensic Medicine and Pathology. 42 (2021) 248–251. https://doi.org/10.1097/PAF.0000000000000669.

[15] D. Gascho, N. Zoelch, H. Richter, A. Buehlmann, P. Wyss, M.J. Thali, S. Schaerli, Heavy metal in radiology: How to reliably differentiate between lodged copper and lead bullets using CT numbers, European Radiology Experimental. 4 (2020) 43. https://doi.org/10.1186/s41747-020-00168-z.

[16] D. Gascho, S. Kottner, A. Buehlmann, D. Schweizer, S.A. Bolliger, M.J. Thali, N. Zoelch, Annular distribution patterns of.357 Magnum fragments in soft tissue simulants after striking hard material that prevented the bullet from exiting, Journal of Forensic and Legal Medicine. 85 (2022) 102286. https://doi.org/10.1016/j.jflm.2021.102286.

[17] L.E. Paulis, J. Kroll, L. Heijnens, M. Huijnen, R. Gerretsen, W.H. Backes, P.A.M. Hofman, Is CT bulletproof? On the use of CT for characterization of bullets in forensic radiology, International Journal of Legal Medicine. (2019). https://doi.org/10.1007/s00414-019-02033-0.

[18] D. Gascho, N. Zoelch, S. Schaerli, Explanation for the contradiction between the results of Diallo et al. (https://doi.org/10.1016/j.jofri.2018.10.001) and Winklhofer et al. (https://doi.org/10.1097/RLI.0000000000000032) in differentiating ferromagnetic from nonferromagnetic bullets by means of the dual-energy index, Journal of Forensic Radiology and Imaging. 19 (2019) 100351. https://doi.org/10.1016/j.jofri.2019.100351.

[19] D. Gascho, N. Zoelch, H. Richter, A. Buehlmann, P. Wyss, S. Schaerli, Identification of bullets based on their metallic components and X-ray attenuation characteristics at different energy levels on CT, American Journal of Roentgenology. 213 (2019) W105–W113. https://doi.org/10.2214/AJR.19.21229.

[20] S. Kottner, L.C. Ebert, G. Ampanozi, M. Braun, M.J. Thali, D. Gascho, VirtoScan: A mobile, low-cost photogrammetry setup for fast postmortem 3D full-body documentations in x-ray computed tomography and autopsy suites, Forensic Science, Medicine and Pathology. 13 (2017) 34–43. https://doi.org/10.1007/s12024-016-9837-2.

[21] S. Kottner, S. Schaerli, M. Fürst, W. Ptacek, M. Thali, D. Gascho, VirtoScan-on-Rails: An automated 3D imaging system for fast postmortem whole-body surface documentation at autopsy tables, Forensic Science, Medicine and Pathology. 15 (2019) 198–212. https://doi.org/10.1007/s12024-019-00095-5.

[22] S. Kottner, M.M. Schulz, F. Berger, M. Thali, D. Gascho, Beyond the visible spectrum—applying 3D multispectral full-body imaging to the VirtoScan system, Forensic Science, Medicine and Pathology. 17 (2021) 565–576. https://doi.org/10.1007/s12024-021-00420-x.

[23] S. Kottner, M.M. Schulz, F. Berger, M.J. Thali, D. Gascho, Multispectral 3D whole-body imaging of dressed and undressed bodies in combination with post-mortem X-Ray computed tomography. In: 2021. https://doi.org/10.15221/21.24.

[24] Ž. Korać, D. Kelenc, J. Hančević, A. Baškot, D. Mikulić, The application of computed tomography in the analysis of permanent cavity: A new method in terminal ballistics, Acta Clinica Croatica. 41 (2002) 205–206. https://hrcak.srce.hr/14732.

[25] M. Große Perdekamp, B. Vennemann, B.P. Kneubuehl, M. Uhl, M. Treier, R. Braunwarth, S. Pollak, Effect of shortening the barrel in contact shots from rifles and shotguns, International Journal of Legal Medicine. 122 (2008) 81–85. https://doi.org/10.1007/s00414-007-0161-y.

[26] G.N. Rutty, P. Boyce, C.E. Robinson, A.J. Jeffery, B. Morgan, The role of computed tomography in terminal ballistic analysis, International Journal of Legal Medicine. 122 (2008) 1–5. https://doi.org/10.1007/s00414-006-0145-3.

[27] S.A. Bolliger, M.J. Thali, M.J. Bolliger, B.P. Kneubuehl, Gunshot energy transfer profile in ballistic gelatine, determined with computed tomography using the total crack length method, International Journal of Legal Medicine. 124 (2010) 613–616. https://doi.org/10.1007/s00414-010-0503-z.

[28] F. Gremse, O. Krone, M. Thamm, F. Kiessling, R.H. Tolba, S. Rieger, C. Gremse, Performance of lead-free versus lead-based hunting ammunition in ballistic soap, PLoS One. 9 (2014) e102015. https://doi.org/10.1371/journal.pone.0102015.

[29] I. Burgos-Díez, F. Zapata, M.J. Chamorro-Sancho, M.J. Ruano-Rando, J.L. Ferrando-Gil, C. García-Ruiz, G. Montalvo, F. Ortega-Ojeda, Comparison between computed tomography and silicone-casting methods to determine gunshot cavities in ballistic soap, International Journal of Legal Medicine. 135 (2021) 829–836. https://doi.org/10.1007/s00414-020-02464-0.

[30] S. Fang, T. Biddlecome, M. Tuceryan, Image-based transfer function design for data exploration in volume visualization. In: Proceedings Visualization '98 (Cat. No.98CB36276), 1998, pp. 319–326. https://doi.org/10.1109/VISUAL.1998.745319.

[31] B.P. Kneubuehl, General wound ballistics. In: Wound Ballistics, Springer, 2011, pp. 87–161.

[32] S. Eggert, R.A. Kubik-Huch, M. Lory, J.M. Froehlich, D. Gascho, M.J. Thali, S.A. Bolliger, The influence of 1.5 and 3 T magnetic resonance unit magnetic fields on the movement of steel-jacketed projectiles in ordnance gelatin, Forensic Science, Medicine and Pathology. 11 (2015) 544–551. https://doi.org/10.1007/s12024-015-9725-1.

[33] D. Gascho, C. Tappero, N. Zoelch, E. Deininger-Czermak, H. Richter, M.J. Thali, S. Schaerli, Synergy of CT and MRI in detecting trajectories of lodged bullets in decedents and potential hazards concerning the heating and movement of bullets during MRI, Forensic Science, Medicine and Pathology. 16 (2020) 20–31. https://doi.org/10.1007/s12024-019-00199-y.

CHAPTER 24

MICRO-CT (μCT)

Sören Kottner, Lars Ebert, Wolf Schweitzer, and Rolf Kaufmann

DOI: 10.4324/9781003413998-26

24.1 INTRODUCTION

X-ray micro-computed tomography or micro-CT (μCT) is an imaging modality that uses X-ray radiation to create 3D cross-sectional images of a physical object. Similar to conventional medical X-ray computed tomography (CT), transmission images are acquired from different angles around a physical object (i.e., the sample) and are used to reconstruct a stack of cross-sectional images (i.e., CT slices). The property that most of all distinguishes μCT from CT is the increased spatial resolution within the micrometer range. Conventional medical CTs nowadays offer scan resolutions up to 0.4 mm^3 [1] whereas μCT offers resolutions up to 0.5 μm^3 (e.g., Zeiss Xradia Versa 600-Series, Carl Zeiss AG, Oberkochen, Germany). However, an increased spatial resolution within the lower micrometer range comes at a cost and results in scan times in the order of 10 hours. Long scanning periods make it especially difficult to prepare and fixate biological and medical samples.

To achieve an increased spatial resolution in the micrometer range, most μCT setups are equipped with so-called microfocus X-ray tubes and area detectors (flat panel detectors, FPD). Microfocus X-ray tubes are designed to establish a small focal spot. Nowadays, conventional microfocus setups allow standard focal spot sizes with a mean diameter of 700 nm [2]. The X-ray beam exiting from the X-ray tube extends in a diverging beam, commonly with a cone or pyramidal shape. Fan beam setups with line detectors (linear diode array detectors, LDA) exist but are less common. All of these setups allow an increased spatial resolution through the principle of geometric magnification (Figure 24.1). Geometric magnification depends on the ratio of the distance between the X-ray source and the detector (focus-detector-distance, FDD), and the distance between the X-ray source and the sample (focus-object-distance, FOD). This means, the closer the sample is located to the X-ray source, the larger the projected image on the detector. As a result, spatial resolution and specimen size depend on each other. To achieve a high spatial resolution, the sample size needs to be comparatively small. In addition to that, with an increased size of the focal spot the so-called penumbra effect occurs which leads to edge blurring in the acquired images (Figure 24.1). Overall, the size of the focal spot together with the magnification defines the best achievable spatial resolution. In combination with microfocus X-ray tubes, an increased spatial resolution can be established with the help of high-resolution image detectors [2].

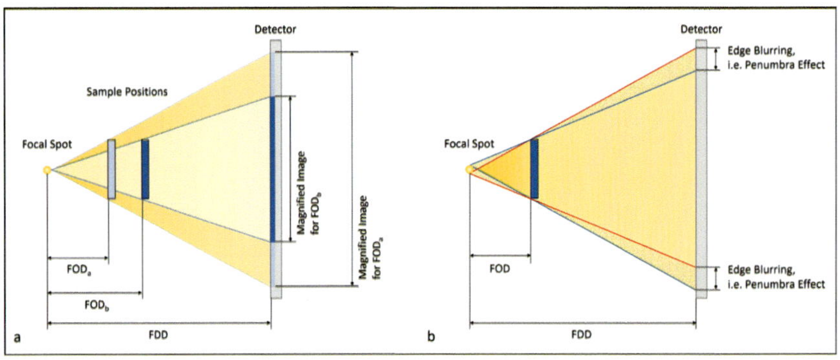

Figure 24.1 Illustration of the principles of geometric magnification a and the penumbra effect b. Image a illustrates that geometric magnification depends on the ratio of the distance between the X-ray source and the detector (focus-detector-distance, FDD), and the distance between the X-ray source and the sample (focus-object-distance, FOD). Image b shows that the size of the focal spot affects the quality of the acquired image. With an increased size of the focal spot, edge blurring (i.e., the penumbra effect) affects the image quality.

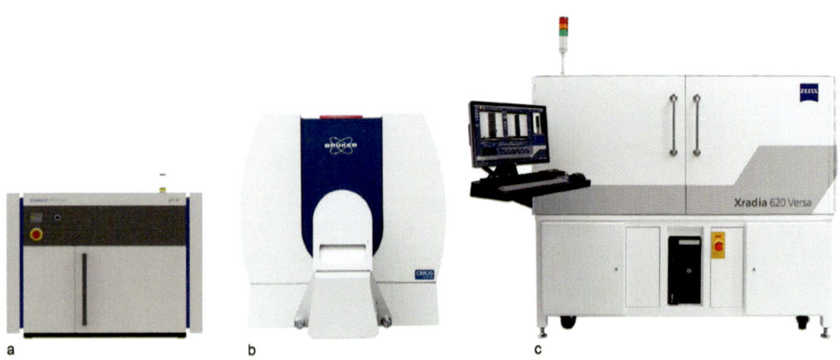

Figure 24.2 Selection of different commercially available μCT systems. Image a shows a fully shielded desktop μCT setup (μCT 45, Scanco Medical AG, Wangen-Brüttisellen, Switzerland). The system features an automated sample changer that can fit up to 20 samples. Image b shows an in-vivo μCT system based on a medical setup (SKYSCAN 1276 CMOS Edition, Bruker BioSpin MRI GmbH, Ettlingen, Germany). The system is specifically designed for fast in-vivo scans of small animals and biological samples and features integrated physiological monitoring devices. Image c shows a high-resolution μCT system based on an industrial setup (ZEISS Xradia 620 Versa, Carl Zeiss GmbH, Oberkochen, Germany). This system offers sub-micron resolutions up to 0.5 μm^3 with a minimum achievable voxel size of 0.04 μm^3.

24.2 MICRO-CT—TYPES AND SETUPS

The market for μCT setups is rather diverse and offers a selection of different systems ranging from setups suitable for multiple applications and highly specialized systems for the analysis of specific samples (Figure 24.2). Therefore, μCT setups come in various shapes and forms with different dimensions and requirements regarding radiation protection. Apart from that, μCT setups can be divided into two groups regarding the setup design (Table 24.1). Either the sample is rotated while the position of the X-ray source and the detector is fixed (industrial setup) or the X-ray source and the detector are rotated around the sample (medical setup). An Illustration of the two setup designs used for μCT imaging is given in Figure 24.3.

Systems based on industrial setups are used for a broad spectrum of applications ranging from industrial to biological and medical applications and research. Industrial setups are generally more versatile in their use due to the build-up of their main components (X-ray source, mechanical axes, sample holder, detector). A common characteristic of industrial setups is that they allow adjusting the spatial resolution according to the sample size. This can be done by modifying the distances between the X-ray source and the sample as well as the distance between the X-ray source and the detector. Apart from that, it is common to find additional axes that allow the positioning of the X-ray source and the detector in relation to the sample. This enables scanning of elongated objects with the highest possible resolution. Due to their broad spectrum of applications, industrial setups come in different sizes and offer a variability of voltages ranging from a few kilovolts up to megavolts. Even though industrial setups can adapt the spatial resolution to the sample size, sample mounting on the other hand can become challenging. This is why these setups are usually not useful for in-vivo scans.

Systems based on a medical setup on the other hand are specifically designed for biological and medical samples and especially for in-vivo scans. These setups are, due to the build-up of the main components, more limited in their functionality. Even though sample mounting is comparably easy and supports in-vivo scans, medical setups commonly exhibit lower spatial resolutions. The spatial resolution is primarily limited by the accuracy of the movement of the source and the detector within the gantry. Since medical setups are primarily developed for biological and medical specimens and in-vivo scans, the range of voltages is commonly much smaller compared to an industrial setup.

Table 24.1 General Differences between Industrial Setups and Medical μCT Setups

	μCT (Industrial Setup)	μCT (Medical Setup)
Rotation	Sample	X-ray source and detector
Magnification	Adjustable according to sample size	Fixed
Hounsfield scale	Not calibrated/not adjusted	Calibrated/adjusted
Sample mounting	Alignment to center of rotation	Alignment to the center of the CT-table/tray
Area of use	Ex-vivo	In-vivo and ex-vivo
Voltage	Large range	Smaller range

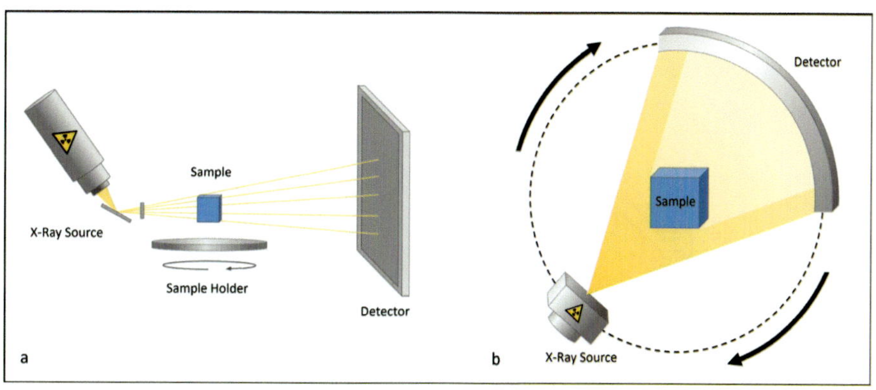

Figure 24.3 Illustration of the two setup designs used for μCT imaging. Image a shows a schematic drawing of an industrial setup, where the sample is rotated while the X-ray source and the detector are fixed. Image b shows a schematic drawing of a medical setup, where the X-ray source and the detector are rotated around a fixed sample.

24.3 MICRO-CTS AND CONVENTIONAL MEDICAL CTS

Spatial resolutions provided by µCTs compared to conventional medical CTs allow visualizing and discriminating materials, tissues, and structures on a much higher level. As an example, skull substructures of a piglet such as mastoid cells and the inner ear cochlea were scanned with a medical CT (Somatom Definition Flash, Siemens Healthineers, Erlangen, Germany) with an anisotropic voxel size of $0.33 \times 0.3 \times 0.6$ mm^3 and a µCT (XtremeCT II, Scanco Medical AG, Wangen-Brüttisellen, Switzerland) with an isotropic voxel size of 35 µm^3. A visual comparison between the axial images of the two scans illustrates the improved delineation of the bone lamellae within the µCT images (Figure 24.4).

Apart from spatial resolution, µCTs and conventional medical CTs present further differences regarding specifications such as scanning duration, sample size, and range of voltages amongst others (Table 24.2). Generally, the duration of a µCT scan is much higher than the duration of a conventional CT scan. Scanning times for a µCT scan range from several minutes to several hours depending on the setup and the sample. Conventional medical CT scanners, however, are built to perform a whole-body scan in a few seconds or minutes. Apart from that, the sample sizes that can be scanned in both systems vary greatly. While the sample sizes for µCTs are commonly in the range of millimeters to centimeters, medical CTs allow scanning up to 2 meters in length with a typical diameter of 50–80 cm. In addition to that, µCT setups are commonly not giving image intensities in Hounsfield units unless the application specifically requires that. Whereas medical CTs are by default calibrated to the Hounsfield scale.

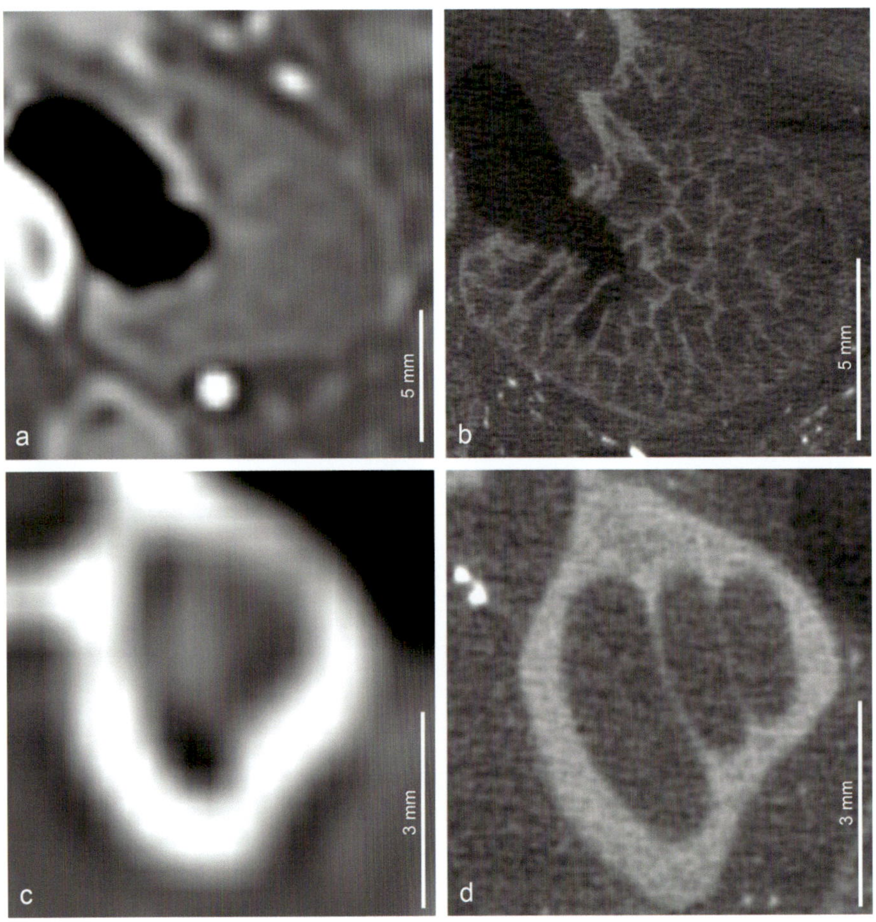

Figure 24.4 Comparison of skull substructures of a piglet from postmortem medical CT and postmortem µCT scans. Images a and b refer to axial images of mastoid cells, whereas images c and d show axial images of the inner ear cochlea of the piglet. Images a and c refer to the scan data from the medical CT (Somatom Definition Flash, Siemens Healthineers, Erlangen, Germany). Images b and d refer to the scan data from the µCT (XtremeCT II, Scanco Medical AG, Wangen-Brüttisellen, Switzerland). The data from the medical CT has an anisotropic voxel size of $0.33 \times 0.33 \times 0.6$ mm^3. The data from the µCT has an isotropic voxel size of 35 µm^3.

Table 24.2 General differences Between μCT Setups and Conventional Medical CT Scanners

	μCT (Industrial Setup)	CT (Medical Setup)
Scanning duration	Up to several hours	Seconds to a few minutes
Sample size	Millimeters to centimeters	Centimeters to a few meters
Hounsfield scale	Commonly not calibrated/not adjusted	Calibrated/adjusted
Spatial resolution	$\geq 0.5\ \mu m^3$	$0.4\ mm^3$
Voltage	40–450 keV	70–150 keV

24.4 APPLICATIONS IN THE FIELD OF FORENSIC MEDICINE

In the field of forensic medicine, the use of µCT was reported early on by Thali et al. in 2003 [3]. In this study, the applicability of µCT for the documentation and analysis of tool marks on bones was presented. In the following years, the use of µCT for the investigation and analysis of sharp force trauma to bones, tool marks, and cut marks as well as false starts on bones have been reported by various authors [4–10]. Since 2003, µCT has been increasingly applied to many fields of forensic medicine. In 2013 Rutty et al. reviewed the research areas and topics that were successfully addressed by the use of µCT [11]. Besides the evaluation of tool marks and cut marks to the bone, µCT has also been used in several studies investigating the distribution of gunshot residue (GSR) in gunshot wounds. In 2011 Cecchetto et al. evaluated the amount and distribution of GSR for different distances between the firearm and samples of human calves [12]. In the following years, Cecchetto and his team studied the impact of decomposition, fire, water, and textiles on the detectability of GSR with the help of µCT [13–16]. µCT has also been used in the assessment of human age based on volume measurements and ratios of different parts of human teeth [17–19]. In some studies, the feasibility and usefulness of µCT for the investigation of strangulations were presented. These studies specifically focused on microstructures and fractures of the hyoid bone and thyroid cartilage [20–24]. Apart from that, µCT has also been used as a tool for non-invasive fetal autopsies [25–28].

Currently, a specialized µCT-Setup to conduct CT angiography of neonates is being developed by Scanco Medical AG, the Swiss Federal Laboratories for Materials Science and Technology (Empa, Dübendorf, Switzerland), the Department of Neonatology of the University Hospital Zürich (USZ, Zürich, Switzerland), and the Zürich Institute of Forensic Medicine of the University of Zürich (IRM-UZH, Zürich, Switzerland). This project is funded by the Commission for Technology and Innovation (CTI). The development of such a system was deemed necessary, as the resolution of clinical CT scanners is not sufficient for the small vessel sizes in neonates. Primarily, the project aims to document malformations specifically of the vascular system in neonates. The scanner itself is a modified Scanco Medical XtremeCT II with an integrated cooling system allowing for longer scanning times. The scanner additionally features a large scanning field as well as the option for dual-energy scans.

References

[1] D. Gascho, M.J. Thali, T. Niemann, Postmortem computed tomography: Technical principles and recommended parameter settings for high-resolution imaging, Medicine, Science and the Law. 58 (2018) 70–82. https://doi.org/10.1177/0025802417747167.

[2] M. Sackewitz, Fraunhofer-Allianz Vision, eds., Leitfaden zur industriellen Röntgentechnik: zerstörungsfreie Prüfung mit Bildverarbeitung, 2., vollständig aktualisierte und erweiterte Auflage, Fraunhofer Verlag, Stuttgart, 2015.

[3] M.J. Thali, U. Taubenreuther, M. Karolczak, M. Braun, W. Brueschweiler, W.A. Kalender, R. Dirnhofer, forensic microradiology: Micro-computed tomography (Micro-CT) and analysis of patterned injuries inside of bone, Journal of Forensic Science. 48 (2003) 2002220. https://doi.org/10.1520/JFS2002220.

[4] C. Capuani, J. Rouquette, B. Payré, J. Moscovici, M.B. Delisle, N. Telmon, C. Guilbeau-Frugier, Deciphering the elusive nature of sharp bone trauma using epifluorescence macroscopy: A comparison study multiplexing classical imaging approaches, International Journal of Legal Medicine. 127 (2013) 169–176. https://doi.org/10.1007/s00414-012-0678-6.

[5] W. Baier, D.G. Norman, J.M. Warnett, M. Payne, N.P. Harrison, N.C.A. Hunt, B.A. Burnett, M.A. Williams, Novel application of three-dimensional technologies in a case of dismemberment, Forensic Science International. 270 (2017) 139–145. https://doi.org/10.1016/j.forsciint.2016.11.040.

[6] G. Pelletti, G. Cecchetto, A. Viero, P. Fais, M. Weber, D. Miotto, M. Montisci, G. Viel, C. Giraudo, Accuracy, precision and inter-rater reliability of micro-CT analysis of false starts on bones: A preliminary validation study, Legal Medicine. 29 (2017) 38–43. https://doi.org/10.1016/j.legalmed.2017.10.003.

[7] G. Pelletti, G. Viel, P. Fais, A. Viero, S. Visentin, D. Miotto, M. Montisci, G. Cecchetto, C. Giraudo, Micro-computed tomography of false starts produced on bone by different hand-saws, Legal Medicine. 26 (2017) 1–5. https://doi.org/10.1016/j.legalmed.2017.01.009.

[8] D.G. Norman, W. Baier, D.G. Watson, B. Burnett, M. Painter, M.A. Williams, Micro-CT for saw mark analysis on human bone, Forensic Science International. 293 (2018) 91–100. https://doi.org/10.1016/j.forsciint.2018.10.027.

[9] D.G. Norman, D.G. Watson, B. Burnett, P.M. Fenne, M.A. Williams, The cutting edge—Micro-CT for quantitative toolmark analysis of sharp force trauma to bone, Forensic Science International. 283 (2018) 156–172. https://doi.org/10.1016/j.forsciint.2017.12.039.

[10] L. Waltenberger, S. Beitel, B. Bernegger, P. Heimel, W. Weninger, F. Kanz, A histological comparison of non-human rib models suited for sharp force trauma analysis, Forensic Science International. 319 (2021) 110661. https://doi.org/10.1016/j.forsciint.2020.110661.

[11] G.N. Rutty, A. Brough, M.J.P. Biggs, C. Robinson, S.D.A. Lawes, S.V. Hainsworth, The role of micro-computed tomography in forensic investigations, Forensic Science International. 225 (2013) 60–66. https://doi.org/10.1016/j.forsciint.2012.10.030.
[12] G. Cecchetto, C. Giraudo, A. Amagliani, G. Viel, P. Fais, F. Cavarzeran, G. Feltrin, S.D. Ferrara, M. Montisci, Estimation of the firing distance through micro-CT analysis of gunshot wounds, International Journal of Legal Medicine. 125 (2011) 245–251. https://doi.org/10.1007/s00414-010-0533-6.
[13] G. Cecchetto, A. Amagliani, C. Giraudo, P. Fais, F. Cavarzeran, M. Montisci, G. Feltrin, G. Viel, S.D. Ferrara, MicroCT detection of gunshot residue in fresh and decomposed firearm wounds, International Journal of Legal Medicine. 126 (2012) 377–383. https://doi.org/10.1007/s00414-011-0648-4.
[14] P. Fais, C. Giraudo, R. Boscolo-Berto, A. Amagliani, D. Miotto, G. Feltrin, G. Viel, S.D. Ferrara, G. Cecchetto, Micro-CT features of intermediate gunshot wounds severely damaged by fire, International Journal of Legal Medicine. 127 (2013) 419–425. https://doi.org/10.1007/s00414-012-0775-6.
[15] P. Fais, C. Giraudo, A. Viero, A. Amagliani, G. Viel, M. Montisci, D. Miotto, G. Cecchetto, Identification of bullet entrance in different type of intermediate firearm wounds through micro-computed tomography analysis, Journal of Forensic Radiology and Imaging. 3 (2015) 147–152. https://doi.org/10.1016/j.jofri.2015.07.004.
[16] C. Giraudo, P. Fais, G. Pelletti, A. Viero, D. Miotto, R. Boscolo-Berto, G. Viel, M. Montisci, G. Cecchetto, S.D. Ferrara, Micro-CT features of intermediate gunshot wounds covered by textiles, International Journal of Legal Medicine. 130 (2016) 1257–1264. https://doi.org/10.1007/s00414-016-1403-7.
[17] F.M. Vandevoort, L. Bergmans, J. Van Cleynenbreugel, D.J. Bielen, P. Lambrechts, M. Wevers, A. Peirs, G. Willems, Age calculation using X-ray microfocus computed tomographical scanning of teeth: A pilot study, Journal of Forensic Science. 49 (2004) 1–4. https://doi.org/10.1520/JFS2004069.
[18] H. Someda, H. Saka, S. Matsunaga, Y. Ide, K. Nakahara, S. Hirata, M. Hashimoto, Age estimation based on three-dimensional measurement of mandibular central incisors in Japanese, Forensic Science International. 185 (2009) 110–114. https://doi.org/10.1016/j.forsciint.2009.01.001.
[19] H. Aboshi, T. Takahashi, T. Komuro, Age estimation using microfocus X-ray computed tomography of lower premolars, Forensic Science International. 200 (2010) 35–40. https://doi.org/10.1016/j.forsciint.2010.03.024.
[20] M. Kettner, S. Potente, B. Schulz, P. Knauff, P.H. Schmidt, F. Ramsthaler, Analysis of laryngeal fractures in decomposed bodies using microfocus computed tomography (mfCT), Forensic Science, Medicine and Pathology. 10 (2014) 607–612. https://doi.org/10.1007/s12024-014-9584-1.
[21] P. Fais, C. Giraudo, A. Viero, D. Miotto, F. Bortolotti, F. Tagliaro, M. Montisci, G. Cecchetto, Micro computed tomography features of laryngeal fractures in a case of fatal manual strangulation, Legal Medicine. 18 (2016) 85–89. https://doi.org/10.1016/j.legalmed.2016.01.001.
[22] W. Baier, C. Mangham, J.M. Warnett, M. Payne, M. Painter, M.A. Williams, Using histology to evaluate micro-CT findings of trauma in three postmortem samples: First steps towards method validation, Forensic Science International. 297 (2019) 27–34. https://doi.org/10.1016/j.forsciint.2019.01.027.
[23] W. Baier, B.A. Burnett, M. Payne, J.M. Warnett, M.A. Williams, Using micro-computed tomography to examine the larynx in cases of suspected strangulation: A comparison of case findings and control images, International Journal of Legal Medicine. 134 (2020) 1441–1450. https://doi.org/10.1007/s00414-019-02194-y.
[24] X. Wang, C. Wang, S. Zhang, W. Wang, X. Li, S. Gao, K. Li, J. Chen, H. Wang, L. Chen, J. Shi, X. Liu, Z.-J. Li, Microstructure of the hyoid bone based on micro-computed tomography findings, Medicine. 99 (2020) e22246. https://doi.org/10.1097/MD.0000000000022246.
[25] J.C. Hutchinson, O.J. Arthurs, N.J. Sebire, Postmortem research: Innovations and future directions for the perinatal and paediatric autopsy, Archives of Disease in Childhood Education and Practice Edition. 101 (2016) 54–56. https://doi.org/10.1136/archdischild-2015-309321.
[26] J.C. Hutchinson, X. Kang, S.C. Shelmerdine, V. Segers, C.M. Lombardi, M.M. Cannie, N.J. Sebire, J.C. Jani, O.J. Arthurs, Postmortem microfocus computed tomography for early gestation fetuses: A validation study against conventional autopsy, American Journal of Obstetrics and Gynecology. 218 (2018) 445.e1–445.e12. https://doi.org/10.1016/j.ajog.2018.01.040.
[27] S.C. Shelmerdine, J.C. Hutchinson, O.J. Arthurs, N.J. Sebire, Latest developments in postmortem foetal imaging, Prenatal Diagnosis. 40 (2020) 28–37. https://doi.org/10.1002/pd.5562.
[28] S.C. Shelmerdine, I.C. Simcock, J.C. Hutchinson, A. Guy, M.T. Ashworth, N.J. Sebire, O.J. Arthurs, Postmortem microfocus computed tomography for noninvasive autopsies: Experience in >250 human fetuses, American Journal of Obstetrics and Gynecology. 224 (2021) 103.e1–103.e15. https://doi.org/10.1016/j.ajog.2020.07.019.

CHAPTER 25

FORENSIC MEDICINE AND 3D PRINTING

Wolf Schweitzer and Lars Ebert

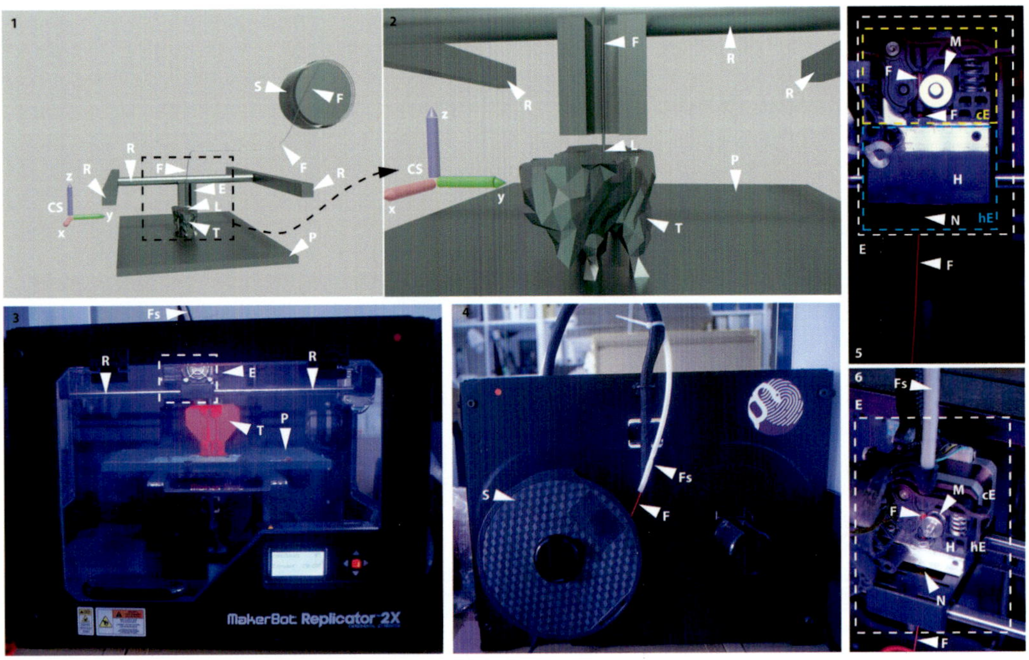

Figure 25.1 An overview of the operational mechanics of a PLA-based 3D printer, taking into account the coordinate system CS depicted in figures 1 and 2, is provided. A uniformly thick PLA filament (F), typically available on a roll and affixed to a spool (S), is directed towards the printer's extruder (E) through a guide tube (Fs). The extruder is composed of a cold section (cE) and a heated section (hE). The cold section houses a motor (M) tasked with pulling the filament, which remains rigid at this stage, into the heated section. The heat within the hot end (H) melts the PLA filament, allowing it to be precisely dispensed through a nozzle (N) at the designated print location. Initially, the molten PLA forms the base layers on the build platform (P), which is gradually lowered in the −Z direction to facilitate the layered construction of the model (T). Additional layers of PLA are applied atop the model, contingent on the cooling and solidification of the preceding layer. Rapid printing may result in warping due to insufficient cooling. The extruder navigates the build platform area via rails (R) along the X and Y axes.—Figures 1 and 2 serve as conceptual diagrams illustrating the PLA 3D printer's mechanism. Figures 3 and 4 showcase the specific 3D printer utilized for producing the model seen in Figure 25.3, originally equipped with dual-ABS extruders but later adapted to include a prototype PLA extruder (E). Figures 5 and 6 reveal the intricate components of the experimental extruder (E), captured after dismantling the temperature-regulating fan for these images.

DOI: 10.4324/9781003413998-27

25.1 INTRODUCTION

3D printing began gaining broader acceptance in medical specialties in the late 1990s [1]. These days, these printers are so popular that exploits of their software security issues have become an issue [2].

In this chapter we will review some of the more popular 3D-printing varieties from our own practical view. We will also review contemporary uses in forensic medicine. The ability of 3D design and 3D printing to achieve rapid prototyping and thus materials—handles, frames, supports, devices—that otherwise may have to be ordered, taking a lot of time and possibly cost, seems attractive. In addition, some radiological findings identified in the deceased may lend themselves to 3D printing, where a physical 3D model may provide an examiner [3] or a student a better understanding than plain images.

25.2 EXPLANATION OF 3D PRINTING

3D printing is an umbrella term that is applied to a variety of methods, and these typically provide a layer-adding method (rather than milling or carving out a shape).

In fused depsotion modeling (FDM), the material is extruded. This is usually done by continuously passing the material through a heater and a small outlet, a nozzle, so that the material is melted before it is deposited. The material is first applied to the empty print bed. After an initial object layer is applied, additional layers are applied on top of previous object layers. For an illustrated explanation, see Figure 25.1 (using our 3D printer, a modified Replicator 2X model originally manufactured by MakerBot Industries, New York City, NY, USA). In the context of FDM, desktop printing of polylactic acid (PLA) or acrylonitrile butadiene styrene (ABS) appears to be the most widely used technology [4–6]. PLA allows for particularly fast 3D printing, and the material can be relatively inexpensive [7]. However, 3D printing is not limited to this. Other methods include photopolymerization (in which a liquid resin is cured by exposure to light), material jetting (in which the light-sensitive material can be introduced into the print bed in the form of droplets before ultraviolet light is used to cure the material), binder jetting, sheet lamination, directed energy deposition, or powder bed fusion [8]. Therefore, the term additive manufacturing is used to describe 3D printing.

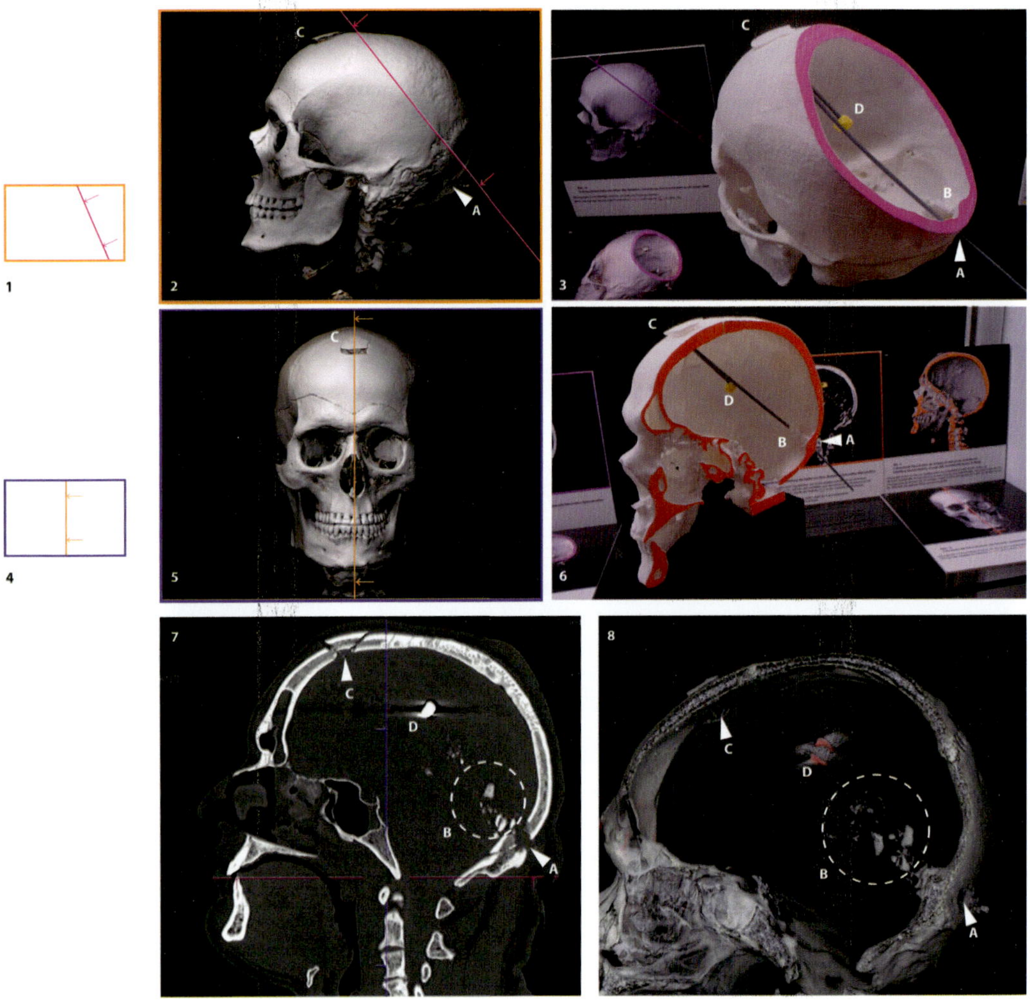

Figure 25.2 Utilizing 3D printing for educational visualization of injuries; this example features two 3D-printed replicas from the same scenario. In the instance of a gunshot wound to the back of the head (9-mm pistol), an entrance defect was observed in the occipital skull (A), alongside apparent bone fragments within the skull (B). A round, funnel-shaped fracture in the frontal bone near the coronal suture suggested an internal ricochet (C), with the bullet's final resting place (D) depicted in figures 7 and 8. The skull was printed twice (figures 3, 6), with the data set up for an oblique occipitoparietal view (figures 1–3) and a sagittal perspective (figures 4–6). The 3D model visually representing the bullet's path (D in figures 3 and 6) was attached to a gray rod to simulate the potential trajectory. Together with the use of symbolic colors (for schematic depiction of image orientations: figures 1 and 4; orthogonal views: figures 2 and 5; 3D models: figures 3 and 6), this approach allows for an exploration of the spatial arrangement between reality and CT images.

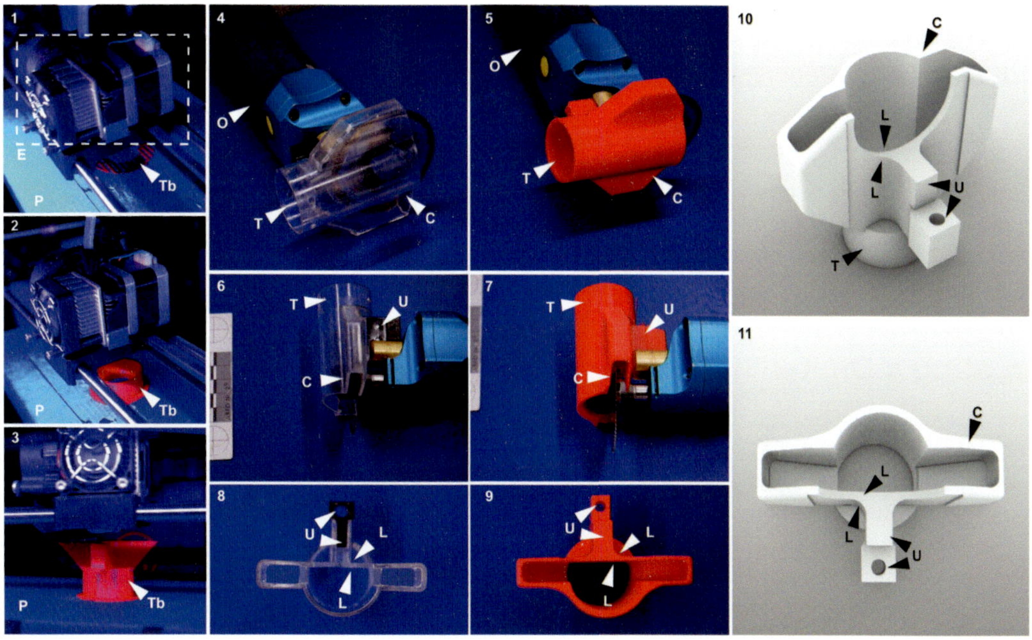

Figure 25.3 Employing 3D-printing techniques, we aimed to design, test, and develop substitute parts for an autopsy tool. Specifically, we use an oscillating saw for cutting bone, notably for skull openings during autopsies (O). This saw is equipped with a plastic shield (the clear element in 4, 6, 8) designed to connect to a suction hose via a circular port (T). The shape of this shield not only directs the airflow but also partially encloses the saw blade (C), enhancing its functionality. The shield is attached to the saw at a connection point (U), which slots into a section of the saw and is secured with a screw. The issue we encountered was the frequent breaking of this plastic shield, particularly at the junction between its connection point (U) and the portion covering the saw blade (L). To address this, we created a 3D-printed prototype part (the red element in 5, 7, 9) as part of a feasibility study. Our design introduces rounded enhancements at the juncture of the connection point (U) to other components, an alteration aimed at reducing breakage (as evidenced by comparing L in images 8 and 9). This modified design is currently on hold, awaiting additional technical enhancements and usage evaluation.—Images 1–3 document the 3D printing of this component (with E representing the extruder, P the build plate, and T the model in progress). Image 1 captures the initial PLA layers bonding to the build plate. The subsequent growth of the part is depicted in images 2 (approximately 15 minutes after image 1) and 3 (around 30 minutes following image 2). The orientation of the 3D model in relation to the build plate is crucial for determining the feasibility of printing the object without support structures.

One significant advantage of this method of object production is its ability to rapidly and cost-effectively print any 3D digital model using processes like PLA or ABS printing. This enables the quick testing of prototypes, fabrication of new items or replacement parts, and generation of 3D objects for straightforward demonstrations.

The phrase “rapid prototyping” is often used interchangeably with 3D printing, yet it refers to a distinct process. It involves the use of 3D printing for the quick, cost-effective, and precise creation of a single physical version of a digital 3D model. This process occurs during the design and refinement stages. The tangible model produced can then undergo a thorough evaluation—it can be physically examined, tested for fit or installation, subjected to stress tests, and modified using tools. Following this, any identified shortcomings and necessary adjustments are documented to enhance the design. Subsequently, the digital 3D model is refined based on this feedback and printed anew. This cycle of testing and improvement may repeat multiple times until the digital 3D model meets the final product specifications. Utilizing inexpensive, readily produced plastic models for visual and functional testing significantly aids in optimizing the design and dimensions of objects. However, this does not imply that the final product will be manufactured in the same manner.

The determination of how to finalize a 3D digital model often occurs only after the rapid prototyping stage. This decision is typically made on an individual basis, evaluating whether prototypes deemed suitable for testing meet the criteria for the final product or if an alternative manufacturing method is required. Upon completing the development phase, the same 3D model can be forwarded for more sophisticated production techniques. For instance, instead of utilizing plastic, a binder jetting method may be selected for crafting metal items, or a laser sintering technique could be employed to produce high-definition acrylic pieces [9].

25.3 STEP-BY-STEP TECHNIQUE

25.3.1 General Object Design and Instance Creation

There are a variety of methods to create an object. These include sheet metal fabrication, molding, casting, machining, hand shaping or carving. A variety of materials may be used across these techniques, including clay, and last but not least, 3D printing. Some of these methods, including 3D printing, can only be employed after a digital 3D model of the object has been created [10, 11].

25.3.2 Creating a Digital 3D Model

If the surface of an object must be digitally defined before manufacturing of a solid object can begin, that 3D model must enclose a volume. In other words, that surface then describes what is termed a solid [12].

There are two main methods that are used to create a 3D model. A first option is to capture the surface of an existing real object (possibly through 3D scanning or photogrammetry, even in forensic medicine [13–15]). A second option is to manually create the 3D model by way of software. Both methods may be combined, such as by first capturing the 3D surface of a real object and then using the resulting 3D model for further digital processing or by first approaching a 3D object through digital creation and subsequent modification of a 3D-printed instance using tools (such as saws, files, or drills). Not all surface angles are suitable for optical 3D surface scanning, so CT scanning has been used to start creating 3D models, sometimes with similar problems related to the initial 3D surface data quality [16, 17].

Surface data quality clearly is an issue. Some 3D digital models initially are too large, too jagged, or too noisy. This may be the case when using 3D surface data from 3D surface scanners, using photogrammetry, or creating 3D surface models from CT data. Subsequently, the 3D models may still need to be modified. This step of modifying the 3D model may include removing non-manifolds, reducing a possibly excessive number of polygons, smoothing (to achieve more convincing visual realism [17]), and defining an actual solid geometry [17, 18].

When using software programs to define 3D models, parametric 3D software (as opposed to free-form models) can allow the user to access selected parameters (rather than the complexity of actual polygon coordinates) of a 3D digital model. Parametric software allows the design, shape, position, size, or orientation of model regions or aspects to be changed more efficiently as a 3D model evolves [19].

Another advantage of using 3D digital models for object creation over the direct creation of physical models is that finite element modeling can be used to investigate the mechanical properties of a particular 3D model or to compare some physical properties of the real object with the properties of the virtual object. The resulting design changes can be achieved more quickly with finite element modeling than with traditional physical testing, especially when the production of a single 3D print requires a lot of time and materials [20].

25.4 ROBUSTNESS AND TESTING

The structural robustness of 3D-printed objects made from PLA or ABS is generally considered somewhat limited, influenced by factors like moisture exposure and demonstrating anisotropic mechanical characteristics [7, 21, 22]. Anisotropy in this context means the material is more likely to fracture along its layers rather than across them, with potential differences in strength up to approximately 50% depending on the direction [7], a phenomenon especially noticeable in fused deposition modeling techniques. Furthermore, creating overhangs often necessitates support structures, which can introduce hard-to-access voids [7], making the removal of these supports a significant challenge. Innovations in support material that can be chemically dissolved are being developed to address this issue.

These aspects are continually being refined. Typically, the exterior of a 3D-printed item is characterized by a specified number of outer layers with designated thickness, while the interior can be filled with a chosen pattern and density of what is known as infill, unless the item is printed as a solid object. The structural integrity of a 3D-printed object can be enhanced through the use of particular infill patterns. Improvements are also being made to the mechanical strength of PLA, such as the development of new composites that include carbon fibers. Surface finishing techniques can involve mechanical or chemical processes [5]. The issue of inherent mechanical anisotropy and its practical implications may require evaluation on a case-by-case basis or through established guidelines, including adjusting the orientation of the 3D digital model in relation to the printing bed and layer configuration [23].

Assessing the educational or illustrative effectiveness of a 3D-printed model compared to flat representations (like those on paper) is complex and likely necessitates interactive feedback with participants [24].

Evaluating 3D models and printed objects often involves a combination of computational modeling and physical tests [20, 21, 25]. In forensic applications, testing might also extend to interactions with fluids, microbes, and the effects of disinfection, sterilization, or repetitive mechanical stress [26].

25.5 APPLICATION DOMAINS IN FORENSIC MEDICINE

25.5.1 Medical Teaching, Demonstration, and Diagnostics

Medical imaging techniques such as postmortem computed tomography (PMCT) provide a solid technical basis for the creation of 3D printed models of skeletal or other body parts [27]. These models can effectively illustrate pathologies, the presence of foreign objects like bullets, or injury details. The use of 3D models in medicine extends to educational purposes, surgical planning, and the specification of implants, offering numerous benefits [28]. For instance, in a case involving a comminuted skull fracture, 3D-printed replicas provided highly realistic representations, particularly when derived from autopsy results [29], which may find application in a whole range of trauma cases (e.g., [30–32]).

In the realm of diagnostics, research has been conducted on the utility of 3D-printed models for distinguishing between multiple sclerosis lesions and nonspecific white matter changes based on supratentorial MRI scans of the brain [3]. Furthermore, 3D printing has facilitated the reconstruction of faces following severe fractures to facial and cranial bones, enhancing the diagnostic usefulness of facial photographs for identification purposes and making the deceased more recognizable for grieving families [33]. The practicality of producing such detailed exhibits has been convincingly shown [17, 34]. Yet the use of these models in legal proceedings remains exceedingly rare.

In a notable example, a 3D-printed model depicting the head injuries of a 6-year-old girl was introduced in court, though specific details about the model's creation were not provided [35].

Another instance reported by Baier et al. [36] involved the use of a 3D-printed model to visually link a complex skull fracture pattern with potential weapons. This model was produced from CT scans using stereolithography with white resin to mimic the color of natural bone, ostensibly achieving a resolution higher than that of the original scans. Notably, this approach allowed court participants, including the jury and judge, to handle and inspect the model without any safety concerns.

Our investigation has focused on the application of 3D printing for visualizing injuries for educational purposes. Specifically, we visualized a fatal gunshot wound to the head in several formats, including two distinct 3D-printed representations (Figure 25.2).

25.5.2 Weapon Models and Models for Biomechanical Tests

A study on the biomechanical simulation of an infant's head for impact analysis utilized a 3D-printed skull model, enhancing bio-fidelity through the use of dual materials via polyjet technology [37].

For surgical simulations of the temporal bone, models were created using varying proportions of several thermoset polymers to replicate the distinct mechanical characteristics of the temporal bone's anatomical substructures, including its vascular system. This research prioritized mechanical properties over visual accuracy, finding that models derived from micro-CT data surpassed those based on standard clinical CT scans in performance [38].

Research into the mechanical efficacy of 3D-printed nylon composites reinforced with continuous Kevlar strands indicated that the material's structural design significantly influences its impact resistance [39].

3D printing's role in weapon manufacturing remains minor but noteworthy. The Liberator, a gun with 3D-printed components, is known to scatter polymer fragments broadly upon discharge, highlighting potential safety and forensic considerations [40, 41].

Furthermore, 3D printing has been applied to producing projectiles, notably for deploying extended-release contraceptives to wildlife. This innovative approach involved mixing the contraceptive hormone progesterone in powder form with PLA pellets, which were then 3D printed into biodegradable projectiles for this specific application.

25.5.3 Instruments, Tools, Component Parts

The development of tools and components for forensic science applications exemplifies the utility of prototyping, testing, and creating objects.

With the emergence of new techniques like postmortem computed tomography angiography (PMCTA), we discovered the absence of commercially available parts necessary for these applications, prompting us to design and manufacture these components in-house [42]. This process led to approximately ten iterations to finalize the design of these parts.

In forensic science, there is also a practice of adapting commercially available materials for specialized purposes, which might not perfectly fit the intended use, such as employing vibratory saws designed for plaster removal [43]. For instance, we identified recurrent breakage in a specific part of a plastic attachment for an oscillating bone saw used in autopsies. Through four rounds of design modifications, we developed a prototype that strengthened this vulnerable area (Figure 25.3).

This field often requires thorough mechanical strain evaluation, physical testing, and documentation of both the efficacy and failures of components, which can be crucial for quality accreditation and control documentation. Additionally, it's important to navigate legal considerations, including copyright, patent issues, or liability concerns.

25.6 DISCUSSION

Deciding whether to develop in-house expertise in 3D printing software and equipment operation or to outsource these tasks is not straightforward. The technical demands of both software and hardware for 3D printing are significant, necessitating careful consideration of whether to invest in personal equipment, collaborate, or utilize a 3D printing service [9]. The prototyping culture within one's business or institution can significantly influence this decision, particularly as 3D printing involves complex procedural and conceptual challenges, with prototyping cycles often exceeding 20 to 50 iterations, unlike traditional methods such as clay modeling [11]. A practical recommendation is to avoid connecting 3D printers to institutional or corporate networks permanently to mitigate security risks [2].

The acceptability of using 3D-printed models for courtroom evidence in forensic cases remains a topic of debate. These models serve as tools to elucidate spatial relationships, which is only one aspect of a broader objective. The skill lies in presenting these relationships without causing distraction and offering various perspectives. The debate over the efficacy of 3D technology extends beyond forensics, for instance, in how filmmakers use stereoscopy to engage viewers without overwhelming them with the narrative [44]. Interestingly, a study showed a preference for 3D-printed skulls over photographs or 2D images among individuals from different legal and non-legal backgrounds [45]. 3D printing's ability to fabricate new parts, components, or structures offers innovative ways to conduct research or enhance products. Essential to this process is the iteration through numerous cycles of design, construction, testing, and refinement.

Ultimately, the direction of 3D model development depends on the intended outcomes. Once a final design is reached, using 3D printing for mass production might not be as viable as other manufacturing methods. A break-even analysis can help determine the most cost-effective approach. While 3D printing may be economical for producing up to 200 units, injection molding becomes more cost-effective at larger volumes [46]. However, scenarios involving frequent design changes, geographically dispersed manufacturing, or crowdfunding could alter this assessment [47]. To date, the large-scale application of 3D printing in forensic science remains limited.

25.7 PRACTICAL CONCLUSIONS

As it appears, 3D printing denotes a part of manufacturing that contains a range of techniques, from digital 3D model creation through prototype testing to obtaining finalized products. Their application in forensic medicine may be in generating anatomical models to exhibit pathology or trauma, to build tool parts, or to generate biomechanical models. All of these are new, however. The technical requirements for software and hardware aspects may be challenging, so using a 3D-printing service instead, or seeking collaboration for 3D printing with a nearby facility may have to be weighed carefully against setting up one's own device. As the development of the relevant software and the physical aspects of making physical objects—3D printing in a narrow sense—are currently ongoing, following these developments is certainly interesting.

ACKNOWLEDGMENTS

The authors express their gratitude to Emma Louise Kessler, MD, for her generous donation to the Zürich Institute of Forensic Medicine, University of Zürich, Switzerland. While the authors have made efforts to ensure clarity and structure, the development of this manuscript occurred under the constraints of available time and timelines. Released under CC BY 4.0. Text and Figures republished in modified and edited form from [48] which had been released under CC BY 4.0. The retrospective use of anonymized data was submitted to and evaluated by the Ethics Committee of the Canton of Zürich (reference number 15–0686).

References

[1] Dimitris Mitsouras, Peter Liacouras, Amir Imanzadeh, Andreas A Giannopoulos, Tianrun Cai, Kanako K Kumamaru, Elizabeth George, Nicole Wake, Edward J Caterson, Bohdan Pomahac, et al. Medical 3d printing for the radiologist. *Radiographics*, 35(7):1965–1988, 2015.

[2] Samuel Moore, Phillip Armstrong, Todd McDonald, and Mark Yampolskiy. Vulnerability analysis of desktop 3d printer software. In *2016 Resilience Week (RWS)*, pages 46–51, Proceedings of a meeting held 16–18 August 2016, Chicago, Illinois, USA, 2016. IEEE.

[3] Braeden D Newton, Katy Wright, Mandy D Winkler, Francesca Bovis, Masaya Takahashi, Ivan E Dimitrov, Maria Pia Sormani, Marco C Pinho, and Darin T Okuda. Three-dimensional shape and surface features distinguish multiple sclerosis lesions from nonspecific white matter disease. *Journal of Neuroimaging*, 27(6):613–619, 2017.

[4] Yichi Song, Yizhuo Li, Wenzhe Song, K Yee, K-Y Lee, and Vito L Tagarielli. Measurements of the mechanical response of unidirectional 3d-printed pla. *Materials & Design*, 123:154–164, 2017.

[5] Eda Hazal Baran and H Yildirim Erbil. Surface modification of 3d printed pla objects by fused deposition modeling: A review. *Colloids and Interfaces*, 3(2):43, 2019.

[6] Jason T Cantrell, Sean Rohde, David Damiani, Rishi Gurnani, Luke DiSandro, Josh Anton, Andie Young, Alex Jerez, Douglas Steinbach, Calvin Kroese, et al. Experimental characterization of the mechanical properties of 3d-printed abs and polycarbonate parts. In *Advancement of Optical Methods in Experimental Mechanics*, Volume 3, 1st ed. ImprintRiver Publishers, 2017.

[7] John Ryan C Dizon, Alejandro H Espera Jr, Qiyi Chen, and Rigoberto C Advincula. Mechanical characterization of 3d-printed polymers. *Additive Manufacturing*, 20:44–67, 2018.

[8] Dimitrios Mitsouras and Peter C Liacouras. 3D printing technologies. In Frank T Rybicki and Gerald T Grant, editors, *3D Printing in Medicine*, pages 5–22. Springer, Cham, Switzerland, 2017.

[9] Marco Wirth and Frederic Thiesse. Shapeways and the 3d printing revolution. In *2014 European-Conference on Information Systems (ECIS)*, pages 1–14, Proceedings of the ECIS held June 9–11, 2014, in Tel Aviv, Israel, 2014. ECIS.

[10] James G Bralla. *Handbook of Manufacturing Processes*. Industrial Press, New York, NY, USA, 2006.

[11] Michael Schrage. Cultures of prototyping. *Bringing Design to Software*, 4(1):1–11, 1996.

[12] Gordon Fisher. *Blender 3D Printing Essentials*. Packt Publishing Ltd, London, United Kingdom of Great Britain, 2013.

[13] W Bru¨schweiler, M Braun, HJ Fuchser, and R Dirnhofer. Photogrammetric evaluation of injuries to skin, soft tissue and bones to determine the weapon—basic aspects. *Rechtsmedizin*, 7:76–83, 1997.

[14] Wolf Schweitzer, Emil Ro¨hrich, Michael Schaepman, Michael J Thali, and Lars Ebert. Aspects of 3d surface scanner performance for postmortem skin documentation in forensic medicine using rigid benchmark objects. *Journal of Forensic Radiology and Imaging*, 1(4):167–175, 2013.

[15] Wolf Schweitzer, Haruki Fukuda, Michael Thali, Stephan Bolliger, and Lars Ebert. Mobile forensic photogrammetry in the field: Conservative approach to non-collaborative surfaces. *Forensic Imaging*, 38:200597, 2024.

[16] Peter C Liacouras, Divya Sahajwalla, Mark D Beachler, Todd Sleeman, Vincent B Ho, and John P Lichtenberger. Using computed tomography and 3d printing to construct custom prosthetics attachments and devices. *3D Printing in Medicine*, 3(1):8, 2017.

[17] Rachael M Carew, Ruth M Morgan, and Carolyn Rando. A preliminary investigation into the accuracy of 3d modeling and 3d printing in forensic anthropology evidence reconstruction. *Journal of Forensic Sciences*, 64(2):342–352, 2019.

[18] Marco Attene, Marcel Campen, and Leif Kobbelt. Polygon mesh repairing: An application perspective. *ACM Computing Surveys (CSUR)*, 45(2):1–33, 2013.

[19] Jorge D Camba, Manuel Contero, and Pedro Company. Parametric CAD modeling: An analysis of strategies for design reusability. *Computer-Aided Design*, 74:18–31, 2016.

[20] Sunil Bhandari and Roberto Lopez-Anido. Finite element analysis of thermoplastic polymer extrusion 3d printed material for mechanical property prediction. *Additive Manufacturing*, 22:187–196, 2018.

[21] Todd Letcher and Megan Waytashek. Material property testing of 3d-printed specimen in pla on an entry-level 3d printer. In *2014 ASME International Mechanical Engineering Congress and Exposition*, volume 46438, pages 1–8, Proceedings of 2014 ASME International Mechanical Engineering Congress and Expositio held November 14–20, 2014, in Montreal, Quebec, Canada, 2014. American Society of Mechanical Engineers.

[22] Eunseob Kim, Yong-Jun Shin, and Sung-Hoon Ahn. The effects of moisture and temperature on the mechanical properties of additive manufacturing components: Fused deposition modeling. *Rapid Prototyping Journal*, 22(6):887–894, 2016.
[23] Sung-Hoon Ahn, Michael Montero, Dan Odell, Shad Roundy, and Paul K Wright. Anisotropic material properties of fused deposition modeling abs. *Rapid Prototyping Journal*, 8(4), 2002. https://doi.org/10.1108/13552540210441166
[24] Eric J Pyle. The role of classroom artifacts in the clinical supervision of science. *NASSP (National Association of Secondary School Principals) Bulletin*, 82(597):70–76, 1998.
[25] M Heidari-Rarani, M Rafiee-Afarani, and AM Zahedi. Mechanical characterization of fdm 3d printing of continuous carbon fiber reinforced pla composites. *Composites Part B: Engineering*, 175:107147, 2019.
[26] Joan Ferra`s-Tarrago´, Oihana Sabalza-Bazta´n, Jose Miguel Sahuquillo-Arce, Manuel A´ ngel Angulo-Sa´nchez, Jose Amaya-Valero, Carolina De-La-Calva Ceinos, and Francisco Baixauli-Garc´ıa. Security of 3d-printed polylactide acid piece sterilization in the operating room: A sterility test. *European Journal of Trauma and Emergency Surgery*:1–6, 2021.
[27] Alessandro Marro, Taha Bandukwala, and Walter Mak. Three-dimensional printing and medical imaging: A review of the methods and applications. *Current Problems in Diagnostic Radiology*, 45(1):2–9, 2016.
[28] Frank T Rybicki and Gerald T Grant. *3D Printing in Medicine*. Springer, Cham, Switzerland, 2017.
[29] Mattias Kettner, Peter Schmidt, Stefan Potente, Frank Ramsthaler, and Michael Schrodt. Reverse engineering—rapid prototyping of the skull in forensic trauma analysis. *Journal of Forensic Sciences*, 56(4):1015–1017, 2011.
[30] Daniel Wittschieber, Laura Beck, Volker Vieth, and Maria L Hahnemann. The role of 3dct for the evaluation of chop injuries in clinical forensic medicine. *Forensic Science International*, 266:e59–e63, 2016.
[31] Vasiliki Chatzaraki, Michael J Thali, Garyfalia Ampanozi, and Wolf Schweitzer. Fatal road traffic vehicle collisions with pedestrian victims: Forensic postmortem computed tomography and autopsy correlation. *The American Journal of Forensic Medicine and Pathology*, 39(2):130–140, 2018.
[32] Wolf Schweitzer and Michael Thali. Computed tomography of scalp embedded gravel: Differentiation between falling and kicking. *Journal of Forensic Radiology and Imaging*, 3(4):200–209, 2015.
[33] Petra Urbanova´, Toma´sˇ Vojt´ısˇek, Jan Frisˇhons, Ondrˇej Sˇandor, Mikola´sˇ Jurda, and Jan Krajsa. Applying 3d prints to reconstructing postmortem craniofacial features damaged by devastating head injuries. *Legal Medicine*, 33:48–52, 2018.
[34] Lars Chr Ebert, Michael J Thali, and Steffen Ross. Getting in touch—3d printing in forensic imaging. *Forensic Science International*, 211(1–3):e1–e6, 2011.
[35] Clare Scott. 3D Printed Skulls Presented as Evidence in Murder Trial, in a First for the British Legal System. 3DPrint. COM-Website, accessed 10 July 2020, https://3dprint.com/133715/ ellie-butler-murder-trial/, 2013.
[36] Waltraud Baier, Jason M Warnett, Mark Payne, and Mark A Williams. Introducing 3d printed models as demonstrative evidence at criminal trials. *Journal of Forensic Sciences*, 63(4):1298–1302, 2018.
[37] Michael Jones, D Darwall, G Khalid, R Prabhu, A Kemp, Owen J Arthurs, and P Theobald. Development and validation of a physical model to investigate the biomechanics of infant head impact. *Forensic Science International*, 276:111–119, 2017.
[38] Austin S Rose, Julia S Kimbell, Caroline E Webster, Ola LA Harrysson, Eric J Formeister, and Craig A Buchman. Multimaterial 3d models for temporal bone surgical simulation. *Annals of Otology, Rhinology & Laryngology*, 124(7):528–536, 2015.
[39] Mackenzie Scrocco, Timothy Chamberlain, Chloe Chow, Logan Weinreber, Edward Ellks, Connor Halford, Pedro Cortes, Brett P Conner, and Youngstown Business Incubator. Impact testing of 3d printed kevlar-reinforced onyx material. In *2018 Annual International Solid Freeform Fabrication Symposium—An Additive Manufacturing Conferenc*, pages 1121–1144, Proceedings of the 29th Annual International Solid Freeform Fabrication, held August 13–15, 2018, in Austin, Texas, USA, 2018.
[40] Hanna Honsberger, Damien Rhumorbarbe, Denis Werner, Fabiano Riva, Matthieu Glardon, Alain Gallusser, and Olivier Dele´mont. How to recognize the traces left on a crime scene by a 3d-printed liberator? Part 1. discharge, exterior ballistic and wounding potential. *Forensic Science International*, 286:245–251, 2018.
[41] Hanna Honsberger, Denis Werner, Damien Rhumorbarbe, Fabiano Riva, Matthieu Glardon, Alain Gallusser, and Olivier Dele´mont. How to recognise the traces left on a crime scene by a 3d-printed liberator? Part 2. elements of ammunition, marks on the weapons and polymer fragments. *Forensic Science International*, 295:137–144, 2019.
[42] Wolf Schweitzer, Markus Enders, and Michael Thali. Very affordable post mortem CT angiography kit: Feasibility study using immersion pump and 3D printed parts. *Journal of Forensic Radiology and Imaging*, 16:11–18, 2019.
[43] Kjell Tore´n and Per Jonsson. Is skull sawing by autopsy assistants overlooked as a cause of vibration-induced white fingers? *Scandinavian Journal of Work, Environment & Health*:227–229, 1996.
[44] Scott Higgins. 3D in depth: Coraline, Hugo, and a sustainable aesthetic. *Film History: An International Journal*, 24(2):196–209, 2012.
[45] Soren Blau, Erin Phillips, Chris O'Donnell, and Greg Markowsky. Evaluating the impact of different formats in the presentation of trauma evidence in court: A pilot study. *Australian Journal of Forensic Sciences*, 51(6):695–704, 2019.
[46] Matthew Franchetti and Connor Kress. An economic analysis comparing the cost feasibility of replacing injection molding processes with emerging additive manufacturing techniques. *The International Journal of Advanced Manufacturing Technology*, 88(9–12):2573–2579, 2017.
[47] Thierry Rayna and Ludmila Striukova. The impact of 3d printing technologies on business model innovation. In *Digital Enterprise Design & Management*, pages 119–132. Springer, 2014.
[48] Wolf Schweitzer, M Thali, E Aldomar, and L Ebert. Overview of the use of 3d printing in forensic medicine. *Rechtsmedizin*:1–8, 2020.

CHAPTER 26

COMPUTATIONAL METHODS

Akos Dobay, Raffael Golomingi, and Lars Ebert

DOI: 10.4324/9781003413998-28

26.1 INTRODUCTION

Image segmentation can be traced back to the emergence of computer vision. The field of computer vision was born in the mid 1960s, and it is traditionally credited to Lawrence Gilman Roberts (1937–2018), an American engineer whose MIT dissertation was entitled "Machine Perception of Three-Dimensional Solids" [1]. Computer vision was developed to gain a higher level of understanding from digital images. Essentially, the question was how to equip a computer with an electronic eye able to see and distinguish what a human does. In the 1970s, several computer vision algorithms were developed that are today part of image segmentation. These are, for instance, edge detections or polyhedral modeling. More recent ones include machine learning and deep learning techniques. Image segmentation is one of the first steps in quantifying information in a digital image. A digital image displayed on a computer screen exists only as a series of ordered pixels. A very simple way to analyze an image would be to count the number of pixels of a certain kind, for instance, pixel intensity or pixel color. In some cases, this information can be enough to derive knowledge about the actual content of an image. Beyond counting and classifying pixel values, it is necessary to delineate borders of the objects represented on the image. This operation can be done manually by labelling each pixel and assigning each pixel to a class of objects. If we consider an image of 2400×1600, the total number of pixels is 3,840,000. Manual labelling is therefore not possible, with constant technological improvements bringing larger and larger image resolutions. This operation can be done only with an algorithm and taking advantage of the speed of computers. As comparison, a processor in the 70s clocked at 1 MHz would have taken approximately 4 seconds to process an image of 2400×1600 pixels in size (1 million instructions per second). Today, the fastest processors are clocked around 5 GHz. The same operation takes only a millisecond.

Image segmentation and labeling are synonymous. Both aim at extracting higher-level information from a digital image. This operation can be facilitated if the contours on the image are very sharp. This is the case when the contrast is high. Therefore, increasing the contrast is in the first step towards identifying classes of objects on a digital image. If this is not sufficient, a more elaborate algorithm can be used to achieve this result. In the next section, we will review the most important techniques used in image segmentation while giving an outlook of possible applications. We will also review finite element techniques.

26.2 UNDERSTANDING 3D SEGMENTATION TECHNIQUES

Understanding 3D segmentation is not very different from understanding digital image segmentation. We can roughly divide image segmentation into two groups: *semantic segmentation* and *instance segmentation*. While semantic segmentation focuses on attributing a pixel value to a class of objects, instance segmentation should be able to recognize all the pixel belonging to the same object. Instance segmentation is therefore more complex than semantic segmentation. The task of identifying pixels that represent a class of objects such as clouds in a sky, animal species in a particular environment, furnitures in a room, or handwritten letters requires knowing the pixel value or tonal distribution of the objects present in the class and being able to subtract it from the overall background information. Identifying individual objects on a digital image requires deeper knowledge that simply analyzing the tone distribution of the pixels' value. In other words, instance segmentation requires information about the boundaries of an object. To find the boundaries of an object it is necessary to delineate the contour of an object. In a digital image with very high contrasts and sharp edges, finding boundaries can be summarized by applying an algorithm comparing pixel values both in color and intensity. When the edges are blurred and both pixel color and intensity are highly variable within an object, segmentation becomes a complex problem. A common problem is when an object is partially hidden by another object, breaking the identity of an object into multiple pieces. In those cases, it is necessary to complement the search with additional information while applying different techniques. The field of computer vision and pattern recognition is an active field of research since there is a need for automating complex tasks to the efficiency level (both in speed and quality) that a human can achieve. The use of machine learning and deep learning techniques are among the most promising approaches to equal human intelligence.

One of the techniques used in digital image segmentation is called *thresholding*. A threshold value is applied to an entire image. The method turns a greyscale image into a bitmap (binary image) format by converting every pixel value below the threshold to one (black pixel value) and every pixel value above the threshold to zero (white pixel value). Since there can be multiple threshold values for a given greyscale image, this operation can be used to generate multiple bitmap images depending on the region of interest. Otsu's method, named after Nobuyuki Otsu, can be applied to search automatically for the optimal threshold value in a greyscale image [2]. Based on pixel intensity counts (how many pixels in the image share the same intensity) an intensity histogram can be drawn together with the probability density of each intensity present in the image. In essence, the method tries to maximize the inter-class variance. This is the value for which the difference in the calculated variance for each class reaches its maximum.

Like Otsu's method, the *balanced histogram thresholding* method tries to divide the image into two classes: the foreground and the background. By removing pixel intensity counts either starting from the high intensities until there is a point along the intensity axis where the histogram is well balanced or from the low intensities until such a point is found, the method should converge toward a point along the intensity axis that is the same irrespective of which side of the intensity axis we start. Methods based on Otsu's approach have been specifically developed for computed tomography imaging [3].

The *K-means clustering algorithm* is an iterative technique to partition an image into clusters. In this method, K pixels are chosen randomly or based on some higher-level information to become the initial center of the clusters. A center represents either a color or an intensity. The algorithm assigns each pixel in the image to a specific cluster based on how similar the pixel value is compared to the pixel value constituting the center of a cluster. At each step the cluster centers are re-computed by averaging the value of all the pixels in a cluster. The value of each center is replaced with the corresponding averaged values. The convergence is attained when the pixels in the image are systematically assigned to the same cluster. Histogram-based methods are usually efficient for locating regions having the same intensity or color. By extracting relevant information in the histogram such as searching for maxima and minima one can locate clusters in the image.

Region-growing methods are efficient when the image contains neighboring pixels with similar values. A pixel is chosen as initial seed and each neighboring pixel is compared to the seed. The similarity threshold decides whether a pixel in the neighborhood will belong to the same region. If the criterion is not satisfied a new region is created.

Edge detection proceeds by locating regions where a change in pixel intensity/color reaches a local maximum. This method also works well if the borders are well defined. To overcome this obstacle, a typical preprocessing consists of smoothing regions that have similar pixel intensity/color. By extending such averaging technique the contrast between regions will be increased, thus allowing for a better detection of the boundaries.

26.3 3D SEGMENTATION FOR MEDICAL IMAGE ANALYSIS

The need for 3D segmentation in medical imaging is motivated by several goals, like assessing the volume and the boundaries of a region of interest. This information can be further used to quantify the weight of an organ for instance or its shape and diagnose a pathology or the cause of death. Typical 3D segmentation is done using an encoder-decoder architecture. Figure 26.1 gives a schematic example of an encoder-decoder architecture.

Brain segmentation is performed on magnetic resonance imaging to identify regions of interest. For instance, for measuring and visualizing the brain's anatomical structures, or for surgical, or, identifying pathological cells such as cancer cells. Detection and segmentation of cells in fluorescence microscopy images.

26.3.1 Finite Element Method

The basic idea behind the finite element method is to partition a problem into small domains so that each domain can be treated separately. The fundamental concept of the finite element method is closely aligned with the approach introduced by Hrennikoff in 1941 [4], while its mathematical foundations were laid by Richard Courant in 1943 [5]. This approach is particularly convenient if you have a system that is highly heterogeneous or has a complex geometry. When modeling the geometry of an object, it is convenient to make a digital model if it can be described using basic geometrical

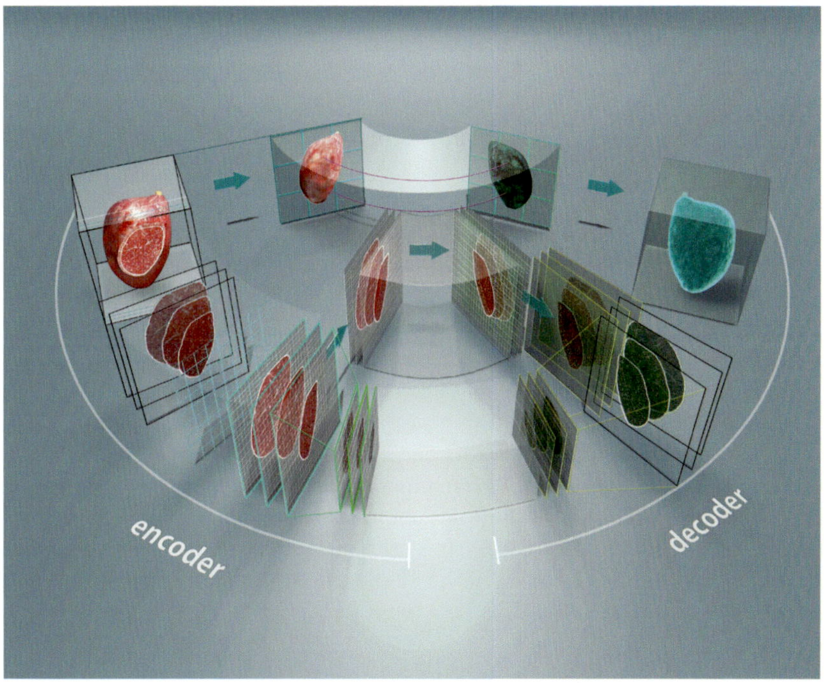

Figure 26.1 A typical encoder-decoder architecture. Information contained in a slice is encoded into a series of feature maps of smaller size. While encoding the slices, intermediary results are set aside to be used as a mask while decoding the slices and help restore the slices to their original size.

Table 26.1 Different Thresholding Methods for Image Segmentation

Name	Effect
Thresholding	Turns a greyscale image into a bitmap format
Balanced histogram thresholding	Search for the optimum threshold
Region-growing	Can be used for color images
Edge detection	Can be used to search for features
Gaussian blurring	Noise reduction

objects or using symmetries. For instance, a cylinder has an axial symmetry, making it easy to describe using a mathematical description. For instance, the base of a cylinder follows the mathematical description of a disc for which the equations are known. Due to the axial symmetry of a cylinder, the third dimension is simply described by the height. Hence, each point of a cylinder is given by a set of parametric equations:

$$\begin{cases} x = \rho\cos\theta \\ y = \rho\sin\theta \\ z = \alpha \end{cases} \tag{1}$$

If the aim is to model a cylinder, then equation (1) can be used for its description. If the shape contains local variations, meaning that there are points that cannot be reach by these equations then the only possibility is to describe the object by making a list of all possible points. Such set of points is of course infinite; therefore, an approximation is needed by selecting only a subset of points. The points that are selected are those that play a critical role in the description of the object and its properties. If the object contains local areas that have an importance, these areas should be better documented. Hence, those regions should contain more points in the subset.

Until now we have focused on the geometry of the object. What about the physical properties of the object? An object is made in a specific material. Each material behaves differently when subjected to various conditions and constraints. For instance, copper and iron have different properties. When subjected to certain constraints such as deformation, the forces that are applied on the object will have different effects. Again, if the object is made of different materials these properties must be described locally. It means that each coordinate has a different physical property. The finite element method helps to overcome this problem. By dividing the object into smaller regions, one can assign different properties to different regions. When these regions are subjected to a constraint, they will respond accordingly.

Finally, the finite element method allows for a global solution to be the sum of local solutions. Using the example of a cylinder, an object can be described as composed of multiple geometries. If this is the case, then the same object can be described as the sum of simpler geometries. Another example of a simple geometry is the sphere using the following parametric equations:

$$\begin{cases} x = r\sin\theta\cos\rho \\ y = r\sin\theta\sin\rho \\ z = r\cos\theta \end{cases} \tag{2}$$

It is possible to describe an object that has a handle surmounted by a sphere as a combination of these two equations. So the finite element method is a convenient way to model real objects that are often far away from simple geometrical shape.

The finite element method is used in engineering to perform tests at a development stage. For instance, how a specific design will respond to various constraints when it is put in use. A typical example is car industry and the need for car industry to perform crash test and how robust a given chassis design is. The shape of a car is obviously complex and far from a simple geometrical shape. Therefore, the finite element method is an appropriate approach to tackle such problems. Moreover, not only do cars have complex shapes, but various parts of the car are made of various materials. Describing a car in a computer is usually made using finite element method software.

26.3.2 Tessellation

Tessellation is a method in computer graphics to describe the volume of an object using vertices. Usually this technique is applied by choosing a set of points (vertices) on a given surface and linking them with edges. The result of this operation is a grid covering the surface. The number of points or vertices on the surface will define the details of the small variations applied to the surface. If the surface is rather flat, for instance, there is no need to have too many points. However, if the surface is very rough with a lot of small bumps, then it is necessary to use many points to take into consideration all the small local variations. Tessellation is the first step when rendering a surface.

26.3.3 Rendering

The rendering technique is used to create a photorealistic view of an object or a scene [6]. The resulting image is called a render. The file containing the rendering information contains information about the geometry, the texture, the lighting, and the shading. The rendering techniques are multiple. One technique is called ray tracing. The ray tracing technique computes the path of light going through a given pixel on the screen. The light goes in a straight line and will hit an object at a certain angle. The light can either be absorbed, reflected, refracted, or eventually fluoresce if the photon is exciting an atom and the atom is emitting another photon. Figure 26.2 shows the flowchart of the various digital processing steps necessary to display photorealistic objects.

26.3.4 Medical Imaging

The use of the finite element method in medical imaging is a logical step. The human body is far from being symmetrical and having a simple geometrical shape. Therefore, the best way to describe the human body in a computer is the use of the finite element method. There are different fields where this method can be applied. We have orthopedics, where we need to design a shape that will replace an existing part of the body, for instance, if we wish to replace a joint. This is typical in surgery. With finite elements, one can take any bone, create a digital copy, and test its resistance in various conditions. This operation would not be possible without a computer.

Finite-element methods allow us to investigate biomechanics of the musculoskeletal system. With finite element methods one can test hypotheses that are difficult to test *in vivo* or *in vitro*.

Usually, the data are prepared before they can be visualized. First, segmentation, as described in the previous section. Here, each pixel or voxel of the images receive a specific description such as bone, cartilage, organs, etc. Two-dimensional (2D) or three-dimensional (3D) reconstruction of the geometry of the model follows. Then comes meshing and the assignment of material properties to various parts of the model.

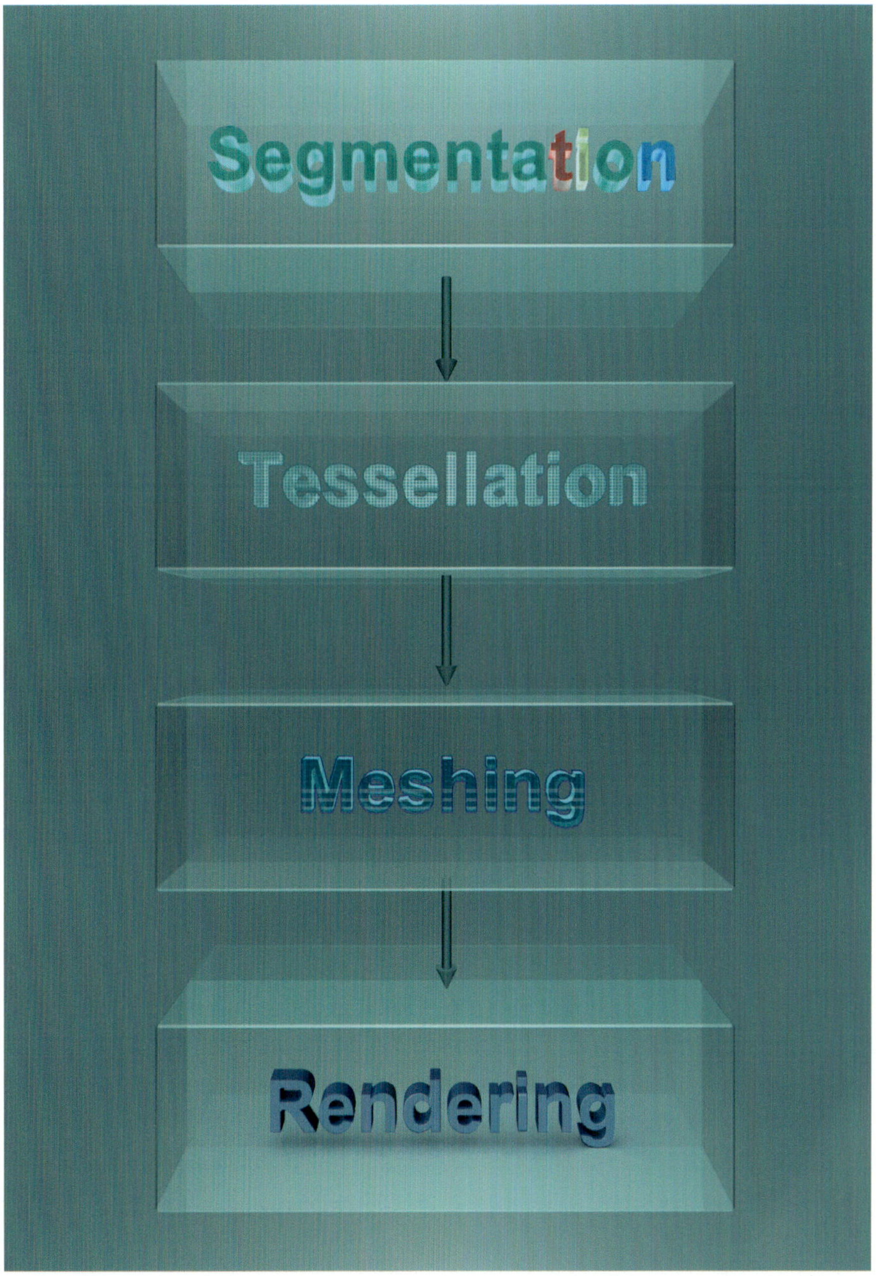

Figure 26.2 Flowchart for displaying a photorealistic object in a digital world. The segmentation will determine the boundaries of the object. The tessellation and the meshing will provide the ground covering the object with a rendering surface.

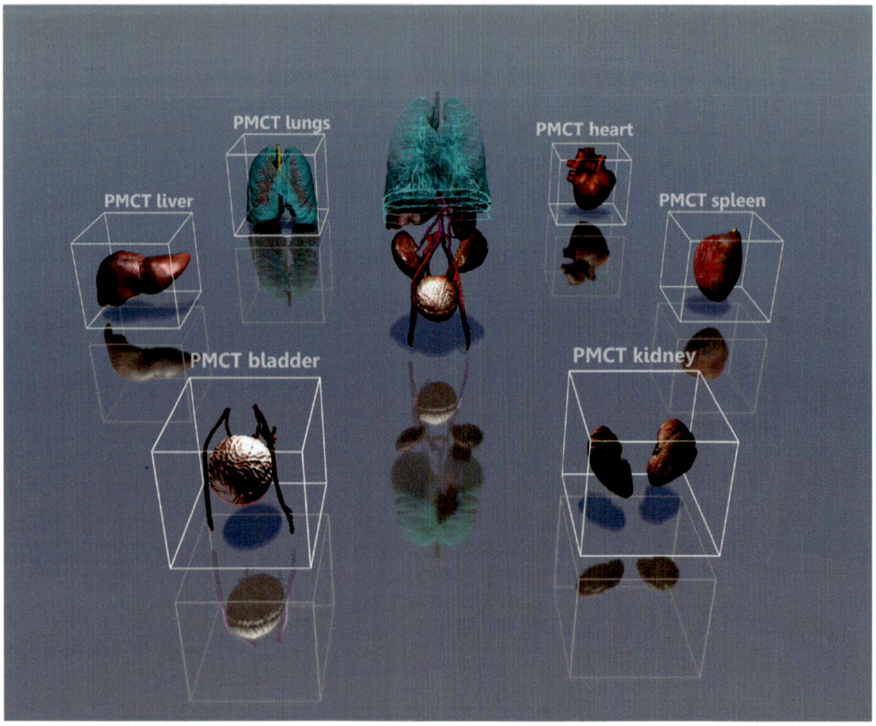

Figure 26.3 Segmentation of the heart, lungs, kidneys, spleen, liver, and bladder. By assembling a collection of such segmented 3D images, it is possible to train a deep learning algorithm and automate the task.

This rendering technique is used in postmortem computed tomography to render the images taken with a scanner. The rendering can be done at multiple levels. One can render the muscles, the organs, or the bones. This technique is used, for instance, in *Syngo.via* (Siemens Healthineers GmbH, Erlangen, Germany). Figure 26.3 displays organs that are relevant for postmortem forensic imaging: the liver, bladder, kidneys, heart, lungs, and spleen.

26.3.5 Statistical Shape Modeling

Statistical shape modeling (SSM) is a technique used in computer vision and medical imaging that involves building a statistical model able to capture the shape variations present in a three-dimensional dataset [7, 8]. The statistical model can subsequently be used to analyze and compare the shapes in the dataset and predict the shape of new objects that belong to the same class.

In forensic anthropology, SSM is used to analyze and compare the shape of skeletal remains. This is particularly useful in cases where only fragments of bones are found, as it allows forensic anthropologists to reconstruct the full shape of the bone using statistical models. SSM can also be used to compare the shape of bones found at different crime scenes and can help to determine whether the bones belong to the same individual.

To use SSM in forensic anthropology, researchers first build a statistical model of the shape variations present in a population of bones. This model is typically built using CT scans or laser scans of the bones, which are then segmented to extract the surface geometry. Once the statistical model is built, it can be used to analyze and compare the shape of bones found at crime scenes. By comparing the shape of these bones to the statistical model, forensic anthropologists can determine the likelihood that the bones belong to a particular individual or population.

Overall, statistical shape modeling is a powerful tool in forensic anthropology, allowing researchers to reconstruct the shape of bones and compare them to known populations. This can help to identify unknown individuals and provide important evidence in criminal investigations.

26.3.6 Minimum and Maximum Intensity Projection of Postmortem CT Volumetric Dataset

Minimum and maximum intensity projections of the postmortem CT volumetric dataset provide information on the state of a body that can be displayed on an image. The technique is described in Ebert et al. [9]. The minimum intensity

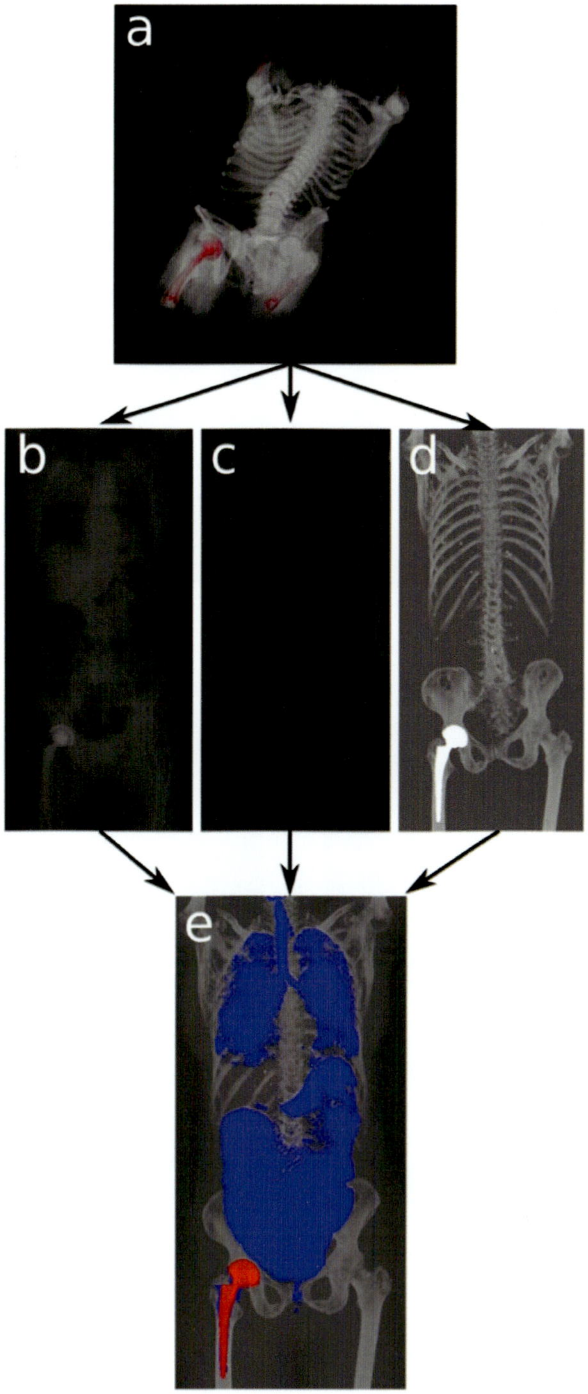

Figure 26.4 (a) Original PMCT volumetric dataset. (b) The average projection. (c) Minimum intensity projection in blue. (d) Maximum intensity projection. (e) Combination of the minimum intensity projection (blue) with the maximum intensity projection (red) to display in the same projection the gas distribution and foreign objects, in this case a hip replacement. Image taken from [9].

projection corresponds to the gas distribution inside the body. The maximum intensity projection gives information about the presence of foreign objects inside the body such as joint replacements, projectiles, etc. These two projections can be combined in a single image. Figure 26.4 shows an example how these two projections can be combined in a practical case to display the gas distribution and a hip replacement [9].

26.3.7 Rib Cage Projection

Figure 26.5 shows the projection of the three-dimensional rib cage on a two-dimensional map using the unfolding tool called CT bone reading [10]. CT bone reading is accessible via a plugin in the software Syngo.via (Siemens Healthineers

GmbH, Erlangen, Germany). With this transformation, we developed a deep learning model called RifNet to detect rib fractures [11]. RifNet reached a precision of 0.89 ± 0.03.

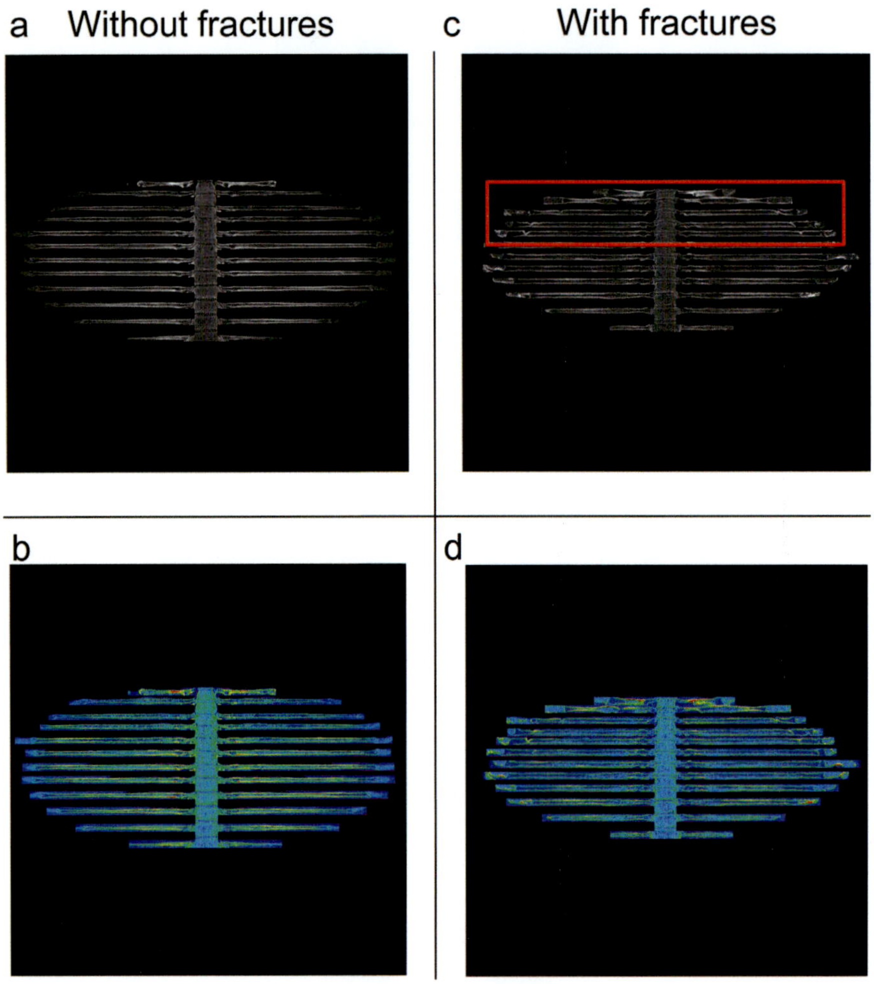

Figure 26.5 Rib cage projection. The greyscale CT image (A and C) of the projection is subsequently transformed into RGB to highlight the contrast using the colormap function nipy_spectral of the Python module Matplotlib.

ACKNOWLEDGMENTS

The authors thank Michael Bolliger for the artwork in Figures 26.1, 26.2, and 26.3.

References

1. Roberts, Lawrence Gillman (1963). Machine Perception of Three-Dimensional Solids, Ph.D. dissertation.
2. Otsu, Nobuyuki (1979). A threshold selection method from gray-level histograms. IEEE Transactions on Systems, Man, and Cybernetics. 9 (1): 62–66.
3. Batenburg, Karel J. and Sijbers, Jan (2009). Adaptive thresholding of tomograms by projection distance minimization. Pattern Recognition. 42 (10): 2297–2305.
4. Hrennikoff, Alexander (1941). Solution of problems of elasticity by the framework method. Journal of Applied Mechanics. 8 (4): 169–175.
5. Courant, Richard (1943). Variational methods for the solution of problems of equilibrium and vibrations. Bulletin of the American Mathematical Society. 49: 1–23.
6. Appel, Arthur (1968). Some techniques for shading machine renderings of solids. AFIPS '68 (Spring) Proceedings of the April 30-May 2, Spring Joint Computer Conference: 37–45.
7. Heimann, Tobias, and Meinzer, Hans-Peter (2009). Statistical shape models for 3D medical image segmentation: A review. Medical Image Analysis. 13 (4): 543–563.
8. Lüthi, Marcel, Gerig, Thomas, Jud, Christoph, and Vetter, Thomas (2018). Gaussian process morphable models. IEEE Transactions on Pattern Analysis and Machine Intelligence. 40 (8): 1860–1873.
9. Ebert, Lars Christian, Seckiner, Dilan, Sieberth, Till, Thali, Michael Joseph, and Franckenberg, Sabine (2021). An algorithm for automatically generating gas, bone and foreign body visualizations from postmortem computed tomography data. Forensic Science Medicine and Pathology. 17(2): 254–261. https://doi.org/10.1007/s12024-021-00363-3. Epub 2021 Apr 27. PMID: 33905073; PMCID: PMC8119247.
10. Ringl, Helmut, Lazar, Matthias, Töpker, Michael, Woitek, Ramona, Prosch, Helmut, Asenbaum, Ulrika, Balassy, Csilla, Toth, Daniel, Weber, Michael, Hajdu, Stefan, Soza, Grzegorz, Wimmer, Andreas, and Mang, Thomas (2015). The ribs unfolded—a CT visualization algorithm for fast detection of rib fractures: Effect on sensitivity and specificity in trauma patients. European Radiology. 25: 1865–1874.
11. Ibanez, Victor, Gunz, Samuel, Erne, Svenja, Rawdon, Eric J., Ampanozi, Garyfalia, Franckenberg, Sabine, Sieberth, Till, Affolter, Raffael, Ebert, Lars Christian, and Dobay, Akos (2022). RiFNet: Automated rib fracture detection in postmortem computed tomography. Forensic Science Medicine and Pathology. 18: 20–29.

CHAPTER 27

RADIOMICS

Carlo Tappero and Wolf Schweitzer

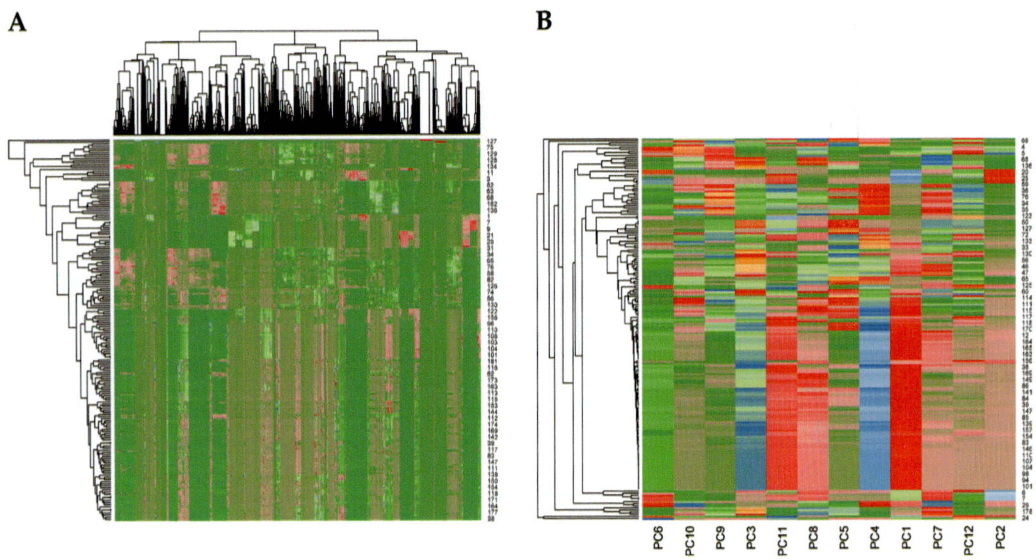

Figure 27.1 Raw data versus principal component analysis of high-dimensional data—Data: over 150 lung regions were encoded with over 1000 single derivative descriptors each; there were attributed to one of eight different diagnostic groups. Raw data plotted as a heatmap (A) shows a graphic depiction of the data (red–green: negative–positive values) with single observations row-wise and features/descriptors across columns; at the same time an attempt was made to cluster the observations so similar observations would be grouped together. After reduction of the dimensions through principal component analysis, the data shows a different structure with relation to numeric values as indicated by the heatmap (B) here, and also in association with the diagnostic groups (see Figure 27.2).

DOI: 10.4324/9781003413998-29

27.1 AIM

The objective of this chapter is to provide a concise overview of radiomics, elucidate its operational mechanism, and ultimately motivate its application in your research endeavors. As implied by the name, radiomics primarily concentrates on the analysis of radiological or medical images. From these images, a wealth of information (referred to as radiomics features) can be extracted through the application of mathematical methods.

27.2 DEFINITION

Radiomics, at its core, represents a cutting-edge approach in medical imaging analysis, harnessing the power of mathematics to extract a wealth of information from radiological images. The fundamental principle underlying radiomics is to delve beyond the surface of images obtained through various medical imaging modalities, such as MRI, CT scans, or PET scans. By employing sophisticated mathematical algorithms, radiomics unveils hidden patterns, subtle textures, and nuanced characteristics within these images that might escape the naked eye. These extracted features, aptly termed radiomics features, encompass a diverse array of quantitative metrics, ranging from intensity histograms to textural descriptors, offering invaluable insights into the underlying biology and pathology of the imaged tissues.

Understanding radiomics begins with recognizing its emphasis on relying on mathematical methods to decode the intricate nuances embedded within medical images. Rather than relying solely on qualitative observations or manual annotations, radiomics enables the quantification and characterization of diverse imaging features. Through the systematic analysis of radiomics features, researchers gain a deeper understanding of pathologic processes, morphological trauma classifiers, and other medicolegally relevant indicators. Moreover, the non-invasive nature of radiomics that allows use on previously captured data makes it particularly attractive for retrospective studies (Figure 27.3).

When confronted with a radiological image, the reader may typically employ two primary modes of interpretation: qualitative and quantitative (Figure 27.4).

Qualitative description involves subjective assessments based on visual inspection, where the observer identifies and describes the presence or absence of specific features, anomalies, or abnormalities within the image. This approach relies heavily on the observer's expertise, experience, and intuition, often resulting in varied interpretations among different individuals. While qualitative assessment offers valuable insights into the overall appearance and gross characteristics of the imaged structures, it may lack precision and consistency, leading to potential diagnostic ambiguity or variability in medicolegal diagnosis or decision-making. Conversely, professional training and quality control are used to ensure an optimal interrater correlation [6].

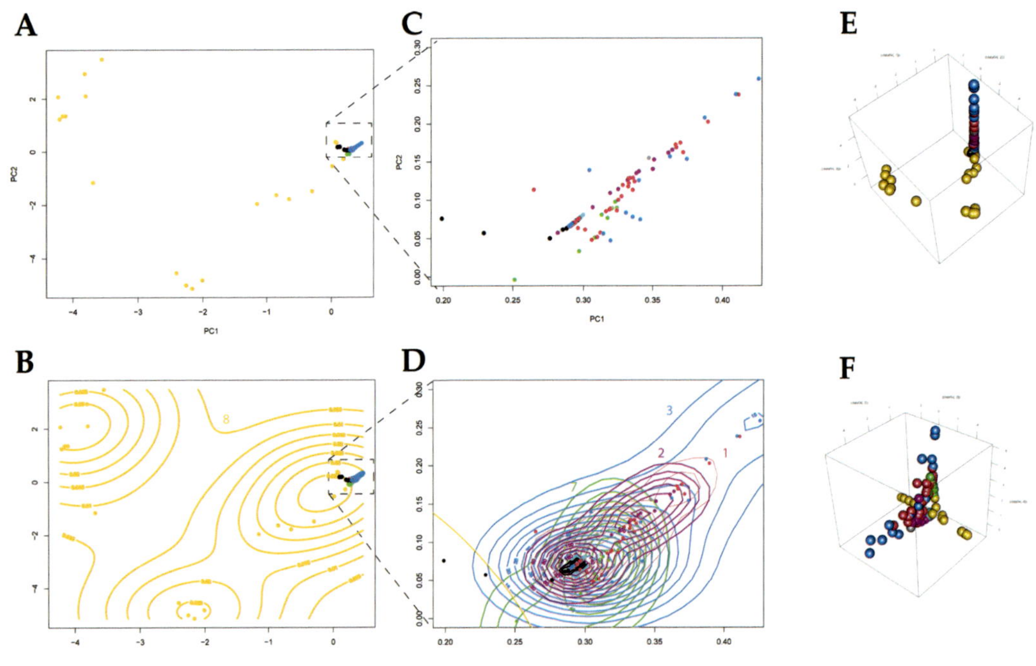

Figure 27.2 Principal component analysis to reduce high dimensionality—Data: over 150 lung regions (each one represented with a dot/sphere in these plots) were encoded with over 1000 single derivative descriptors each; there were attributed to one of 8 different diagnostic groups. Then, principal component analysis was performed. These plots (A, C: scatter plots; B, D: scatter plots with added kernel density estimates of the plotted data; E, F: three-dimensional plots; all plots: principal component analysis data; A-D with the first two, E-F with the first three principal components) visualize a dataset of initially highly dimensional radiomics data. Single observations (each case is a dot in these plots) are colored with their colors corresponding to their diagnostic group (A, B: overview, C, D: zoomed detail; colors: diagnostic groups). With exception of one particular diagnostic category (8, yellow), the others seem to provide relatively densely clustered data, whereas groups overlap (see detail enlargements in C and D). The question there will be whether the groups are numerically distinct from each other, and whether any statistical model built on this data provides good predictions. If one considers the first two principal components (A–D), then some diagnostic groups seem impossible to distinguish from their data alone. But if one adds the third principal component (E) or even considers a plot of the second to fourth principal component (F), the group configurations may be visualized better.

Figure 27.3 First order statistics, second-order statistics. To approach an explanation of the difference between first- and second-order statistic, one can think about a box full of balls of different colors. If you perform a first-order analysis, you may count the ball colors and arrive at a result. A second-order analysis may quantify the interaction with the nearby structure, yielding an entirely different type of evaluation (source: own 3D Blender model [1–3]).

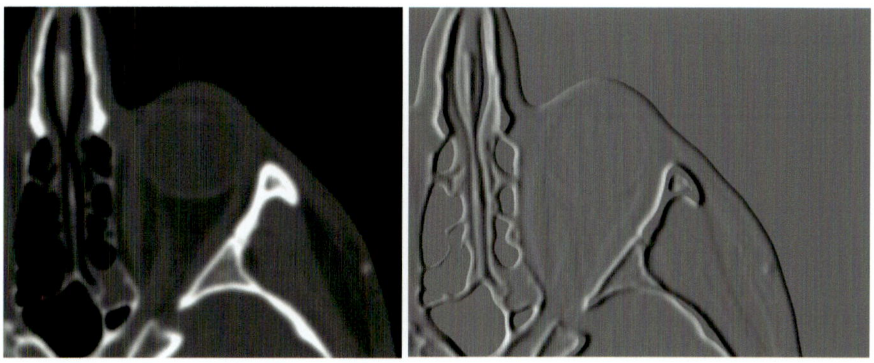

Figure 27.4 Derivative statistics—2D. The left image is the CT of a head. The soft tissues show a small particle. The right image shows a horizontal gradient image (Sobel filter) of the image on the left. It allows for a better distinction of data that constitutes a candidate for the geometric property of an edge. This is an example of a mathematical derivative of an original image.

In contrast, *quantitative description* entails the objective measurement and analysis of various image attributes using numerical metrics or mathematical algorithms [7]. This approach aims to quantify specific features, such as size, shape, density, or texture, within the radiological image. By quantifying imaging biomarkers and extracting precise measurements, quantitative analysis enhances reproducibility, accuracy, and reliability in diagnostic interpretation. Moreover, quantitative metrics provide valuable data for statistical analysis, trend monitoring, and outcome prediction, contributing to evidence-based medicine and medicolegal research endeavors. While quantitative analysis complements qualitative observation by offering objective measurements and enhancing diagnostic precision, it also poses challenges related to choice of methods and parameters, algorithm development, validation, and implementation in medicolegal practice.

While qualitative assessment relies on subjective interpretation and visual inspection to identify salient features and bases on professional training and experience, quantitative analysis employs (however much subjectively chosen and attuned) objective measurements and mathematical techniques to quantify imaging characteristics. By integrating both approaches, radiologists and researchers can attempt to harness the complementary strengths of qualitative observation and quantitative analysis to enhance diagnostic accuracy, improve autopsy preparation and ancillary or supportive postmortem imaging diagnoses, and advance our understanding of postmortem imaging morphology (Figure 27.5).

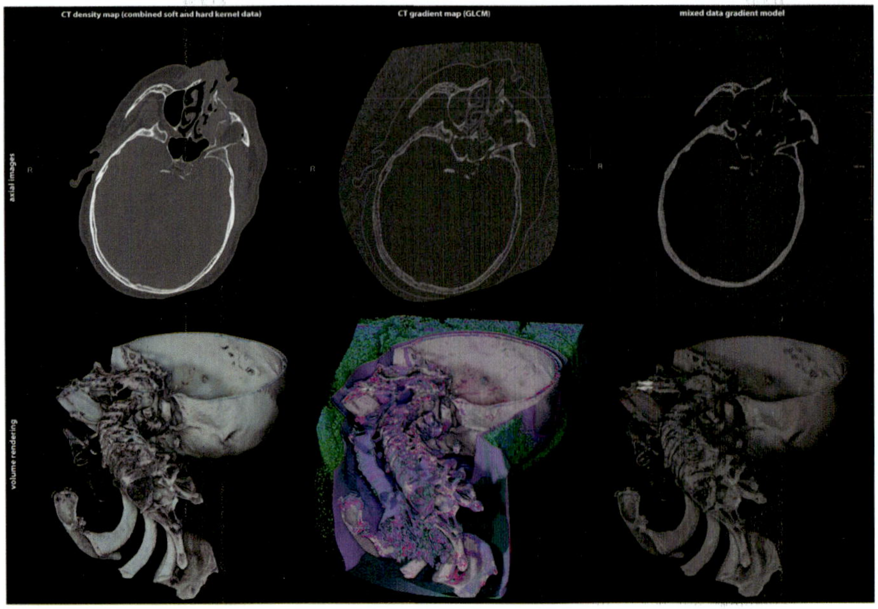

Figure 27.5 First-order statistics of 3D volume data of a head CT. The left column shows the CT of a head, conventionally as density map. The middle column shows a kernel-based gradient which allows for a better distinction of the actual contrast aspect. The right column shows a mixed data model that combines both. This is an applied example of a mathematical derivative of original 3D volume data. Its use has effective advantages for thin bone reconstruction (see Figure 27.6) (reprint from [4] with permission).

27.3 TECHNICAL BACKGROUND

Digital radiological images yield extensive numerical data beyond the visual representation, constituting vital information distinct from the image itself. These derived numbers, although not visually apparent in the image, serve as quantitative descriptors of underlying tissue properties, spatial relationships, and imaging characteristics. On basis of these numerical values, researchers employ statistical methods to analyze patterns, correlations, and trends within the imaging data. Ultimately, these statistical analyses may enhance diagnostic accuracy, validate hypotheses, and uncover medicolegally relevant imaging biomarkers [8].

One may ask where such large numbers not related to the image presentation itself come from all of a sudden—because 30 years ago, radiomics apparently was not a thing. However, the field may have started out with a different designation: image texture analysis [9–11]. Thereby, radiomics continues previously introduced methods to uncover and interpret hidden information within digital (and radiological) images through advanced texture and shape analysis techniques, revealing a "radiological texture" that offers deeper insights into the tissue of interest than previously recognized. This approach includes established analytical methods such as texture, entropy, and Haralick features, combined with modern computational capabilities, to extract valuable data from images and correlate these measurements for medical imaging analysis [12].

Radiomics has been described as a research domain that studies the extraction and analysis of large amounts of advanced quantitative imaging features [7, 13]. It is simply a radiology application of multivariate statistics with data of a very high dimensionality. This has always posed peculiar problems, that are also present in radiomics. A considerable number of the mathematically obtained image derivatives, that reside at the radiomics concept core center, appear to stem from a particular computational ease with which they are acquired—but not always from a particularly high evidence for their actual biological significance. In fact, many of the numbers that end up in data tables intended to be used as radiomics classifiers could be regarded as having no known or at least only obscure biological significance (Figure 27.7).

27.3.1 Parameters Other Than Diagnostic Value

Hundreds of variables may be generated from a single image or from a single region of interest. Certain features extracted from medical images may not solely reflect the underlying pathology or trauma being investigated. Factors such as imaging acquisition parameters, equipment variations, and patient-specific characteristics can introduce confounding effects, potentially distorting statistical analyses. For instance, parameters such as slice thickness, resolution, and imaging protocols, can influence the texture and intensity characteristics of the images, thereby impacting the extracted radiomics

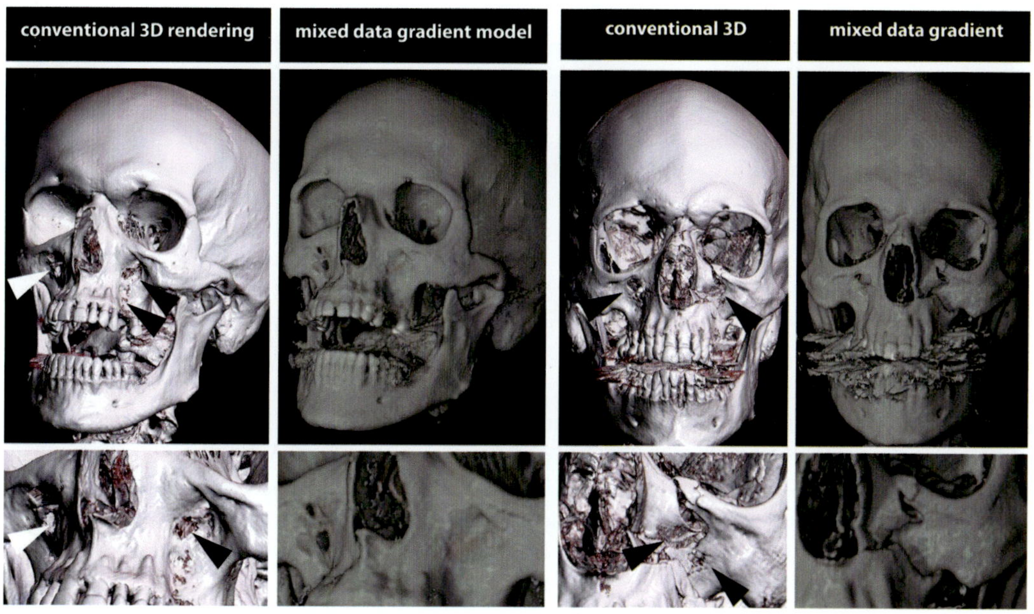

Figure 27.6 First-order statistics to reduce thin bone related artifacts in 3D (head CT). A comparison between conventional and mixed data gradient volume rendering techniques for visualizing facial bones, both intact and with fractures, is presented. The conventional 3D-renderings depict typical volume rendering inherent artificial defects in the thin anterior wall of the maxillary sinuses, as indicated by arrows. Notably, the case on the left demonstrates an absence of injuries when reconstructed accurately. The case on the right provides insight into how the mixed data gradient model effectively delineates fracture lines along the anterior wall of the left maxillary sinus from artificial defects (reprint from [4] with per-mission).

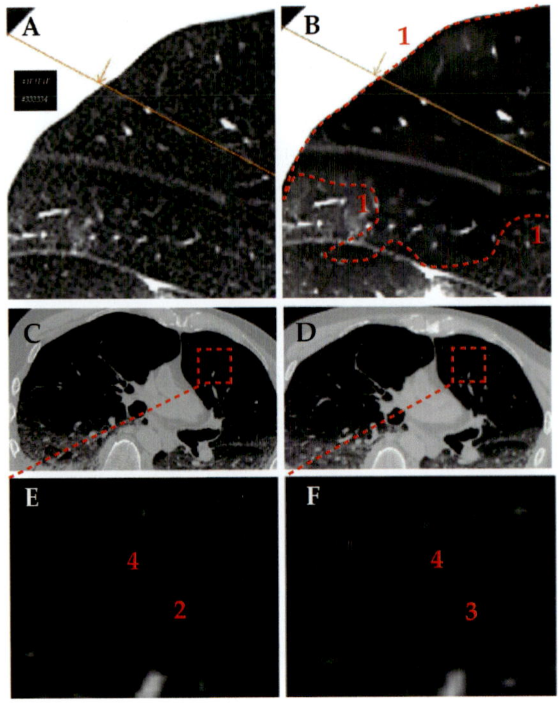

Figure 27.7 Beyond radiomics: PMCT without and with added artificial intelligence smoothing filter. A: conventional CT-scanner (Somatom Definition Flash)–based lung CT (120 kV, 400 mAs, BR60 kernel). B: typical morphology of a typical AI-based filtering result (Topaz Photo AI) which one may also find in some new or advanced CT-scanner postprocessing systems to improve image display of otherwise overly noisy data [5]. In comparison, consider another lung PMCT (120 kV, 400 mAs, BR60 kernel: 120 kV, 400 mAs, BR60 kernel, D: photon counting CT, 120 kV, 275 mAs, Q4 kernel, no further user postprocessing). 1: uncanny low density and absence of substructures as result of unexplainable steps taken by the algorithm. 2, 3: emphysematous bulla, 4: lung substructures. Users of modern technology must be able to visually recognize the hallmarks of AI image filtering.

features. These acquisition-related variations subsequently can lead to errors or biases in subsequent statistical analyses, potentially obscuring the radiomics features of the targeted pathology or trauma. With that, differences in imaging hardware and software across imaging facilities may further contribute to variability in radiomics features, complicating the interpretation and generalization of results. To mitigate these challenges, standardization, quality control measures, and numerical techniques such as normalization may have to be considered. By accounting for and adjusting acquisition-related variations, researchers can enhance the robustness and reliability of radiomics-based biomarkers for disease diagnosis, prognosis, and treatment response assessment. A powerful method to transcend machine or scanner related issues may be to evaluate data from various scanners or medical centers [14]. Overall, understanding and addressing the potential influence of acquisition methods on radiomics features are critical for ensuring the validity and reproducibility of radiomics studies, also in medicolegal research and practice settings [15, 16].

27.3.2 Descriptive Statistics and Data Exploration

Radiomics may provide added value if extracted features enhance the data beyond the conventionally visible, but not noise or meaningless content such as artifacts [8, 17].

Explore and visualize data first. Exploring and understanding data before embarking on predictive modeling endeavors may provide relevant angles for the examiner. This preliminary step allows practitioners to gain insights into the intricacies of the data, look for underlying patterns, and identify potential challenges or biases. Moreover, the selection of appropriate methods for predictive modeling may hinge on the characteristics uncovered during the data exploration and visualization phase. Thus, conducting a thorough examination of the data from multiple perspectives may be helpful to inform subsequent modeling decisions effectively.

However, datasets boasting hundreds or thousands of features or descriptors present unique challenges in terms of visualization and analysis. The sheer volume of variables can overwhelm traditional visualization techniques, making it difficult to discern meaningful relationships or patterns visually. Consequently, practitioners may need to employ methods such as dimensionality reduction techniques or feature selection algorithms, to effectively navigate and distill the essential information embedded within the dataset. By prioritizing thorough data exploration, practitioners can enhance the efficacy and accuracy of predictive modeling efforts [18].

Example: In Figure 27.1, data is shown as typically encountered in radiomics. In Figure 27.2, a first approach reduces the dimensions of data visualized in Figure 27.1 by use of principal component analysis, and one may visualize the data so one can investigate the problems that the data poses with greater ease. In this example, the groups appear to be densely grouped with exception of one group (group 8) which shows a wider scatter of single group members. With regard to data density estimates, there is considerable overlap between the different groups, so the error for predictions based on this data may be relatively high.

Describe statistical moments and relevant data properties. Evaluating the distribution within a dataset serves as more than just a statistical observation; it serves as a pivotal cue for guiding subsequent analytical decisions. Such distributions may inform the selection of statistical tests and methodologies in the context of normality, in the context of scedasticity, which pertains to the homogeneity of variance across the data and also in the context of tail behavior. These aspects may inform the choice of appropriate statistical models and inferential techniques [19, 20].

Consider data in its morphological context. When examining data, the initial observation often involves visual scrutiny, where one seeks to discern apparent patterns or correlations that may hint at underlying relationships or structures. This visual inspection serves as a crucial step in uncovering the intricate nuances embedded within the dataset, allowing researchers to glean insights into its inherent morphology and characteristics. By scrutinizing the data's visual representation, researchers may detect patterns, trends, or anomalies that prompt further exploration and analysis. Beyond simply identifying visual correlations, this initial inspection may unveil potentially relevant correlations or differences within the dataset. These aspects could manifest as distinctive patterns, clusters, or trends that hold significance in the context of the research objectives or domain-specific knowledge. Upon identifying these possibly relevant functions or aspects, researchers can proceed to encode them into formal representations or mathematical models. This encoding process involves translating the observed patterns or correlations into quantifiable variables, parameters, or features that can be systematically analyzed and manipulated. By encoding these insights into the data, researchers can attempt to extract meaningful information, derive predictive models, or generate actionable insights that contribute to further understanding or decision-making within the respective domain. Thus, from the initial visual inspection to the subsequent encoding of relevant functions or aspects, this process forms the foundation for data-driven exploration and analysis, driving insights and discoveries across various fields of research and application.

27.3.3 Generating and Selecting Features or Descriptors for Predictive Modeling

Defining characteristics, features, and descriptors. In the realm of predictive modeling, the selection of discriminating features plays a pivotal role in determining the model's efficacy and predictive performance. However, not all features are created equal; for a feature to be deemed useful in predictive modeling, it must transcend mere statistical significance and instead be genuinely representative of underlying hypotheses or observations. In other words, the feature should encapsulate meaningful information that aligns with the research question or domain-specific knowledge, thus contributing substantively to the predictive capacity of the model. To ascertain the suitability of a feature for predictive modeling, researchers must evaluate its relevance and interpretability within the context of the research objectives. This entails considering whether the feature reflects meaningful patterns, trends, or relationships that align with existing hypotheses or observational insights. Additionally, the feature should possess discriminatory power, meaning it should exhibit discernible differences or variability across different classes or conditions of interest, thereby enabling the model to differentiate and classify instances effectively. Furthermore, the selection of discriminating features involves a nuanced balance between statistical significance and substantive relevance. While features may demonstrate statistical significance in terms of their association with the outcome variable, their true utility lies in their ability to contribute to the model's predictive accuracy and generalizability. Thus, features that are not only statistically significant but also conceptually meaningful and representative of underlying phenomena are more likely to enhance the predictive performance and interpretability of the model. In summary, discriminating features in predictive modeling should not only exhibit statistical significance but also be genuinely representative of hypotheses or observations within the research domain. By prioritizing features that align with substantive insights and possess discriminatory power, researchers can construct more robust and interpretable predictive models that yield actionable insights and facilitate informed decision-making [21].

What has previously drawn skepticism within the scientific community are so-called "fishing expeditions", where researchers adopt an exploratory or data-driven approach without a priori hypotheses guiding their inquiry. Such endeavors may involve indiscriminate analysis of data in search of unexpected patterns or associations, often leading to spurious findings or false positives. While exploratory research can yield serendipitous discoveries and generate hypotheses for further investigation, the absence of a guiding hypothesis may compromise the rigor and interpretability of the findings [22]. Conversely, any extensive search for the actual meaning of a newly-found classifier in radiomics-like multivariate statistics has also been discouraged by other authors [17]. In reality, however, not all assumedly different morphological groups turn out to exhibit significantly different radiomics features. Then, a firm theoretical grasp on the classifiers one uses, particularly also if they fail at discriminating groups, may be particularly paramount from the assumption that the null-hypothesis could actually be the better explanation.

Example: if the stature [body length] of a body is estimated from the length of all bones in a multivariate statistic, one may propose that all bone measurements that contribute directly to the make of the stature should be an integral part of the classifiers, and not removed from the equation for any reason; if then one finds through statistical procedures, that the legs' long bone lengths do not correlate with the stature in the given sample, then it may still be wrong to remove the femur or tibia lengths from the equations based on theoretical grounds and prior research [23]. Instead, the better explanation may be that the actual sample may not be representative for a larger population.

Data conditioning. Data conditioning, often regarded as a relevant pre-processing step in predictive modeling, entails the meticulous refinement and preparation of datasets to enhance their suitability for analysis and modeling tasks. At its core, data conditioning aims to optimize the performance and accuracy of predictive models by mitigating the detrimental effects of noise, irrelevant variables, and spurious correlations within the dataset. This process involves systematically identifying and filtering out extraneous or redundant information, thus streamlining the dataset and focusing on the most informative features. One of the primary objectives of data conditioning is to eliminate noise, which refers to random fluctuations or irregularities in the data that do not convey meaningful information. By reducing or eliminating noise through various techniques such as smoothing, filtering, or outlier removal, researchers can enhance the signal-to-noise ratio within the dataset, thus improving the model's ability to discern genuine patterns and relationships. Moreover, data conditioning involves identifying and removing irrelevant variables that do not contribute substantively to the predictive task at hand. These unnecessary variables may introduce complexity, increase computational burden, or confound the analysis, ultimately hindering the model's performance. Through feature selection or dimensionality reduction techniques, researchers can prune the dataset of extraneous variables, focusing on those that are most relevant and informative for the predictive task. Additionally, data conditioning aims to address the issue of co-correlation, where certain variables may exhibit high interdependence or redundancy, potentially inflating model complexity and multicollinearity issues. By identifying and addressing co-correlating variables through techniques such as principal component analysis (PCA) or correlation analysis, researchers can reduce redundancy and improve the model's interpretability and generalizability. In essence, data conditioning serves as a critical preparatory step in predictive modeling, facilitating the construction of robust and accurate models by refining the dataset to focus on the most salient and informative features. By eliminating noise, removing irrelevant variables, and addressing co-correlation, researchers can enhance the efficacy and re-liability of predictive models, ultimately advancing our ability to extract actionable insights and make informed decisions from data [17].

Feature selection: data proposed to be used as classifying features may fall into the categories irrelevant, redundant, weakly relevant but non-redundant, and, relevant. Irrelevant features refer to variables or attributes within the dataset that have little to no bearing on the classification task. These features lack discriminatory power and do not contribute meaningfully to distinguishing between different classes or categories. As such, including irrelevant features in the modeling process can introduce noise and unnecessary complexity, potentially impairing the performance of the classifier. Redundant features, on the other hand, exhibit high levels of interdependence or correlation with other features in the dataset. While redundant features may individually possess discriminatory power, the inclusion of all of them in the modeling process can lead to multicollinearity issues, where multiple variables convey redundant information. Removing redundant features helps streamline the dataset and mitigate issues associated with model complexity and overfitting. Weakly relevant but non-redundant features occupy a middle ground between irrelevant and relevant features. These features may exhibit some degree of association with the target variable but lack strong discriminatory power on their own. While they may contribute marginally to the classification task, their inclusion in the model may not significantly enhance predictive performance. Nonetheless, these features may still offer valuable insights when considered in conjunction with other relevant features or through feature engineering techniques. Finally, relevant features are those that possess substantive discriminatory power and directly contribute to distinguishing between different classes or categories of significance to the research hypothesis. These features capture meaningful patterns, relationships, or characteristics within the data that are relevant to the classification task. By focusing on relevant features, researchers can construct more accurate and interpretable classifiers that effectively discriminate between classes and generalize well to unseen data. The ways to reduce high dimensions and at the same time throw out irrelevant data are manifold and subject of much debate—including that of information leak [17].

Feature subgroups. Radiomic features may be categorized into the following subgroups (shape features, first-order statistics, second-order statistics, high-order statistics). *Example:* In Figure 27.3, the idea behind radiomic features is explained using an easy to understand visual model; see also the paper by Lubner et al. (2017) [24].

Shape features. When considering geometric properties in the context of classifying features for predictive modeling, a range of metrics related to shape and size can offer valuable insights into the underlying characteristics of objects or entities within the dataset. Examples of geometric properties include volume, maximum diameter along different orthogonal directions, maximum surface area, tumor compactness, and sphericity [25]. *Volume:* The volume of an object represents the amount of space it occupies in three-dimensional space. In medical imaging, such as MRI or CT scans, the volume of a tumor can serve as a discriminatory feature for classification tasks. For instance, larger tumor volumes may indicate more advanced disease stages or higher malignancy grades, thus influencing the classification of tumors into different categories based on severity. *Maximum diameter along different orthogonal directions:* This metric describes the maximum distance between two points on the surface of an object along perpendicular axes. In the context of tumor classification, variations in the maximum diameter along different directions can reflect the irregularity or asymmetry of tumor shapes. For example, tumors with elongated or irregular shapes may exhibit larger disparities in maximum diameters along orthogonal directions com-pared to spherical or symmetrical tumors. *Maximum surface area:* The maximum surface area of an object corresponds to the largest possible area of its external boundary. In tumor classification, the surface area-to-volume ratio can provide insights into the tumor's morphology and spatial distribution. For instance, tumors with complex or irregular surface contours may have higher surface area-to-volume ratios compared to more compact or spherical tumors of similar volume. *Tumor compactness:* Tumor compactness quantifies the degree of packing or density of tumor cells within the volume occupied by the tumor. Higher compactness values indicate a denser packing of tumor cells, which may correlate with more aggressive tumor phenotypes or poorer treatment responses. Thus, tumor compactness can serve as a discriminative feature for classifying tumors into different prognostic or risk categories. *Sphericity:* Sphericity measures the degree to which an object approaches a perfect sphere in shape. In tumor classification, sphericity can provide in-sights into the regularity or symmetry of tumor shapes. For example, tumors with high sphericity values are more likely to exhibit spherical or rounded shapes, whereas tumors with lower sphericity values may have more irregular or non-spherical morphologies.

In summary, geometric properties such as volume, maximum diameter, surface area, tumor compactness, and sphericity offer a diverse range of features that can be exploited for predictive modeling in various domains, including medical imaging analysis, material science, and engineering. By incorporating these geometric metrics as classifying features, researchers can enhance the discriminatory power and interpretability of their models, thus facilitating more accurate and insightful classification tasks. *First-order statistics features* serve as fundamental descriptors of the distribution of voxel values within medical imaging data, devoid of considerations for spatial relationships among voxels. These features provide insights into the overall intensity characteristics of the image, offering valuable information about the voxel intensity distribution's central tendency and variability. One commonly utilized approach involves histogram-based analysis, which summarizes the frequency distribution of voxel intensities across the image. Within the realm of first-order statistics, various metrics are employed to characterize the distribution of voxel values comprehensively. These metrics encompass essential statistical properties such as the mean, median, maximum, and minimum voxel intensities, offering insights into the typical range and variability of intensity values present in the image. Additionally, measures of skewness and kurtosis provide information about the asymmetry and peakedness of the intensity distribution, respectively, shedding light on the distribution's shape and departure from normality. Other first-order statistics features include measures of uniformity and randomness, which assess the homogeneity and complexity of voxel intensity patterns within the image. Uniformity quantifies the degree of uniformity or consistency in voxel intensities, while randomness, often quantified using entropy, reflects the level of disorder or unpredictability in intensity distributions. These metrics are particularly valuable as they capture aspects of image texture and complexity that may be indicative of underlying tissue characteristics or pathological features. One notable advantage of first-order statistics features is their robustness to non-diagnostic acquisition-related factors, such as variations in imaging protocols or equipment settings. By focusing on aggregate statistics of voxel intensities, first-order features are less sensitive to spatial variations or artifacts within the image, making them more reproducible and resilient to potential confounding factors. As a result, first-order statistics features have emerged as valuable tools in radiomics analysis, facilitating standardized and reliable characterization of medical imaging data for various medicolegal applications and research endeavors [26].

Second-order statistics features, often referred to as textural features, offer insights into the spatial relationships and arrangements of voxel intensities within medical imaging data. These features capture the statistical inter-relationships between neighboring voxels, providing a quantitative measure of texture and heterogeneity within the image. By analyzing patterns of voxel intensities at different spatial scales and orientations, second-order statistics features enable characterization of complex structures, such as lesions or tissues, based on their internal texture properties. One common approach to deriving textural features is through the use of matrices that quantify voxel intensity relationships within the image. The grey-level co-occurrence matrix (GLCM) is a widely utilized method that captures the frequency of occurrence of pairs of voxel intensity values at a specified distance and in a predetermined direction. By analyzing the GLCM, various texture descriptors can be derived, such as contrast, homogeneity, and entropy, which reflect different aspects of spatial arrangement and heterogeneity within the image. Another matrix-based approach is the grey-level run-length matrix (GLRLM), which quantifies the occurrence of consecutive voxels with the same intensity value along fixed directions within the image. From the GLRLM, textural features such as run-length non-uniformity, short-run emphasis, and long-run emphasis can be computed, providing additional insights into the spatial distribution and continuity of voxel intensity patterns. Additionally, other second-order statistics features include the gray-level size zone matrix (GLSZM) and the neighborhood gray-tone difference matrix (NGTDM). The GLSZM characterizes the distribution of connected voxel clusters of similar intensities within the image, offering information about the size and spatial extent of homogeneous regions. On the other hand, the NGTDM quantifies the local variations in voxel intensity between neighboring regions, capturing textural properties related to subtle changes in intensity gradients and transitions. Overall, second-order statistics features play a vital role in radiomics analysis, facilitating the quantitative characterization of texture and spatial heterogeneity within medical imaging data. By employing matrices and statistical metrics to capture the complex interplay of voxel intensities, these features enable the extraction of valuable information about underlying tissue properties, lesion characteristics, and disease states, ultimately contributing to improved diagnosis, prognosis, and treatment planning in medicolegal practice [27, 28].

Higher-order statistics features represent a sophisticated approach to extracting intricate information from medical imaging data by applying advanced statistical methods following image preprocessing steps. These features aim to uncover complex patterns, suppress noise, and enhance the visibility of details within the image, ultimately offering valuable insights into underlying structures and textures. One common strategy involves applying filters or mathematical transforms to the images to manipulate the intensity values and highlight specific features of interest. Fractal analysis is one such technique used to characterize the self-similar or fractal-like properties of structures within the image. By quantifying the complexity and irregularity of patterns at different scales, fractal analysis provides a means to assess the spatial heterogeneity and organization of tissues or lesions. This approach is particularly useful for capturing the intricate details of anatomical structures or pathological features that exhibit fractal-like patterns. Another powerful method is the wavelet transform, which decomposes the image into multiple levels of detail and approximation coefficients across different spatial frequencies. By analyzing the wavelet coefficients at various scales and orientations, researchers can extract texture features that capture both fine and coarse details within the image. This enables the detection of subtle variations in texture and the delineation of structural boundaries with high precision. Further-more, Laplacian transforms of Gaussian-filtered images offer a means to enhance texture patterns and highlight regions of interest by accentuating edges and discontinuities in intensity values. By convolving the image with a Gaussian filter to smooth out noise and then applying the Laplacian operator to detect intensity gradients, this technique can extract areas with increasingly coarse texture patterns, facilitating the identification of relevant features within the image. Overall, higher-order statistics features can be used to uncover intricate details and spatial relationships within medical imaging data.

Model creation—testing: the robustness and reliability of a statistical model are often evaluated by testing it with the same dataset used for training. This process typically yields favorable results, as the model has essentially memorized the patterns that are within the data. However, to assess the model's generalizability and performance in real-world scenarios, more rigorous validation methods are necessary. One such method is cross-validation, where the dataset is split into training and test sets multiple times, allowing the model to be trained on different subsets of data and tested on the remaining portions. This approach provides a more comprehensive evaluation of the model's performance and helps mitigate the risk of overfitting to the training data. Taking another step towards ensuring model robustness involves using test sets that are disjoint from the training data, meaning they contain samples that the model has not been exposed to during training. This approach, known as holdout validation, helps assess the model's ability to generalize to unseen data and provides a more realistic evaluation of its performance. By evaluating the model on independent test sets, researchers can gauge its performance in diverse settings and ascertain its applicability beyond the specific dataset used for training. An even more rigorous validation approach involves testing the model on datasets acquired from different sources or using different imaging modalities, such as CT scanners. This approach, known as external validation or generalization testing, assesses the model's performance across heterogeneous datasets, thereby validating its robustness and applicability in real-world medicolegal settings. By testing the model on disjoint datasets acquired from different sources, researchers can evaluate its performance under varying conditions and ensure its reliability across diverse patient populations and imaging protocols. In summary, while testing a statistical model with the same data used for training provides initial insights into its performance, more rigorous validation methods, such as cross-validation, holdout validation, and external validation, are necessary to assess its generalizability and robustness. By systematically evaluating the model on independent test sets and disjunct datasets acquired from different sources, researchers can gain confidence in its performance and applicability in real-world medicolegal practice [15].

Model creation—application: The practical application of model predictions extends beyond the confines of academic studies, often necessitating a broader understanding of the data within its real-world context. While a model may yield accurate predictions within the scope of a research study, factors that were previously considered confounding may suddenly become crucial when the model is deployed in real-life scenarios. For instance, in medical diagnosis, a predictive model may accurately classify patients based on certain features, but real-world prevalences of underlying conditions or comorbidities may significantly impact the model's performance and utility. Moreover, the interpretation of prediction accuracy can vary depending on the application and industry standards. While academic guidelines may categorize prediction accuracies below 50% as poor, 50–75% as moderate, 75–90% as good, and greater than 90% as excellent, these benchmarks may not always align with real-world expectations. For example, in financial fraud detection or cybersecurity, a 97% success rate may be deemed inadequate due to the potentially severe consequences of false negatives. In such high-stakes scenarios, even a small margin of error can have significant implications, necessitating more stringent performance standards. In industrial manufacturing, the pursuit of quality has led to the implementation of rigorous quality control programs aimed at minimizing defects to higher levels. For instance, some manufacturing processes strive to achieve defect rates as low as 3 or 4 parts per million opportunities (PPM), indicating a level of precision and reliability that by far exceeds academic appreciation levels. Achieving such low defect rates requires not only meticulous data gathering and analysis but also the implementation of robust quality assurance measures throughout the whole production process. By using advanced statistical techniques and process optimization strategies as well as qualitative improvements, manufacturers can minimize variability, enhance product quality, and ultimately improve customer satisfaction. In summary, the practical use of model predictions demands a nuanced understanding of the data's context and real-world implications. Interpretations of prediction accuracy may vary across different industries and applications, with benchmarks such as academic guidelines often falling short of real-world standards. By aligning predictive models with industry-specific requirements and quality standards, organizations can ensure the reliability, effectiveness, and success of their predictive analytics initiatives in diverse operational settings [29].

27.4 MEDICOLEGAL AND FORENSIC BACKGROUND

Forensic medicine has a strong basis in morphology. With accumulating CT-scan data and other digital information, the basis for radiomics analysis certainly is present. For instance, radiomics can aid in the identification of specific tissue signatures associated with traumatic injuries or the detection of pathological changes indicative of underlying diseases.

From there on out, there appear to be two intrinsically different ways of exploiting morphology for the benefit of diagnostic predictions using computers:

- exploiting *any derived numerical features*: justify the extraction of a particular numerical feature on an assumed or established relevance for the hypothesis under examination. Each numerical feature is carefully selected based on its presumed or established relevance to the hypothesis being investigated. The justification for extracting a particular numerical feature lies in its potential to provide valuable insights into the underlying biological or pathological processes associated with the condition under examination. By scrutinizing these numerical features, researchers aim to uncover meaningful patterns, correlations, or biomarkers that can inform diagnostic predictions and aid in medicolegal decision-making.
- exploiting *whole morphological data* (i.e., images, volume data, surface data). This approach involves leveraging the entirety of morphological data, including images, volumetric data, and surface data, without prior extraction of specific numerical features. Instead of focusing on individual features, researchers analyze the holistic morphological characteristics of the dataset. By examining the complete morphological landscape, researchers aim to capture complex spatial relationships, structural variations, and subtle nuances that may not be apparent when considering isolated numerical features alone. This approach allows for a comprehensive exploration of the morphological data, enabling the identification of patterns, anomalies, and diagnostic clues that may otherwise remain obscured.
- When COVID-19 became a focus, diagnostic aspects were a big issue also in forensic medicine [30, 31]. Radiomics was applied for a range of applications when evaluating lung CT scans. In one study, lung volumes were segmented, before extracting shape, intensity and texture as radiomics features in order to predict poor prognosis [32]. In another study, radiomics was used to classify chest X-ray images into healthy, COVID-19 and non–COVID-19 pneumonia [33].

The more interesting and challenging option obviously is the first one, where one derives numerical features, before starting data analysis, rather than feeding raw image data. As outlined above, hypothesis-driven research may benefit from an understanding-based classifier definition, rather than suffering classifier removal at the hands of a pure statistical evaluation. From the application of multivariate statistics in general and radiomics in particular, it has become clear that obvious diagnoses, instances where anyone can see a diagnostic difference, do not lend them-selves that much for multivariate analysis and radiomics (unless one wants to study the behavior of the method under unanimous data). At any rate, one tends to use these methods for situations where a phenotype overlap is assumed to lie somewhere between partial and complete.

27.4.1 Overlaps of Morphology

Morphological features within a particular category, group, diagnosis, or finding can exhibit a wide spectrum of variation, ranging from stark differences to remarkable similarities, or even a blend of both. The intricacies of morphological characteristics may vary significantly depending on the underlying condition or pathology under investigation. In some instances, morphological features may align closely between different categories or diagnoses, presenting *overlapping patterns or manifestations*. These are particularly interesting areas for the application of radiomics. Conversely, distinct differences in morphological features may distinguish one condition from another, contributing to diagnostic specificity and accuracy.

In the realm of radiomics, the ability of statistical analysis to provide enhanced discrimination between different diagnoses or diagnostic groups hinges largely on the nature and quality of the data being utilized. While radiomics-based statistics hold promise for uncovering previously hidden or subtle patterns within imaging data, their efficacy in providing further discrimination may vary depending on factors such as dataset composition, imaging protocol variability, as well as sample size and adequacy of selected radiomics features. Moreover, imaging limitations, such as contrast, resolution, and specificity of visualization, may impact the accuracy and reliability of diagnostic conclusions, highlighting the importance of judicious interpretation and multidisciplinary collaboration in medicolegal practice. Consequently, the utility of radiomics-based discrimination must be carefully evaluated within the wider context of the specific dataset and diagnostic task at hand. It's essential to acknowledge that certain diagnoses cannot be solely reliant on radiology data but necessitate complementary examinations and tests for accurate assessment and data validation if not model generation. The integration of multiple diagnostic modalities, including laboratory tests, medicolegal assessments, and imaging studies, is often indispensable for comprehensive diagnostic evaluation.

Typical topics that may be subjected to medicolegal disambiguation attempts with a possible, at least partial, phenotypical overlap, where a statistical approach may be interesting, may be given as examples:

- Estimation of time since death based on PMCT: the estimation of postmortem interval (PMI) has traditionally relied on observing sequential changes in the body over time, influenced by various external, internal, and environmental factors. However, complexities in death scenes often make it challenging to accurately account for these factors, leading to potential compromises in PMI estimation accuracy. As one body may exhibit the same postmortem alterations after a certain PMI compared to another body with a different PMI, there certainly exists phenotype overlap or a certain diagnostic error for PMI estimation. First studies aimed to assess the potential of postmortem CT (PMCT) radiomics for distinguishing between early and late PMI stages [34, 35].
- The presence of both subdural and retinal hemorrhages, while indicating potential non-accidental injury, must be differentiated from inborn disorders, as there can be morphological resemblances between these distinct medical and legal categories [36–38]. The morphologic differences also between different inborn disorders may be subtle or absent in individual cases, while distinct radiological features were described. While genetic testing confirmation is essential, precise medicolegal and radiological phenotyping remains the cornerstone of the diagnostic framework [39]. Assessments for suspected non-accidental trauma may involve seeking advice from genetic and metabolic specialists to examine patients for genetic disorders that may imitate or worsen child abuse. Therefore, conditions such as osteogenesis imperfecta (OI) and glutaric aciduria type 1 (GA1) might be subject to particular scrutiny during legal proceedings. Presently, there are no established societal protocols for the genetic or metabolic evaluation of suspected cases [40].
- In evaluating skull fractures, the medicolegal expert may have to decide whether they are caused by a fall or by a blow [41, 42]. The medicolegal differentiation between falls and blows is a common yet challenging task. One frequently employed criterion for this purpose is the hat brim line (HBL) rule, which suggests that injuries above the HBL are unlikely to be fall-related. However, some studies have questioned the relevance of this rule. One study examined the causes, number, and locations of fractures on the skulls and trunks of 400 individuals who underwent CT scans after trauma. The goal was to aid in interpreting such injuries in cases where soft tissues are absent, such as skeletonized or heavily decomposed bodies. By combining multiple criteria and assessing their predictability, the authors aimed to enhance the accuracy of distinguishing between falls and blows. Parameters such as anatomical location and the number of fractures per region may offer valuable insights for statistically distinguishing between falls and blows. In another evaluation of that data, random forest methods were used with a 83% success rate [43, 44].
- Firearm injuries pose classification problems with the question of manner of death [45, 46]. Here, morphological differentiation has been attempted using statistical evaluation; thereby, suicide had higher mortality with more severe injuries, and more head/neck/facial injuries than homicide, and the majority of suicides were with handguns [47]. Radiomics so far has not been used for this differentiation. However radiomics had been used to predict head trauma outcome [48, 49].

27.5 APPLICATION

From a view on their medicolegal or case related application, these analyses are still in their infancy. So far, they appear to be not widely used in forensic practice. Still, there are some promising studies in medicolegal radiology, particularly in the oncological field, and some first developments also cover the forensic field.

27.5.1 Forensic Imaging Experience

Forensic imaging studies that touch on the principle of multivariate statistics using shape descriptors, if not radiomics in a more narrow sense, so far covered research into predictive classification of skin injuries (using discriminant analysis [50]), hypothermia-related lung findings (using a simple general linear model [51]), obstructive asphyxia-related lung findings (using several models [52]) and hemorrhagic pericardial effusions (using deep learning [53]).

Even the inclusion of one texture or radiomics related feature (a type of 3D gradient) to enhance conventional volume rendering has paved the way for significantly better volume rendering [4]. The single different layers for this path are visually not as interesting for the application as they are to illustrate the method (Figure 27.5). Whereas the final result seems impenetrable with regard to giving away the mathematics behind it, like, how it was really generated, use of radiomics for targeted improvement can be a powerful tool for visualization (see Figure 27.6).

27.5.2 Medicolegal Use

For example, Sandrasegaran et al. used CT-data based texture analysis (CTTA) to assess 60 patients with nonmetastatic pancreatic ductal adenocarcinoma before treatment and the found that of 16 radiomics feature analyzed only two showed a statistical significance. Low kurtosis of CT-data based regions of interest correlated with poor overall survival and that higher mean of the positive pixels (MPP) (threshold of >29.4) was associated with better progression-free survival. However, they also found that subjective radiological observation of local vascular invasion (which would make the tumor unresectable) and the presence of metastatic disease were more strongly associated with the prognosis [54]. Li et al. and Fu et al. used CTTA to analyze the survival rate of HCC patient comparing TACE with surgery or TACE plus Sorafenib [55, 56]. Other studies show how CTTA can identify renal masses and also their subtypes malignancies grade according to Fuhrman grade [57, 58]. Other studies are focusing on lung tumor differentiation, histological correlation with EGFR expression and so on [59, 60].

27.5.3 Legal Aspect of Case-Based Application

The European General Data Protection Regulation (GDPR) outlines specific requirements regarding the use of algorithms in decision-making processes that affect individuals. According to the GDPR, individuals subject to algorithm-based decisions are entitled to receive information on various aspects of the decision, including insight into how the result was generated. While there is ongoing debate concerning the extent and specificity of this transparency in practical applications, the underlying principle affirms the necessity of informed engagement. This expectation applies not only to the subjects of such decisions but also to the professionals deploying the relevant technologies. In this regard, it has been argued that the GDPR presents an opportunity for computer scientists to lead the development of algorithmic systems and evaluation frameworks that promote fairness, avoid discrimination, and facilitate interpretability [61–63]. From a clinical perspective, sound medicolegal practice in radiology increasingly emphasizes the need for both informed consent and adherence to quality guidelines, especially when algorithmic tools are involved [64]. In this context, it is essential to recognize the potential implications of auto-mated decision-making. Artificial intelligence (AI), for example, may produce outputs that exhibit atypical visual characteristics. This is illustrated in Figure 27.7, where the filtered image (B, D, F in Figure 27.7) displays conspicuously flat regions with unusually low attenuation values. In such cases, the system manufacturer would be expected to provide a step-by-step, technically detailed justification as to why the altered image can be considered diagnostically valid when compared with the original, unfiltered version (A, C, E in Figure 27.7). As this field evolves, it is foreseeable that the integration of such technologies will demand significant attention and resources. Yet, it is also worth acknowledging that in some cases, improvements in fundamental hardware reliability may be far more impactful for medicolegal utility in forensic death investigation than overreliance on postprocessing algorithms in the face of the hardware limitations of new developments that may easily not deliver on the hope and promise [65].

27.6 CONCLUSION

Radiomics is a fascinating field that allows one to add significant abstract layers of evaluation to pragmatic and typical visual radiology image reading. However, there are relevant technical pitfalls to look out for and consider.

ACKNOWLEDGMENTS

The authors express their gratitude to Emma Louise Kessler, MD, for her generous donation to the Zürich Institute of Forensic Medicine, University of Zürich, Switzerland. Output of AI tools was critically considered for use in the generation of this chapter (Instatext, Chat Generative Pre-trained Transformer). Figures 27.5 and 27.6 reprinted from [4], copyright 2023, with explicit permission from Elsevier (Elsevier licenses 572360001448 and 5724591327348). Released under CC BY 4.0. While the authors have made efforts to ensure clarity and structure, the development of this manuscript occurred under the constraints of available time and timelines. The authors thank Jakob Heimer for his contributions to this text.

References

[1] Blender Online Community. Blender—a 3D modelling and rendering package. Blender Institute, Amsterdam; 2025. Available from: http://www.blender.org.
[2] Ebert LC, Franckenberg S, Sieberth T, Schweitzer W, Thali M, Ford J, et al. A review of visualization techniques of postmortem computed tomography data for forensic death investigations. International Journal of Legal Medicine. 2021;135(5):1855–1867.
[3] Schweitzer W. swisswuff.ch—Balls (3D) by Wolf Schweitzer [CC BY 4.0]; 2025. Available from: https://www.swisswuff.ch/wordpress/?p=2204.
[4] Schweitzer W, Thali M, Aldomar E. Virtopsy visualisation: Mixed data gradient model for more accurate thin bone visualization in 3D rendering. Forensic Imaging. 2023;33:200529.
[5] Huber NR, Ferrero A, Rajendran K, Baffour F, Glazebrook KN, Diehn FE, et al. Dedicated convolutional neural network for noise reduction in ultra-high-resolution photon-counting detector computed tomography. Physics in Medicine & Biology. 2022;67(17):175014.
[6] Alpert HR, Hillman BJ. Quality and variability in diagnostic radiology. Journal of the Amer-ican College of Radiology. 2004;1(2):127–132.
[7] Lambin P, Rios-Velazquez E, Leijenaar R, Carvalho S, Van Stiphout RG, Granton P, et al. Radiomics: Extracting more information from medical images using advanced feature analysis. European Journal of Cancer. 2012;48(4):441–446.
[8] Heimer J. Statistical methods in radiomics; 2021. Master Thesis, Swiss Federal Institute of Technology Zürich.
[9] Haralick RM. Statistical and structural approaches to texture. Proceedings of the IEEE. 1979;67(5):786–804.
[10] Nailon WH. Texture analysis methods for medical image characterisation. Biomedical Imaging. 2010;75:100.
[11] Pietikainen MK. Texture analysis in machine vision. vol. 40. World Scientific; 2000.
[12] Parekh V, Jacobs MA. Radiomics: A new application from established techniques. Expert Review of Precision Medicine and Drug Development. 2016;1(2):207–226.
[13] Kumar V, Gu Y, Basu S, Berglund A, Eschrich SA, Schabath MB, et al. Radiomics: The process and the challenges. Magnetic Resonance Imaging. 2012;30(9):1234–1248.
[14] Fournier L, Costaridou L, Bidaut L, Michoux N, Lecouvet FE, de Geus-Oei LF, et al. Incorporating radiomics into clinical trials: Expert consensus endorsed by the European Society of Radiology on considerations for data-driven compared to biologically driven quantitative biomarkers. European Radiology. 2021;31:6001–6012.
[15] Mackin D, Fave X, Zhang L, Fried D, Yang J, Taylor B, et al. Measuring CT scanner variability of radiomics features. Investigative Radiology. 2015;50(11):757.
[16] Lu L, Ehmke RC, Schwartz LH, Zhao B. Assessing agreement between radiomic features computed for multiple CT imaging settings. PloS One. 2016;11(12):e0166550.
[17] Kuhn M, Johnson K. Feature engineering and selection: A practical approach for predictive models. CRC Press; 2019.
[18] Lowe J, Matthee M. Requirements of data visualisation tools to analyse big data: A structured literature review. Responsible Design, Implementation and Use of Information and Communication Technology. 2020;12066:469.
[19] Andrews D, Gnanadesikan R, Warner J. Transformations of multivariate data. Biometrics. 1971:825–840.
[20] Greselin F, Ingrassia S, Punzo A. A more informative approach to compare scedasticity under the assumption of multivariate normality. Rapporti di Ricerca del Dipartimento di Metodi Quantitativi per le Scienze Economiche ed Aziendali. 2009:2009.
[21] Almuallim H, Dietterich TG, Citeseer. Learning with many irrelevant features. AAAI 91: Proceedings of the Ninth National Conference on Artificial Intelligence [Anaheim Convention Center, Anaheim, CA, July 14–19, 1991]. 1991;2:547–552.
[22] Gelman A, Loken E. The garden of forking paths: Why multiple comparisons can be a problem, even when there is no "fishing expedition" or "p-hacking" and the research hypothesis was posited ahead of time. Department of Statistics, Columbia University. 2013:348.
[23] Hauser R, Smoliński J, Gos T. The estimation of stature on the basis of measurements of the femur. Forensic Science International. 2005;147(2–3):185–190.
[24] Lubner MG, Smith AD, Sandrasegaran K, Sahani DV, Pickhardt PJ. CT texture analysis: Definitions, applications, biologic correlates, and challenges. Radiographics. 2017;37(5):1483–1503.

[25] Limkin EJ, Reuzé S, Carré A, Sun R, Schernberg A, Alexis A, et al. The complexity of tumor shape, spiculatedness, correlates with tumor radiomic shape features. Scientific Reports. 2019;9(1):4329.
[26] Traverso A, Wee L, Dekker A, Gillies R. Repeatability and reproducibility of radiomic features: A systematic review. International Journal of Radiation Oncology* Biology* Physics. 2018;102(4):1143–1158.
[27] Van Griethuysen JJ, Fedorov A, Parmar C, Hosny A, Aucoin N, Narayan V, et al. Computational radiomics system to decode the radiographic phenotype. Cancer Research. 2017;77(21):e104–e107.
[28] Wang Y, Li C, Yin G, Wang J, Li J, Wang P, et al. Extraction parameter optimized radiomics for neoadjuvant chemotherapy response prognosis in advanced nasopharyngeal carcinoma. Clinical and Translational Radiation Oncology. 2022;33:37–44.
[29] Chakrabarty A, Tan KC. The current state of six sigma application in services. Managing Service Quality: An International Journal. 2007;17(2):194–208. https://doi.org/10.1108/09604520710735191.
[30] Fitzek A, Sperhake J, Edler C, Schröder A, Heinemann A, Heinrich F, et al. Evidence for systematic autopsies in COVID-19 positive deceased: Case report of the first German investigated COVID-19 death. Rechtsmedizin (Berlin, Germany). 2020;30(3):184.
[31] Schweitzer W, Ruder T, Baumeister R, Bolliger S, Thali M, Meixner E, et al. Implications for forensic death investigations from first Swiss postmortem CT in a case of non-hospital treatment with COVID-19. Forensic Imaging. 2020;21:200378.
[32] Wu Q, Wang S, Li L, Qian W, Hu Y, Li L, et al. Radiomics analysis of computed tomography helps predict poor prognostic outcome in COVID-19. Theranostics. 2020;10(16):7231.
[33] Hu Z, Yang Z, Lafata KJ, Yin FF, Wang C. A radiomics-boosted deep-learning model for COVID-19 and non-COVID-19 pneumonia classification using chest x-ray images. Medical Physics. 2022;49(5):3213–3222.
[34] De-Giorgio F, Ciasca G, Fecondo G, Mazzini A, Di Santo R, De Spirito M, et al. Post mortem computed tomography meets radiomics: A case series on fractal analysis of post mortem changes in the brain. International Journal of Legal Medicine. 2022;136(3):719–727.
[35] Klontzas ME, Leventis D, Spanakis K, Karantanas AH, Kranioti EF. Postmortem CT radiomics for the prediction of time since death. European Radiology. 2023:1–9.
[36] Hartley L, Khwaja O, Verity C. Glutaric aciduria type 1 and nonaccidental head injury. Pediatrics. 2001;107(1):174–175.
[37] Bishop FS, Liu JK, McCall TD, Brockmeyer DL. Glutaric aciduria type 1 presenting as bilat-eral subdural hematomas mimicking nonaccidental trauma: Case report and review of the literature. Journal of Neurosurgery: Pediatrics. 2007;106(3):222–226.
[38] Gago LC, Wegner RK, Capone Jr A, Williams GA. Intraretinal hemorrhages and chronic subdural effusions: Glutaric aciduria type 1 can be mistaken for shaken baby syndrome. Retina. 2003;23(5):724–726.
[39] Sudhakar SV, Muthusamy K, Arunachal G, Shroff M. Genomics and radiogenomics in inher-ited neurometabolic disorders–a practical primer for pediatricians. The Indian Journal of Pediatrics. 2019;86:923–938.
[40] Hildebrandt C, Wilson CR, Kritzer A. Standardizing genetic and metabolic consults for non-accidental trauma at a large pediatric academic center. Child Abuse & Neglect. 2022;125:105480.
[41] Kremer C, Sauvageau A. Discrimination of falls and blows in blunt head trauma: Assessment of predictability through combined criteria. Journal of Forensic Sciences. 2009;54(4):923–926.
[42] Bilo RA, Robben SG, Van Rijn RR. Forensic aspects of pediatric fractures: Differentiating accidental trauma from child abuse. Springer Science & Business Media; 2010.
[43] Henriques M, Bonhomme V, Cunha E, Adalian P. Blows or falls? Distinction by random forest classification. Biology. 2023;12(2):206.
[44] Henriques M, Saliba-Serre B, Martrille L, Blum A, Chaumoître K, Donato P, et al. Discrimina-tion between falls and blows from the localization and the number of fractures on computed tomography scans of the skull and the trunk. Forensic Sciences Research. 2023:owad006.
[45] Desinan L, Mazzolo G. Gunshot fatalities: Suicide, homicide or accident? A series of 48 cases. Forensic Science International. 2005;147:S37–S40.
[46] Molina DK, Wood LE, DiMaio VJ. Shotgun wounds: A review of range and location as pertaining to manner of death. The American Journal of Forensic Medicine and Pathology. 2007;28(2):99–102.
[47] Foote CW, Doan XL, Vanier C, Cruz B, Sarani B, Palacio CH. Suicide versus homicide firearm injury patterns on trauma systems in a study of the National Trauma Data Bank (NTDB). Scientific Reports. 2022;12(1):15672.
[48] de la Rosa E, Sima DM, Vyvere TV, Kirschke JS, Menze B. A radiomics approach to trau-matic brain injury prediction in CT scans. In: 2019 IEEE 16th International Symposium on Biomedical Imaging (ISBI 2019). IEEE; 2019. pp. 732–735.
[49] Zhang L, Zhuang Q, Wu G, Yu J, Shi Z, Yuan Q, et al. Combined radiomics model for prediction of hematoma progression and clinical outcome of cerebral contusions in traumatic brain injury. Neurocritical Care. 2021:1–11.
[50] Röhrich E, Thali M, Schweitzer W. Skin injury model classification based on shape vector analysis. BMC Medical Imaging. 2012;12(1):1–12.
[51] Schweitzer W, Thali M, Giugni G, Winklhofer S. Postmortem pulmonary CT in hypothermia. Forensic Science, Medicine, and Pathology. 2014;10(4):557–569.
[52] Schweitzer W, Thali M. Fatal obstructive asphyxia: Transpulmonary density gradient characteristic as relevant identifier in postmortem CT. Journal of Forensic Radiology and Imaging. 2019;19:100337.
[53] Ebert LC, Heimer J, Schweitzer W, Sieberth T, Leipner A, Thali M, et al. Automatic detection of hemorrhagic pericardial effusion on PMCT using deep learning-a feasibility study. Forensic Science, Medicine and Pathology. 2017;13(4):426–431.

[54] Sandrasegaran K, Lin Y, Asare-Sawiri M, Taiyini T, Tann M. CT texture analysis of pancreatic cancer. European Radiology. 2019;29(3):1067–1073.
[55] Li M, Fu S, Zhu Y, Liu Z, Chen S, Lu L, et al. Computed tomography texture analysis to facilitate therapeutic decision making in hepatocellular carcinoma. Oncotarget. 2016;7(11):13248.
[56] Fu S, Chen S, Liang C, Liu Z, Zhu Y, Li Y, et al. Texture analysis of intermediate-advanced hepatocellular carcinoma: Prognosis and patients' selection of transcatheter arterial chemoembolization and sorafenib. Oncotarget. 2017;8(23):37855.
[57] Deng Y, Soule E, Samuel A, Shah S, Cui E, Asare-Sawiri M, et al. CT texture analysis in the differentiation of major renal cell carcinoma subtypes and correlation with Fuhrman grade. European Radiology. 2019;29(12):6922–6929.
[58] Bhandari A, Ibrahim M, Sharma C, Liong R, Gustafson S, Prior M. CT-based radiomics for differentiating renal tumours: A systematic review. Abdominal Radiology. 2021;46(5):2052–2063.
[59] Ganeshan B, Goh V, Mandeville HC, Ng QS, Hoskin PJ, Miles KA. Non–small cell lung cancer: Histopathologic correlates for texture parameters at CT. Radiology. 2013;266(1):326–336.
[60] Xiong J, Li X, Lu L, Schwartz LH, Fu X, Zhao J, et al. Implementation strategy of a CNN model affects the performance of CT assessment of EGFR mutation status in lung cancer patients. IEEE Access. 2019;7:64583–64591.
[61] Brkan M. Do algorithms rule the world? Algorithmic decision-making and data protection in the framework of the GDPR and beyond. International Journal of Law and Information Technology. 2019;27(2):91–121.
[62] Hacker P. Teaching fairness to artificial intelligence: Existing and novel strategies against algorithmic discrimination under EU law. Common Market Law Review. 2018;55(4).
[63] Goodman B, Flaxman S. European Union regulations on algorithmic decision-making and a "right to explanation". AI Magazine. 2017;38(3):50–57.
[64] Deigert F, Cherewick T, Gunn W. "Lawsuit-proof" your radiation oncology practice. Oncology Issues. 2009;24(6):19–22.
[65] Willemink MJ, Persson M, Pourmorteza A, Pelc NJ, Fleischmann D. Photoncounting CT: Technical principles and clinical prospects. Radiology. 2018;289(2):293–312.

PART 3

TRAUMA DIAGNOSIS

CHAPTER 28

FORENSIC IMAGING OF MECHANICAL TRAUMA

Wolf Schweitzer, Carlo Tappero, and Garyfalia Ampanozi

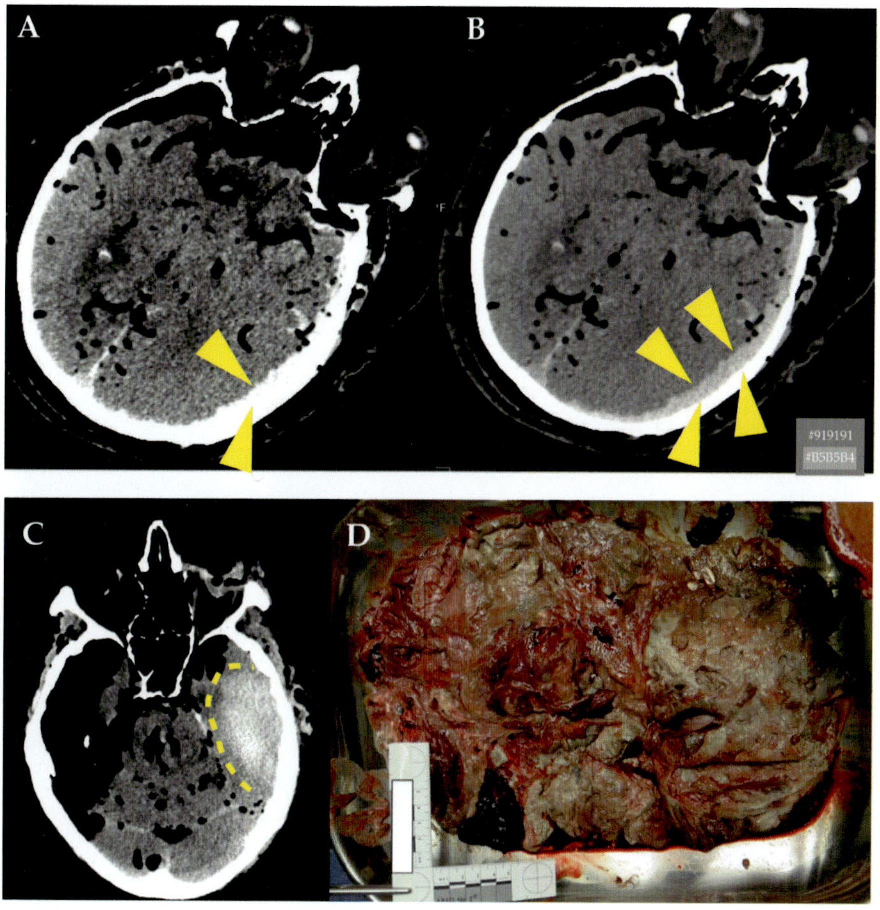

Figure 28.1 PMCT can be helpful in identifying and localizing brain hemorrhages, particularly in cases of putrefaction. PMCT can be helpful in identifying and localizing brain hemorrhages, particularly in cases of putrefaction. Proper windowing is paramount for PMCT reading. Images A and B show a subdural hematoma (yellow arrows) before (A) and after (B) applying windowing commands when using the image viewing software. These commands provide results adequate for discerning the CT density of brain from blood, even before proper alignment of the head within the CT viewing software. To demonstrate the difficulty of diagnosing and properly localizing intracranial hemorrhages in putrefied bodies during autopsy, consider PMCT in C, which shows a large intracranial hemorrhage (yellow dashed outline). Making this diagnosis may be far more difficult even during, and certainly after, removal of the liquefied brain (D). Given the blurred margins of the PMCT hemorrhage correlate and the brain level, it must be considered that the blood may have altered position during putrefaction due to gravity. Image re-used from [1] (Open Access, CC BY 4.0 Deed—Attribution 4.0 International—Creative Commons).

DOI: 10.4324/9781003413998-31

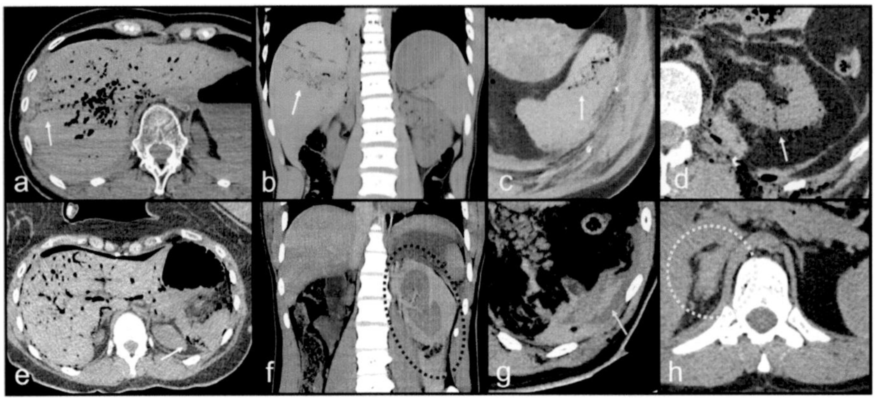

Figure 28.2 Use of PMCT to identify abdominal hemorrhages. PMCT multiplanar reconstructions (MPR) in different cases with true positive abnormal findings. PMCT axial slice through the abdomen (a); true positive abnormal gas distribution not following the vascular tree within the liver. PMCT coronal slice through the abdomen (b); true positive abnormal gas distribution within the liver not following the vascular tree. PMCT axial slices through the spleen (c) and left kidney (d); true positive abnormal gas distributions within a spleen and a kidney. PMCT axial slice through the abdomen (e); true positive abnormal gas distribution within the spleen. Note the prominent but normally distributed gas in the liver within the vascular tree. Liver was true negative in this case. PMCT coronal slice through the abdomen (f); extended parenchymal disruption of the left kidney with sentinel clot sign. PMCT axial slice through the spleen (g); sentinel clot sign. PMCT axial slice through the right adrenal gland (h); blood around the right adrenal gland (note the difference with the left one) as true positive sign on PMCT. Image re-used from [2], which was published as Open Access with a CC BY 4.0 license. Deed—Attribution 4.0 International—Creative Commons.

28.1 INTRODUCTION

Mechanical trauma occurs when an external force damages, injures, or alters the body, including organs, bones, and other tissues. It is an interaction that leaves a discernible mark, such as a bruise or a break, caused by an external agent. This can be direct (e.g., a punch, impact, perforation, stab, or shot) or indirect (e.g., a radiating fracture or contrecoup brain injury caused by inertia). Subsequent complications or vital reactions may include hemorrhage, embolism of fat, thrombi, foreign bodies or gas/emphysema, inflammation or infection. Foreign bodies, such as fragmented or intact projectiles from a firearm or gunshot residue, can be relevant as they may be linked to the weapon used [4, 5].

In forensic medicine, mechanical trauma is categorized, with each category tending to exhibit specific characteristics. These categories represent the medicolegal textbook approach to clear classification.

Sharp Force Trauma: This occurs when a relatively sharp edge slices through skin, tissues, bones, organs, or vasculature. Knives, axes, and broken glass can cause this type of damage. This is considered a focused force, in contrast to the more diffused nature of blunt trauma; however, this distinction is relative. This trauma can sever tissues and blood vessels, leading to bleeding. Cuts may be clean or roughly hacked, potentially resulting in bone fragments.

Gunshot Trauma: This is when a relatively small projectile pierces the body at a relatively high velocity. A projectile may enter with a focused, relatively intense energy. Depending on specifics, the entrance wound may be small, with injured bone mostly but not always pushed inwards, possibly creating beveling. If the projectile has enough energy, it passes through the body, thus creating an exit wound, and also here, bone beveling may be seen. The wound track may be inferred from looking at entrance and exit wounds and any injuries in between. Temporary cavitation may cause injuries in some distance from an actual trajectory. A projectile may cause a relatively straight trajectory, or it may also be deflected and then cause a trajectory that is not straight [6, 7]. A number of studies have considered the diagnostic value of postmortem imaging in gunshot injuries [8–11].

Blunt Force Trauma: This is when a relatively wide, dull surface hits the body. One thinks of a fist, a club, a rock, as object used as a weapon, but also a collision of a person with inside (e.g., as driver or passenger) or outside (e.g., as pedestrian or bicycle rider or motorbike rider) car, bus, train or truck structures, or a collision with a ground or floor structure (fall from standing position, fall from great height). It is the kind of force that may crush and deform. Relatively to a gunshot, blast or explosive trauma, blunt trauma tends to be a slower kind of damage, a force pressing down on the bone, allowing more time for it to bend before it breaks. One might see bruises, scrapes, dislocation, and fractures. There may be foreign materials transferred onto or into the body as a consequence [12]. The fractures can be depressed, meaning the bone was pushed inward. Sometimes, the bone fragments will chip at the edges as they rub against each other, also

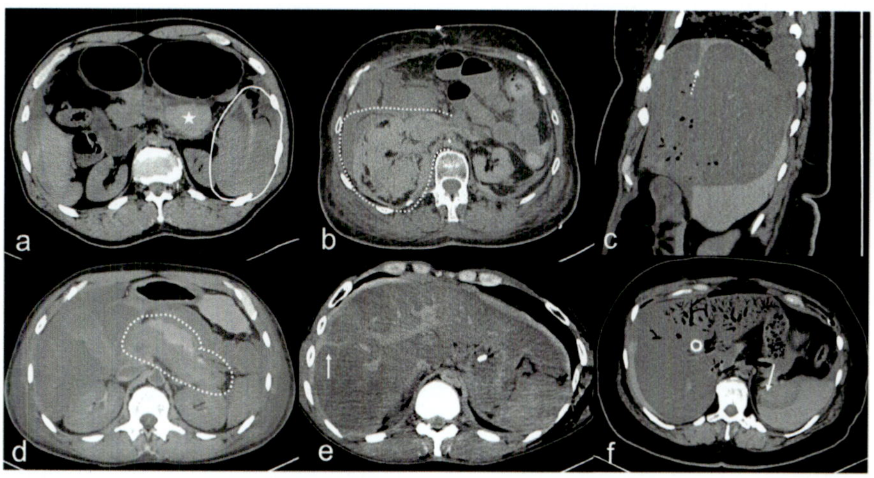

Figure 28.3 Use of PMCT to identify abdominal hemorrhages. PMCT multiplanar reconstructions (MPR) in different cases with false positive (a–d) and one case with true positive (e) abnormal findings. a PMCT axial slice through the abdomen; false positive sedimented blood around the pancreas (asterisk) in a case of spontaneous splenic artery rupture (white line). b PMCT axial slice through the abdomen; false positive sentinel clot for the right kidney in a case of aortic rupture. c PMCT sagittal slice through the abdomen of the same case with f (details in f). d PMCT axial slice through the abdomen; false positive sentinel clot sign around the pancreas, in a case where actually the liver and the spleen were injured. e PMCT axial slice through the abdomen of a case with true positive parenchymal disruption of the right liver lobe and sentinel clot sign around the liver. f PMCT-axial slice through the abdomen of a case with false positive parenchymal disruption of the liver (misinterpretation of a hepatic groove—dashed arrow in c). Note the true positive parenchymal disruption of the spleen (white arrow in f) with sedimented blood around the organ. Image re-used from [2], which was published as Open Access with a CC BY 4.0 license. Deed—Attribution 4.0 International—Creative Commons.

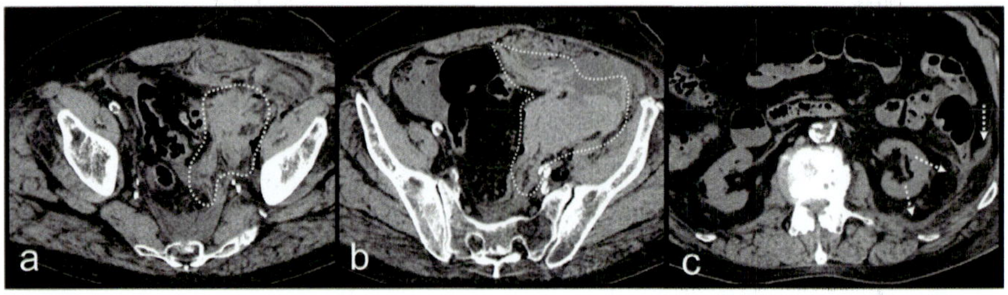

Figure 28.4 Use of PMCT to identify abdominal hemorrhages. PMCT of a case with true positive findings for acute rupture of the left common femoral artery PMCT axial slices through the pelvis (a, b) and the retroperitoneal spaces at renal level (c). Note the accumulation of hyperdense fluid at the level of the artery in the pelvis (a and b), whereas the expanded fluid in the retroperitoneal spaces (c) is visually more hypodense. Image re-used from [2], which was published as Open Access with a CC BY 4.0 license. Deed—Attribution 4.0 International—Creative Commons

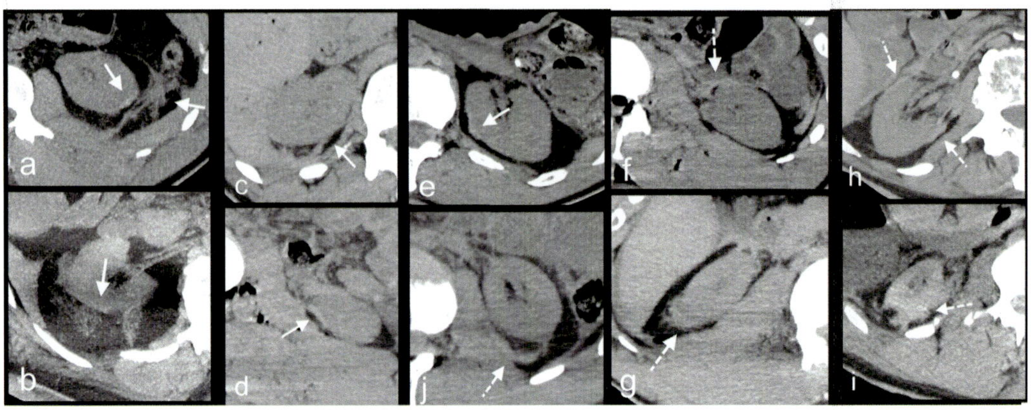

Figure 28.5 Use of PMCT to identify abdominal hemorrhages. a–e PMCT axial slices through true positive findings of kidney ruptures (true positive findings with white arrows). f–j PMCT axial slices through false positive findings of suspected kidney ruptures (false positive findings with white dashed arrows). Image re-used from [2], which was published as Open Access with a CC BY 4.0 license. Deed—Attribution 4.0 International—Creative Commons.

from repeated hits. Blunt trauma is a wide category; it is not a strict term, however much it is set off against sharp force trauma, explosions or blasts, and gunshot trauma.

Blast and Explosive Trauma: Types or categories of mechanical trauma, though they may overlap in any particular given case or stand out, exhibiting specific diagnostic criteria in any other given case, all also have their own characteristics. In one particular example that highlights the difficulty in assigning a textbook trauma category to a real case, one may start by acknowledging that a razor blade may be relatively blunt with regard to its edge or edges, as it may have been never what one calls razor sharp or not all that sharp any longer, and, when moved over skin, no cuts may be caused, just abrasions, as when scraping with the face across a rough surface. Looking at such an injury without being informed of a razor having been moved across that skin surface, one might then be tempted to guess that blunt trauma was behind that injury when in fact a razor blade, which would be, by convention, termed sharp force injury, caused it. That is not to say these categories are entirely useless: as another example, no blunt object such as a perfectly spherical marble of e.g. 3 cm diameter has been seen to reliably cause a slit-like sharp force injury cleanly penetrating e.g. a rib, whereas a thin sharp metallic knife blade may be much better able to do so. Despite the appearance of an existence of overlaps and vague boundaries, not all seems lost with regard to simple textbook categories of trauma.

In a more modern understanding of reconstructing trauma, the distinctions of blunt and sharp force trauma may be seen as residing on a metric or morphological continuum [13]. This may become apparent when juxta positioning various penetrating skull fractures as in Figures 28.11 or 28.8.

28.1.1 Limitations of PMCT in Diagnosing and Documenting Mechanical Trauma in PMCT

Clearly, PMCT has strengths and weaknesses when comparing it to other methods of examining a body (e.g. [14–16]).

Reliable postmortem CT trauma diagnostics hinge on carefully optimized image acquisition and interpretation. Detecting subtle injuries demands low image noise, high contrast and adequate spatial resolution. On helical CT scanners this is best

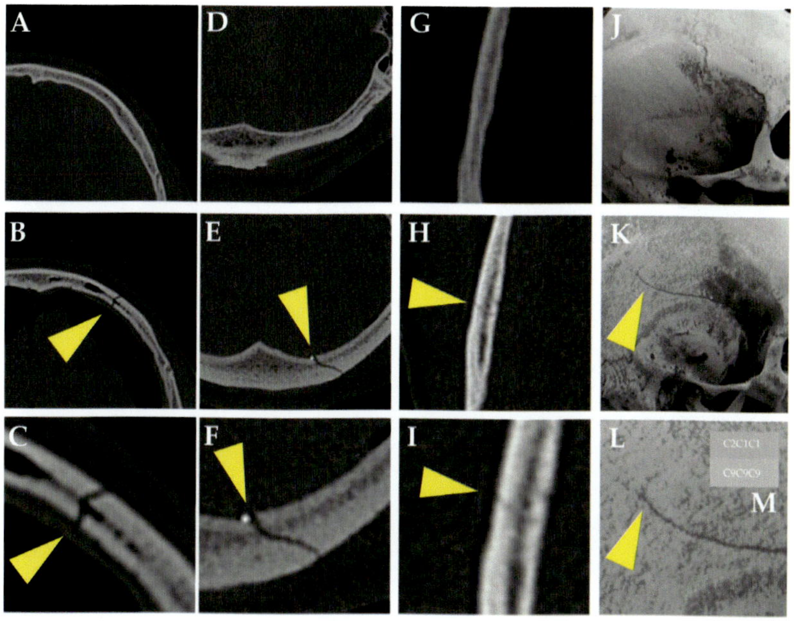

Figure 28.6 Relevance of PMCT to investigate skull fractures. Fracture identification on PMCT may be done on axial (or sagittal, coronal or MPR) slices. Here, we compare a non fractured (top row, included/shown for reference) and a fractured (middle row, detail of findings in middle row: bottom row) skull with regard to burst fracture lines. The technical aspect to consider when viewing the data is to first sufficiently enlarge the anatomical area on the screen and then to check for gray value or density absences. In order to be able to do that, proper windowing is required. While some fracture lines may be seen in any overview without difficulty (e.g., B, E), thinner fracture lines (e.g. H) may benefit from using zoom function (as in I). As a consequence of a firearm injury, the fracture shown in E and F also contains an adjacent particle of high CT density. The thin fracture line in H and I also is visible in the 3D rendering of that same finding (K and L). When comparing the findings with a normal skull (middle and bottom versus top row), the findings may be easier to discern. M: if these two gray areas are similar, wherever you read this, then you may not be able to see the fracture line in H and I well.

achieved with minimal pitch, maximal tube current time product (mAs), and maximal rotation time; settings around a 1 s rotation, pitch ≈ 0.35 and 1200–1800 mAs routinely outperform low-dose clinical protocols by pushing soft-tissue contrast toward that of T1-weighted MRI [17]. Recent photon-counting scanners may not yet match the speed or dose robustness of conventional energy-integrating systems under these high-load conditions, and many hospital scanners may lack the tube output or cooling capacity for sustained very-high-dose operation. Consequently, every CT scanner intended for high-quality forensic work should undergo cautious protocol and phantom testing before considering it for routine use.

When the hardware and scan protocols are appropriate, PMCT may be relatively good at visualizing skeletal fractures, foreign bodies, airway injuries, intra- and extra-vascular gas, and a spectrum of vascular lesions [18]. Pulmonary CT patterns potentially identifying obstructive asphyxia remain under investigation, but conventional autopsy is not perfect here either [19, 20]. Intravenous or intraluminal contrast media can outline wound tracks and vascular breaches [21]. Higher-grade pulmonary fat embolism is only rarely recognizable in CT, as are fat emboli in limb veins [22, 23] and thromboembolism may always require autopsy for validation [24].

PMCT's main weakness is a soft-tissue contrast. Even very high-dose acquisitions do not match the naked-eye or microscopic resolution of an autopsy, especially for parenchymal lacerations, cardiac wounds, stab or bullet paths where hemorrhage is absent because circulation had already ceased [18]. The method therefore may simultaneously surpass and fall short of autopsy morphology, albeit in different ways [14–16]. How far investigators push beyond simply ruling out legally relevant third party involvement in any given death examination depends on legal mandate, resources and also on communication between pathologists, police and judiciary [25].

If PMCT leaves the cause of death or other relevant aspects uncertain, a medicolegal autopsy with ancillary studies is definitely recommended; conversely, in clear-cut cases, further dissection may justifiably be omitted [14]. Many aspects of PMCT practice and its research base within this text will, however, at least for time being, remain incomplete, untouched, or not mentioned for valid reasons [26, 27].

28.1.2 Goal of This Text

The goal of this text is not to reproduce all of forensic medicine or forensic pathology and thus not an attempt to summarize years of board certification related training and relevant experience [28]. Instead, we approach what appear to be specific imaging aspects of forensic examination of mechanical trauma.

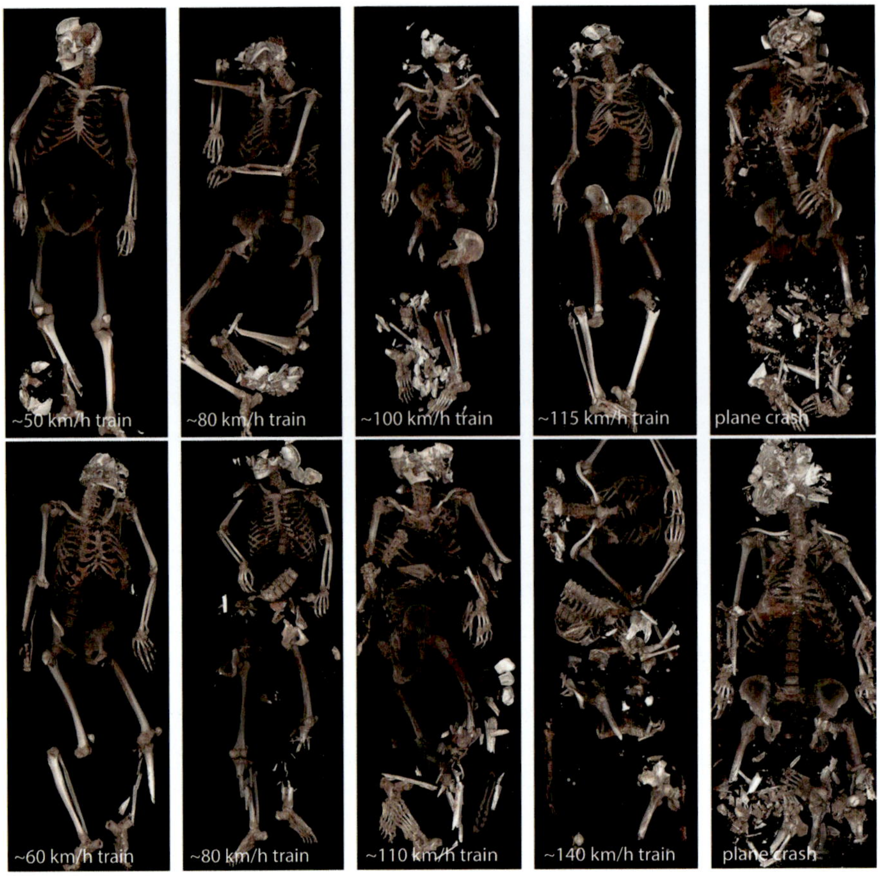

Figure 28.7 Relevance of PMCT to investigate skeletal fracture patterns. High-grade trauma and body fragmentation—The images show, from left to right, an apparent progression of lower collision speeds to higher collision speeds of a vehicular structure against a body. On the left are cases where a person had themselves be run over by a train (starting from about 50 km/h, see image labels), progressing towards the right to pedestrian/train collisions up to about 140 km/h; the rightmost images show plane crash victims. Here, the fracture pattern is of relevance, whereas identifying the fractures appears to pose no particular problems. That impression, on occasion, may be misleading, particularly when the forensic question is that of delineating other forms of trauma from the apparent high grade blunt trauma [3]. Deed—Attribution 4.0 International—Creative Commons.

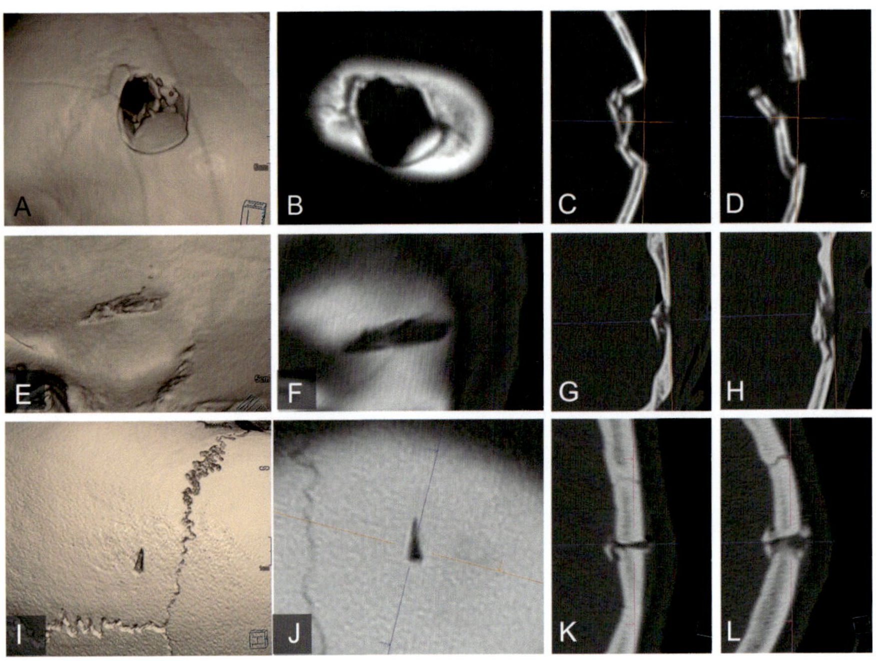

Figure 28.8 Relevance of PMCT in investigating bone defect fracture shapes. Series of penetrating skull fractures.—Top row: impalement with a blunt object. Middle row: knife. Bottom row: billhook.—The relative aspects of blunt vs. sharp can be viewed as ranging on a metric continuum.

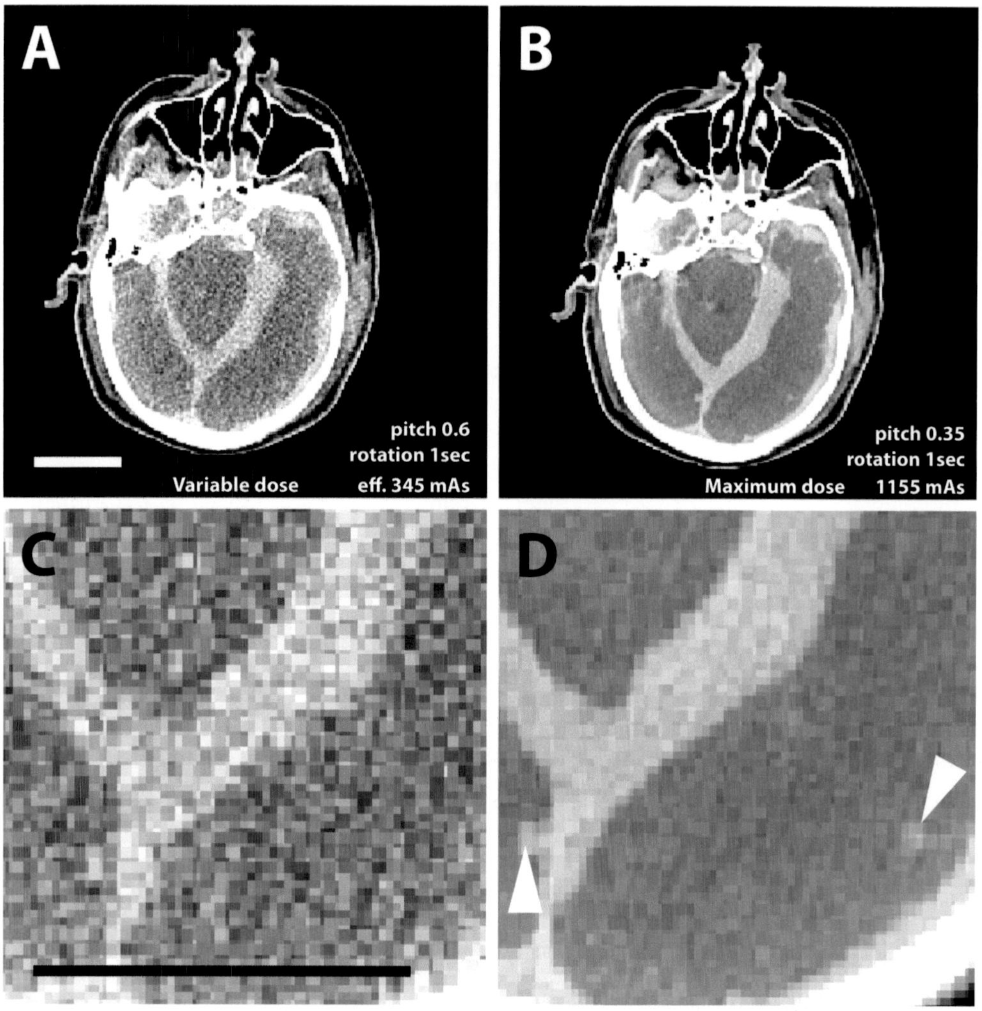

Figure 28.9 Relevance of CT data quality for brain hemorrhage identification. CT of the brain at variable dose (overview: A, detail: C, variable dose setting (CareDose, Siemens, Erlangen, Germany) with an effective dose of 345 mAs, spiral pitch factor (ratio of the table increment over the detector collimation) of 0.6, rotation time 1 sec, tube voltage of 120 kVp.) allows one to easily delineate a subdural hemorrhage. The same scanner was used at higher dose (overview: B, detail: D—fixed CT dose of 1155 mAs, spiral pitch factor (ratio of the table increment over the detector collimation) of 0.35, rotation time 1 sec, tube voltage 120 kVp) and a number of additional regions exhibiting blood density become apparent: not only does it appear to be a lot easier to ascertain left temporal subdural and subarachnoidal hemorrhage, also small areas of intracerebral hemorrhage (see D: arrows) can now be discerned better. Bar 4cm.

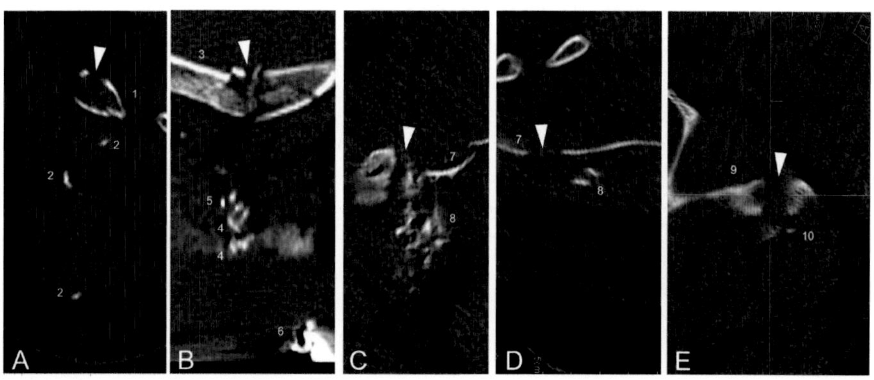

Figure 28.10 Relevance of PMCT to investigate the direction of a shot in gunshot injury cases (A–E) from top to bottom. Pistol, caliber 7.65 mm (A, B); pistol, caliber 9 mm (C–E). When a bone such as a rib (A, B) or scapula (C–E) is hit (region approximated by arrow), bone fragments (2, 4, 8, 10) and even fragments of metal density (5) may be identified along the track as it extends, including the remainder of the projectile (6), thus indicating the direction of the shot. Bones: 1 rib, 3 rib, 7 scapula, 9 scapula.

28.2 MORPHOLOGICAL ELEMENTS OF MECHANICAL TRAUMA

The technical specifics of trauma identification using PMCT data as they relate to forensic medicine or pathology usually can be addressed by considering the detection of small, hard to see findings.

Once these can be identified, putting the remaining findings along the morphological spectrum in place appears to be mostly a matter of practicing forensic medicine.

28.2.1 Identifying Traumatic Hemorrhage

Small traumatic hemorrhages may benefit from good soft tissue contrast with low noise (Figure 28.9). Better specification of suspicious findings on native PMCT may be achieved with MRI [29–32]. Overall, diagnosis requires high soft tissue contrast with low noise, optimal use of the CT scanner's resolution and proper use of CT viewing software (see also Figure 28.1).

In general, PMCT, possibly enhanced with contrast staining of the vasculature [33], but also without extra contrast, appears to be relatively good at identifying traumatic intracranial hemorrhage [34].

It is particularly helpful in cases of putrefaction, where it may provide useful diagnostic documentation also when compared to autopsy [1, 29]. Figure 28.1 demonstrates a case of putrefaction and subdural hemorrhage. There, it may become clear just how thinner subdural bleeds may get missed [34], which, however, may in part be attributed to CT viewing technique. At any rate, putrefaction does constitute a significant technical challenge for proper autopsy of brain related pathology and trauma, also of intracranial hemorrhage [35].

Fatal acute thoracic and abdominal hemorrhage [36], and to some degree, gastrointestinal hemorrhage [37]), may be detected with PMCT, particularly if the volume of blood is sufficiently large.

Perihepatic blood may be difficult to identify, particularly if the amount is small. Sentinel clots that highlight splenic injury may have to be delineated against normal finding which can be a challenge when there is no or only little fat surrounding the spleen. Abnormal gas distribution within solid organs and combinations of PMCT signs (e.g., sentinel clot plus parenchymal disruption) may offer the a higher specificity and sensitivity for identifying true bleeding sources, particularly in short survival intervals. The sentinel clot sign is highly specific but prone to false positives—especially around the spleen and right kidney—when the actual source is else-where (often the liver), and it is more common in surgical or natural deaths with survival beyond one hour. Be cautious of sedimented or surrounding blood alone as a marker; it is frequent in post-resuscitation changes and shows lower diagnostic accuracy without corroborating signs [2]. See Figures 28.2, 28.3, 28.4, and 28.5 for detailed illustration of PMCT appearances in relation to trauma with abdominal hemorrhages.

28.2.2 Identifying Gas

Fatal gas embolism may be detected using PMCT, particularly when the body is not putrefied/decomposed [38–40]. Assessing gas and gas distribution may help shaping a path to PMCT interpretation and subsequent workup in instances with what one may believe to be significant gas particularly in the right heart and pulmonary arteries [41].

Gas in tissues such as in soft tissue emphysema may highlight a direct sequel of relevant trauma, also in conjunction with laryngeal fractures [42] but also penetrating wounds. Gas in the abdominal cavity may indicate intestinal perforation.

28.2.3 Identifying Non Biological Materials

Identification of materials such as (but not restricted to) glass, metal, including projectile fragments or gunshot residue, and documentation and interpretation with regard to their presence or distribution has been a mainstay of forensic PMCT interpretation also in the context of trauma (e.g., [4, 5, 12, 43, 44]).

Technically, any analysis and documentation bases on the CT density difference of such materials compared to human tissues. This basically means that CT viewing software has to be used accordingly.

28.2.4 Identifying Bone Fractures

Fracture identification: A practical issue may be that of identifying fractures. That is obviously not a particular difficulty once fractures are sufficiently large or displaced, such as evidenced in Figure 28.7.

In order to approach the arguably more difficult to identify subtle thin fractures, consider Figure 28.6. There, we placed relatively subtle typical trauma findings against similar anatomical regions of a non fractured skull for better comparison.

There are limitations concerning resolution of clinical CT scanners typically used for PMCT. Correct diagnosis of subtle fractures of small structures such as laryngeal fractures will need a higher resolution, whereas sufficiently resolving small bones of around 1 to 2 mm width such as the hyoid bone or thyroid cartilage will typically require a resolution of around 0.05 mm for a reliable diagnosis which requires a device such as a micro CT scanner, whereas a clinical photon counting CT scanner is not expected to yield such [45–47].

Fragments: The presence of a hard to detect bone fracture as such is one aspect of skeletal injury. Bone fragments and their distribution may be another aspect that has the capacity to supplement autopsy. PMCT can be relatively good at exhibiting the directional relation of fracture, bone fragment distribution or scatter, subtle soft tissue hemorrhage, and thus interpreting findings, e.g. as to shot direction [48]. Figure 28.10 shows examples of fragments in relation to shot direction, but consider also Figure 28.14 with scattered bone fragments.

Specific fracture aspects: Fracture aspects that are typical for forensic pathology, such as, for example, related to firearm shot direction (Figure 28.10) or morphology of penetrating skull fractures across blunt and sharp trauma, neurosurgery and firearm injuries (Figure 28.8 and 28.11) may be sought and identified, diagnosed and documented using PMCT.

Applying the general principle elsewhere: In the same way as explained above, the software allows one to identify and demonstrate any other skeletal fractures, across shoulder, spinal, rib, pelvic or extremity findings. In other words, the path to that type of morphology documentation lies along the similar technical underlying principles demonstrated here. This may be illustrated and expanded in a later version of this text.

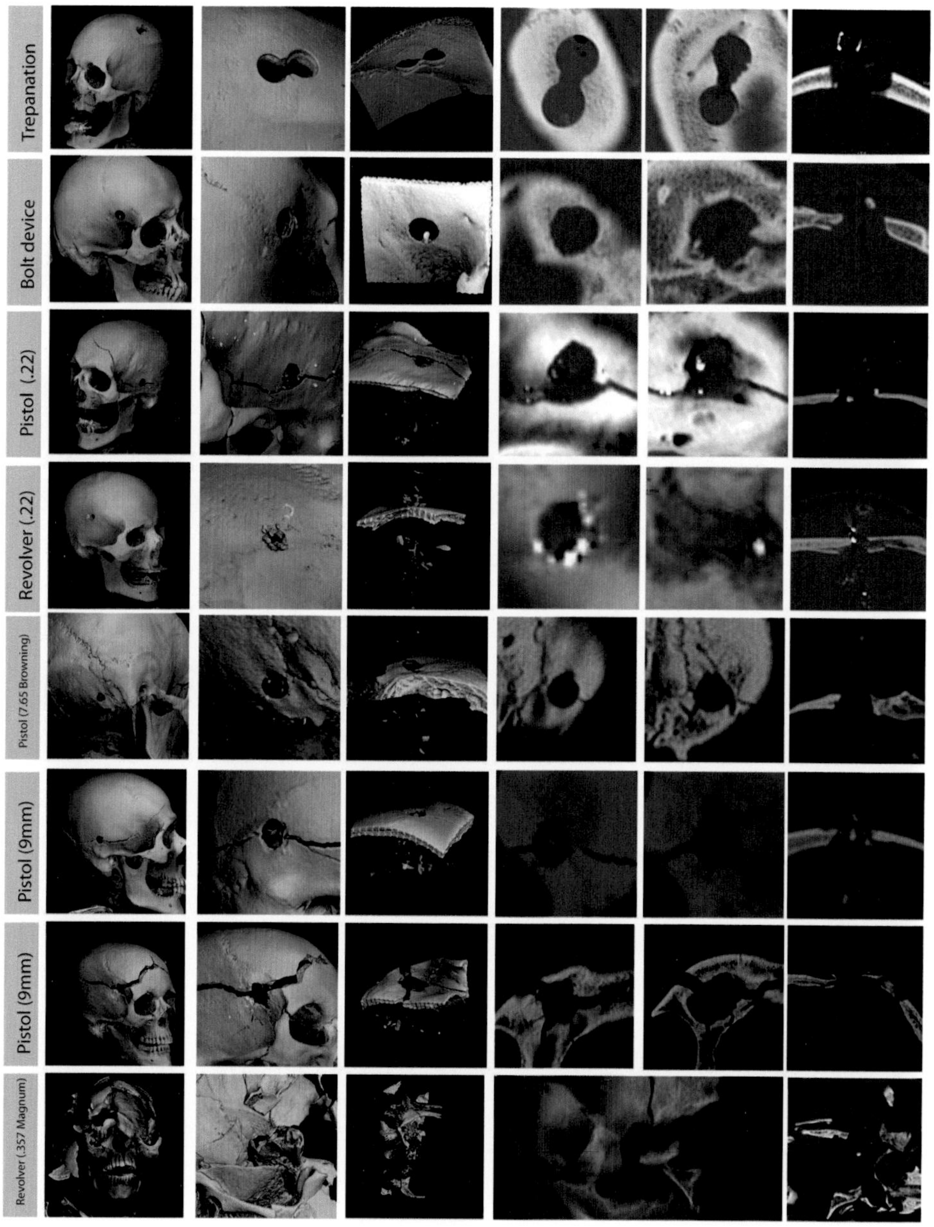

Figure 28.11 Relevance of PMCT in investigating bone defect fracture shapes. First row: neurosurgical trepanation. Second row: stun gun/bolt device. Third row: pistol caliber.22. Fourth row: revolver caliber.22. Fifth row: pistol caliber 9mm. Sixth and seventh row: pistol caliber 9 mm. Eighth row: revolver caliber.357 magnum. The image series illustrates a side-by-side positioning of roundish skull perforations.

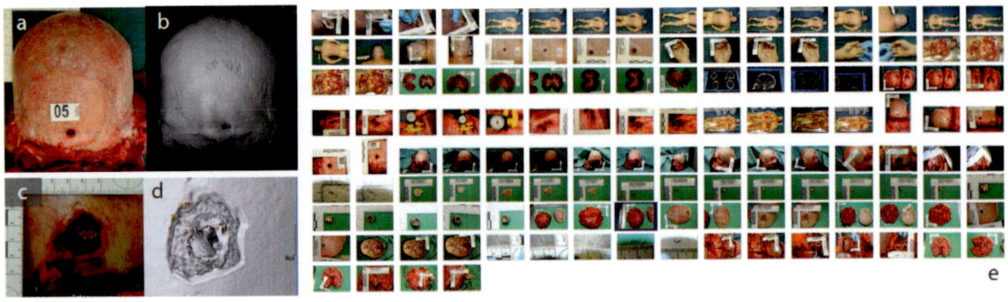

Figure 28.12 Relevance of PMCT in allowing for continuous overlap and redundancy in 3D. Gunshot injury to the head, autopsy photos. While the photos, in this case of a fatal pistol shot to the back of the head, are numerous and apparently complete, none provide an actual visual understanding of the way the shot was carried out.

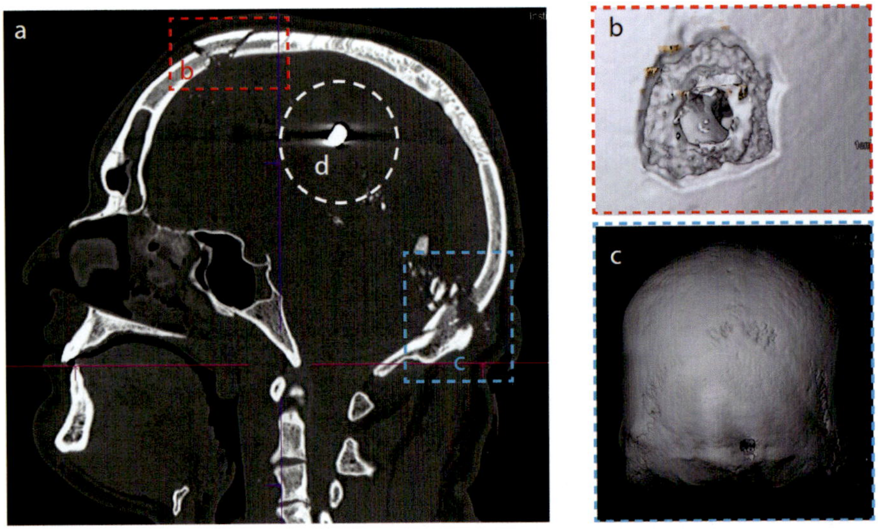

Figure 28.13 Relevance of PMCT in allowing for continuous overlap and redundancy in 3D: gunshot injury to the head. With a single adequately placed MPR image in this case (a), entrance wound to the skull (c), fracture due to ricochet (b), and final position of projectile (d) are shown.

28.3 SIGNIFICANCE OF PMCT SCANNING FOR FORENSIC MEDICINE CASE DOCUMENTATION

While the inclusion of PMCT scanning brings about a range of known advantages in forensic medicine, two seem to stand out as particularly significant. One is that redundant overlapping or supplementing visuals may be used in the attempts to deliver a better understanding of what appears to be relevant morphology. The other one is that PMCT can be used to validate autopsy findings.

28.3.1 Providing Redundant Visuals for Better Understanding

In order to render, deliver, convey, transmit, or give reasonably understandable insight into what constitutes the relevant aspects of the work delivered, when documenting a death from medicolegal perspective, the PMCT images allow to reach or at least accompany semantic, morphological and narrative clarity in what appears to be a relevant form (see Figures 28.12, 28.13, and 28.14).

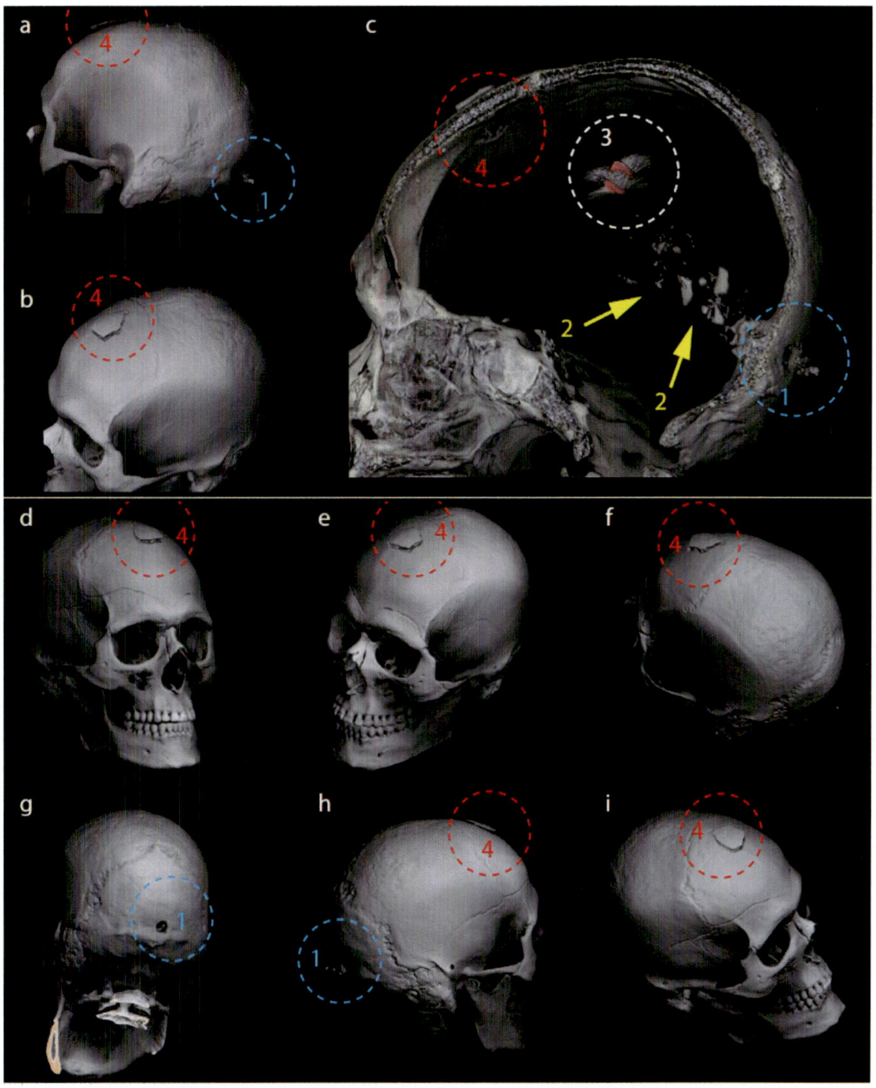

Figure 28.14 Relevance of PMCT in allowing for continuous visual overlap and redundancy in 3D:gunshot injury to the head. Forensic medicine typically tries to understand injury patterns or constellations that are a combination of single features, aspects or details. Is a head injury caused by the shot from a firearm? Was there preceding blunt trauma due to blows or a fall? Partly redundant views (a through i, labels: 1 skull entrance wound, 2 bone fragments, 3 projectile, 4 fracture due to ricochet) help the viewer understand the combination of absence and presence of findings, enabling them to also reach an own conclusion.

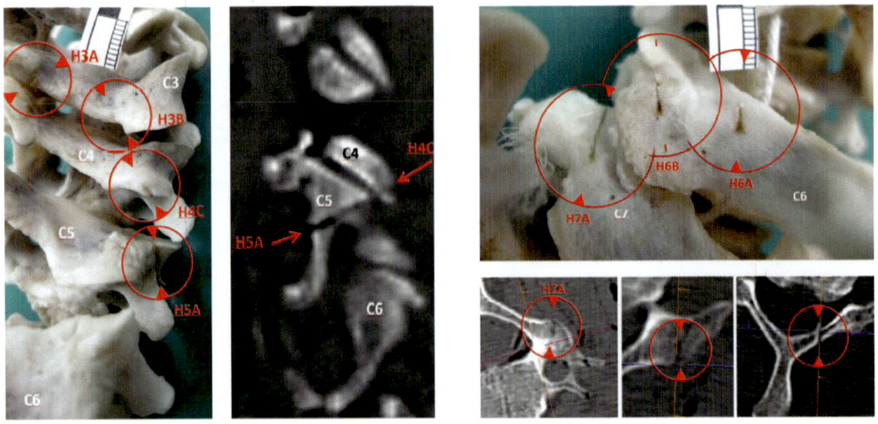

Figure 28.15 Relevance of PMCT in showing cervical spine injuries to have been present before autopsy. When arguing that cervical spine injuries may be attributed to knife stabs, the prior removal of that cervical spine with a knife during autopsy may raise the question of just how sure one can be that the cervical spine was not injured then. The above images show photos of a cervical spine after removal and after maceration, exhibiting a range of injuries (labels: H3A, H3B, H4C, H5A, H6A, H7A, H7B) of a range of cervical vertebrae (labels: C3 through C7). Side by side documentation of the macroscopic findings next to the PMCT data that was obtained before autopsy helps to answer such a question. In this case, numerous cervical spine injuries were documented over 24 pages for clarification and documentation.

Redundant and partially overlapping images in forensic case documentation, particularly in trauma-focused forensic postmortem imaging may, depending on case specifics, provide a useful basis when approaching a reconstructive understanding. A single image can be informative and relevant, but sometimes the single image remains a claim. Redundant images may help test that claim for coherence. Such a principle of presenting partly redundant images was used in Figures 28.6 and 28.11.

28.3.2 Validating Autopsy Findings through preceding Touch-Free 3D Documentation

When a case goes to autopsy, dissection will cause cuts that then may raise question as to whether a tear or cut, stab or similar injury or defect could be ascertained to have resulted from prior injury. Comparing PMCT and macroscopic finding as shown in Figure 28.15 may then help to provide a basis to ascertain that an injury inside a deceased person was present before autopsy was started.

28.4 SUMMARY

Mechanical trauma may cause PMCT to capture relevant morphology. Resulting data could then be used to identify, document, and demonstrate or show visual representation of what appear to be relevant findings. The choice of CT scanner and protocols may have a significant impact on resolution, contrast and resulting cost versus benefit.

ACKNOWLEDGMENTS

The authors express their gratitude to Emma Louise Kessler, MD, for her generous donation to the Zürich Institute of Forensic Medicine, University of Zürich, Switzerland. Output of AI tools was critically considered for use in the generation of this chapter (e.g., Instatext, Chat Generative Pre-trained Transformer). Figures 28.1 re-used from [1], Open Access, CC BY 4.0. Figures 28.2, 28.3, 28.4 and 28.5 re-used from [2], Open Access, CC BY 4.0 license. Released under CC BY 4.0. While the authors have made efforts to ensure clarity and structure, the development of this manuscript occurred under the constraints of available time and timelines. The retrospective use of anonymized data was submitted to and evaluated by the Ethics Committee of the Canton of Zürich (reference number 15–0686).

References

[1] Tappero C, Thali MJ, Schweitzer W. The possibility of identifying brain hemorrhage in putrefied bodies with PMCT. Forensic Science, Medicine and Pathology. 2020;16(4):571–576.

[2] Chatzaraki V, Thali MJ, Ampanozi G. Diagnostic accuracy of postmortem computed to- mography for bleeding source determination in cases with hemoperitoneum. International Journal of Legal Medicine. 2021;135:593–603.

[3] Schweitzer W, Thali MJ, Ampanozi G. Planned complex suicide combining pistol head shot and train suicide and Virtopsy examination. Forensic Imaging. 2022;28:200485.

[4] Stein KM, Bahner M, Merkel J, Ain S, Mattern R. Detection of gunshot residues in routine CTs. International Journal of Legal Medicine. 2000;114:15–18.

[5] Schweitzer W, Verster J, Aldomar E, Ebert L, Bolliger SA, Thali MJ, et al. Forensic volumetric visualization of gunshot residue in its anatomic context in forensic post mortem computed tomography: Development of transfer function preset. Forensic Imaging. 2021;25:200451.

[6] Kerkhoff W, Bolck A, Alberink I, Mattijssen E, Hermsen R, Riva F. Pistol bullet deflection through soft tissue simulants. Forensic Science International. 2018;289:270–276.

[7] Lux C, Kettner M, Federspiel JM, Ramsthaler F, Verhoff MA. Atypical wound trajectory after a tangential pistol shot. International Journal of Legal Medicine. 2023;137(2):595–600.

[8] Van Kan R, Haest I, Lahaye M, Hofman P. The diagnostic value of forensic imaging in fatal gunshot incidents: A review of literature. Journal of Forensic Radiology and Imaging. 2017;10:9–14.

[9] Giorgetti A, Giraudo C, Viero A, Bisceglia M, Lupi A, Fais P, et al. Radiological investigation of gunshot wounds: A systematic review of published evidence. International Journal of Legal Medicine. 2019;133:1149–1158.

[10] Pinto A, Russo A, Reginelli A, Iacobellis F, Di Serafino M, Giovine S, et al. Gunshot wounds: Ballistics and imaging findings. In: Seminars in ultrasound, CT and MRI. vol. 40. Elsevier; 2019. p. 25–35.

[11] Cascini F, Polacco M, Cittadini F, Paliani GB, Oliva A, Rossi R. Postmortem computed tomography for forensic applications: A systematic review of gunshot deaths. Medicine, Science and the Law. 2020;60(1):54–62.

[12] Schweitzer W, Thali M. Computed tomography of scalp embedded gravel: Differentiation between falling and kicking. Journal of Forensic Radiology and Imaging. 2015;3(4):200–209.

[13] Passalacqua NV, Rainwater CW. Skeletal trauma analysis: Case studies in context. John Wiley & Sons; 2015.

[14] Chatzaraki V, Heimer J, Thali M, Dally A, Schweitzer W. Role of PMCT as a triage tool between external inspection and full autopsy–case series and review. Journal of Forensic Radiology and Imaging. 2018;15:26–38.

[15] Chatzaraki V, Thali MJ, Ampanozi G, Schweitzer W. Fatal road traffic vehicle collisions with pedestrian victims: Forensic postmortem computed tomography and autopsy correlation. The American Journal of Forensic Medicine and Pathology. 2018;39(2):130–140.

[16] Ampanozi G, Halbheer D, Ebert LC, Thali MJ, Held U. Postmortem imaging findings and cause of death determination compared with autopsy: A systematic review of diagnostic test accuracy and meta-analysis. International Journal of Legal Medicine. 2020;134:321–337.

[17] Flach PM, Gascho D, Schweitzer W, Ruder TD, Berger N, Ross SG, et al. Imaging in forensic radiology: An illustrated guide for postmortem computed tomography technique and protocols. Forensic Science, Medicine, and Pathology. 2014;10(4):583–606.

[18] Adelman A, Vasserman M, Graziani G, Kugel C, Meir K, Bdolah-Abram T, et al. Postmortem computed tomography compared to medico-legal autopsy–pathologies in the torso and limbs. Journal of Forensic Radiology and Imaging. 2018;12:43–49.

[19] Schweitzer W, Thali M. Fatal obstructive asphyxia: Trans-pulmonary density gradient characteristic as relevant identifier in postmortem CT. Journal of Forensic Radiology and Imaging. 2019;19:100337.

[20] Jakobsen SR, Schellerup L, Boel LWT, Hansen K. Lung densitometry in postmortem computed tomography-comparison across different fatal asphyxia groups. Forensic Science, Medicine and Pathology. 2024;p. 1–7.

[21] Bolliger S, Preiss U, Gläser N, Thali M, Ross S. Radiological stab wound channel depiction with instillation of contrast medium. Legal Medicine. 2010;12(1):39–41.

[22] Chatzaraki V, Heimer J, Thali MJ, Ampanozi G, Schweitzer W. Approaching pulmonary fat embolism on postmortem computed tomography. International Journal of Legal Medicine. 2019;133(6):1879–1887.

[23] Chowdhary V, Mehta V, Bajaj T, Scheiner J. Rare imaging of a known entity: Fat embolism seen on CT in lower extremity vein after trauma. Radiolog Case Reports. 2017;12(3):488–490.
[24] Ampanozi G, Held U, Ruder TD, Ross SG, Schweitzer W, Fornaro J, et al. Pulmonary thromboembolism on unenhanced postmortem computed tomography: Feasibility and findings. Legal Medicine. 2016;20:68–74.
[25] Kawasumi Y, Hosokai Y, Usui A, Sato M, Takane Y, Saito H, et al. What information in postmortem computed tomography should radiologist provide to forensic doctor? In: European congress of radiology. Vienna, Austria: European Congress of Radiology; March 3–7, 2011. p. C–1799.
[26] Nolte KB. Research issues in forensic pathology: A survey of academic institutions employing forensic pathologists. Human Pathology. 2004;35(5):532–535.
[27] Gulland A. Doctors cite lack of time as greatest barrier to research. British Medical Journal Publishing Group. 2016;352:i1488.
[28] Mangin P, Bonbled F, Väli M, Luna A, Bajanowski T, Hougen HP, et al. European Council of Legal Medicine (ECLM) accreditation of forensic pathology services in Europe. International Journal of Legal Medicine. 2015;129:395–403.
[29] Tschui J, Jackowski C, Schwendener N, Schyma C, Zech WD. Postmortem CT and MR brain imaging of putrefied corpses. International Journal of Legal Medicine. 2016;130(4):1061–1068.
[30] Bauer M, Berger C, Gerlach K, Scheurer E, Lenz C. Post mortem evaluation of brain edema using quantitative MRI. Forensic Science International. 2022;337:111376.
[31] Ringger R, Schwendener N, Klaus J, Jackowski C, Zech WD. Postmortem CT and MRI diagnosis of acute cerebral hemorrhage in a putrefied corpse. Forensic Science International: Reports. 2023;8:100345.
[32] Schreckenbauer L, Klaus J, Schwendener N, Fridle C, Hartmann C, Zech WD. T1 and T2 signal appearance of different age stages of intracranial hemorrhage in postmortem MRI. Forensic Science International. 2025:112506.
[33] Becker KJ, Baxter AB, Bybee HM, Tirschwell DL, Abouelsaad T, Cohen WA. Extravasation of radiographic contrast is an independent predictor of death in primary intracerebral hemorrhage. Stroke. 1999;30(10):2025–2032.
[34] Legrand L, Delabarde T, Souillard-Scemama R, Sec I, Plu I, Laborie J, et al. Comparison between postmortem computed tomography and autopsy in the detection of traumatic head injuries. Journal of Neuroradiology. 2020;47(1):5–12.
[35] MacKenzie JM. Examining the decomposed brain. The American Journal of Forensic Medicine and Pathology. 2014;35(4):265–270.
[36] Hessmann M, Hofmann A, Kreitner KF, Lott C, Rommens P. The benefit of multislice CT in the emergency room management of polytraumatized patients. Acta Chirurgica Belgica. 2006;106(5):500–507.
[37] Leitman IM, Paull DE, Shires 3rd G. Evaluation and management of massive lower gastrointestinal hemorrhage. Annals of Surgery. 1989;209(2):175.
[38] Jackowski C, Sonnenschein M, Thali MJ, Aghayev E, Yen K, Dirnhofer R, et al. Intrahepatic gas at postmortem computed tomography: Forensic experience as a potential guide for in vivo trauma imaging. Journal of Trauma and Acute Care Surgery. 2007;62(4):979–988.
[39] Schweitzer W, Marti M, Majcen R, Steinmann P, Thali M, Fornaro J, et al. Lethal lung tear and coronary artery gas embolism in shallow water apnea diving exercise–PMCT, PMMRI, autopsy and histology. Journal of Forensic Radiology and Imaging. 2014;2(4):199–204.
[40] Aquila I, Pepe F, Manno M, Frati P, Gratteri S, Fineschi V, et al. Scuba diving death: Always due to drowning? Two forensic cases and a review of the literature. Medico-Legal Journal. 2018;86(1):49–51.
[41] Egger C, Vaucher P, Doenz F, Palmiere C, Mangin P, Grabherr S. Development and validation of a postmortem radiological alteration index: The RA-Index. International Journal of Legal Medicine. 2012;126:559–566.
[42] Schulze K, Ebert LC, Ruder TD, Fliss B, Poschmann SA, Gascho D, et al. The gas bubble sign—a reliable indicator of laryngeal fractures in hanging on postmortem CT. The British Journal of Radiology. 2018;91(1084):20170479.
[43] Bolliger SA, Oesterhelweg L, Spendlove D, Ross S, Thali MJ. Is differentiation of frequently encountered foreign bodies in corpses possible by Hounsfield density measurement? Journal of Forensic Sciences. 2009 Sep;54(5):1119–1122.
[44] Flach P, Ampanozi G, Germerott T, Ross S, Krauskopf A, Thali M, et al. Shot sequence detection aided by postmortem computed tomography in a case of homicide. Journal of Forensic Radiology and Imaging. 2013 Apr;1(2):68–72.
[45] Tarifi AA, Sarhan SZ, Amer MA, Al Safarini AD, Mustafa DF, Altabaa KS, et al. Radiological morphometric study of the hyoid bone using three-dimensional computed tomography (3D-CT) scans. International Journal of Morphology. 2023;41(2).
[46] Kettner M, Potente S, Schulz B, Knauff P, Schmidt PH, Ramsthaler F. Analysis of laryngeal fractures in decomposed bodies using microfocus computed tomography (mfCT). Forensic Science, Medicine, and Pathology. 2014;10:607–612.
[47] Thomsen FSL, Horstmeier S, Niehoff JH, Peña JA, Borggrefe J. Effective spatial resolution of photon counting CT for imaging of trabecular structures is superior to conventional clinical CT and similar to high resolution peripheral CT. Investigative Radiology. 2022;57(9):620–626.
[48] Makhlouf F, Scolan V, Ferretti G, Stahl C, Paysant F. Gunshot fatalities: Correlation between postmortem multi-slice computed tomography and autopsy findings: A 30-months retrospective study. Legal Medicine. 2013;15(3):145–148.

CHAPTER 29

INTOXICATION

Sabine Franckenberg

During external forensic examination of a deceased at a crime scene, there are findings that might already raise the suspicion of an intoxication. The location and surrounding of the body (such as a drug addict supportive housing, drug utensils), obtained background information (for example, known drug abuse of the deceased) and needle marks/scarred subcutaneous veins on the body might be indicative of an intoxication case (Figure 29.1).

In the further investigation of the deceased, postmortem computed tomography (PMCT) is nowadays often performed either as an adjunct or in special cases even as a replacement for autopsy.

While it is not possible yet to actually prove a lethal intoxication as cause of death in PMCT alone, there are indicative findings to consider when reading a whole-body PMCT. Especially cerebral edema, pulmonary edema, aspiration and/or distended urinary bladder are seen more frequently in intoxication cases than in other causes of death (1–4) (Figure 29.1 and 29.2). Yet, to confirm (or exclude) suspected intoxication, blood and urine samples still have to be taken and analyzed. Sampling can either be performed during regular autopsy, or minimally invasive transcutaneous with a needle either freehand or CT-guided with the Virtobot (5). This way, if the findings from external examination, PMCT and toxicological analysis already allow for diagnosis of cause of death, autopsy might be redundant in specific cases.

Listed below are the most common indicative findings for intoxication in PMCT.

DOI: 10.4324/9781003413998-32

29.1 NEUROCRANIUM

Already in clinical computed tomography (CT), findings such as cerebral edema or specific hemorrhage patterns can be found as an expression of the toxicity of drugs or their derivatives (6–8).

Cerebral edema—cerebral edema in general is a very unspecific post mortal finding and is therefore quite often seen in intoxication cases in general, but it is most often seen in PMCT of opioid and cocaine fatalities (1). (Figure 29.2a).

Brain hemorrhage—Already in non-lethal intoxications, brain hemorrhage can be seen as a common complication, such as for example in amphetamines (Figure 29.3). Symmetrical hemorrhage in the basal ganglia is the most common, and often lethal complication in methanol intoxication (9, 10). Interestingly, a study from Iran observed a significant increase in methanol intoxications during the COVID-19 pandemic due to the misbelief that the consumption of alcohol-containing sanitizers or homemade methanol-containing beverages might actually protect against the virus (11).

29.2 THORAX

Pulmonary edema—Intoxications in general are one of the major causes for neurogenic pulmonary edema (NPE) following an acute, severe compromise of the central nervous system (12). Additionally, a direct toxic effect on the alveolocapillary membrane has been discussed in methadone/heroin intoxication, causing increased capillary permeability (13). In computed tomography, pulmonary edema can be found already in non-fatal methadone and heroin intoxication (13, 14). Acute pulmonary (and sometimes hemorrhagic) edema is most often seen in postmortem examinations in lethal opioid, cocaine, benzodiazepine and psychotropic drug intoxication (15). Especially in lethal opioid intoxications, pulmonary edema was the finding seen most often in PMCT (in 95 % of the cases) (1) (Figure 29.2b).

Aspiration—Aspiration of stomach content as a finding in PMCT is usually interpreted as a vital reaction of an unconscious person in agony and therefore very unspecific. Since lethal intoxications often come along with a prolonged agony, aspiration is a frequent finding. In lethal opioid intoxication, aspiration has been described in 2/3 of the cases (1).

29.3 ABDOMEN

Gastrointestinal tract—Drugs (tablets, body packs) might be visible in PMCT inside the GI tract as hypo-, iso- or hyperdense lesions (Figure 29.4). Hyperdense lesions within the stomach content were for example seen in oral medication intoxication cases (16). Therefore, such findings might also be indicative of a possible intoxication case.

Distended Urinary Bladder—Fatal intoxication in general shows a correlation with urinary bladder distension, with a cut-off of > 330 ml for intoxication cases (without consideration of the separate substances; sensitivity/specificity 25%/97%) in PMCT (2). Urinary bladder distention is especially seen in lethal opioid, cocaine, benzodiazepine and psychotropic drug intoxication (4).

In clinical CT a cut-off of >416 ml in mono-intoxication with alcohol could be shown (sensitivity/specificity 50.9%/76.3%) (17). In mono-intoxication with opioids, a distended urinary bladder was found in only 1/3 of the cases (1).

29.4 COMBINATION OF FINDINGS

Often, a combination of cerebral edema, pulmonary edema, aspiration and/or distended urinary bladder is seen in lethal intoxication cases (1–3). Especially the combination of lung edema, brain edema and a full urinary bladder is considered a highly specific constellation (specificity = 100 %) for lethal opioid and cocaine intoxication (Figure 29.2a–c) (1).

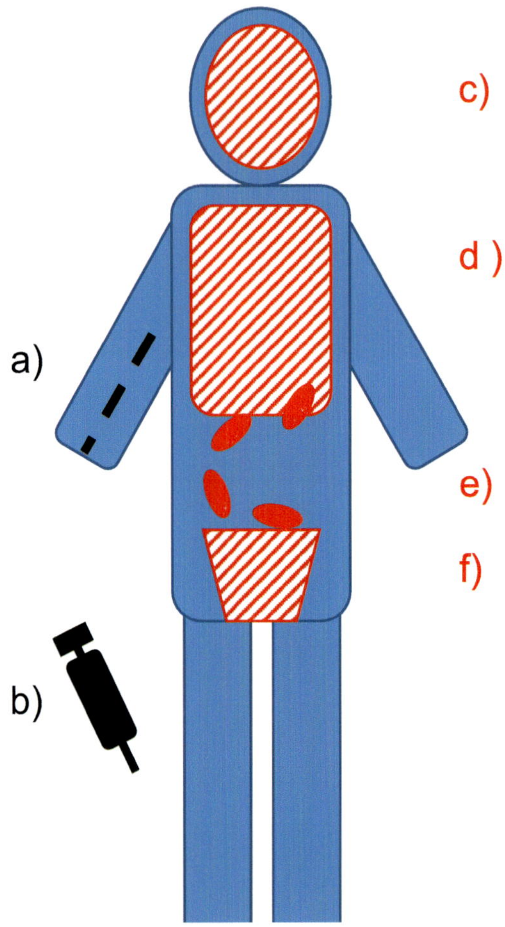

Figure 29.1 External findings indicative of intoxication in black: a) needle marks/scarred subcutaneous veins, b) drug utensils. Internal findings both in PMCT and at autopsy indicative of intoxication in red: c) brain edema, d) pulmonary edema, e) tablets, body packs, f) bladder distension (1).

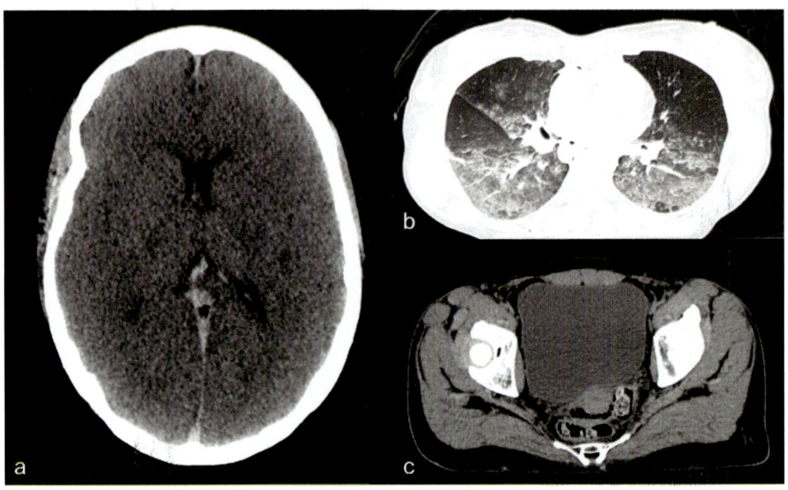

Figure 29.2 The typical, highly specific constellation in lethal opioid intoxication in unenhanced PMCT: a) brain edema, b) pulmonary edema and c) bladder distention.

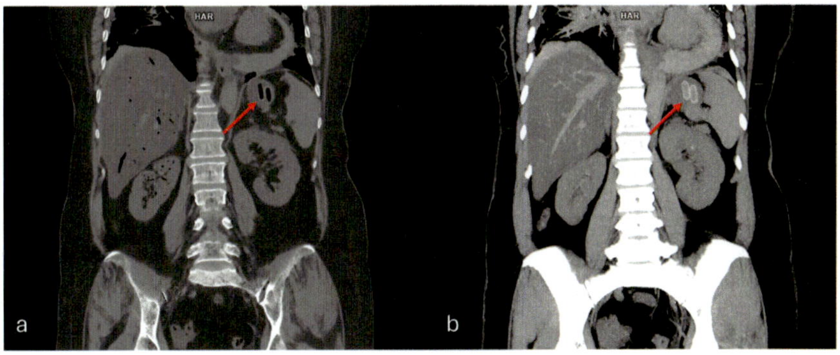

Figure 29.3 Axial unenhanced CT of the brain with bilateral hemorrhage in a clinical case of amphetamine intoxication. Image courtesy of PD Dr. med. S. Winklhofer, University Hospital Zürich.

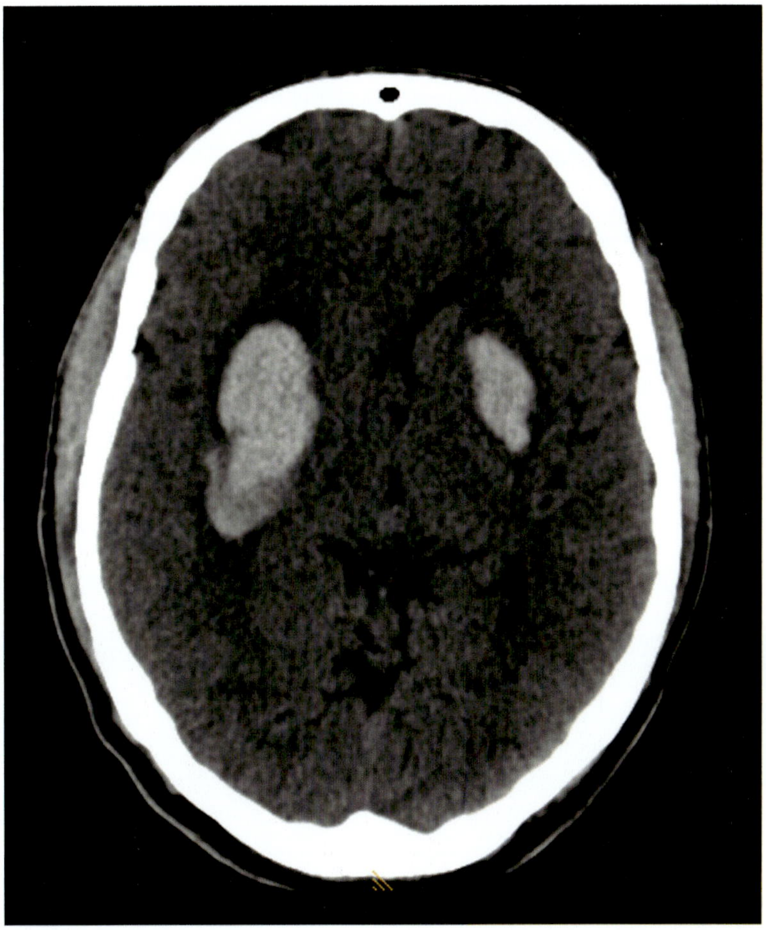

Figure 29.4 Coronal unenhanced PMCT of the abdomen in a soft tissue window, b) MIP reconstruction of the same case. Two rounded, oblong structures (red arrows) can be seen in the stomach, which are highly suspicious of body packs. As additional finding, there is already some gas in the liver veins as a sign of beginning decomposition.

References

1. Winklhofer S, Surer E, Ampanozi G, Ruder T, Stolzmann P, Elliott M, Oestreich A, Kraemer T, Thali M, Alkadhi H, Schweitzer W (2014) Postmortem whole body computed tomography of opioid (heroin and methadone) fatalities: Frequent findings and comparison to autopsy. Eur Radiol.24(6):1276–82.
2. Rohner C, Franckenberg S, Schwendener N, Oestreich A, Kraemer T, Thali MJ, Hatch GM, Ruder TD (2013) New evidence for old lore—urinary bladder distension on postmortem computed tomography is related to intoxication. Forensic Sci Int.225(1–3):48–52.
3. Soon LP, Subramaniam K, Madon NF, Mohamad Noor MH, Chainchel Singh MK, Mahmood MS (2019) Postmortem CT scan in intoxication cases: A necessity or just an indulgence. Indian J Radiol Imaging.29(3):305–9.
4. Murdzinski J, Franckenberg S, Ebert L, Fliss B, Thali MJ, Blüthgen C (2023) Correlation of a distended urinary bladder volume with lethal intoxication in general and for different substance subgroups in particular: A retrospective study. Toxicologie Analytique et Clinique.35(2):138–42. https://doi.org/10.1016/j.toxac.2022.11.002
5. Staeheli SN, Gascho D, Ebert LC, Kraemer T, Steuer AE (2017) Time-dependent postmortem redistribution of morphine and its metabolites in blood and alternative matrices-application of CT-guided biopsy sampling. Int J Legal Med.131(2):379–89.
6. Geibprasert S, Gallucci M, Krings T (2010) Addictive illegal drugs: Structural neuroimaging. AJNR Am J Neuroradiol. 31(5):803–8.
7. Büttner A, Mall G, Penning R, Weis S (2000) The neuropathology of heroin abuse. Forensic Sci Int.113(1–3):435–42.
8. Jamshidi F, Sadighi B, Aghakhani K, Sanaei-Zadeh H, Emamhadi M, Zamani N (2013) Brain computed tomographic scan findings in acute opium overdose patients. Am. J Emerg Med.31(1):50–3.
9. Kim HJ, Na JY, Lee YJ, Park JT, Kim HS (2015) An autopsy case of methanol induced intracranial hemorrhage. Int J Clin Exp Pathol.8(10):13643–6.
10. Taheri MS, Moghaddam HH, Moharamzad Y, Dadgari S, Nahvi V (2010) The value of brain CT findings in acute methanol toxicity. Eur J Radiol.73(2):211–4.
11. Simani L, Ramezani M, Roozbeh M, Shadnia S, Pakdaman H (2022) The outbreak of methanol intoxication during COVID-19 pandemic: Prevalence of brain lesions and its predisposing factors. Drug Chem Toxicol.45(4):1500–3.
12. Finsterer J (2019) Neurological perspectives of neurogenic pulmonary edema. Eur Neurol.81(1–2):94–102.
13. Frand UI, Shim CS, Williams MH, Jr. (1972) Methadone-induced pulmonary edema. Ann Intern Med.76(6):975–9.
14. Steinberg AD, Karliner JS (1968) The clinical spectrum of heroin pulmonary edema. Arch Intern Med.122(2):122–7.
15. Püschel KSA (2003) Drogennot- und -todesfälle. Madea BBB, editor. Berlin: Springer Verlag; 2003. 1344 p.
16. Aghayev E, Jackowski C, Christe A, Thali M (2010) Radiopaque stomach contents in postmortem CT in suicidal oral medication intoxication: Report of three cases. J Forensic Leg Med.17(3):164–8.
17. Gümbel D, Schneidler F, Frank M, Bockholdt B, Hinz P, Napp M, Spitzmüller R, Ekkernkamp A, Langner S (2017) Urinary bladder volume measured in whole-body CT scans is a useful marker for alcohol intoxication. Alcohol (Fayetteville, NY).65:45–50.

PART 4

DISEASE DIAGNOSIS

CHAPTER 30

CARDIOVASCULAR SYSTEM

Garyfalia Ampanozi, Carlo Tappero, and Wolf Schweitzer

Natural causes of death from the cardiovascular system may form a considerable portion of the cases encountered in a forensic institute.

This may also be because sudden deaths are reportable deaths in many jurisdictions (1)—cardiac disease was alleged to have sudden death even as its first clinical manifestation in 19–26% of cases (2).

In one study that evaluated natural death related data from a British coroner's office, causes of death attributed to the cardiovascular system (ischemic heart disease: 58,7%, non-ischemic heart disease: 7,5%, aortic aneurysm: 3,4%) surpassed those attributed to respiratory disease (17,7%), central nervous system disease (4,3%), gastrointestinal hemorrhage (2,2%) and alcohol related (2,8%) (3). However, because structural cardiac findings are common and readily demonstrable at autopsy, there is a systematic tendency to attribute death to "cardiac arrest" or cardiovascular disease, even though the terminal mechanism may merely reflect the final common pathway rather than the true precipitating cause. Comprehensive investigation of toxicological, metabolic, pulmonary, neurological, and multiorgan contributors remains comparatively underutilized, despite evidence that a substantial proportion of presumed sudden cardiac deaths are ultimately reclassified after detailed autopsy and toxicological work-up.

Conversely, cardiovascular diagnoses may also be identified in non-natural death, for example after blunt force trauma (4), sharp force trauma (5), firearm injury (6), in cases of intoxication (7) or after surgery, blunt trauma or immobilization (8, 9). Deaths due to a myocardial infarction or aortic dissection shortly after consulting with a doctor because of chest pains, for example, may be investigated as possibly non-natural death, as medical negligence and malpractice deaths are not considered natural deaths. Generally, any cardiovascular diagnosis may be morphologically impressive but etiologically non-specific [Hanzlick R. The medical cause of death manual: instructions for writing cause of death statements for deaths due to natural causes. College of American Pathologists 1994.].

Findings on postmortem imaging, suggesting a *potential* involvement of a diagnosis related to the cardiovascular system in the cause of death are numerous (10). Calcification of coronary arteries and valves, stents, bypass vessels, heart devices and even calcifications of the myocardium on the ground of old heart infarctions or aneurysmatic enlargements of the myocardium may be visible on native PMCT, allowing to at least suspect heart pathology, without, however, the ability to provide correct and specific attribution to the cause of death. If traumatic causes of death, third party involvement and intoxication can be excluded, these findings could suggest however a *possible* cause of death from the cardiovascular system.

Clinical imaging of the heart may be performed with methods of CT angiography, echocardiography, nuclear medicine or magnetic resonance imaging (MRI) (11). Forensic settings typically approach postmortem imaging with a native postmortem computed tomography scan (PMCT). The imaging of vasculature and the heart after death is greatly aided by postmortem CT angiography (PMCTA) and postmortem MRI (PMMR) (12, 13).

In numerous instances, postmortem imaging techniques may not fully or adequately characterize significant diseases affecting the cardiovascular system. Therefore, whenever a comprehensive evaluation of such conditions is necessary, we strongly recommend the performance of an autopsy to ensure accurate diagnosis and assessment.

DOI: 10.4324/9781003413998-34

30.1 CARDIOMEGALY

Cardiomegaly holds significant relevance as a finding in cases of sudden deaths (14).

Heart weight, as a correlate for cardiomegaly, can be estimated both on PMCT and PMMR.

- The cardiothoracic ratio may be determined on PMCT in order to predict cardiomegaly (15, 16).
- Assessing the circumferential area of the left heart ventricle on axial images may allow for a rough estimate of the heart weight (17, 18).
- Measurements of heart wall thickness and some of the valve circumferences by both modalities (PMCT, PMMR) have been found to be statistically significant different (19, 20), whereas cardiac PMMR assessment of myocardial wall thickness tends to overestimate the actual autopsy measurements (19, 20).

Normal values referring to postmortem imaging cardiac measurements have not yet been determined, as postmortem imaging measurements of normal hearts differ from normal value tables based on autopsy measurements.

In order to understand why there is cardiomegaly, it may be important to discriminate between different possible reasons. The differential diagnosis includes hypertrophic cardiomyopathy, which involves abnormal thickening of the heart muscle, hypertensive heart disease caused by high blood pressure, adaptive changes in the heart size seen in athletes due to physical training, and other conditions that can lead to heart enlargement, such as aortic stenosis or ventricular septal defect. Usually, the relevant morphological criteria to identify these diagnoses are most cost-efficiently examined at autopsy (10, 21, 22).

30.2 ISCHEMIC HEART DISEASE

Coronary artery calcification may be identified on native PMCT.

It is that easy to detect once a certain amount of calcification is present, that it seems to be a sufficiently impressive finding to possibly imply coronary artery atherosclerosis as cause of death, whereas however coronary calcification amount does not strictly correlate with the degree of stenosis (23). Much rather, one should avoid succumbing to the specificity paradox (24) where the impressively visible finding is mistaken to be particularly relevant or specific: an age-sex matched study comparing 55 cases of fatal opioid poisoning and 55 non-toxicological cases (mostly natural deaths) found that coronary artery atherosclerosis was present in similar incidences in both groups (poisoning group: 36%, control group: 29%; difference not statistically significant) (25).

Severe coronary artery stenosis and occlusions may be identified with added contrast, so PMCTA has been shown to be a valuable tool for coronary artery stenosis assessment after death (26). In native PMMR, Ruder et al. described the presence of chemical shift artefact as an indicator for patency and paired dark bands as a finding related to coronary stenosis (27).

Myocardial infarction is a frequently encountered finding of the cardiovascular system, whereas it may be seen as immediate or as a contributory cause of death (28).

And yet, it is barely and rarely depicted by unenhanced PMCT. Indirect signs of possibly existing indicators for heart disease, such as coronary artery calcification, past bypass operation, stents in the coronary arteries, or calcified scars of previous myocardial infarctions, are not substantial evidence of a relevant myocardial infarction (Figure 30.1). Rare findings correlated with cardiac ischemia, such as myocardial bridges (29) or coronary artery anatomic anomalies (10, 30, 31), can be revealed by postmortem imaging. Figure 30.2 depicts a thrombosed aneurysm of the right coronary artery with total occlusion of the vessel.

There is strong evidence in the literature that native PMMR can yield relevant documentation of this pathological entity, either of the heart in-situ or after removal at autopsy (32, 33).

PMMR not only enables identification of the infarction area, but also provides basis for infarction age estimation. The distinction between acute, subacute and chronic infarction may be based on T2-weighted images. There, central hypointense areas with peripheral hyperintense margins, and hyperintense, and hypointense areas, respectively, have been

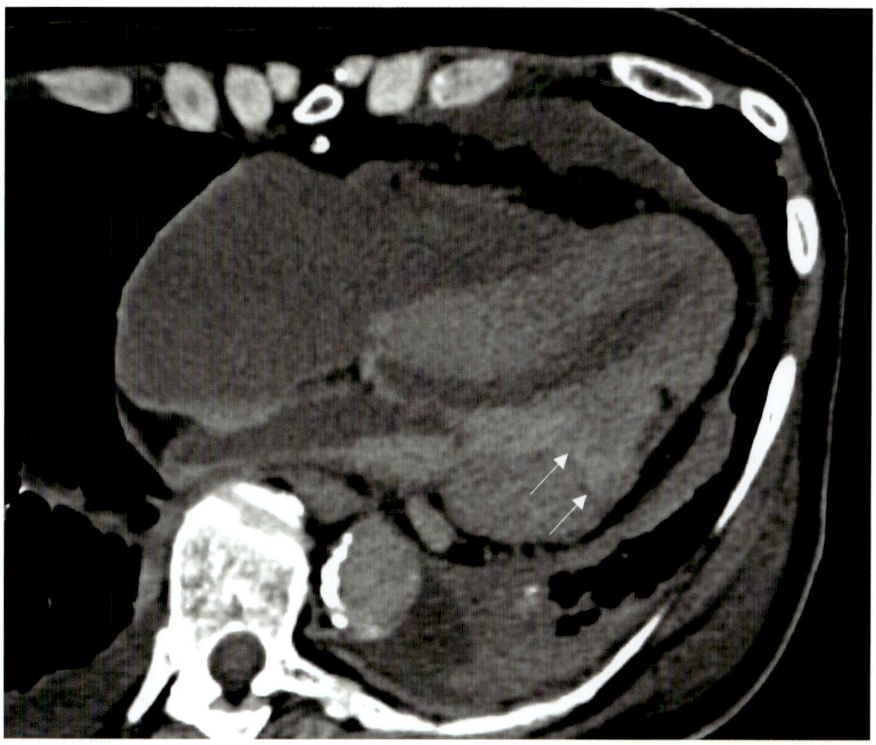

Figure 30.1 Unenhanced postmortem computed tomography, four-chamber view of the heart. Note the hyperdensity of the myocardium (white arrows), corresponding to an old infarction scar with peripheral hemorrhagic components. An acute hemorrhagic infarct could look similar.

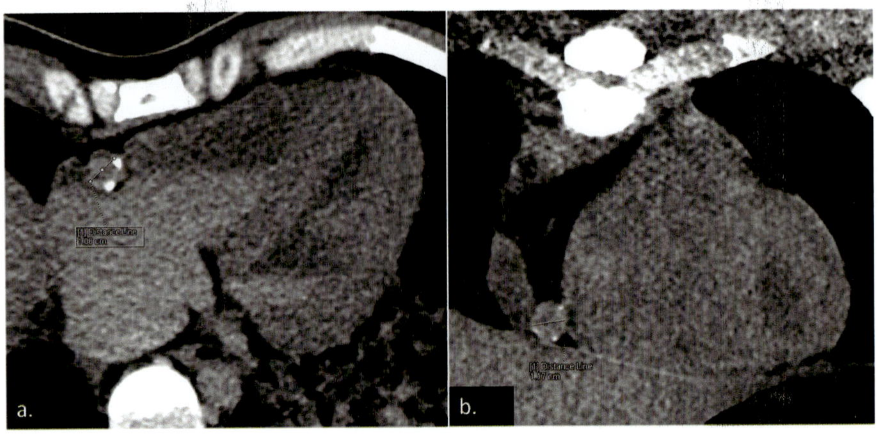

Figure 30.2 Unenhanced postmortem computed tomography of the heart, a) axial and b) coronal view. Aneurysma of the right coronary artery, with a maximal diameter of ca. 1.17 cm. It was found to be thrombosed and occluded during autopsy.

Table 30.1 MRI Findings Related to Infarct Age and Sequence According to (35)

MRI and Myocardial Infarction Stages (According to (35), Figure 30.5)	Peracute (0-~6h)	Acute (~1 day to ~1 week)	Subacute (~1 week—~2 months)	Chronic (> ~2 months)
Necrotic center of infarction	T1 isointense T2 possibly hypointense	T1 possibly hyperintense T2 hypointense	T1 isointense T2 hyperintense	T1 hypointense T2 hypointense
Marginal region of infarction	T1 possibly hypointense T2 possibly hyperintense	T1 hypointense T2 hyperintense	T1 isointense T2 isointense	T1 isointense T2 isointense

found to correlate with diagnostic criteria (34, 35). Even (per)acute myocardial infarctions that are not identifiable macroscopically were proposed to be diagnosable on PMMR as hypointense regions (34–37). Knowing their localization before autopsy can guide targeted histological sampling during autopsy, or guide biopsy. Ischemic heart disease can also be depicted by PMCTA (either whole body or targeted) in terms of pathological myocardial enhancement (38).

In addition to T1 and T2 sequences, adequate cardiac PMMR imaging may benefit from considering also a susceptibility weighted sequence. A more technical prospective approach may interprete T2 increased signal combined with indication of blood on susceptibility weighted sequences as anything between a well capillarized scar and a hemorrhagic infarction. With that in mind, cardiac MRI may indicate regions of particular interest for histology but not necessarily identify candidates as acute cause of death with certainty.

Currently, the combination of PMMR and guided biopsy of the myocardium provides the highest accuracy in the pre-autopsy investigation of acute or chronic myocardial ischemia (39).

30.3 HEMOPERICARDIUM

Hemopericardium can be identified on native PMCT. Additional findings including signs of increased intrapericardial pressure such as flattening of the anterior surface of the heart and compression of the coronary sinus or the pulmonary trunk suggest pericardial tamponade as a possible cause of death.

The presence of blood in the pericardial sac (hemopericardium) is clearly identifiable not only on PMMR but also on unenhanced PMCT (see Figures 30.3 and 30.4). While diagnosis of pericardial tamponade is primarily clinical, postmortem signs indicating a fatal hemopericardium and thus a pericardial tamponade have been described. The "hyperdense armored heart sign" (HAH) is a concentric or contour-following dense ring around the epicardial heart contour that was described by Shiotani et al. (40) as the most frequently observed PMCT finding in cases of aortic dissection leading to hemopericardium. Watanabe et al. (41) proposed HAH as a sign of a fatal pericardial tamponade, since the origin of its shape was be suggested by a beating myocardium. Other than hyperdense armored heart sign (HAH), other PMCT signs of fatal heart tamponade are the flattening of the right ventricular wall and coronary sinus and pulmonary artery compression, and, to a lesser extent, hepatic and renal vein distension (42). However, PMCT images alone may be misleading, as the HAH may be absent, but visible on PMMR images.

Although PMCTA and PMMR are the methods of choice in identifying or at least narrowing down the cause of the hemopericardium, unenhanced PMCT can very often be suggestive of it, particularly for the differentiation between ruptured myocardial infarction and ruptured aortic dissection (43), which are the most common etiologies for hemopericardium (26).

Differentiation between aortic dissection and ruptured myocardial infarction, the two most often encountered causes of pericardial tamponade, may be possible on unenhanced PMCT images as the examples in Figures 30.3 and 30.4 depict.

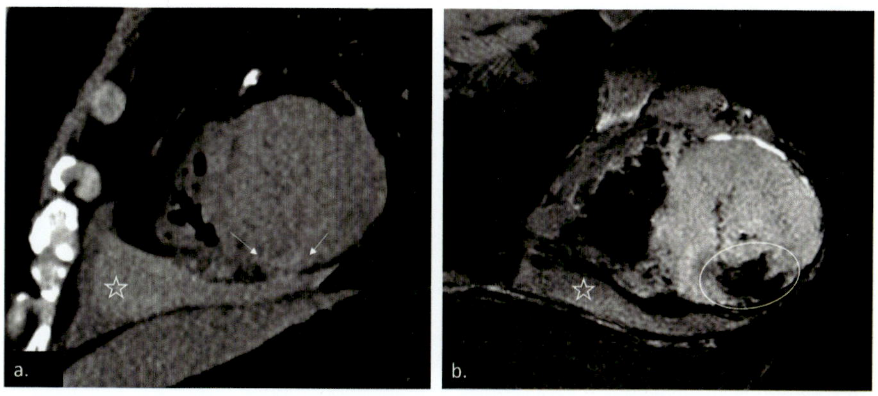

Figure 30.3 a) Unenhanced postmortem computed tomography, short axis view of the heart. Note the hyperdensity of the myocardium (white arrows) and the hemopericardium (white star). b) Postmortem magnetic resonance imaging, T2 weighted FFE Hemo sequence. Hemorrhagic myocardial infarction and rupture (encircled) with resulting hemopericardium (white star).

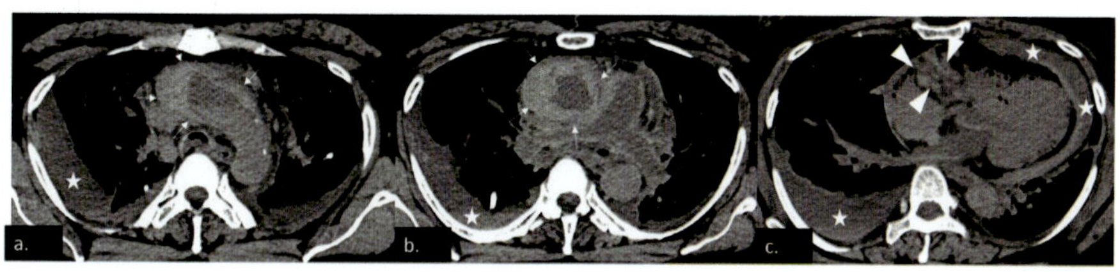

Figure 30.4 Unenhanced postmortem computed tomography, axial images of the thoracic cavity depicting a) the aortic arch, b) the pulmonary artery trunk and the ascending aorta and c) the origin of the aorta from the left ventricle. Note the thickening of the aortic wall (white arrows). There was blood infiltration of the epicardial adipose tissue (arrowheads) as well as blood in the pericardial sac and the right pleural cavity (white stars).

30.4 PULMONARY THROMBOEMBOLISM

Pulmonary thromboembolism is frequently encountered in forensic institutes since it is considered a sudden, unexpected death because of the acute onset of symptoms.

Pulmonary thromboembolism is a challenging diagnosis on unenhanced PMCT.

A definitive diagnosis of this entity on postmortem imaging is usually impossible (26). The main reason is that postmortem formation of clots results in native PMCT findings that are practically impossible to reliably discern from pulmonary thromboembolism. Centrally located hyperdense material in the pulmonary trunk and in the main pulmonary arteries may raise the suspicion and can be mistaken as a thrombotic embolus by inexperienced or experienced readers alike. PMCTA will reveal contrast medium filling defects in both cases (postmortem clots and thromboembolism), so only PMCT-guided biopsy with subsequent histology or autopsy will finally allow a diagnosis (26).

Indirect findings have been proposed as indicators of pulmonary thromboembolism. Perivascular edema of the lower extremities as sign of leg vein thrombosis, and a distinct irregular shape of the pulmonary trunk and artery content (44) (Figure 30.5) in combination with venous vessel congestion (45) are rated as strongly suggestive of pulmonary thromboembolism on unenhanced PMCT. PMMR has also been suggested as an adjunct in the preautopsy diagnosis of pulmonary thromboembolism (46).

Although different radiological signs have been described, autopsy remains necessary for this diagnosis. Furthermore, the source of venous emboli typically is important for a complete case documentation; native PMCT (or PMCTA) seems not very good at securing the diagnosis of lower or upper extremity vein thrombosis, thromboses in pelvic blood vessels, or of thrombosis located in the heart (47).

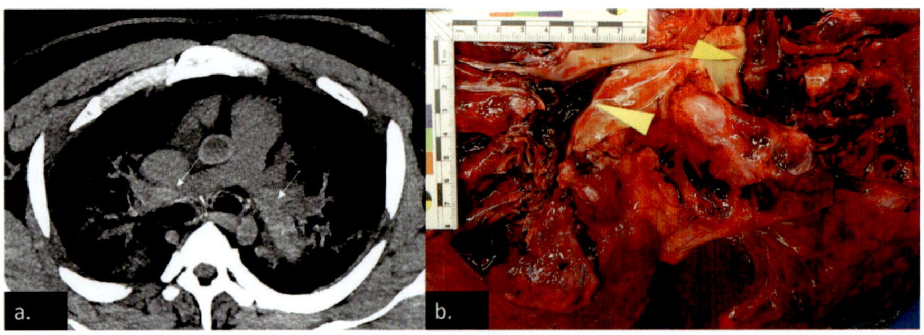

Figure 30.5 a) Postmortem computed tomography, axial image of the pulmonary artery trunk and the right and left pulmonary arteries. Inhomogeneous content in both pulmonary arteries (arrows), indicative of pulmonary artery thromboembolism. b) Autopsy photograph of the dissected pulmonary artery trunk and main pulmonary arteries, with thromboembolism (yellow arrowheads).

30.5 AORTIC DISSECTION, AORTIC RUPTURE

Aortic dissection and rupture are described in persons with atherosclerosis, bicuspid aortic valve, and connective tissue disorders (48). As clinical detection of non-traumatic aortic dissection or rupture cases can be elusive, they were termed "great masqueraders"; this explains that they may frequently be the focus of medical malpractice lawsuits (49). Traumatic aortic rupture may be due to blunt chest trauma, as in automobile accidents or falls from height (50).

Detection of traumatic aortic rupture, or of arterial injury in general, is best performed using PMCTA (51).

Aortic aneurysms are identified as enlargements of the diameter of the aorta. They can lead to massive blood loss in cases of rupture (Figure 30.6). PMCT is the method of choice for identifying this pathology as the cause of death, with PMCTA possibly enabling the exact localization of the rupture site (52). This may be of significance in cases of triage, or for planning of the autopsy (52).

Aortic dissection may be identified on unenhanced PMCT with specific findings such as dislocated aortic wall calcifications, a visible intimomedial flap (as a result of the postmortem aortic wall hyperattenuation (53)), or the double sedimentation sign (54). Both pathological entities are well depicted on postmortem imaging. Dislocated calcification towards the aortic lumen, intimomedial flap depiction (without calcification) and the double sedimentation sign (uneven sedimentation level between true and false aortic lumen) are identifiable on unenhanced PMCT and allow for the diagnosis of aortic dissection (54). Involvement of the coronary arteries or rupture of the adventitia and blood exsanguination into the pericardial sac (Figure 30.4), the pleural or the abdominal cavity are the most frequently encountered causes of death. PMCTA allows for a definitive diagnosis and documentation of this pathological finding, and PMMR also clearly demonstrates the dissection membrane.

Case-related data published and used under Cantonal Ethics Committee of Canton of Zürich waiver 15–0686.

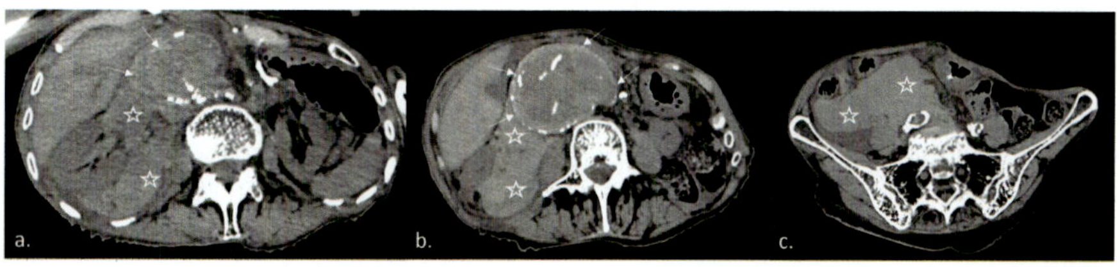

Figure 30.6 Postmortem computed tomography, axial images of the abdomen. Abdominal aortic aneurysma (arrows), with rupture and a large retroperitoneal hemorrhage (white stars).

References

1. Dwyer AJ, Visser P, Russell L. Interpreting the coroners act at the bedside: How do junior doctors know they are doing it correctly? Med J Australia. 2013;199(1):25–6.
2. Marrugat J, Elosua R, Gil M. Epidemiology of sudden cardiac death in Spain. Rev Esp Cardiol. 1999;52(9):717–25.
3. Thomas AC, Knapman PA, Krikler DM, Davies MJ. Community study of the causes of natural sudden-death. Brit Med J. 1988;297(6661):1453–6.
4. Abdolrahimi SA, Sanati HR, Ansari-Ramandi MM, Heris SO, Maadani M. Acute myocardial infarction following blunt chest trauma and coronary artery dissection. J Clin Diagn Res. 2016;10(6):OD14–5.
5. Bartoloni G, Trio F, Bartoloni A, Giorlandino A, Pucci A. A fatal stab wound causing selective injury to the left anterior descending coronary artery, myocardial infarction and delayed cardiac tamponade. Forensic Sci Int. 2013;229(1–3):e16–8.
6. Kumar S, Moorthy N, Kapoor A, Sinha N. Gunshot wounds: Causing myocardial infarction, delayed ventricular septal defect, and congestive heart failure. Tex Heart Inst J. 2012;39(1):129–32.
7. Mittleman MA, Mintzer D, Maclure M, Tofler GH, Sherwood JB, Muller JE. Triggering of myocardial infarction by cocaine. Circulation. 1999;99(21):2737–41.
8. Vonknorring J. Postoperative myocardial-infarction: A prospective-study in a risk group of surgical patients. Surgery. 1981;90(1):55–60.
9. Michota FA. Prevention of venous thromboembolism after surgery. Cleve Clin J Med. 2009;76(10):45–52.
10. Michaud K, Grabherr S, Jackowski C, Bollmann MD, Doenz F, Mangin P. Postmortem imaging of sudden cardiac death. Int J Leg Med. 2014;128(1):127–37.
11. de Roos A, Higgins CB. Cardiac radiology: Centenary review. Radiology. 2014;273(2 Suppl):S142–59.

12. Schweitzer W, Ebert L, Thali MJ. Comparison of very affordable immersion pump and heart-lung-machine for post mortem CT angiography in forensic pathology: A case-control study. Foren Imag. 2022;30.
13. Ruder TD, Hatch GM, Ebert LC, Flach PM, Ross S, Ampanozi G, et al. Whole body postmortem magnetic resonance angiography. J Forensic Sci. 2012;57(3):778–82.
14. Tavora F, Zhang Y, Zhang M, Li L, Ripple M, Fowler D, et al. Cardiomegaly is a common arrhythmogenic substrate in adult sudden cardiac deaths, and is associated with obesity. Pathology. 2012;44(3):187–91.
15. Winklhofer S, Berger N, Ruder T, Elliott M, Stolzmann P, Thali M, et al. Cardiothoracic ratio in postmortem computed tomography: Reliability and threshold for the diagnosis of cardiomegaly. Forensic Sci Med Pathol. 2014;10(1):44–9.
16. Jotterand M, Faouzi M, Dédouit F, Michaud K. New formula for cardiothoracic ratio for the diagnosis of cardiomegaly on postmortem CT. Int J Leg Med. 2020;134(2):663–7.
17. Hatch GM, Ampanozi G, Thali MJ, Ruder TD. Validation of left ventricular circumferential area as a surrogate for heart weight on postmortem computed tomography. J Forens Radiol Imaging. 2013;1(3):98–101.
18. Ruder TD, Stolzmann P, Thali YA, Hatch GM, Somaini S, Bucher M, et al. Estimation of heart weight by postmortem cardiac magnetic resonance imaging. J Forens Radiol Imaging. 2013;1(1):15–8.
19. Ampanozi G, Hatch GM, Flach PM, Thali MJ, Ruder TD. Postmortem magnetic resonance imaging: Reproducing typical autopsy heart measurements. Leg Med (Tokyo). 2015;17(6):493–8.
20. Chatzaraki V, Thali MJ, Schweitzer W, Ampanozi G. Left myocardial wall measurements on postmortem imaging compared to autopsy. Cardiovasc Pathol: The Offici J Soc Cardiovasc Pathol. 2019;43:107149.
21. Cohle SD, Balraj E, Bell M. Sudden death due to ventricular septal defect. Pediatr Devel Pathol. 1999;2(4):327–32.
22. Kragel AH, Roberts WC. Sudden-death and cardiomegaly unassociated with coronary, valvular, congenital or specific myocardial-disease. Am J Cardiol. 1988;61(8):659–60.
23. Michaud K, Grabherr S, Doenz F, Mangin P. Evaluation of postmortem MDCT and MDCT-angiography for the investigation of sudden cardiac death related to atherosclerotic coronary artery disease. Int J Cardiovasc Imaging. 2012;28(7):1807–22.
24. Hanzlick R. *The medical cause of death manual: Instructions for writing cause of death statements for deaths due to natural causes*. Northfield, IL: College of American Pathologists. 1994. MMS ID 997914243406676; ISBN 9780930304621 (pbk.); ISBN 0930304624 (pbk.); NLM Unique ID 9506457.
25. Winklhofer S, Surer E, Ampanozi G, Ruder T, Stolzmann P, Elliott M, et al. Postmortem whole body computed tomography of opioid (heroin and methadone) fatalities: Frequent findings and comparison to autopsy. Eur Radiol. 2014;24(6):1276–82.
26. Michaud K, Jacobsen C, Basso C, Banner J, Blokker BM, de Boer HH, et al. Application of postmortem imaging modalities in cases of sudden death due to cardiovascular diseases-current achievements and limitations from a pathology perspective: Endorsed by the association for european cardiovascular pathology and by the international society of forensic radiology and imaging. Virchows Arch. 2023;482(2):385–406.
27. Ruder TD, Bauer-Kreutz R, Ampanozi G, Rosskopf AB, Pilgrim TM, Weber OM, et al. Assessment of coronary artery disease by postmortem cardiac MR. Eur J Radiol. 2012;81(9):2208–14.
28. Madsen M, Davidsen M, Rasmussen S, Abildstrom SZ, Osler M. The validity of the diagnosis of acute myocardial infarction in routine statistics: A comparison of mortality and hospital discharge data with the Danish MONICA registry. J Clin Epidemiol. 2003;56(2):124–30.
29. Ampanozi G, Gascho D, Hatch G, Schulze C, Thali MJ, Ruder TD. What is unsought will go undetected: Myocardial bridging on postmortem computed tomography. J Forens Radiol Imaging. 2014;2(1):5–8.
30. Martinez RM, Flach PM, Ebert LC, Bartsch C, Thali MJ, Ampanozi G. Anomalous left coronary artery origin on postmortem imaging in correlation with autopsy. J Forens Radiol Imaging. 2014;2(3):146–8.
31. Takei H, Sano R, Takahashi Y, Takahashi K, Kominato Y, Tokue H, et al. Usefulness of coronary postmortem computed tomography angiography to detect lesions in the coronary artery and myocardium in cases of sudden death. Leg Med (Tokyo). 2018;30:46–51.
32. Jackowski C, Schweitzer W, Thali M, Yen K, Aghayev E, Sonnenschein M, et al. Virtopsy: Postmortem imaging of the human heart in situ using MSCT and MRI. Forensic Sci Int. 2005;149(1):11–23.
33. Cafarelli FP, Macarini L, Cipolloni L, Maglietta F, Guglielmi G, Sessa F, et al. Ex situ heart magnetic resonance imaging and angiography: Feasibility study for forensic purposes. Foren Imag. 2021;25.
34. Jackowski C, Schwendener N, Grabherr S, Persson A. Postmortem cardiac 3-T magnetic resonance imaging: Visualization of sudden cardiac death? J Am Coll Cardiol. 2013;62(7):617–29.
35. Jackowski C, Christe A, Sonnenschein M, Aghayev E, Thali MJ. Postmortem unenhanced magnetic resonance imaging of myocardial infarction in correlation to histological infarction age characterization. Eur Heart J. 2006;27(20):2459–67.
36. Ruder TD, Ebert LC, Khattab AA, Rieben R, Thali MJ, Kamat P. Edema is a sign of early acute myocardial infarction on postmortem magnetic resonance imaging. Forensic Sci Med Pathol. 2013;9(4):501–5.
37. Gascho D, Von Allmen A, Landsmann A, Hünermund T, Tappero C, Thali MJ, et al. Diagnostic value of T1- and T2-weighted 3-Tesla MRI for postmortem detection and age stage classification of myocardial infarction. Forensic Sci Med Pat. 2023.
38. Sabatasso S, Vanhaebost J, Doenz F, Palmiere C, Michaud K, Dedouit F, et al. Visualization of myocardial infarction in postmortem multiphase computed tomography angiography: A feasibility study. Am J Forensic Med Pathol. 2018;39(2):106–13.

39. Wagensveld IM, Blokker BM, Pezzato A, Wielopolski PA, Renken NS, von der Thüsen JH, et al. Diagnostic accuracy of postmortem computed tomography, magnetic resonance imaging, and computed tomography-guided biopsies for the detection of ischaemic heart disease in a hospital setting. Eur Heart J Cardiovasc Imaging. 2018;19(7):739–48.
40. Shiotani S, Watanabe K, Kohno M, Ohashi N, Yamazaki K, Nakayama H. Postmortem computed tomographic (PMCT) findings of pericardial effusion due to acute aortic dissection. Radiat Med—Med Imaging Radiat Oncol. 2004;22(6):405–7.
41. Watanabe S, Hyodoh H, Shimizu J, Okazaki S, Mizuo K, Rokukawa M. Classification of hemopericardium on postmortem CT. Leg Med. 2015;17(5):376–80.
42. Filograna L, Laberke P, Ampanozi G, Schweitzer W, Thali MJ, Bonomo L. Role of postmortem computed tomography (PMCT) in the assessment of the challenging diagnosis of pericardial tamponade as cause of death in cases with hemopericardium. Radiol Med. 2015;120(8):723–30.
43. Ampanozi G, Flach PM, Ruder TD, Filograna L, Schweitzer W, Thali MJ, et al. Differentiation of hemopericardium due to ruptured myocardial infarction or aortic dissection on unenhanced postmortem computed tomography. Forensic Sci Med Pathol. 2017;13(2):170–6.
44. Ampanozi G, Held U, Ruder TD, Ross SG, Schweitzer W, Fornaro J, et al. Pulmonary thromboembolism on unenhanced postmortem computed tomography: Feasibility and findings. Leg Med (Tokyo). 2016;20:68–74.
45. Mueller SL, Thali Y, Ampanozi G, Flach PM, Thali MJ, Hatch GM, et al. Distended diameter of the inferior vena cava is suggestive of pulmonary thromboembolism on unenhanced postmortem CT. J Forens Radiol Imaging. 2015;3(1):38–42.
46. Jackowski C, Grabherr S, Schwendener N. Pulmonary thrombembolism as cause of death on unenhanced postmortem 3T MRI. Eur Radiol. 2013;23(5):1266–70.
47. Tadlock MD, Chouliaras K, Kennedy M, Talving P, Okoye O, Aksoy H, et al. The origin of fatal pulmonary emboli: A postmortem analysis of 500 deaths from pulmonary embolism in trauma, surgical, and medical patients. Am J Surg. 2015;209(6):959–68.
48. Sherk WM, Khaja MS, Williams DM. Anatomy, pathology, and classification of aortic dissection. Tech Vasc Interv Rad. 2021;24(2).
49. Elefteriades JA, Barrett PW, Kopf GS. Litigation in nontraumatic aortic diseases: A tempest in the malpractice maelstrom. Cardiol. 2008;109(4):263–72.
50. Ledbetter S, Stuk JL, Kaufman JA. Helical (spiral) CT in the evaluation of emergent thoracic aortic syndromes: Traumatic aortic rupture, aortic aneurysm, aortic dissection, intramural hematoma, and penetrating atherosclerotic ulcer. Radiol Clin N Am. 1999;37(3):575–+.
51. Schoder M, Prokop M, Lammer J. Traumatic injuries: Imaging and intervention of large arterial trauma. Eur Radiol. 2002;12(7):1617–31.
52. Chatzaraki V, Heimer J, Thali M, Dally A, Schweitzer W. Role of PMCT as a triage tool between external inspection and full autopsy: Case series and review. J Forens Radiol Imaging. 2018;15:26–38.
53. Shiotani S, Kohno M, Ohashi N, Yamazaki K, Nakayama H, Ito Y, et al. Hyperattenuating aortic wall on postmortem computed tomography (PMCT). Radiat Med. 2002;20(4):201–6.
54. Ampanozi G, Flach PM, Fornaro J, Ross SG, Schweitzer W, Thali MJ, et al. Systematic analysis of the radiologic findings of aortic dissections on unenhanced postmortem computed tomography. Forensic Sci Med Pathol. 2015;11(2):162–7.

PART 5

OTHER DIAGNOSES

CHAPTER 31

NORMAL POSTMORTEM CHANGES

Garyfalia Ampanozi, Carlo Tappero, and Wolf Schweitzer

Knowledge of what may be expected as changes which normally take place in a human body after death and that are without any further diagnostic significance is a requirement for a successful identification of pathology or trauma. These postmortem changes are specific, so a clinical radiologist should know them in order to avoid pitfalls when reading postmortem computed tomography (PMCT) images.

Adding to this, the mere absence, reduction, or any excess of postmortem alterations anticipated as normal, can bear diagnostic significance. As an example, the absence of livor mortis (i.e. blue-red discoloration of the skin) may indicate fatal hemorrhage, and along the same lines, also absence of hypostatic blood settled in the lungs may be found in a case with fatal hemorrhage.

DOI: 10.4324/9781003413998-36

31.1 EARLY VS. LATE POSTMORTEM CHANGES

Postmortem decay of the body begins right after death and is influenced by the condition of the body as well as ambient factors such as temperature. The changes are categorized as "early" and "late" changes. Early changes comprise supravital reactions, livor mortis, rigor mortis and convergence of the body temperature to the ambient temperature, which in most practical situations will be a process of cooling of the body over time. That early time after death may extend over one or two days. Later changes or decomposition then may contain putrefaction, mummification, as well as animal predation and skeletonization, and that period may cover the time interval starting at around one to three days after death [1, 2].

31.2 EARLY POSTMORTEM CHANGES

Earliest postmortem changes in PMCT are well identifiable in the vasculature, heart and lungs. While extensive gas formation is a clear hallmark of late postmortem changes, gas may be present earlier.

31.2.1 Gas

Gas formation can occur as a common normal postmortem change. It may be encountered relatively soon after death and it may be visible earliest in the liver vessels (Figure 31.1). When reporting gas on PMCT, it may be a good idea to also report the time interval since death or estimated time of death, and, the overall distribution of gas to possibly discern various types of gas such as intravascular air by injection or an injury, bacterial gas from an infection, putrefactive or decompressive gas.

One particular pitfall that may occur when not considering the normal postmortem nature may consist in attributing excessive significance as to trauma or pathology to such a finding. Another pitfall is to overlook it given the actual postmortem interval, as gas also is a marker of some significance other than being a normal postmortem change: it may indicate fatal gas embolism and as that, it is present in the right heart and pulmonary vasculature at death already. And it may indicate bacterial infection, upon which gas formation may be observed to progress more rapidly than in cases with no bacterial infection. As a large number of factors play into postmortem gas formation, there currently are no ready-to-use tables as reliable guide as to just how much, where and when there is gas to be expected—but gas embolism and bacterial infection may constitute relevant differential diagnoses to normal postmortem gas formation.

In cases of bacterial infection, gas may appear more rapidly, so putting the presence of gas in the context of known or estimated time since death may give a clue as to the possible presence of a bacterial infection, particularly in the early postmortem period [3]. As gas formation may become excessive as part of late postmortem changes, at some stage, no differential diagnosis can be made any more. At some stage, one might make the time of PMCT an urgency as to avoid scanning after an unnecessarily long delay. Delaying PMCT for hours or days will obstruct gas interpretation in cases where the body is admitted a mere hours after death.

With intraosseous access lines in place, increased gas was found as early as 8 hours after death [4].

The delineation of findings suggestive of gas embolism, particularly fatal gas embolism, from normal postmortem gas can be challenging. To approach this question, one research group defined a radiological alteration index (RAI); according to their study, a grade III finding of gas in the heart may be attributed to fatal gas embolism if a case is attributed less than 50 points on that scale [5].

31.2.2 Postmortem Blood Clots

Postmortem clot formation both in the great vessels and in the heart cavities is normal, and their visibility on PMCT thus constitutes a normal postmortem alteration. The differential diagnosis however is that of thrombosis and embolism.

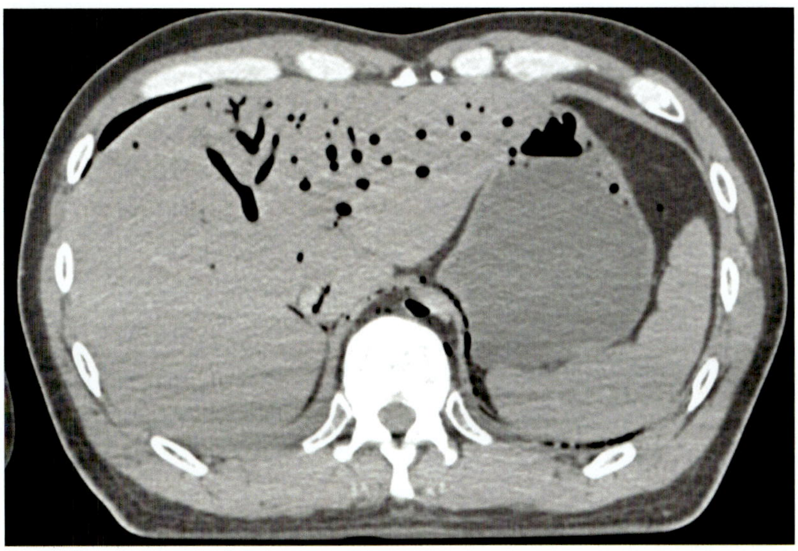

Figure 31.1 Postmortem computed tomography, axial image through the abdomen. Gas in the liver vessels as an early decomposition sign.

Especially blood clots depicted in the pulmonary artery are of great interest, since any true antemortem pulmonary artery thromboembolism is a particularly relevant finding.

Differentiation between postmortem clots and true thromboembolism is currently not possible in a fully reliable way on unenhanced PMCT and only some indirect signs can be used [6].

Kasagawa et al. described four types (block-like, cast-like, fluid level-like and unclear) of hyperdense formations in the heart cavities in PMCT [7] and found out that only the first two of them corresponded to blood clots at autopsy [7].

PMCT Angiography as well as postmortem magnetic resonance imaging (PMMR) could prove helpful in such cases [8, 9]. Usually, we recommend autopsy to disambiguate such findings.

31.2.3 Blood Sedimentation and Hypostasis

Livor mortis may become evident on the skin of the deceased within 30 minutes to 2 hours after death, but it may be absent in cases of fatal hemorrhage or severe anemia. Cutaneous livor mortis can also be depicted on PMCT as thickening of the skin and hyperdensity of the subcutaneous fat of the dependent body regions [10]. Internal correlates for livor mortis are typically seen in the lungs, where the appearance of hypostatic blood accumulation tends to be that of ground glass opacities. These are typically clearly distinguishable from the opposite regions of the lung (Figure 31.2) [11–13].

A diagnostic criterium for normal postmortem hypostasis tends to be that of a relatively straight demarcation line, which can help to discriminate hypostatic blood accumulations in the lungs from findings in viral pneumonia. So, a pitfall for clinical radiologists may be to overinterpret hypostasis in the lungs as indicative of pneumonia. Differentiation of normal internal livor mortis from pathological lung alterations such as pneumonia may be facilitated using the techniques of external inflation of the lungs (postmortem ventilation [14]).

Conversely, a normal appearance of lungs after death always contains hypostasis. A reduced or absent density difference in lungs between the positionally dependent areas and their opposite regions, in a sense of reduced or absent hypostasis, may be correlated with trauma or pathology. One differential diagnosis there is mechanical asphyxia [15], and similar findings of low or absent transpulmonary density gradient may be identified after fatal hemorrhage [16] or in cases of fatal hypothermia [17, 18].

However, many pulmonary postmortem CT findings may only indicate the possibility of such options, and autopsy will be required to clarify or validate findings, and to arrive at a comprehensive diagnosis.

As the capillaries tend to lose their capacity to serve as containment with increasing postmortem interval, a visual discrimination of hypostatic blood settlement in the lungs with a clear demarcation line may disappear after a few days.

In the great vessels and the heart cavities, blood may exhibit sedimentation of red blood cells (Figure 31.3). Hypostasis or gravity dependent settling of blood is a normal postmortem phenomenon, which starts immediately after cessation of heart activity and blood circulation. The blood is settling down to the lowest body parts, thus following gravity. This will depend on the position of the body at time of death. Livor mortis may be visible during external inspection of the deceased as a purplish-red discoloration of the skin on the dependent parts of the body, particularly if the skin has only relatively little pigmentation, and, depending on ambient temperature and conditions, typically for the first one or two days after death. Later, discoloration in context of later postmortem changes may obfuscate the appearance of typical livor mortis.

Also, hemorrhage—extravasated blood—may appear with a clear sedimentation line on PMCT. So, blood collections in the sense of a hemorrhage—be it in the soft tissues as part of decollement, hemothorax, hemopericardium, hemoperitoneum or hemorrhage in the skull cavity—may all exhibit sedimentation levels.

This sedimentation has been correlated to high fibrinogen concentrations antemortem, which could be interpreted as an indication for the cause of death [19]. An expression of this phenomenon is also the denser dorsal part of the superior sagittal sinus (in cadavers in a supine position), which should not be confused with antemortem thrombosis (Figure 31.4) [20, 21].

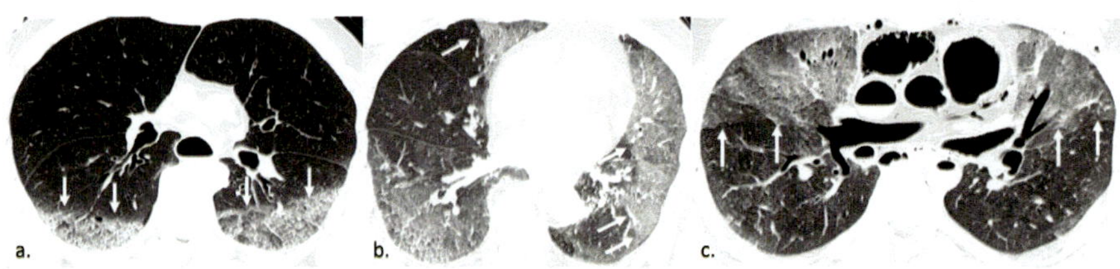

Figure 31.2 Postmortem computed tomography, axial images through the thoracic cavity. Internal livores (arrows) in the lungs corresponding to a) supine, b) initial left sided and c) prone position of the deceased. https://creativecommons.org/licenses/by-nc-nd/4.0/

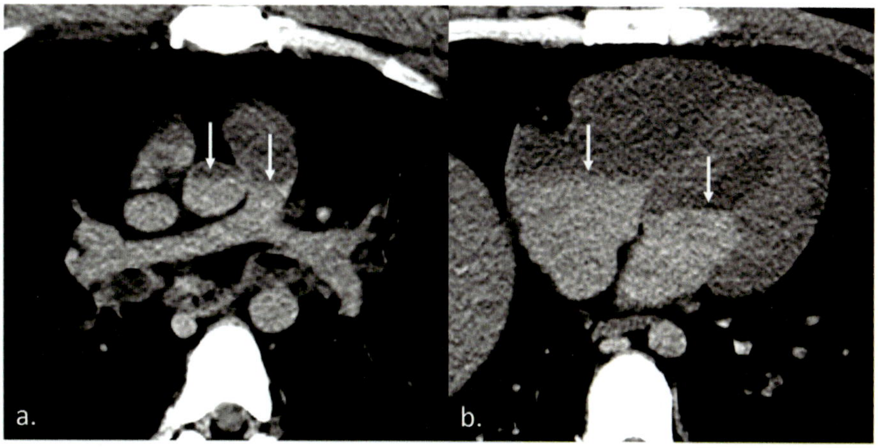

Figure 31.3 Postmortem computed tomography, axial views through a) the great vessels and b) the heart. Sedimentation of blood in the lumen of the vessels and in the heart cavities (arrows). https://creativecommons.org/licenses/by-nc-nd/4.0/.

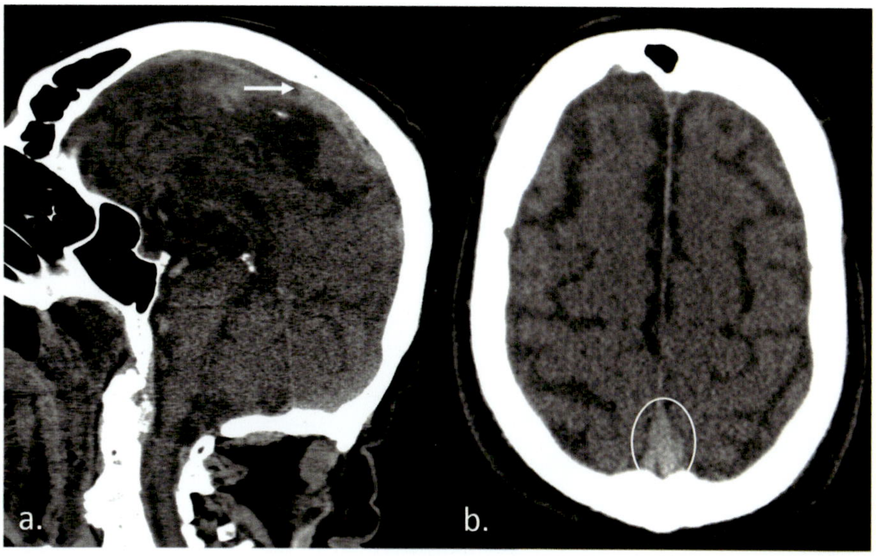

Figure 31.4 Postmortem computed tomography, a) sagittal and b) axial view of the head. Note the denser content of the dorsal part of the superior sagittal sinus (arrow in a, encircled in b). This should not be mistaken as an antemortem thrombosis. https://creativecommons.org/licenses/by-nc-nd/4.0/.

31.2.3.1 Rigor Mortis

Rigor mortis is, in part, a temperature dependent phenomenon and typically describes the state of skeletal muscles. It is a phenomenon that may have an onset at around 1–2 hours after death and may endure for around 1–2 days. Rigor mortis may obstruct a proper body position in the CT scanner which is why a larger gantry diameter and a longer scanning table are of greatest advantage for PMCT.

Stiffness of all body muscles may also affect the heart and there, it may yield a hypertrophy-like appearance of the human heart [22]. This should not be mistaken as proof of pathology: studies have shown that heart wall measurements on postmortem imaging often overestimate the actual myocardial thickness [23, 24] and that PMCT heart measurements are significantly greater than AMCT measurements [25].

31.2.4 Other Density and Volume Differences

Hyperattenuating aortic wall is a further normal PMCT finding. This phenomenon was first described by Shiotani et al. [26] who attributed it to two main factors. First, a decrease in the aortic diameter results in thickening of the aortic wall and increased CT attenuation. Second, hemodilution occurs due to fluid infusions during cardiopulmonary resuscitation (CPR), leading to a decrease in the usual blood CT attenuation within the aortic lumen. Additionally, absence of motion artifacts further contributes to the observed hyperattenuation [26]. Moreover, Okuma et al. conducted a comparison between antemortem and postmortem CT scans of the same individuals and discovered a significant thickening of the aortic wall in the postmortem scans [27]. Furthermore, Hyodoh et al. demonstrated that the aortic diameter decreases and the abdominal aorta loses its typical round shape, particularly noticeable in younger individuals without calcification [28].

Radiodensity of cerebrospinal fluid and vitreous humor may change postmortem, even allowing to approximate an estimation of time since death [29, 30]. Increase of the above-mentioned densities seems to take place independently of the temperature, and at least cerebrospinal fluid radiodensity correlates linearly to postmortem interval [29].

Gray and white matter PMCT densities: there is a decrease of the Hounsfield units (HUs) density of the brain grey matter rapidly after death followed by an increase of the white matter HU [31]. Gray-white matter differentiation is gradually lost on postmortem CT images of the brain with increasing time since death [21].

Organs' densities may change after death; a study has shown that thyroid gland's CT values may be decreased compared to antemortem measurements on same individuals [32].

Organs' volumes may also be affected; the volume of liver, spleen and kidneys all decreased significantly on postmortem CTs compared to antemortem CTs [30, 33]. Adrenal glands were also decreased in volume in early postmortem interval, as compared to antemortem measurements [34]. On the other hand, bowel volume increased over time as was expected because of decompositional gas formation [30].

31.3 LATE POSTMORTEM CHANGES

In a state of putrefaction, main findings to be expected are:

- Gas accumulation into soft tissues, such as fat and muscle, as well as organ parenchyma and body cavities, may increase with increasing postmortem interval and increasing degree of decomposition; Egger et al. developed an index, the radiological alteration index (RAI [5]), for enabling quantification of gas in cadavers undergone PMCT.
- Free and partly sedimented fluid in the pleural, abdominal and cranial cavity, accompanied with liquification of the organs (Figure 31.5) [35], often with a CT-density of fat or oil.
- Loss of CT-typical organ or tissue structure and delineation, loss of consistency and liquification. In cases of progressed decomposition and mummification the internal organs may not be any more identifiable [10] and density of soft tissues may increase considerably. If brain parenchyma is altered due to putrefaction, postmortem imaging including PMCT and PMMR, especially the latter one, may be superior to autopsy regarding assessing brain morphology [36]. Still, fatal hemorrhages may be easier to find on PMCT than at autopsy [37]. The eye lenses may settle to the gravity-dependent side (Figure 31.5) and so may the organs, which will also progressively shrink.

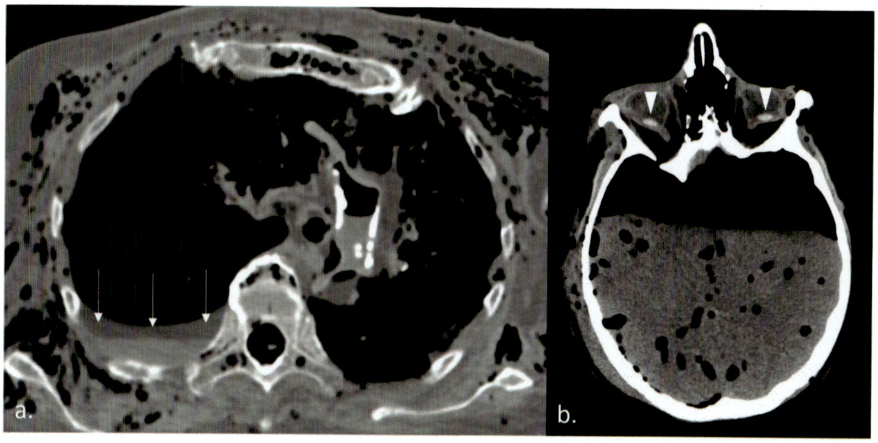

Figure 31.5 Postmortem computed tomography, axial image through a) the thoracic cavity b) the head. Free fluid, with sedimentation, in the right thoracic cavity (arrows). Putrefaction gas in soft tissues and great vessels. Liquification of the brain. Settling back of the eye lenses (arrowheads). https://creativecommons.org/licenses/by-nc-nd/4.0/.

Case-related data published and used under Cantonal Ethics Committee of Canton of Zürich waiver 15–0686.

References

1. Madea B. History of forensic medicine. Handbook of Forensic Medicine. 2014:1–14.
2. Brinkmann B, Madea, B. Handbuch gerichtliche Medizin 1. Brinkmann BMB, editor: Springer; 2004.
3. Baumeister R, Gauthier S, Schweitzer W, Thali MJ, Mauf S. Small—but fatal: Postmortem computed tomography indicated acute tonsillitis. Journal of Forensic Radiology and Imaging. 2016;6:52–6.
4. Hillewig E, Aghayev E, Jackowski C, Christe A, Plattner T, Thali MJ. Gas embolism following intraosseous medication application proven by postmortem multislice computed tomography and autopsy. Resuscitation. 2007;72(1):149–53.
5. Egger C, Vaucher P, Doenz F, Palmiere C, Mangin P, Grabherr S. Development and validation of a postmortem radiological alteration index: The RA-Index. International Journal of Legal Medicine. 2012;126(4):559–66.
6. Ampanozi G, Held U, Ruder TD, Ross SG, Schweitzer W, Fornaro J, et al. Pulmonary thromboembolism on unenhanced postmortem computed tomography: Feasibility and findings. Legal Medicine (Tokyo). 2016;20:68–74.
7. Kasagawa A, Usui A, Kawasumi Y, Funayama M, Saito H. Block-like and cast-like hyperdense areas in the right heart cavities on postmortem CT strongly suggest the presence of intracardiac blood clots at autopsy. European Radiology. 2021;31(12):8879–86.

8. Burke MP, Bedford P, Baber Y. Can forensic pathologists diagnose pulmonary thromboembolism on postmortem computed tomography pulmonary angiography? The American Journal of Forensic Medicine and Pathology. 2014;35(2):124–31.
9. Flach PM, Franckenberg S, Gascho D, Ampanozi G, Thali MJ, Fliss B. Detection of pulmonary thrombembolism and postmortem clotting on postmortem magnetic resonance imaging. Forensic Imaging. 2024;36:200574.
10. Charlier P, Carlier R, Roffi F, Ezra J, Chaillot PF, Duchat F, et al. Postmortem abdominal CT: Assessing normal cadaveric modifications and pathological processes. European Journal of Radiology. 2012;81(4):639–47.
11. Hyodoh H, Shimizu J, Watanabe S, Okazaki S, Mizuo K, Inoue H. Time-related course of pleural space fluid collection and pulmonary aeration on postmortem computed tomography (PMCT). Legal Medicine (Tokyo). 2015;17(4):221–5.
12. Michiue T, Ishikawa T, Oritani S, Kamikodai Y, Tsuda K, Okazaki S, et al. Forensic pathological evaluation of postmortem pulmonary CT high-density areas in serial autopsy cases of sudden cardiac death. Forensic Science International. 2013;232(1):199–205.
13. Shiotani S, Kohno M, Ohashi N, Yamazaki K, Nakayama H, Watanabe K, et al. Non-traumatic postmortem computed tomographic (PMCT) findings of the lung. Forensic Science International. 2004;139(1):39–48.
14. Germerott T, Preiss US, Ebert LC, Ruder TD, Ross S, Flach PM, et al. A new approach in virtopsy: Postmortem ventilation in multislice computed tomography. Legal Medicine (Tokyo). 2010;12(6):276–9.
15. Schweitzer W, Thali M. Fatal obstructive asphyxia: Trans-pulmonary density gradient characteristic as relevant identifier in postmortem CT. Journal of Forensic Radiology and Imaging. 2019;19.
16. Schober D, Schwendener N, Zech WD, Jackowski C. Postmortem CT: Hounsfield unit profiles obtained in the lungs with respect to the cause of death assessment. International Journal of Legal Medicine. 2017;131(1):199–210.
17. Schweitzer W, Thali M, Giugni G, Winklhofer S. Postmortem pulmonary CT in hypothermia. Forensic Science, Medicine and Pathology. 2014;10(4):557–69.
18. Zeng YW, Zhang XY, Yoshizumi I, Zhang Z, Mizuno T, Sakamoto S, et al. Deep learning-based diagnosis of fatal hypothermia using post-mortem computed tomography. Tohoku Journal of Experimental Medicine. 2023;260(3):253–61.
19. Ishida M, Gonoi W, Hagiwara K, Takazawa Y, Akahane M, Fukayama M, et al. Hypostasis in the heart and great vessels of non-traumatic in-hospital death cases on postmortem computed tomography: Relationship to antemortem blood tests. Legal Medicine. 2011;13(6):280–5.
20. Ishida M, Gonoi W, Okuma H, Shirota G, Shintani Y, Abe H, et al. Common postmortem computed tomography findings following atraumatic death: Differentiation between normal postmortem changes and pathologic lesions. Korean Journal of Radiology. 2015;16(4):798–809.
21. Wagensveld IM, Blokker BM, Wielopolski PA, Renken NS, Krestin GP, Hunink MG, et al. Total-body CT and MR features of postmortem change in in-hospital deaths. PloS One. 2017;12(9):e0185115.
22. Klein WM, Bosboom DGH, Koopmanschap DHJLM, Nievelstein RAJ, Nikkels PGJ, van Rijn RR. Normal pediatric postmortem CT appearances. Pediatric Radiology. 2015;45(4):517–26.
23. Ampanozi G, Hatch GM, Flach PM, Thali MJ, Ruder TD. Postmortem magnetic resonance imaging: Reproducing typical autopsy heart measurements. Legal Medicine (Tokyo). 2015;17(6):493–8.
24. Chatzaraki V, Thali MJ, Schweitzer W, Ampanozi G. Left myocardial wall measurements on postmortem imaging compared to autopsy. Cardiovascular Pathology: The Official Journal of the Society for Cardiovascular Pathology. 2019;43:107149.
25. Okuma H, Gonoi W, Ishida M, Shintani Y, Takazawa Y, Fukayama M, et al. Heart wall is thicker on postmortem computed tomography than on antemortem [corrected] computed tomography: The first longitudinal study. PloS One. 2013;8(9):e76026.
26. Shiotani S, Kohno M, Ohashi N, Yamazaki K, Nakayama H, Ito Y, et al. Hyperattenuating aortic wall on postmortem computed tomography (PMCT). Radiation Medicine. 2002;20(4):201–6.
27. Okuma H, Gonoi W, Ishida M, Shintani Y, Takazawa Y, Fukayama M, et al. Greater thickness of the aortic wall on postmortem computed tomography compared with antemortem computed tomography: The first longitudinal study. International Journal of Legal Medicine. 2014;128(6):987–93.
28. Hyodoh H, Sato T, Onodera M, Washio H, Hasegawa T, Hatakenaka M. Vascular measurement changes observed using postmortem computed tomography. Japanese Journal of Radiology. 2012;30(10):840–5.
29. Koopmanschap DH, Bayat AR, Kubat B, de Bakker HM, Prokop MW, Klein WM. The radiodensity of cerebrospinal fluid and vitreous humor as indicator of the time since death. Forensic Science, Medicine and Pathology. 2016;12(3):248–56.
30. Klein WM, Kunz T, Hermans K, Bayat AR, Koopmanschap DHJLM. The common pattern of postmortem changes on whole body CT scans. Journal of Forensic Radiology and Imaging. 2016;4:47–52.
31. Nishiyama Y, Kanayama H, Mori H, Tada K, Yamamoto Y, Katsube T, et al. Whole brain analysis of postmortem density changes of grey and white matter on computed tomography by statistical parametric mapping. European Radiology. 2017;27(6):2317–25.
32. Ishida M, Gonoi W, Hagiwara K, Takazawa Y, Akahane M, Fukayama M, et al. Postmortem changes of the thyroid on computed tomography. Legal Medicine (Tokyo). 2011;13(6):318–22.
33. Takahashi N, Yajima K, Otaki M, Yoshikawa Y, Ishihara A, Sato Y, et al. Postmortem volume change of the spleen and kidney on early postmortem computed tomography: Comparison with antemortem computed tomography. Japanese Journal of Radiology. 2019;37(7):534–42.

34. Ishida M, Gonoi W, Hagiwara K, Okuma H, Shirota G, Shintani Y, et al. Early postmortem volume reduction of adrenal gland: Initial longitudinal computed tomographic study. Radiology Medicine. 2015;120(7):662–9.
35. Cartocci G, Santurro A, Neri M, Zaccagna F, Catalano C, La Russa R, et al. Postmortem computed tomography (PMCT) radiological findings and assessment in advanced decomposed bodies. Radiology Medicine. 2019;124(10):1018–27.
36. Tschui J, Jackowski C, Schwendener N, Schyma C, Zech WD. Postmortem CT and MR brain imaging of putrefied corpses. International Journal of Legal Medicine. 2016;130(4):1061–8.
37. Tappero C, Thali MJ, Schweitzer W. The possibility of identifying brain hemorrhage in putrefied bodies with PMCT. Forensic Science, Medicine and Pathology. 2020;16(4):571–6.

CHAPTER 32

STAY COOL!—FREEZING AND THAWING IN FORENSIC MEDICINE

Dominic Gascho and Niklaus Zoelch

DOI: 10.4324/9781003413998-37

32.1 INTRODUCTION

The freezing of a homicide victim and the night-time disposal of the dead body in secluded places by the perpetrator have been described in many crime novels. When the corpse is found and still frozen inside, this condition provides crucial clues for the investigator, as told, for example, in the crime novel *Tell's Söhne* by Res Perrot [1]. Although rarely described in the scientific literature, the freezing of murder victims and subsequent efforts to dispose of the body also occur in real life [2,3]. A famous example of this is the story of the serial killer and professional hitman Richard Kuklinski [4,5]. "The Iceman" was arrested in 1986 after investigators finally got on his trail by finding a partially frozen body. For a long time, however, he succeeded in confusing investigators by freezing a body for a long period of time and disposing of it after it thawed. Failure to recognize that a dead body was previously frozen may result in an incorrect estimate of the time since death because decomposition changes, rigor mortis, and livor mortis, which are included in such estimates, behave differently after freezing and thawing. A previously frozen body presents an alteration of the natural decomposition process, namely an "outside-in" manner of decomposition instead of a naturally occurring "inside-out" decomposition process [6,7]. This means that the decomposition rate inside the body is slowed down after freezing, which can be explained by the fact that a body thaws from the outside to the inside and also by the fact that the viability of some bacteria species decreases due to freezing. This change in the usual decomposition process results in an altered appearance in terms of skin discoloration and bloating of the body and thus decomposition changes detected during an external examination could lead to an incorrect estimate of the time since death. Hardening of the skin and skeletal muscles is another effect after freezing [8], which may lead to confusion regarding rigor mortis. Diffusion of hemoglobin into skin tissue caused by intravascular hemolysis can turn skin color to a reddish appearance [9], which in turn may lead to confusion with livor mortis. If the external appearance of the body raises suspicion that it may have been frozen and thawed, specific characteristics are sought to confirm such suspicion. Macroscopic findings noted during autopsy of thawed cadavers are described in the literature. Such post freezing findings may resemble soft tissue injuries, so-called pseudo-bruising, and such pseudo-bruising is likely to result from leakage of red blood cells from autolytically engorged venous plexuses as hypostatic pressure progressively increases [10]. Dark red fluid that may also be found in the body cavities of thawed bodies is associated with accumulated extracellular fluid, with the dark red color coming from hemoglobin from lysed red blood cells [11]. Additional histological examinations may reveal micro-morphological changes resulting from freezing and subsequent thawing of biological tissue [8]. Such post-freezing histological artifacts have been described as long and parallel-shaped empty areas [8], chains of pseudo-bubbles [12], or columnar bubble-like spaces [13]. However, it is considered difficult to distinguish such results from ordinary decomposition processes. In summary, the macroscopic and microscopic findings described are too inconclusive to provide definitive evidence that a body was previously frozen. Therefore, efforts were made to evaluate additional imaging examination techniques with regard to the detection of artifacts after freezing. In the following, it will be discussed which changes of the tissue can be detected by computed tomography (CT), magnetic resonance tomography (MRI) and magnetic resonance spectroscopy (MRS).

32.2 WHAT TISSUE CHANGES CAN BE SEEN ON CT AFTER FREEZING AND THAWING?

Frozen soft tissue presents a contrast to unfrozen soft tissue on computed tomography, and the transition between frozen and unfrozen tissue is usually a steep gradient, resulting in a relatively sharp demarcation on the CT image (Figure 32.1).

Computed tomography is also sensitive to the post-freezing artifacts which are described histologically as empty areas or bubbles. A previously frozen and now thawed body shows tiny and randomly arranged bubble-like patterns in the brain which can be detected by CT (Figure 32.2) [13]. The formation of these bubbles is not fully understood, but can probably be attributed to the rupture of cell membranes after freezing and during thawing. At the beginning of a freezing process extracellular water starts to freeze and forms ice crystals, which can lead to direct damage to the cell membrane on the one hand, but also to indirect damage because it also leads to an increased concentration of solutes in the fraction of the extracellular fluid that is not yet frozen, which then leads to intracellular fluid diffusing out of the cell due to increased osmotic pressure, which eventually leads to osmotically induced cell dehydration or cell shrinkage, and finally, upon thawing, the cell membrane may rupture due to the net flow of fluid returning into the cells [8,11,14].

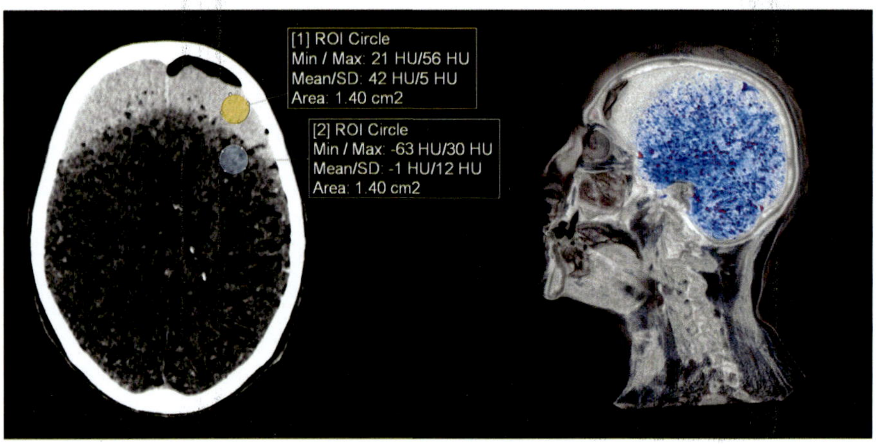

Figure 32.1 Axial CT slice (left image) and 3D volume rendering (right image) of a deceased person who was partially frozen. Region-of-interest (ROI) measurements show the quantitative differences in CT numbers in Hounsfield units (HU) between non-frozen (ROI 1) and assumingly still frozen (ROI 2) brain tissue. Visually, there is an impressive contrast between still frozen brain tissue and already slightly thawed brain tissue.

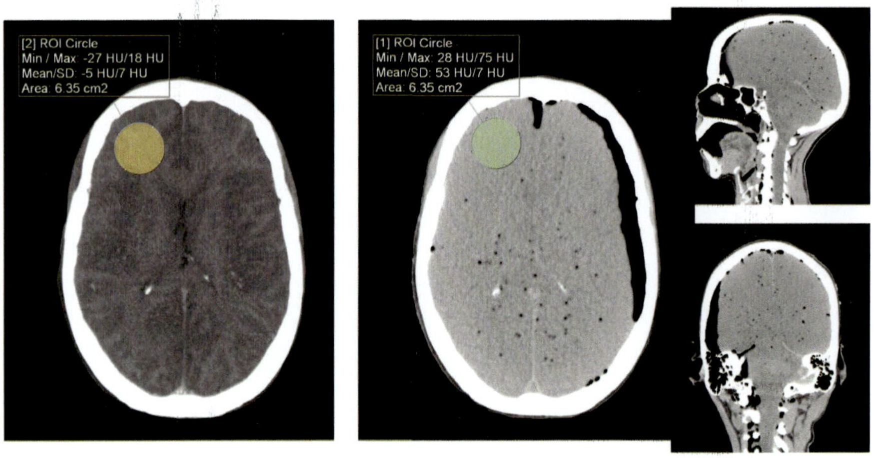

Figure 32.2 Head CT of a frozen body (left image) and after complete thawing about 60 hours later (images on the right). After thawing, randomly distributed bubble-like hypodense structures can be seen in the brain tissue. Note also the circular region-of-interest (ROI) measurement of the CT attenuation in Hounsfield units (HU). The assumingly still frozen brain has a very low mean value of −5 HU (SD: 7 HU) compared to the relatively high mean value of 53 HU (SD: 7 HU) in the thawed brain.

To a lesser extent, these bubbles are also seen in the thoracic and abdominal organs. However, such post-thawing bubble-like patterns on CT images are temporarily unstable and thus difficult to distinguish from decomposition-induced gases that form during ongoing postmortem intervals. Thus, only the clear recognition of a bubble-like pattern in the early decomposition stage is indicative of a previously frozen and now thawed body, but its absence is not an exclusion.

The radiodensity of tissue also shows differences after freezing and thawing. However, in this context, increased CT attenuation values, expressed in Hounsfield units (HUs), are to be expected in thawed tissue as demonstrated in human muscle tissue of amputated arms from body donors which showed strong changes in the measured Hounsfield units before freezing and after complete thawing [15]. Note also the mean value of 53 HU in the thawed brain tissue illustrated in Figure 32.1. A healthy brain of a living human has a mean value of about 35 HU [16]. The muscle tissue study [15] further showed a difference in whether the amputated arm muscle sample was thawed at a cool temperature of 2°C over a period of 48 hours or at room temperature of 19°C over a period of 24 hours. The increase in HU values was stronger for the muscle tissue thawed at room temperature, and this increased HU value remained constant over repeated freeze-thaw cycles. These increased HU values correspond to the description of hardening of skeletal muscles. Thus, a postmortem CT measurement shows changes caused by freezing and thawing in the form of bubbles in the organs and increased CT attenuation values.

32.3 WHICH FREEZE-THAW EFFECTS CAN BE DETECTED WITH MRI AND MRS?

MRI is even more sensitive than CT for depicting the aforementioned bubble-like patterns in the tissue. In addition, quantitative detection of freezing and thawing changes in tissue are also possible with magnetic resonance imaging, as has been demonstrated, for example, in studies on meat in food science. Nuclear magnetic resonance (NMR) imaging was used to study previously frozen meat samples by imaging-based relaxation time measurements [17–19]. Research in the field of cryobiology has used NMR spectroscopy to detect metabolic changes related to organ preservation techniques using pig kidneys and rat livers [20,21]. Inspired by these approaches, measurements of relaxation times and metabolites were performed on sheep legs in a forensic context to detect the changes in tissue induced by the freezing process [22]. In this study on the severed legs of seven sheep, one severed sheep leg was frozen at −20°C and thawed (study group), and the other severed sheep leg was stored at room temperature of +20°C (control group). At specific time intervals, repeated proton MRS single voxel measurements in muscle tissue and bone marrow were performed with a clinical 3 Tesla scanner and a corresponding thigh coil.

In this controlled setting, T_1 of water was shorter, while T_2 appeared longer in muscle tissue after freezing and thawing. However, these changes were considered too small to detect an earlier freezing state in a forensic context.

The ^{1}H-MRS measurements in the bone marrow showed a previously unknown effect. The ratio between the signal from bulk methylene appearing at 1.3 ppm (*lip1.3*) and the sum of detected signal peaks appearing at 5.3 and 5.2 ppm (*lip5*) was significantly lower after freezing and thawing (Figure 32.3). This change in the *lip1.3/lip5* ratio indicates an effective decrease in the bulk methylene proton concentrations in the measured volume. The number of bulk methylene protons is related to the length of the carbon chains of the fatty acids, and it can be assumed that certain processes may alter the fatty acid chains during freezing. The methylene protons may be bound to other compounds, so that their signal contributes to other resonances or is no longer detectable due to a much shorter T_2 due to for instance reduced mobility. Some methylene protons might also have leaked from the measured volume. The exact cause of the change in the *lip1.3/lip5* ratio has not yet been clarified. Moreover, it is not yet clear whether this effect is also observed in human bone

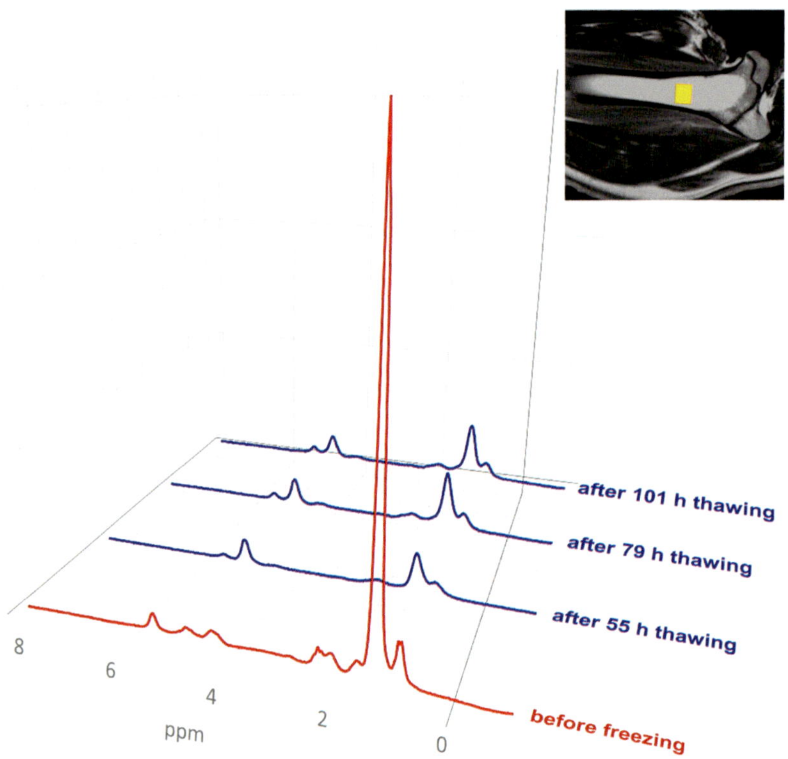

Figure 32.3 1H-MRS spectrum obtained from bone marrow measurements in a sheep leg before freezing and approximately 55 hours, 79 hours, and 101 hours after thawing at room temperature (20°C). The signal from bulk methylene appearing at 1.3 ppm (*lip1.3*) is considerably reduced after freezing and thawing. The ratio between the signal from *lip1.3* and the sum of detected signal peaks appearing at 5.3 and 5.2 ppm (*lip5*) was significantly lower after freezing and thawing.

marrow and whether it actually becomes visible only at sub-zero temperatures or already at low temperatures above 0°C, in which case it could be an indication of hypothermia. Therefore, further research is needed to provide robust evidence for an earlier freezing and thawing of a found body.

References

[1] R. Perrot, Tells Söhne: Ein recht extremer Fall für Wachtmeister Grossenbacher, Gmeiner-Verlag, 2014.

[2] L. Fanton, H. Yappo-Ette, C. Vianey-Saban, D. Malicier, Homicide followed by freezing and burning: Usefulness of measuring SCHAD (short-chain 3-hydroxyacyl-CoA dehydrogenase) activity, Journal of Forensic and Legal Medicine. 13 (2006) 339–340. https://doi.org/10.1016/j.jcfm.2006.06.005.

[3] T.B. Olsen, P.M. Leth, Homicide victims concealed in a freezer, Forensic Science, Medicine and Pathology. (2018) 1–4. https://doi.org/10.1007/s12024-018-9988-4.

[4] F.T. Zugibe, J.T. Costello, The iceman murder: One of a series of contract murders, Journal of Forensic Sciences. 38 (1993) 1404–1408. https://doi.org/10.1520/JFS13544J.

[5] S. Bonn, Why We Love Serial Killers: The Curious Appeal of the World's Most Savage Murderers, Skyhorse Publishing, Inc., 2014.

[6] M.S. Micozzi, Experimental study of postmortem change under field conditions: Effects of freezing, thawing, and mechanical injury, Journal of Forensic Science. 31 (1986) 953–961. https://doi.org/10.1520/JFS11103J.

[7] L.G. Roberts, G.R. Dabbs, A taphonomic study exploring the differences in decomposition rate and manner between frozen and never frozen domestic pigs (Sus scrofa), Journal of Forensic Science. 60 (2015) 588–594. https://doi.org/10.1111/1556-4029.12717.

[8] A.Th. Schäfer, J.D. Kaufmann, What happens in freezing bodies? Experimental study of histological tissue change caused by freezing injuries, Forensic Science International. 102 (1999) 149–158. https://doi.org/10.1016/S0379-0738(99)00043-2.

[9] N. Tabata, M. Morita, J. Azumi, A frozen newborn infant: Froth in the air-passage after thawing, Forensic Science International. 108 (2000) 67–74. https://doi.org/10.1016/S0379-0738(99)00199-1.

[10] M.S. Pollanen, S.D.C. Perera, D.J. Clutterbuck, Hemorrhagic lividity of the neck: Controlled induction of postmortem hypostatic hemorrhages, The American Journal of Forensic Medicine and Pathology. 30 (2009) 322–326.

[11] W.D. Roe, B.D. Gartrell, S.A. Hunter, Freezing and thawing of pinniped carcasses results in artefacts that resemble traumatic lesions, The Veterinary Journal. 194 (2012) 326–331. https://doi.org/10.1016/j.tvjl.2012.03.028.

[12] J.L. Pechal, C.J. Schmidt, H.R. Jordan, M.E. Benbow, Frozen: Thawing and its effect on the postmortem microbiome in two pediatric cases, Journal of Forensic Science. 62 (2017) 1399–1405. https://doi.org/10.1111/1556-4029.13419.

[13] S.A. Bolliger, D. Tomasin, J. Heimer, H. Richter, M.J. Thali, D. Gascho, Rapid and reliable detection of previous freezing of cerebral tissue by computed tomography and magnetic resonance imaging, Forensic Science, Medicine and Pathology. 14 (2018) 85–94. https://doi.org/10.1007/s12024-018-9955-0.

[14] P.L. Steponkus, D.G. Stout, J. Wolfe, R.V.E. Lovelace, Possible role of transient electric fields in freezing-induced membrane destabilization, Journal of Membrane Biology. 85 (1985) 191–198. https://doi.org/10.1007/BF01871513.

[15] A.C. Klop, M.E.M. Vester, K.L. Colman, J.M. Ruijter, R.R. Van Rijn, R.-J. Oostra, The effect of repeated freeze-thaw cycles on human muscle tissue visualized by postmortem computed tomography (PMCT), Clinical Anatomy. 30 (2017) 799–804. https://doi.org/10.1002/ca.22917.

[16] L.A. Cala, G.W. Thickbroom, J.L. Black, D.W. Collins, F.L. Mastaglia, Brain density and cerebrospinal fluid space size: CT of normal volunteers, American Journal of Neuroradiology. 2 (1981) 41–47.

[17] S.D. Evans, Kevin.P. Nott, A.A. Kshirsagar, L.D. Hall, The effect of freezing and thawing on the magnetic resonance imaging parameters of water in beef, lamb and pork meat, International Journal of Food Science & Technology. 33 (1998) 317–328. https://doi.org/10.1046/j.1365-2621.1998.00165.x.

[18] L.D. Hall, S.D. Evans, K.P. Nott, Measurement of textural changes of food by MRI relaxometry, Magnetic Resonance Imaging. 16 (1998) 485–492. https://doi.org/10.1016/S0730-725X(98)00116-7.

[19] K.P. Nott, S.D. Evans, L.D. Hall, Quantitative magnetic resonance imaging of fresh and frozen-thawed trout, Magnetic Resonance Imaging. 17 (1999) 445–455. https://doi.org/10.1016/S0730-725X(98)00189-1.

[20] P.W. So, A.L. Busza, B.J. Fuller, Metabolic effects of citrate in liver during cold hypoxia studied by 1H NMR spectroscopy, Cryobiology. 36 (1998) 225–235. https://doi.org/10.1006/cryo.1998.2082.

[21] J. Nath, T.B. Smith, K. Patel, S.R. Ebbs, A. Hollis, D.A. Tennant, C. Ludwig, A.R. Ready, Metabolic differences between cold stored and machine perfused porcine kidneys: A 1H NMR based study, Cryobiology. 74 (2017) 115–120. https://doi.org/10.1016/j.cryobiol.2016.11.006.

[22] D. Gascho, H. Richter, D.C. Karampinos, J. Heimer, S. Schaerli, M.J. Thali, N. Zoelch, Noninvasive in situ proton MRS in muscle tissue and bone marrow as a novel approach to identify previous freezing in a completely thawed cadaver, NMR in Biomedicine. 33 (2020) e4220.

PART 6

RADIOLOGIC IDENTIFICATION

CHAPTER 33

IDENTIFICATION OF HUMAN REMAINS

Wolf Schweitzer, Rilana Baumeister, and Jakob Heimer

DOI: 10.4324/9781003413998-39

33.1 MEDICOLEGAL AND FORENSIC BACKGROUND

The reason that identification of human remains is an issue in forensic medicine relates to the flip side of another reality: the cases where general practitioners or hospital doctors do sign a death certificate. There, a requirement for such cases is that the identification of the deceased is established and known with certainty.

The identity of a body may be in doubt in cases where the facial and bodily features are otherwise well recognizable but for some reason cannot be matched with ease to antemortem data; these may include (but not restricted to) homeless people, tourists, migrants, or people where antemortem data in form of an outdated photo conflicts with a markedly different appearance, possibly as consequence of lifestyle, aging, or disease. The identity of a body will most certainly pose technical challenges when there are no immediately recognizable identifying features; these cases include those of postmortem decomposition (Figure 33.1), particularly when the body is found outside, and those cases where a relatively high degree of injury or fragmentation is present; to give a few examples, these may be victims of workplace accidents, falls from great heights, collisions of pedestrians with a train or motor vehicle, or other victims of road traffic accidents (Figure 33.2).

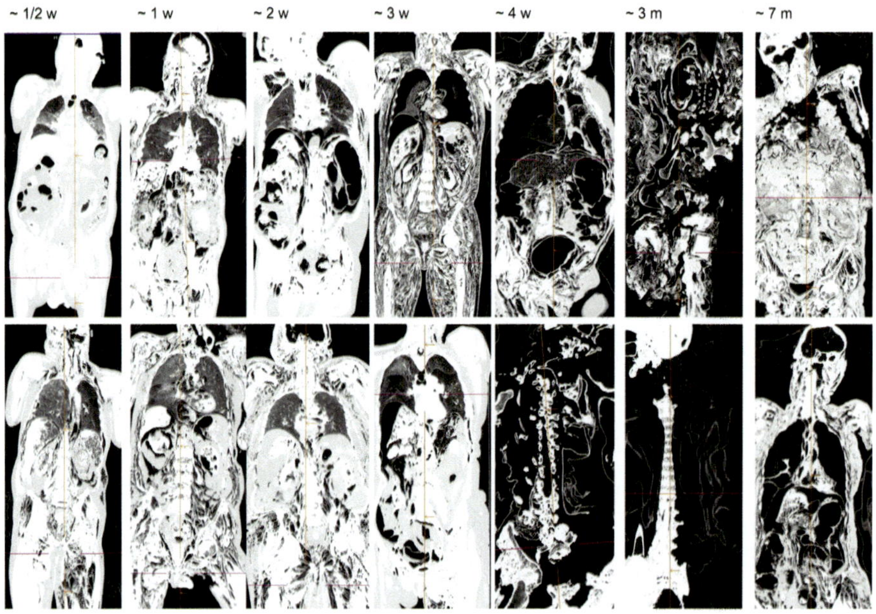

Figure 33.1 Postmortem decomposition. The images show, from left to right, an apparent progression of postmortem decomposition, from about half a week (left, labeled "~ 1/2w") to about 7 months (right, labeled "~7 m"). Gas may be built up all the while soft tissues tend to degrade and possibly attain a higher density than in living people or in the early postmortem interval. As individual case details influence the rate of gas development and both nature and speed of decomposition, the variation over time during the postmortem interval tends to be relatively wide. This image series is to show how examining a decomposed body using PMCT still can reveal relevant anatomy and pathology, at least to a degree.

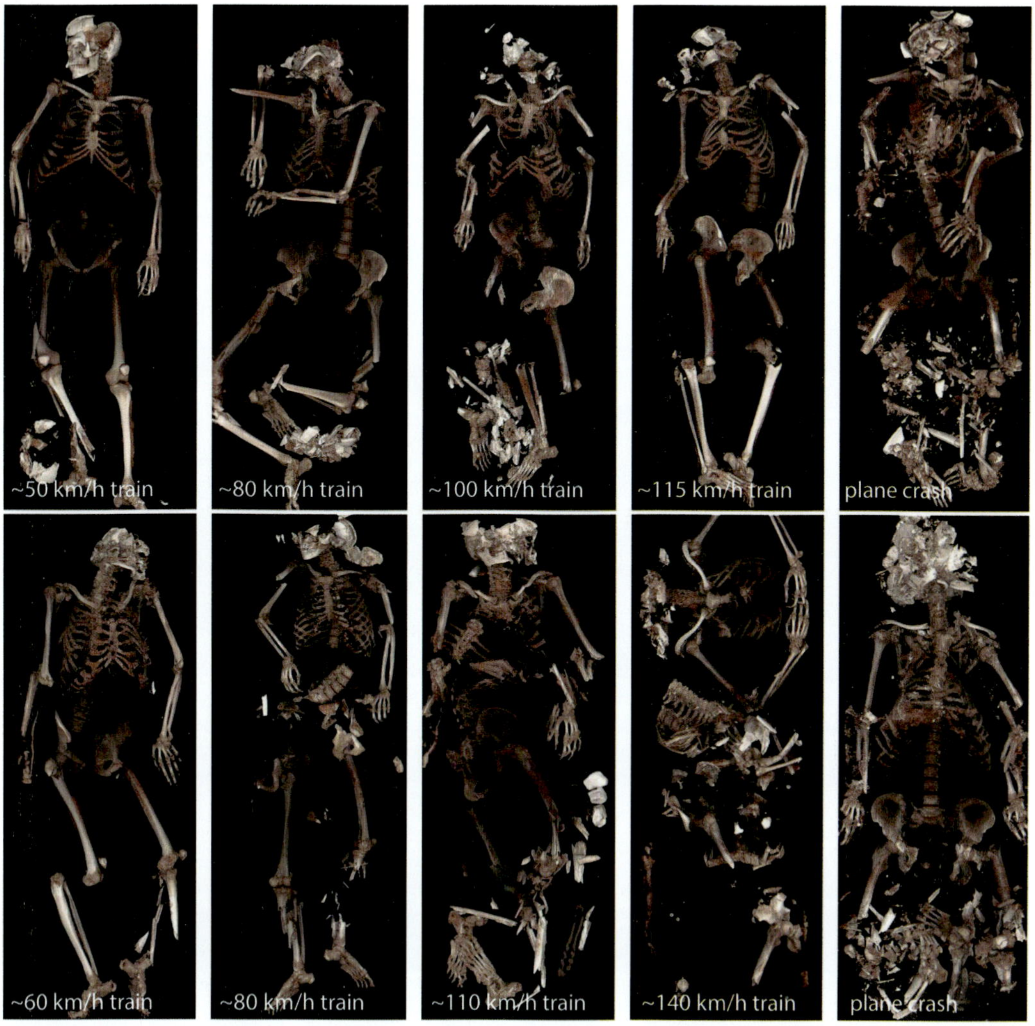

Figure 33.2 High grade trauma and body fragmentation. The images show, from left to right, an apparent progression of lower collision speeds to higher collision speeds of a vehicular structure against a body. On the left are cases where a person was run over by a train (starting from about 50 km/h, see image labels), progressing towards the right to pedestrian/train collisions up to about 140 km/h; the rightmost images show plane crash victims.

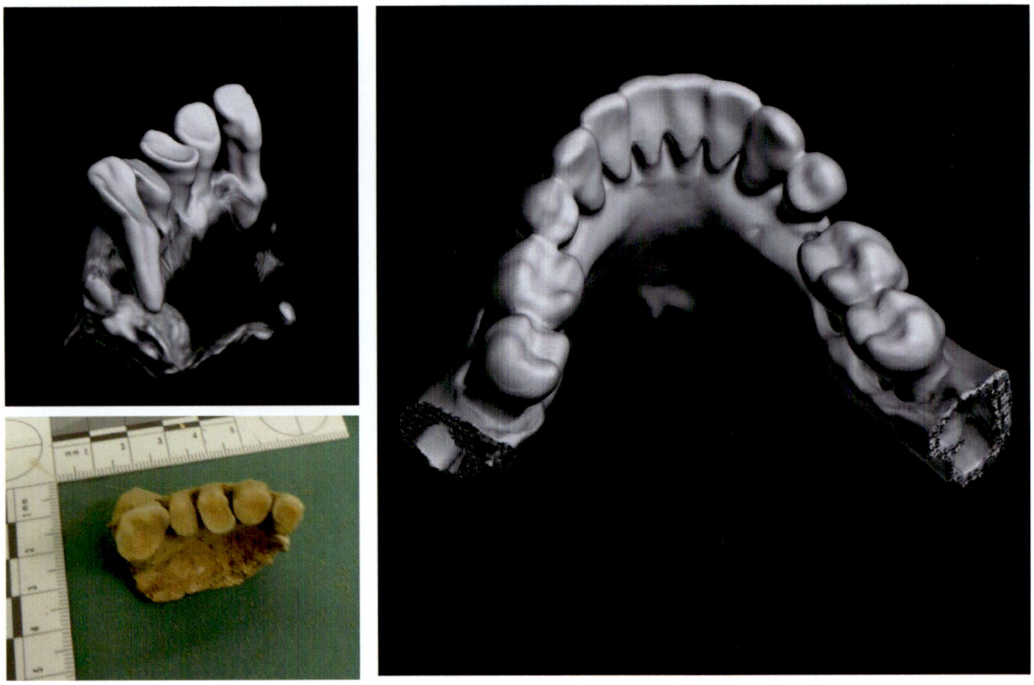

Figure 33.3 Dental status and time since death. The images on the left (3D reconstruction PMCT, photo) show the highly abraded teeth of a person that in all likelihood was buried sometime between the 6th and 9th century (AD [early middle ages]) in the then-cemetery of a Swiss rural town. Bones were later found during construction work. Right side: 40-year-old man, death in 2021.

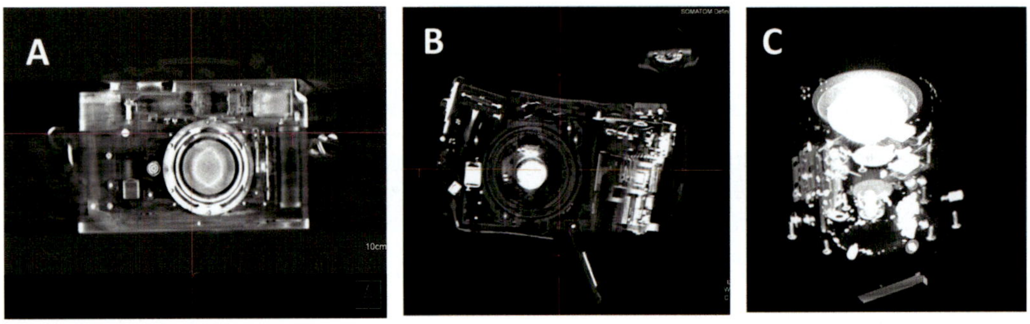

Figure 33.4 Digital or analog consumer devices as part of debris after a plane crash. PMCT scanning may be extended to scanning technical remains identified within the debris (A–C: standard ubiquitous camera parts as found after the crash), possibly of reconstructive value to the investigators.

33.2 RADIOLOGICAL/TECHNICAL BACKGROUND

There are two domains of forensic medicine that warrant a particular introduction to postmortem radiology, that seem relevant for the subject of postmortem identification.

Postmortem decomposition is an aspect that needs a special introduction, in that clinical radiology does not cover any such reality. Thereby, it is relevant to point out that a body may exhibit postmortem decomposition along a continuum, it is not a simple binary state. The bodily tissues degrade over time. That process typically is accompanied by the formation of gas; the images in Figure 33.1 show how over time, gas increases. Nevertheless, various diagnoses may still be made; these may include fatal intracranial hemorrhage [1], vascular injury or pathology [2] and trauma [3]. Due to various decomposition related aspects including fluid accumulation or tissue density increases, the use of PMCT to identify trauma, pathology and anatomy also for comparison with antemortem data may require dedicated attention and work.

High degree of fragmentation may occur in context of falls from great height or accidents with vehicles, airplanes or other machinery such as agricultural or construction work. The images in Figure 33.2) are exemplary and show that even in relatively high degrees of fragmentation, performing a detailed radiological and medicolegal examination is not only possible but necessary. The attentive search for identifying features may require particular important technical steps including windowing, pan, zoom and MPR alignment in these two groups of cases.

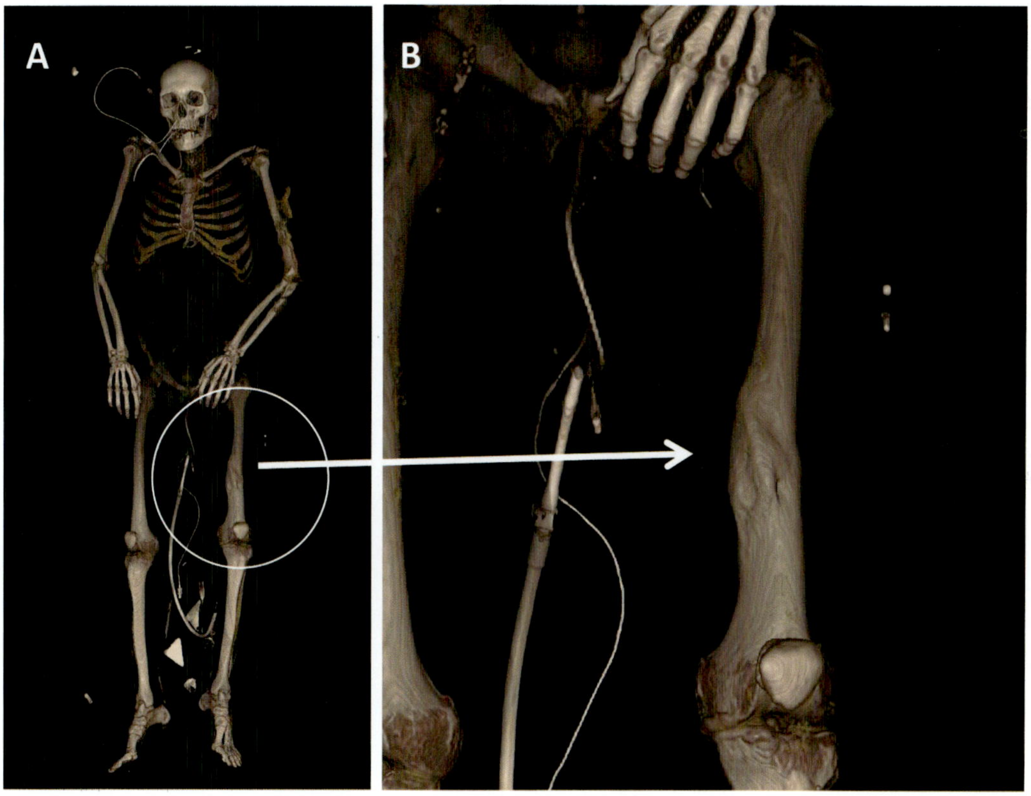

Figure 33.5 First screening of PMCT scan after a plane crash 1/2. Scanning victims of a plane crash may reveal a largely intact skeleton, whereas an old femur fracture (1) may assist in approaching at least preliminary identification.

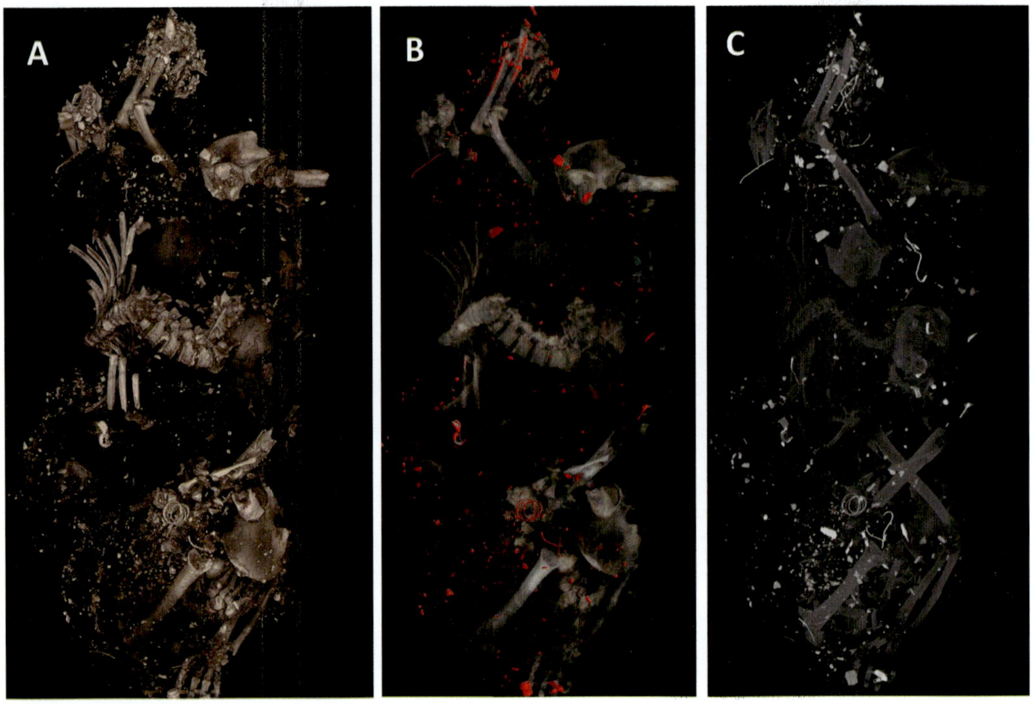

Figure 33.6 First screening of PMCT scan after a plane crash 2/2. PMCT scanning of high grade injuries may be complicated by possible commingling of remains. There, macroscopic and radiological examination of the scanned data for identifiable body parts may be a starting point that greatly benefits from the highest data resolution that is available. Dental restorations and orthopedic implants may be a lot harder to find as also, all sorts of other materials—plane fragments, cargo, soil, rocks, gravel—may be intermingled with the fragmented bodies. An anatomical rendition using VRT (A) may improve anatomical feature identification whereas a metal density highlighting (VRT in B, MIP in C) helps to find implants, wires, rocks, belt buckles and other items of possible metal density.

33.3 FORENSIC IMAGING AND RADIOLOGICAL IDENTIFICATION

In the field of forensic imaging, identifying a deceased may be performed for two fundamentally different situations, whereas in situation "A", no clue initially exists for a presumed identity of the deceased, and, "B", descriptors of a deceased are compared against descriptors of one or several missing persons specifically in order to exclude or confirm a given, known identity.

33.3.1 "A": Situation When No Identity of the Deceased Is Known at the Time of the Examination

This may pertain to human remains found anywhere, and in any state, from a recently deceased body without any injuries or postmortem decomposition, to any high degree of fragmentation, and the possibility of commingling of one or more human bodies. Commingling of human remains can significantly complicate reconstructive analyses and may arise in a wide range of contexts that are not always readily predictable or exhaustively enumerable, including, for example, following aviation accidents, explosive events, flooding, large-scale fires, or the excavation of mass graves.

33.3.1.1 Hazards

Hazards may have to be considered as a basic aspect. As examples, infectious disease, radiation or asbestos could be suggested as possibilities [4].

33.3.1.2 Human or Non-Human

Distinguishing human from non-human remains should be performed. However, this may also require morphological evaluation and DNA testing.

33.3.1.3 Bagging and Commingling, Labeling

If the remains are not attached to each other but separate parts, one may have to approach the situation as if these were remains from more than one individual. Un-attached or separate remains collections or single remains may need to be bagged, labeled, and processed individually [5]. The degree of fragmentation of a body may depend on a range of factors, including postmortem interval and environment, and the type and degree of trauma (Figure 33.2).

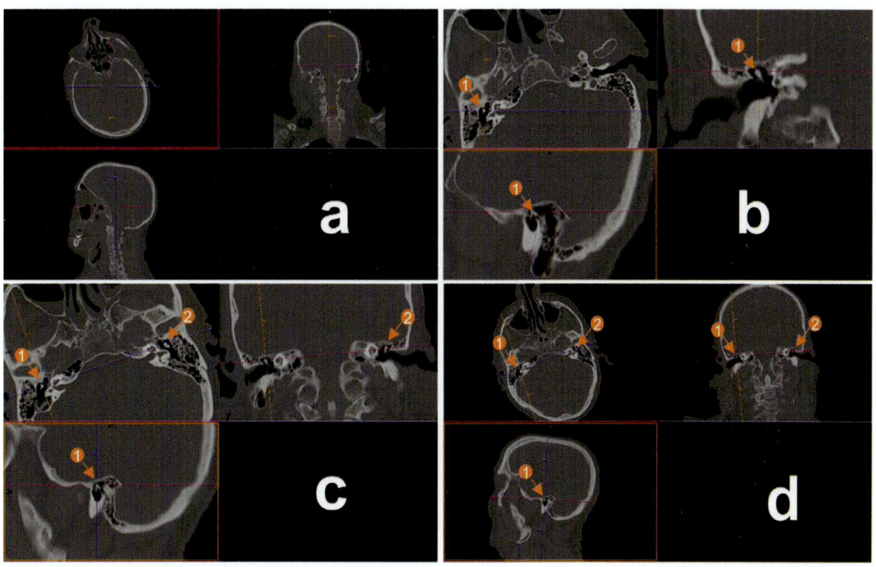

Figure 33.7 Using paranasal sinuses for radiological identification 1/2. Multiplanar reconstruction (MPR) usually uses a cursor (point, defined as the intersection of three lines, each representing the other image plane). These three lines (and the intersection point) are also shown here (a). The cursor is first placed on the left malleus (1, caput mallei dexter) (b). Then the horizontal axis is tilted through the right malleus (2, caput mallei sinister) (c). *Continued in* Figure 33.8.

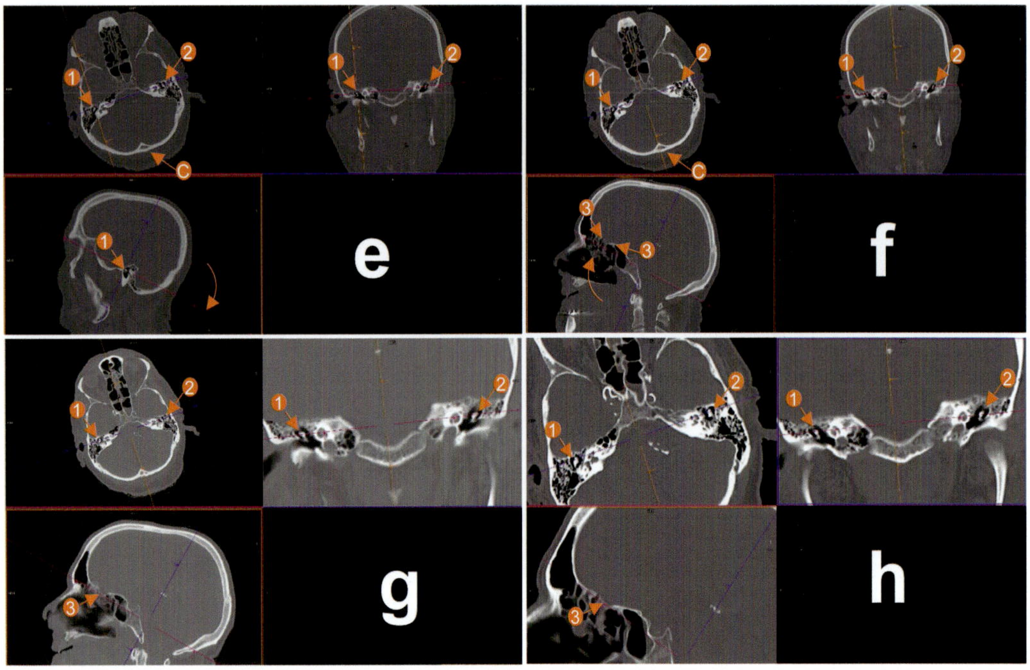

Figure 33.8 Using paranasal sinuses for radiological identification 2/2. *Continued from* Figure 33.8. Now, the anterior-posterior cursor line is tilted backward so that the crista occipitalis interior (C, shown in e) becomes visible. Then, the sagittal plane of the cursor is shifted to it from the left, so that it comes to rest on the most prominent part of the crista (f). Finally, the anterior-posterior cursor line—i.e., the axial image—is rotated through the hammerhead axis, in front of it, and tilted upward from below to the anterior cranial fossa in such a way that the latter is just touched or cut in the sagittal view (3, in Figure g and h). Software: Syngo Via (Siemens, Germany). The same result is easily achieved with exactly the same procedure in any other MPR program.

33.3.1.4 Postmortem CT Scanning and Postmortem Examination

Postmortem CT scanning requires a handling, labeling, or distribution of the bagged and labeled remains on the CT scanner that allows a later sure matching of scan data and remains.

A particular aspect there may be redundancy in documentation. Once one has synchronized the photo camera time with the CT-scanner time, then one may take photos of all CT scan table layouts before a scan was performed, and be able to add these to the individual case documentation with better reliability. The photos need to clearly show size, shape, and label inscription of each bag.

Subsequent postmortem information then needs to be informed about the content of the scans, most notably about foreign materials, possibly identifying features including indications regarding sexual dimorphism of bones, dental restoration, implants, or other identifying features. The survey of a PMCT scan in highly fragmented bodies may differ from case to case (compare Figures 33.5 and 33.6).

33.3.1.5 Time Since Death

At least roughly estimating a time since death based on the overall state of the remains given the place where they were found may be a third step.

Skeletal status may indicate early into an investigation that skeletal remains identified at a construction site may relate to an old cemetery from the early middle ages as shown in Figure 33.3. The PMCT-appearance of a body one or several weeks after death may differ greatly also with regard to the ability to differentiate organs or the amount of gas (Figure 33.1).

33.3.1.6 Sex

Sexual dimorphisms and at least a rough age estimation may be of relevance; there, a large amount of anthropological literature is present to assist whereas postmortem imaging is used to enact the principles of forensic anthropology.

33.3.1.7 Retrieval of Tissue for DNA Extraction

Retrieval of tissue samples for DNA extraction may be a fifth step.

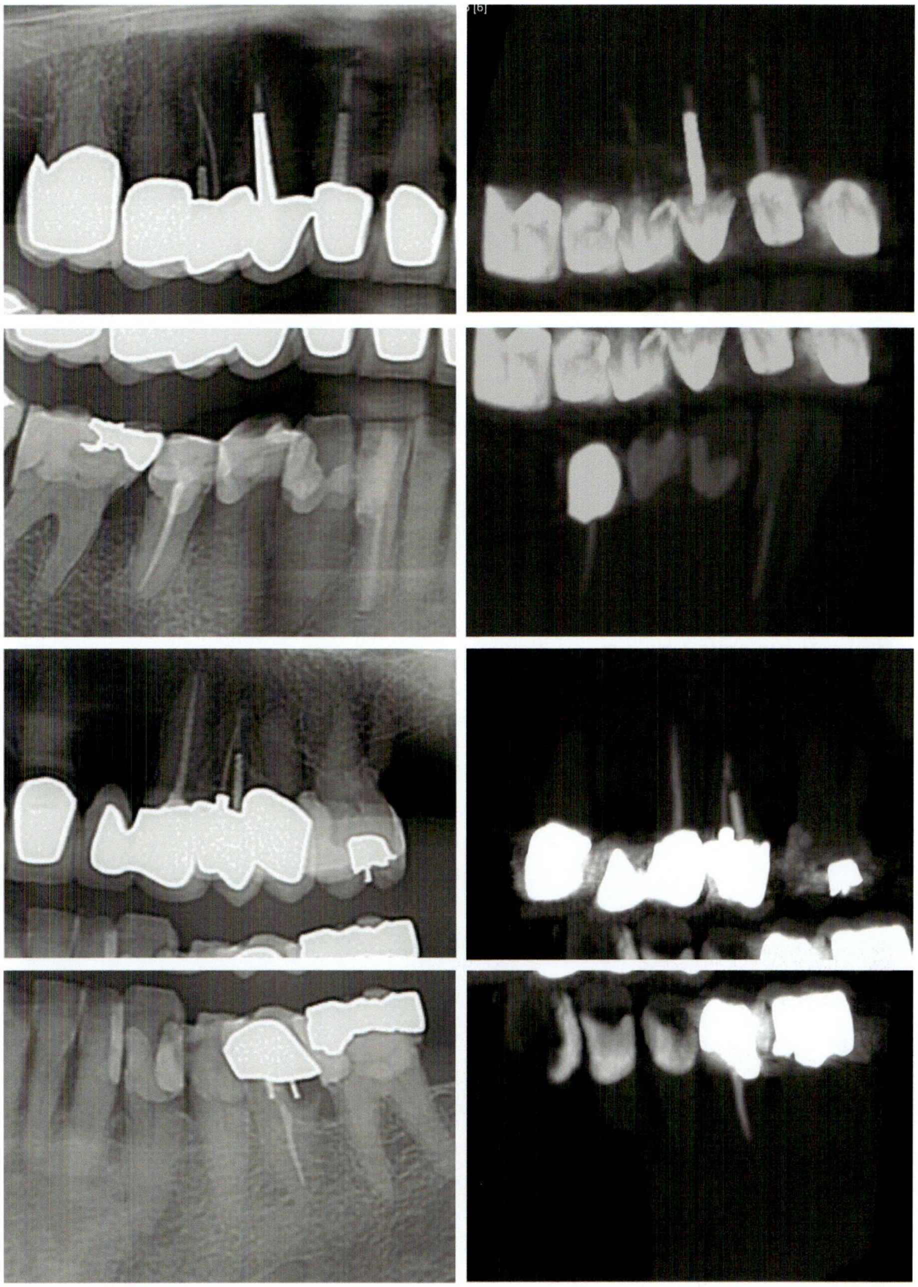

Figure 33.9 Radiologic identification comparing dental restorations. Antemortem dental radiographs (all images on the left) and PMCT of the same location (all images on the right) with MIP (maximum intensity projection) of the teeth of an 80-year old female deceased with presumed but initially unconfirmed identity.

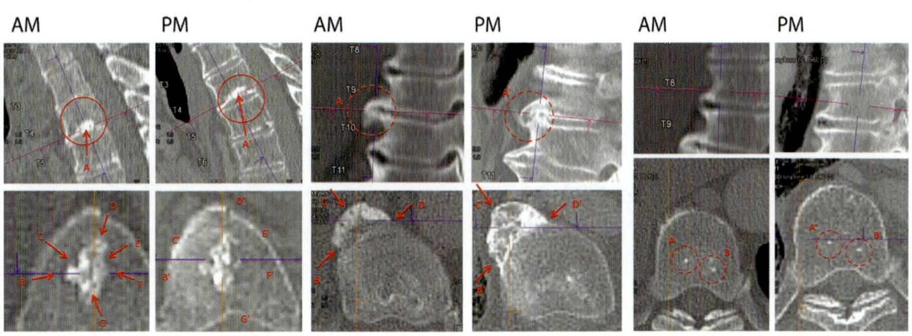

Figure 33.10 Radiologic identification comparing vertebral morphology. Antemortem (AM) and postmortem (PM) images in a 70-year-old deceased. The matching findings are marked with arrows and labels. Edit—image markings.

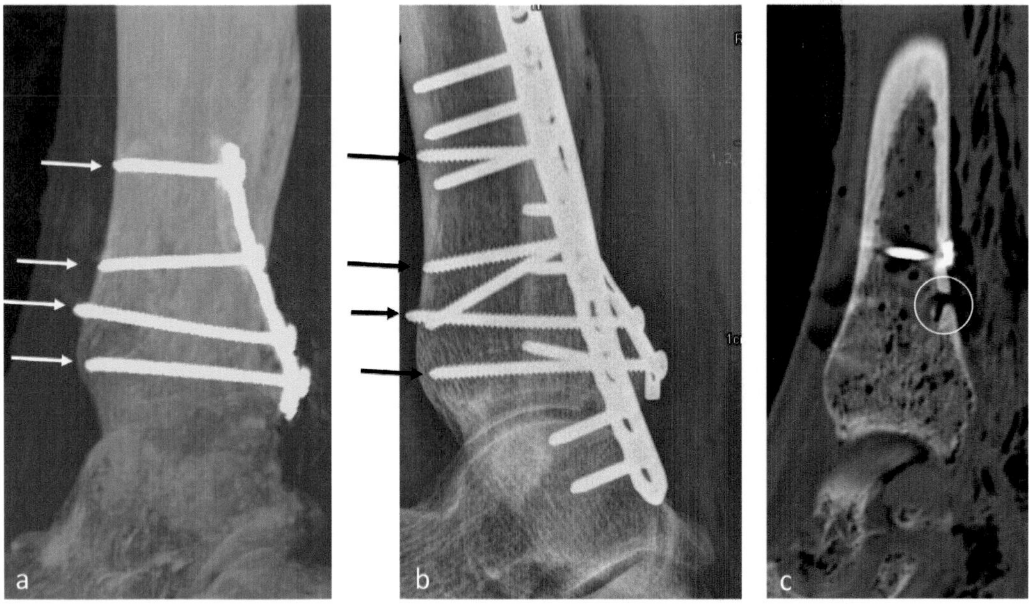

Figure 33.11 Radiologic identification comparing ankle morphology. PMCT (a) with maximum-intensity projections (MIP), antemortem x-ray (b) and postmortem MPR (c) of the right ankle. Removed screws between antemortem and postmortem imaging with borehole from removal (circle) and persistent orthopedic material (white and black arrows). This particular difference between ante- and postmortem is not contradiction to a positive match in this case but not regarded as sufficient for positive identification.

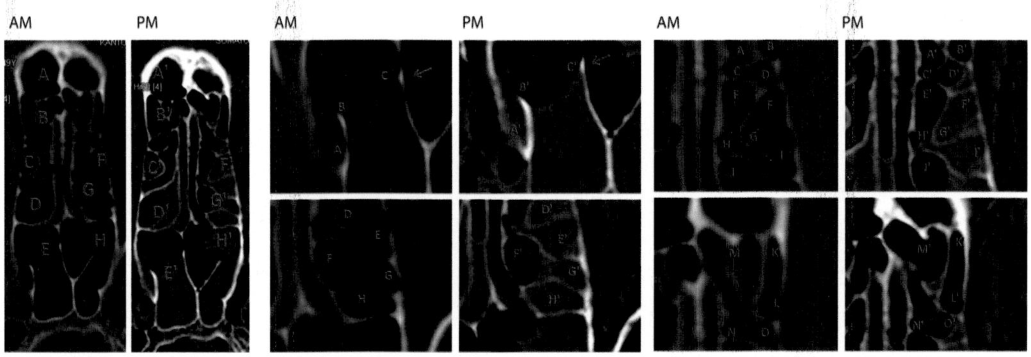

Figure 33.12 Radiologic identification comparing paranasal sinus morphology 1/2. Comparison of antemortem (AM) and postmortem (PM) images after standard plane alignment using MPR (see Figure 33.7 and 33.8). "Single" features labeled.

33.3.1.8 Pathology and Trauma

Pathology, injury and fracture analysis may ensue, whereas differentiating trauma from natural disease is one task, and differentiating antemortem from postmortem changes may not always be easy to solve.

33.3.1.9 Facial Reconstruction

Reconstructive attempts may even include facial soft tissue estimation, in order to produce a visual, an image, of a face to indicate how that person might have looked like.

33.3.1.10 Also Consider CT-Scanning Other Scattered Objects Found at the Scene

Relevant clues for the investigators may be gleaned from debris and contained high-tech objects, such as apparently **intact** (= Figure 33.4 A, C) or **damaged** (= Figure 33.4 B) optical devices.

33.3.1.11 Searching for Missing Persons

Possibly identifying features may have to be listed, verbally described, and visualized, then communicated, circulated, and published to motivate dentists or the general public to propose missing persons to the investigators.

The remains and their descriptions will have to be kept, stored, and then compared against missing persons' records.

Only once potential matches have been found, one may proceed as suggested or outlined in the next section with regard to comparison of antemortem and postmortem findings.

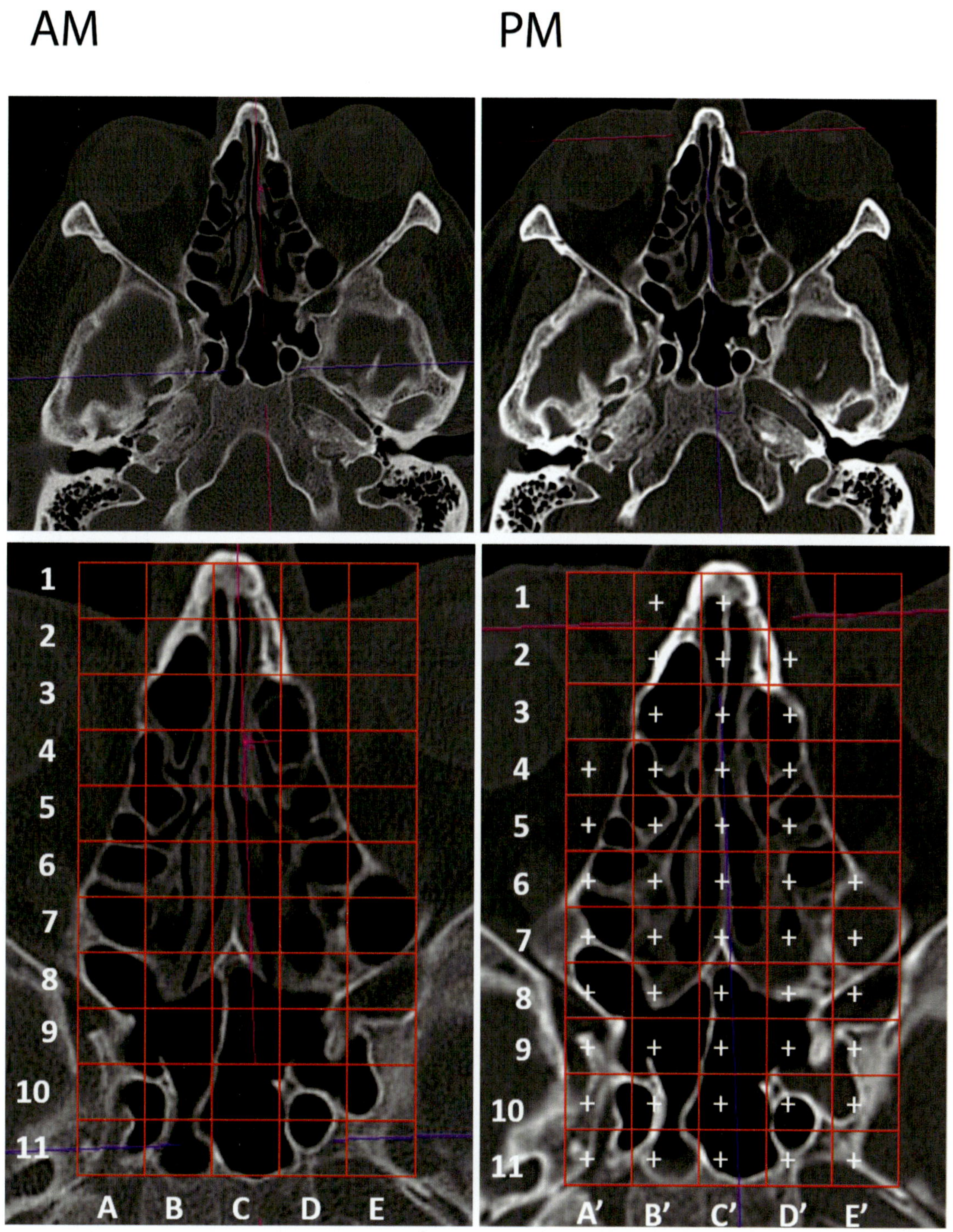

Figure 33.13 Radiologic identification comparing paranasal sinus morphology 2/2. Comparison of antemortem (AM) and postmortem (PM) images after standard plane alignment using MPR (see Figure 33.7 and 33.8). Here a grid method was employed to allow easier comparison of what appear to be homologue areas of anatomy. A positive visual match was declared on all grid squares marked with a "+". A total of 46 matching square regions are identified here.

33.3.2 "B": Situation When the Presumed Identity of the Deceased Is Known, Antemortem Data Is Present, and It Has to Be Matched with Postmortem Data

33.3.2.1 Introduction

Identification of deceased persons is the process of confirming an identity based on their human remains by matching antemortem with postmortem characteristics.

A feature may be defined as any type of isolated, delineated, separate, and single attribute or aspect of the body that can be measured or observed, whether AM or PM. A "match" of a feature may be established when the "same" feature is encountered in both the PM and AM data. Any such identification also relies on the presumed identity of the person documented in the AM data as the correct one, and as it is provided with or on the AM data. The validation of the correct identity pertaining to AM documentation, however, lies outside the scope of this chapter.

Finding a presumed identity may be obvious: if a body is found inside a locked apartment, the identity that may be considered first is the person that was registered at that address.

When human remains are found without any kind of initial circumstantial evidence, the presumed identities must be determined through some search process. Then, presumed identities may be generated from existing missing person lists or databases or through the publication of individual findings—e.g., dental findings in a dental journal.

Once a presumed identity is found, their AM data may be acquired. Then, it is important to note that the process to validate the congruence *between the presumed identity and its associated AM data* lies outside the scope of the postmortem examination by forensic pathologists, dentists or radiologists. If this relationship is wrong, either accidentally or intentionally, the actual identification will also fail, and not even for reasons of wrong AM and PM data matching. For example, an X-ray image in a hospital database might be assigned to the wrong patient. In this case, of course, identification based on a feature on the X-ray will also reproduce this error. Or a person may have faked an identity for the past years of their lives, and, accordingly, the AM records of these past years also carry this aspect; then they die and the presumed identity itself remains matched to the faked identity and thus in fact will be wrong.

If one confirms a match between AM and PM data, it may be relevant to describe what name and birth date the AM data was assigned to, and that the match of radiological data is one that allows one to say that the person behind the AM data also was the person in the PM data—but before asserting an actual identification it may be important to consider a disclaimer with regard to the above said.

For all practical purposes, "radiological identification" only describes the process of matching AM and PM data—it does not deal with the police or administrative work necessary to ascertain the authenticity and validity of AM data with regard to establishing their true identity.

33.3.2.2 Statistical Approach Using Likelihood Ratios

In general, a concordant match of what one may term being a single feature in both the AM and PM records is indicative of a correct identification, but the sheer number of matching features alone may not provide evidence of a true identification [6].

Instead, the value of a feature match for identification depends on the frequency of the feature in the source population. The rarer a feature is, the higher the probability that both the AM and PM data belong to the same individual.

For a feature to be considered useful for identification, it must be unusual and ideally unique. Established identification methods such as DNA or fingerprints make use of a relative trait or feature rarity and/or uniqueness, and, the combination of several or many single traits or features.

The contribution of a feature match to an identification may be quantified mathematically by likelihood ratios (LR). The LR is the ratio between the probability of the presence of a feature given the identification is correct and the probability of the presence of the feature given the identification is incorrect [7]:

$$LR = \frac{P(\text{feature present} \mid \text{ID is correct})}{P(\text{feature present} \mid \text{ID is incorrect})} \quad (33.1)$$

The relationship between feature matches, feature frequencies, and the calculation of likelihood ratios is best illustrated by examples.

Example 1: Assume that an unidentified body shows a sternal foramen (feature 1) on a postmortem computed tomography scan (postmortem data). The assumed identity also shows a sternal foramen visible on a CT scan from several years ago (antemortem data). If the identification is correct, the probability of finding a foramen sternum on the antemortem CT images is 1 (100%). If the identification is incorrect, the probability of finding a sternal foramen on the antemortem

CT images is equal to the prevalence of sternal foramina in the source population minus 1 person. It is known from a dedicated database that about 4.5% of people have a sternal foramen [8]. The first example of a LR is then estimated as:

$$\widehat{\mathrm{LR1}} = \frac{1}{0.045} = 22.2 \tag{33.2}$$

Example 1 thus exemplifies that it is ~22.2 times more likely to observe this match if the identification is correct than if it were incorrect. That said, we add that for practical purposes, the one correctly identified person being among those that make up the prevalence number of 4.5% was not subtracted in this example seeing as if that one person among the whole population is a marginal figure.

Example 2: It may be further noted that the deceased is male (feature 2, postmortem data). The presumed identity is also male (antemortem data). If the identification is correct, the probability that the presumed identify being male is 1 (100%). If the identification is incorrect, the probability of the presumed identify being male is about 0.5 (50%), as approximately 50% of the source population are male. The computed second example of a LR is estimated as:

$$\widehat{\mathrm{LR2}} = \frac{1}{0.5} = 2 \tag{33.3}$$

Given statistically independent characteristics and thus likelihood ratios, they can be multiplied by the product rule. In the example, the multiplied $\widehat{\mathrm{LR}}$ estimate is computed as:

$$\widehat{\mathrm{LR1}} \times \widehat{\mathrm{LR2}} = 22.2 \times 2 = 44.4 \tag{33.4}$$

It is to be noted that the resulting figure of an $\widehat{\mathrm{LR}}$ of 44.4 is in itself an estimate or approximation, because the data basis is an approximation or estimate.

If more concordant matches of single features based on population statistics are established, this product can quickly grow to a very high number, approximating certainty. However, it is important to note that there seems to be no widely accepted threshold at what point a LR is sufficiently high [9].

Example 3—extreme rarity corollary:

Assume an extreme example E, where the deceased has a truly unique appearance. There, the prevalence r of that appearance is only "that one person in the world's population w" (for argument's sake, take that as one person in the universe), so $\mathrm{r} = \frac{1}{w}$ Then, the determination according to Eq. 33.2 and 33.5 as $\widehat{\mathrm{LR}_\mathrm{E}} \approx \frac{1}{r}$ appears to yield a slightly incorrect estimate (see Eq. 33.1: denominator is the probability of features present given incorrect identification). A more precise approach would be as shown in Eq. 33.6:

$$\widehat{\mathrm{LR}_\mathrm{E}} = \frac{1}{\frac{1}{\mathrm{r}}} = \frac{1}{\frac{1}{w}} \tag{33.5}$$

$$= \mathrm{w} = \mathrm{r}$$

$$\mathrm{LR}_\mathrm{E} = \frac{1}{\frac{\mathrm{r} \cdot w - 1}{w - 1}} = \infty \tag{33.6}$$

This can easily be tested with extreme and obviously hypothetical examples: if the planet is inhabited by just three people, one of which has a glass eye (but definitely none of the others), then $\mathrm{r} = \frac{1}{3}$ The likelihood of a deceased person found with a glass eye being correctly identified thereby then would be underestimated by proposing $\widehat{\mathrm{LRE}} = \frac{1}{0.33333} = 3$, whereas a realistic determination of how likely it is that this particular deceased is in fact who we think it is $\mathrm{LR}_\mathrm{E} = \frac{1}{\frac{\frac{1}{3} \cdot 3 - 1}{3 - 1}} = \infty$; in other words, there is simply no way this is another person. This becomes even clearer if we propose another even more extreme and superbly hypothetical example whereas there is only one known or registered person on a planet, and an initially unidentified person is found dead there by, say, a non-identity-non-entity from elsewhere. This may lead to two interesting considerations. One is that when obtaining the likelihood under assumed probabilities of two different hypotheses H0 and H1, these must not intersect ($\mathrm{H0} \cap \mathrm{H1} = \varnothing$). So a person to be identified can be logically treated as a member of the correctly identified group under H1, but they cannot simultaneously be part of the group that is assumed to constitute the incorrectly identified under H0. The other is that not properly subtracting a present individual presumed as correctly identified from the probability that this is not the presumed person may risk to possibly underestimate the likelihood ratio.

33.3.2.3 Approximating a Rule of Thumb

Population statistics of any single characteristics, and reliable figures of mutual dependency of these characteristics, seem difficult to implement in reality.

Assumedly, one knee prosthesis will not occur independently of a second knee prosthesis, and several osteophytes of the spine may certainly not be present independently of each other. Since no reliable figures exist about exact statistical associations, one should approach this problem pragmatically and, if possible, use features from deliberately different domains for identification. For example, if one were to toy with the concept of likelihood ratios of what one perceives as single features, one could assume a better independence of features in the combination of a sternal foramen, a knee prosthesis, and a dental implant, than in the combination of several osteophytes.

The prevalence of any single morphological feature may be given as p(f), and if a number n of independently occurring features f(1) . . . f(n) are used, their chance of co-occurrence may be assumed as their product Π p(f). In simplified terms, this means that one may need at least seven independent features with $p(f) \leq \frac{1}{10}$, or four independent features with $p(f) \leq \frac{1}{200}$ to reach a total $\Pi p(f) \leq \frac{1}{1E6}$, as a rough estimate. This appears to be a safe guess, also since analog shapes may easily be split further into more shape sub-features such as their size and location, curvature aspects, and so on, but with possibly increasing interdependency. So one may easily reach relatively high statistical estimates, just by selecting at least around ten different and independent analog shape features for identification, provided each occurs less frequently than $\sim \frac{1}{200}$, or more.

The Swiss German of Forensic Medicine (SGRM) proposes 12 matching and characteristic traits in order to provide a reliable dental identification, whereas the presence of 6–11 traits allows one to conclude a probable identification, and the presence of 1–6 traits suggest a possible identification [10].

33.3.2.4 Morphological Approach—Analog Comparisons, or Quantification of Shape Mismatch

A single-feature based attempt in radiological identification may draw on the experiences of the forensic application of DNA matching, and also on expectations one may base on the DNA method for any reason. But technically, DNA comparison bases on chemically identified *discrete base pairs*, which by nature of pre-encoded *discrete* base data, allows for rather well based mathematical calculations.

In stark conceptual contrast, the *shapes of the human body are not discrete figures*. The analog and almost fractal continuous nature of morphological shapes in forensic imaging typically evades such a relatively easy well based mathematical description and capture.

So far the best technical approach to achieving quantification of paired bone shapes in match pairs appears to be to *digitally capture and mathematically encode the whole shape, and then to evaluate adequate comparison samples for quantified shape differences*, while both temporal difference between the AM and PM data with aging and other changes (such as surgery or disease) or also significant postmortem alterations may cause problems. This has been performed by Decker et al. (2019), who presented a very convincing example of dealing with analog shapes in a digital world under the light of shape matching for the purpose of identification [11]. Most notably, the whole shapes rather than single features were encoded, and a cut-off for the match error to delineate a negative match was identified.

33.3.2.5 Mismatch of Findings

While calculating and multiplying the likelihood ratio of matches, or appreciating the shapes themselves in their overall and detailed appearance, may provide approximate confidence of correct identification, it is also important to emphasize the importance of feature mismatches.

Because regardless of how many feature matches are present, an unexplainable mismatch precludes correct identification. An unexplainable mismatch occurs for example when the deceased has an intact skull, but the AM data clearly contain a healed borehole from a prior intracranial pressure sensor. This mismatch seems virtually impossible, and the identification must be rejected on that basis.

If conversely the deceased clearly exhibits a healed borehole finding on the skull, and the AM radiology with a skull CT from 15 years ago does not show that, that would not per se constitute a contradiction, as within a time frame of 15 years, the surgical intervention could easily have occurred and the bone healed. There are numerous other easily explainable mismatches that typically increase with the temporal difference between the AM and PM data set dates. For example, a loss of teeth or an increase in arterial calcifications are considered explainable mismatches that do not contradict the identification. The assessment of explainability of mismatches requires expertise of temporal changes in the human body.

33.3.2.6 Radiologic Identification—Antemortem Data Availability

Radiological identification has been described as early as 1927 [12] and also with relation to forensic imaging methods [13]. In case antemortem material is present and obtainable and sufficiently rare features are observable, radiological identification may be a cost-efficient and fast identification method. With multiplanar reconstructions, identical cross sections can be achieved quickly in both datasets, which can display identification features. In case of convincingly rare features, a correct identification can be declared within minutes.

When matching AM and PM data, one attempts to examine a possible identification by matching radiologic features, including computed tomography, X-ray, and magnetic resonance imaging as modalities. For this, the features must be adequately visualized in a postmortem scan, and they must be depicted in the antemortem scan as well.

The most important practical limitation of radiologic identification seems to be obtaining AM imaging data of the missed person—either they are not available because their location is unknown, or they simply do not exist. In theory, the existence of AM images should be positively correlated with age and with pathology. Which brings us to the issue of taking identification a step further: if neither a DNA match nor radiological identification solves the problem in any given case, it may be important to consider performing an autopsy, and if only to look for additional findings for identification.

33.3.2.7 Radiologic Identification—Practical Aspects Regarding Reconstructions and Visualizations

Not every AM image in a case of matching identities guarantees a successful identification, since some AM images do not contain any unique, unusual or rare features. In practice, this may typically occur, for example, in plain radiographs of extremities, chest or spine. For a good visual comparison, data must be prepared adequately. The step of aligning 3D data by use of 3D renderings or by use of multiplanar reconstructions (MPR) can be relatively time consuming.

With that, providing a side-by-side visualization that allows for description and measurement of features, to appreciate their shape and if present shape similarities, and that also allows one to an appreciation of a structural independence of one feature from another, to furthermore visualize their continuous shape analogy between antemortem and postmortem imagery or data, appears to be the best way to document a particular image juxtaposition. Alternatively, use of matching grids to overlay the compared images may allow for better feature comparison. There needs to be a possibility to describe plausible alteration causes when a mismatch is encountered, and discuss, for example, how a particular finding does not rule out a correct match because mismatches also are a reality. Also, there needs to be a point in time when a detected mismatch, or the inability to perform a proper match, is declared and decided. That may be a required step before other means of identification can be sought.

For decades, the main medicolegal findings were frontal sinus patterns, skull suture and vascular groove patterns, bone comparisons (particularly in the light of pathology), as well as all sorts of medical implants and foreign bodies [14]. That has not changed much since then. Most preferable structures for identification are particularly rare or individual custom made surgical implants or procedures, pathological or anatomical findings. There, osseous and CT-dense structures are favored. If there are no such features apparent, then, nasal and paranasal sinuses as well as mastoid cells and dental scans or X-rays appear to be most convenient to use [13]. Especially in disaster victim identification (DVI), dental scans or X-rays, and individual nasal and paranasal sinuses for radiological identification (RadID) are useful REF, and with the use of PMCT, the radiologist is also becoming an integral part of the mortuary identification team REF.

In our experience, these features are most useful:

- paranasal sinuses (technique for standard plane: Figures 33.7 and 33.8; for cases, see Figures 33.12 and 33.13);
- mastoid cells;
- atypical bone findings, including peculiar osteophytes or ossifications in spongious bone, as in Figure 33.10;
- medical implants, as in Figure 33.11;
- dental records (see Figure 33.9).

33.3.2.8 Practical Example: Paranasal Sinuses and Multiplanar Reconstruction

We identified speed as relevant aspect of skull comparison. Usually the postmortem heads are not aligned to the table and thus, the alignment has to be performed manually. As we focused on paranasal sinuses always, getting these aligned always was an issue. At first, our radiologists easily spent 10 or 20 minutes per case, trying to get these planes aligned between antemortem and postmortem data. There are a few fast ways our team subsequently identified and described over the years, that may be used to orient multiplanar reconstructions, in order to yield paranasal sinuses in a comparable way. One particularly fast method is described in Figures 33.7 and 33.8. Therefore, the first line of a standard plane may be placed through the inner ear ossicles [15]. Another relatively rapid method that built on this first method uses the labyrinth as first orienting structures (for details of the method, see [16, 17]).

33.4 SUMMARY

1. **Radiological identification is a powerful, non-invasive method.** Postmortem computed tomography allows for rapid and reliable identification, even in cases of decomposition or fragmentation, when dental, skeletal, or implant-related features are available.
2. **Two identification scenarios must be distinguished.** One must differentiate between (A) cases without a presumed identity and (B) cases with a known presumed identity requiring confirmation. Each pathway involves distinct forensic workflows and evidence standards.
3. **Likelihood ratios (LRs) quantify the strength of feature matches.** The LR expresses how much more likely a feature is if the identification is correct compared to if it is incorrect. This is the cornerstone of statistical radiologic identification.
4. **The rarity of a feature matters more than the number of matches.** A single rare or unique feature contributes more evidential weight than multiple common traits. Feature prevalence directly influences LR magnitude.
5. **Multiplying LRs assumes independence—which is rarely guaranteed.** Mathematical multiplication of LRs is only valid if the features are statistically independent. Many traits (e.g., degenerative changes) exhibit interdependence, requiring careful evaluation.
6. **Avoid calling Bayesian-style reasoning "frequentist."** Estimating identity strength via feature prevalence and LR is a Bayesian approach. Mislabeling it as frequentist undermines statistical accuracy and forensic credibility.
7. **An unexplainable mismatch invalidates identification.** Even numerous matching features cannot compensate for one contradiction that lacks plausible explanation. This is a hard exclusion criterion.
8. **Explainable mismatches require contextual justification.** Discrepancies between antemortem and postmortem findings (e.g., surgeries, tooth loss) can be acceptable if explainable within the known clinical or temporal context.
9. **Radiologic identification depends on available antemortem data.** The greatest limitation of PMCT-based identification is the absence or inaccessibility of prior medical imaging. Without AM data, identification is often not feasible.
10. **More research into shape match quantification and shape variation statistics is needed.** At the moment, much of radiological identification is done qualitatively, taking the above remarks into consideration. That, however, is not satisfactory, as a more scientific basis is highly desirable.

DISCLAIMER

For all practical purposes, "radiological identification" as used in this chapter only describes the process of matching antemortem and postmortem data—it does *not* deal with the police or administrative work necessary to ascertain the authenticity and validity of AM data with regard to establishing their true identity. It is up to the examiner to provide adequately worded reports so as not to inadvertently misinform the reader with regard to the level of evidence given.

ACKNOWLEDGMENTS

The authors express their gratitude to Emma Louise Kessler, MD, for her generous donation to the Zürich Institute of Forensic Medicine, University of Zürich, Switzerland. Released under CC BY 4.0. While the authors have made efforts to ensure clarity and structure, the development of this manuscript occurred under the constraints of available time and timelines. Output of AI tools was critically considered for use in the generation of this chapter (Instatext, Chat Generative Pretrained Transformer). The retrospective use of anonymized data was submitted to and evaluated by the Ethics Committee of the Canton of Zürich (reference number 15–0686).

References

[1] Tappero C, Thali MJ, Schweitzer W. The possibility of identifying brain hemorrhage in putrefied bodies with PMCT. Forensic Science, Medicine and Pathology. 2020;16(4):571–576.

[2] Franckenberg S, Flach PM, Gascho D, Thali MJ, Ross SG. Postmortem computed tomography-angiography (PMCTA) in decomposed bodies: A feasibility study. Journal of Forensic Radiology and Imaging. 2015;3(4):226–234.

[3] Thali MJ, Yen K, Schweitzer W, Vock P, Ozdoba C, Dirnhofer R. Into the decomposed body—forensic digital autopsy using multislice-computed tomography. Forensic Science International. 2003;134(2–3):109–114.

[4] Leditschke J, Collett S, Ellen R. Mortuary operations in the aftermath of the 2009 Victorian bushfires. Forensic Science International. 2011;205(1–3):8–14.

[5] Bassed R, Leditschke J. Forensic medical lessons learned from the Victorian Bushfire Disaster: Recommendations from the Phase 5 debrief. Forensic Science International. 2011;205(1–3):73–76.

[6] Acharya AB, Taylor JA. Are a minimum number of concordant matches needed to establish identity in forensic odontology? Journal of Forensic Odontostomatology. 2003 Jun;21(1):6–13.

[7] McGee S. Simplifying likelihood ratios. Journal of General Internal Medicine. 2002;17(8):647–650.

[8] Christensen AM, Hatch GM. Quantification of radiologic identification (RADid) and the development of a population frequency data repository. Journal of Forensic Radiology and Imaging. 2016;7:14–16.

[9] Steadman DW, Adams BJ, Konigsberg LW. Statistical basis for positive identification in forensic anthropology. American Journal of Physical Anthropology: The Official Publication of the American Association of Physical Anthropologists. 2006;131(1):15–26.

[10] Arbeitsgruppe Forensische Bildgebung. Forensisch-radiologische Zahnuntersuchung an Leichen, SGRM; 2018.

[11] Decker SJ, Ford JM. Forensic personal identification utilizing part-to-part comparison of CT-derived 3D lumbar models. Forensic Science International. 2019;294:21–26.

[12] Culbert WL, LAW FM. Identification by comparison of roentgenograms: Of nasal accessory sinuses and mastoid processes. Journal of the American Medical Association. 1927;88(21):1634–1636.

[13] Hatch GM, Dedouit F, Christensen AM, Thali MJ, Ruder TD. RADid: A pictorial review of radiologic identification using postmortem CT. Journal of Forensic Radiology and Imaging. 2014;2(2):52–59.

[14] Curran W, McGarry A, Petty C, editors. F. A. Davis Company, Philadelphia, PA, USA; 1980.

[15] Schweitzer W, Thali M, Ruder T. Rechtsmedizinische Bildgebung: Single-Plane-Charakterisierung von Schädel-CT als Strategie zur effizienteren formvergleichenden Identifizierung, Rechtsmedizin; 2014.

[16] Ruder TD, Brun C, Christensen AM, Thali MJ, Gascho D, Schweitzer W, et al. Comparative radiologic identification with CT images of paranasal sinuses: Development of a standardized approach. Journal of Forensic Radiology and Imaging. 2016;7:1–9.

[17] Brun CN, Christensen AM, Kravarski M, Gorincour G, Schweitzer W, Thali MJ, et al. Comparative radiologic identification with standardized single CT images of the paranasal sinuses: Evaluation of inter-rater reliability. Forensic Science International. 2017;280:81–86.

PART 7

VISUALIZATION AND DOCUMENTATION FOR REPORTING

CHAPTER 34

WHAT SOFTWARE TO USE IN FORENSIC IMAGING

Lars Ebert

DOI: 10.4324/9781003413998-41

34.1 SOFTWARE FOR MEDICAL VISUALIZATION

A variety of different software packages exist for working with 2D and 3D medical and non-medical data. Some of this software is proprietary, but there are often open source or even free software alternatives with comparable performance and functionality. Open source and free software is an affordable means to start with forensic imaging, as proprietary software is often costly or comes with expensive subscription models, which can be a considerable entry threshold. In this chapter, we provide a list of software packages that are in use at the IRM-UZH. This list is by no means complete and only intended as a starting point (Table 34.1). If possible, we try to present open source alternatives to proprietary software.

34.2 2D IMAGES AND MEDICAL DATA

To create image reports, it is essential to generate, process, and annotate images [1].

For the generation of images from medical 2D data, a Picture Archiving and Communication System (PACS) workstation can be used. Usually, PACS are provided by the manufacturers of CT and MRI scanners (i.e. syngo.via from Siemens in our case). These PACS workstations allow for 2D visualizations such as MPR, MIP, minIP, curved MPR and others and are the main tool of a forensic radiologist [2]. A free alternative for a PACS workstation with integrated PACS server is Horos, a branch of the much more limited free OsiriX Lite. Unlike all of the other software discussed in this chapter, Horos only runs on MacOS and is not available under Windows.

Image processing involves the application of image filters, rescaling, extraction of objects and features, adding transparency, layering, changing of contrast, cropping and other means of changing an already existing image. While proprietary software such as Photoshop are powerful tools, they come with a subscription model that can be highly priced, especially when multiple computer systems have to be equipped. Alternatives to this are GIMP and ImageJ. While GIMP is meant as a free software alternative to Photoshop with a similar set of tools and workflows, ImageJ is often used in medical and science visualization due to the fact that it is easily programmable, can also be used for image analysis and allows for the easy creation of macros to automate tasks, though it requires some limited programming knowledge.

Combining multiple images and annotating them is the prerequisite for the creation of meaningful image reports, as they are supplied to the state attorneys at the IRM-UZH. To do this, Microsoft PowerPoint has proven to be a useful tool, even though it is a bit of an "off-label" use of the program. It allows for the combination of multiple images as well as the addition of text and vector graphics such as filled polygons or arrows to highlight specific findings. A limitation of PowerPoint is the fact that the created graphics can only be exported at a fixed resolution, which might not be sufficient for all purposes. A direct free software alternative to PowerPoint would be LibreOffice Impress, which has comparable functionality. An alternative is Inkscape, which is a vector graphics software that also allows the integration of bitmaps. It is especially useful for creating figures for publication, as the created images can be exported at any resolution, thus maintaining the quality of the original images. While Inkscape is more powerful than PowerPoint when it comes to creating a final graphic, it is also more complex and less intuitive to use.

34.3 3D MEDICAL

3D medical visualization in the form of volume rendering technique (VRT) images is usually covered by any modern PACS workstation [2]. Similar to 2D medical visualizations, Horos can be used as a free software alternative. When it comes to more complex visualizations, additional software such as 3D Slicer or MeVisLab might be required. MeVisLab in particular is aimed at developers and requires some level of programming knowledge, but is highly customizable to specific needs. Even though 3D Slicer and MeVisLab are more complex to use, here it is possible to mix volumetric rendering with polygonal data and 2D resliced planes to further refine the visual output. In addition, MeVisLab has a powerful path tracing algorithm, that is to a certain extent simulating the behavior of photons and their interactions with matter. This allows for realistic looking renderings of volumetric data comparable to Siemens Cinematic Rendering.

34.4 SEGMENTATION

Segmentation is the task of separating certain structures within a dataset. In forensic medical imaging, this can be used as a basis for creating polygon models of these structures to be processed in non-medical software, i.e. to further perform anthropological analyses, perform reconstructions or 3D print parts of the anatomy [3–7]. A variety of different proprietary software for segmentation exists, i.e. Amira and Mimics. They offer a variety of manual, semi-automatic and automatic tools for segmentation, as well as pre-and postprocessing filters and conversion algorithms. Just like for the other software, proprietary segmentation software can be expensive, often with subscription models. A free alternative to the proprietary segmentation software is 3D Slicer. The basic functionality already has a variety of different segmentation tools and it can be extended through a plugin mechanism. 3D Slicer does not offer dedicated support for large datasets such as Micro-CT and can already be slow to use with large conventional 3D datasets, such as high resolution full body PMCT. Also 3D Slicer can be used to generate 3D polygon models from segmented data for further processing.

34.5 3D POLYGON + VIRTUAL REALITY

3D incident reconstructions, where data from a variety of different surface scanning procedures (such as photogrammetry, structured light scanning or laser scanning) as well as medical scanning (i.e. PMCT or PMMR) are combined to answer a specific forensic question, usually are generated in animation software such as 3D Studio Max or Maya [8]. Both of this software is proprietary, with free educational licenses available to be used for a limited time. These software packages cover 3D modeling aspects, rigging (the task of fitting a bone system for animation), animation, alteration of camera parameters and lighting and finally rendering, using techniques such as raytracing for the calculation of correct reflections and refractions [9]. Blender is a free alternative to the other animation software packages, with comparable functionality. All these software packages have in common that, while very powerful, the amount of functions available can be overwhelming and they possess a steep learning curve. Visualizations and animations created in Maya, Blender or 3D Studio Max can be visualized in proper 3D using Virtual Reality as well [10,11]. Multiple game engines such as Unreal or CryEngine are available, an engine called Unity is most commonly used for forensic visualizations. This is mainly due to its relative simplicity and license model for educational purposes, which means there are a variety of different tutorials and plugins available online. This makes Unity an interesting tool for universities. Unity can be extended using C#, provided programming knowledge is present; however, for simple visualizations, this is not required.

34.6 VIDEO

In cases where video footage has to be cut, there are a variety of options such as FinalCut Pro (MacOS only), or Adobe Premiere. They allow for cutting, overlays, color grading and transitions between video and audio as well as overlays of text or other markers. In addition, object tracking is possible, i.e. to highlight a moving object automatically without the necessity for manual tracking. At the IRM-UZH we use Davinci Resolve, which is proprietary but free in the basic version. It contains sufficient functionality to cover all our needs and is well suited to handle RAW video stream data from our Blackmagic cameras.

Table 34.1 Different Software Packages to Be Used in Forensic Imaging

Type	Software	Licence	URL
2D	Adobe Photoshop	proprietary	https://www.adobe.com
2D	GIMP	free/open source	https://www.gimp.org/
PACS	Syngo.via	proprietary	https://www.siemens-healthineers.com/
PACS	HOROS	free/open source	https://horosproject.org/
Annotation	PowerPoint	proprietary	https://www.microsoft.com/
2D Vector	Inkscape	free/open source	https://inkscape.org/de/
3D Visualization	MeVisLab	proprietary	https://www.mevislab.de/
3D Visualization/Segmentation	3DSlicer	free/open source	https://www.slicer.org/
3D Visualization/Segmentation	Amira	proprietary	https://www.thermofisher.com/
3D Visualization/Segmentation	Mimics	proprietary	https://www.materialise.com/
Animation, Modelling, Rendering	3D Studio Max	proprietary	https://www.autodesk.eu/
Animation, Modelling, Rendering	Maya	proprietary	https://www.autodesk.eu/
Animation, Modelling, Rendering	Blender	free/open source	https://www.blender.org/
Game Engine	Cryengine	proprietary	https://www.cryengine.com/
Game Engine	Unreal	proprietary	https://www.unrealengine.com/en-US/
Game Engine	Unity	proprietary	https://unity.com/
Video	DaVinci Resolve	Proprietary	https://www.blackmagicdesign.com/

CHAPTER 35

READING AND DOCUMENTATION

Wolf Schweitzer, Garyfalia Ampanozi, and Carlo Tappero

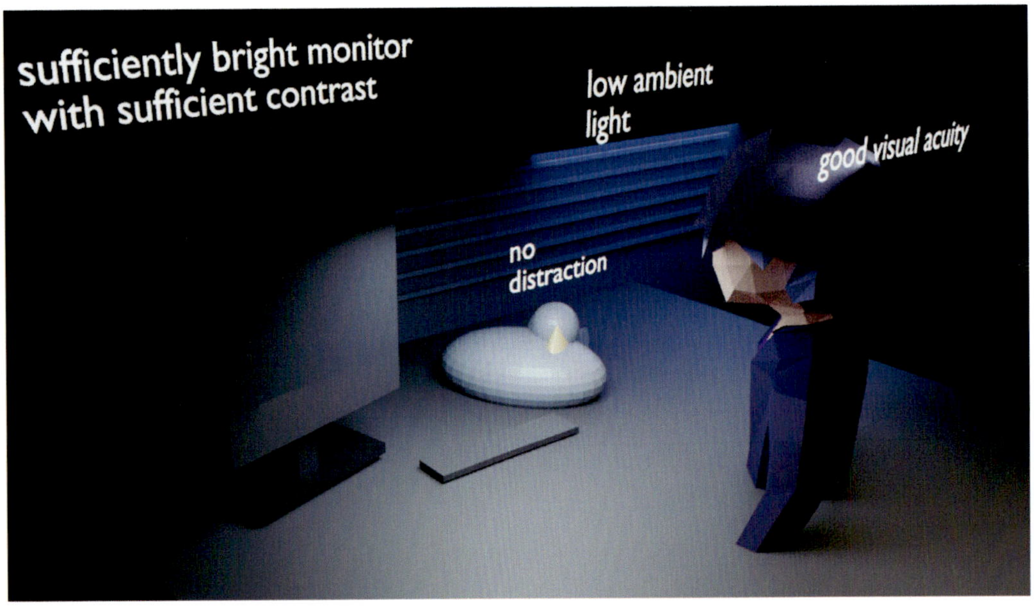

Figure 35.1 Reading radiology images: general setup. As an engram, remember that low ambient light and a sufficiently bright and well resolved (pixel, contrast) monitor (symbolic illustration, rectangular structure to the left) in conjunction with good visual acuity of the reader (symbolic illustration, figure to the right) and no distraction (symbolic illustration of a duck that quietly watches) are best suited for a general setup [source: own 3D Blender model [1–4]].

DOI: 10.4324/9781003413998-42

35.1 BACKGROUND

Reports in forensic medicine that describe the results of an examination are usually structured. With that, they aim to cover content of both so-called positive and negative findings.

Usually, it is expected to provide a report that is complete. From any logical understanding of completeness, however, it does not seem possible to cover *all* humanly thinkable hypothetical angles of any given case—simply because there exists no terminally limiting logic to determine what has to be contained.

What can be targeted for a complete coverage is a type of morphological, anatomical or regional completeness, possibly with focus on certain aspects. There, the body or specimen that is examined may be split into anatomical areas—such as head and neck, thorax, abdomen and extremities. Alternatively, any examination result may be split into grid-like subregions [6, 7].

While some forensic pathologists may focus on what they believe to be relevant, neither the state attorney, legal representative of the bereaved or defense lawyer may share that focus. So a more comprehensive or wider angle on the subject of reporting may help here. And while one could always write a footnote or disclaimer stating that more data could be analyzed, the relevant quote here is "quod not legitur not creditur", what cannot be read cannot be believed, and "quod non est in actis non est in mundo", what is not in the files is not in the world.

What generally should be avoided, both on a level of case reading and on a level of providing adequate staffing, is the assumption that reading and documenting PMCT is a matter of a few minutes.[1] Performing a PMCT scan may require anywhere between 10 to 30 minutes, having data exported, transferred or sent to the reading system may take another 5 to 45 minutes, depending on specifics, and a full PMCT reading and documentation should be allowed to take 1.5 to 2 hours [9].

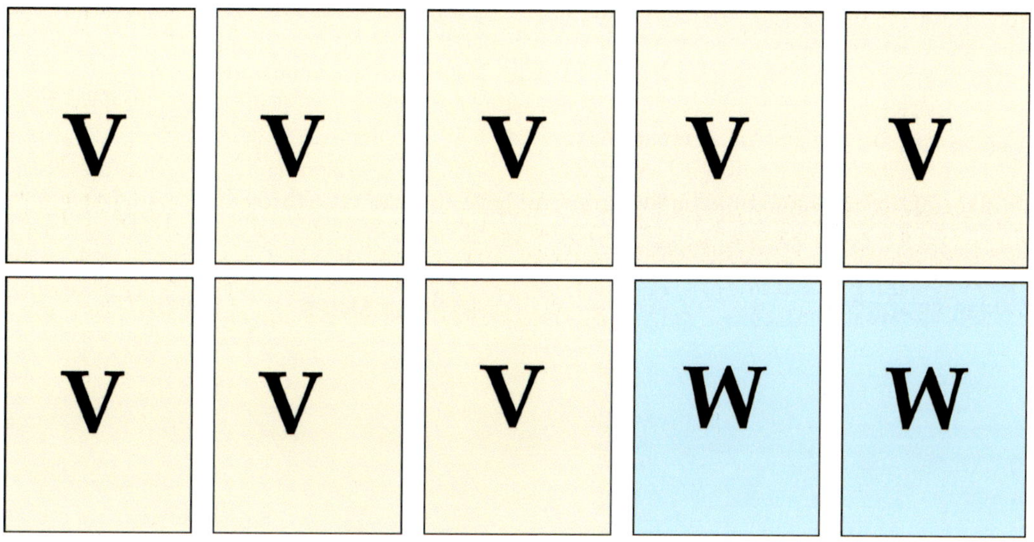

Figure 35.2 Visual and written report—recommended coverage: this schematic overview symbolically illustrates the extent of visual (V) and written (W) report pages that we may recommend to consider for routine use. Actual extent, visual and text content details will have to be prepared if not negotiated with the investigators and pathologists.

35.2 OUR APPROACH TO READING FORENSIC IMAGING

35.2.1 Before the Report—PMCT Scan-Related Aspects for the Advanced PMCT Reader

As some aspects cannot be changed after the PMCT scan, it may be relevant as forensic pathologist or radiologist to see to it that materials such as rocks or stones and metal materials such as jewelry are removed from the body as much as possible. It is to be regarded as a part of the preparatory process of the person with the leading responsibility to see to this. It may be relevant in certain settings to set up a proper process description in the quality manual.

Preparing the body inside the body bag for best positioning may be relevant and be regarded as part of the responsibility of the autopsy technicians or forensic pathologists. Therefore, the arms should be positioned conveniently to create not overly extensive artifacts, particularly for a detailed capturing of thorax and abdomen—whether above the head, anteriorly or elsewhere. Ultimately, the forensic pathologist, radiologist or whoever else reads the images will have a clear interest in getting optimal quality images.

The arrangement and metric spread, in terms of occupying table width, of single parts, objects, items or bones, may also be regarded as the responsibility of the forensic pathologist, radiologist or whoever else is the lead examiner.

That metric spread of single specimen should be kept minimal, as then scan can occur in the center region of the gantry and reconstruction may be performed with a minimal field-of-view.

With any field-of-view f [in millimeter], the in-plane pixel size p with a given reconstruction matrix of m × m will always be $p = \frac{f}{m}$. That means that a higher resolution may be achieved by proper reconstruction. Once the field-of-view is known, slice thickness for scanning can be selected accordingly, and slice thickness z for reconstruction can be set to $z \sim p$ in order to yield least approximately isotropic data. Any scientific reporting of what can be seen, or, more importantly, not seen, on PMCT definitely requires full disclosure of CT device used, kV, mA, rotation time and pitch. As PMCT scanning allows for dose maximization for the benefit particularly of more precise physical soft tissue structure delineation, not exploiting that for any reason usually explains lossy soft tissue contrast.

As not all people performing postmortem CT scans are radiology technicians, it may also be relevant for the forensic pathologist to know that for a helical CT scanner, maximizing CT dose, increasing rotation time and minimizing spiral pitch generally help generate better results [10]. A pitch below 1 denotes overlapping helical acquisition with z-axis oversampling, while higher gantry rotation times further improve signal-to-noise ratio through increased photon statistics.

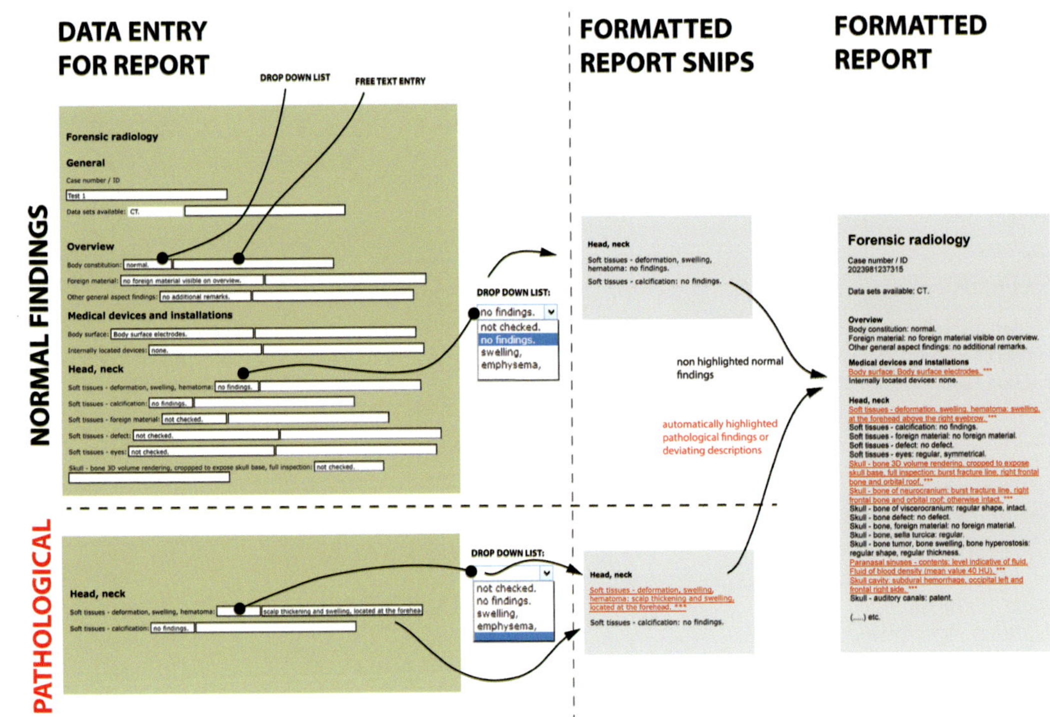

Figure 35.3 Written report: this schematic overview symbolically illustrates the possibility of a structured data entry or text capture system one could design and propose in order to build a structured report from formatted report parts. Ideally, positive findings could be highlighted in the formatted forensic imaging report in a color (here: red and underlined). Figure republished from [5] with permission.

In instances where in an image objects are available as part of the examination such as scattered cameras or mobile devices after an airplane crash, these may be examined in detail and findings reported.

Last but not least, the aspects of various data reconstructions once the PMCT has been acquired and raw data saved should be known to the person in charge of the case. Different kernels and reconstruction techniques may adversely influence the image quality—be that too much noise or an uncannily high degree of smoothing.

35.2.2 Before the Report—Setup for Image Reading

Radiological images are read ideally in a room that is not too bright, using computer screens with a sufficient brightness, pixel and contrast resolution. The ambient room brightness, the type of computer display or monitor and its particular setting with regard to brightness, contrast, as well as a possibly more refined calibration, may be regulated by law, regulation or quality requirements (e.g., [11]). Also, the visual acuity of the person reading the images will have to be sufficiently high [12]. Distraction of any auditory, visual or even olfactory type is to be avoided (Figure 35.1). Any workflow distraction may be inefficient, in that it extends overall work effort, in one study by a significant extra time cost of up to 13 percent [13], as well as increasing the likelihood of an error [14].

35.2.3 Reading Images and Reporting—General Remarks

Quod non est in actis, non est in mundo.[2]

Any report is preceded by reading the PMCT images (or other modalities) under and ideally conforming to something like the aforementioned setup. The reader may ideally have a sufficiently capable computer screen that may be calibrated to the requirements, low ambient light, undisturbed time and no distraction. The list of checked items on any PMCT may vary among jurisdictions and readers. If a PMCT scan has to be particularly sensitive so as not to miss even subtle injuries, a proper protocol may have to be used. Typically, a higher contrast (as brought about by a higher CT dose as quantified by the mAs value) and a higher pixel resolution (by reconstructing a smaller field-of-view and matching that higher axial inplane resolution with a similar slice distance to approximate isotropic voxels) are of particular help, as are attempts to reduce artifacts or improve reconstruction, such as placing single bones in a water bed [15, 16]. Quite possibly, the subsequent reading of the acquired PMCT data may have to be thorough and comprehensive. To that end, the reader may take advantage of the available visualization modalities such as the ones mentioned in the visualization chapter of this book, and such as warranted by anatomical region, type of trauma and forensic question. One should put any possibly normal postmortem findings, such as related to blood settling down according to gravity after cessation of circulation (as in hypostasis), hyperdense blood vessels, gas and emphysema, in context of the specific case. One should make sure to consider artifacts, such as beam hardening artifacts, as explanation for a visual finding, as they may be typically present in the region of brain stem and cerebellum when there are dental restorations or implants. Artifacts may also occur when the arms are scanned in a position that is attached to the torso, thereby possibly making a reliable CT-density reading of thoracic or abdominal findings impossible.

Structured reporting and including visuals have been a relevant consideration in our team since years [5, 17]. We currently provide both a visual report alongside a structured written report, with a page count distribution similar to what symbolically is illustrated in Figure 35.2. These visuals serve as support, helping to demonstrate the findings, clarify e.g. the extent of trauma, and potentially also identify normal findings or absence of trauma in cases where such was expected. The provision of structured reporting and the provision of visual documentation in this field is underpinned by significant advancements in medical imaging technologies and digital computer systems, which have evolved to address the complex needs of modern users and their increasing demands. Incorporating visual evidence not only augments the textual report but also provides a type of proof that can be critical for further proceedings. Vismann argues that the Latin maxim *quod non est in actis, non est in mundo* (what is not in the file is not in the world) that comes up in literary and judicial texts referring to the written foundation for Roman court proceedings summarizes the performative operation of the law in constructing reality" (quoted from [18]).

35.2.4 Written Report

The written report may reflect findings across all anatomic regions or cover the whole scanned area or volume. It should describe findings in necessary detail but also so the reader can relate these to their medico-legal meaning and significance. Vague or suspected findings may be described as such, as a particular finding does not always provide clear or unanimous evidence. It may be important to differentiate between the ability to describe a finding once it is there, which depends on various aspects and parameters, and the ability to exclude a finding such as a hemorrhage or fracture, which may depend on different parameters. For example, it may be wrong to state that "a laryngeal fracture is excluded" based on a PMCT with a normal clinical CT-scanner [19], and but it may be fair to state that "given the actual device and resolution, no signs of a fracture were identified".

Structured reporting in the realm of forensic imaging may take a relevant role by enhancing the clarity and comprehensibility but also a certain minimal coverage if not completeness of findings [5, 20–22]. The role of a structure to guide reporting also lies in extending the reader's search past and beyond the apparently first or relevant appearing diagnosis [23, 24]. The technical system to be used for a structured report ideally provides the recipient of the report with an easy-to-read text. For example, one could devise a structured report template setup that allows an item-per-item data entry, possibly a combined drop-down menu with pre-configured text selection choices and a free text field for each finding, to allow for different or unexpected findings to be described. The software to be used for such a template based system may be freely chosen by the reader and their clients, possibly also after consulting with specialists for programming, data base, text form creation or similar informatics system devices and programs. If such a reporting system also feeds into a database, while generating the written report, later data mining and analysis may be facilitated [25].

There are reasons also not to overly focus on structured reporting [24]. Rigid adherence to structured report templates can disrupt the visual search patterns of the reader. This disruption can lead to a phenomenon known as "eye dwell", where readers may find themselves focusing more on the report template than on the images themselves. Such shifts in focus can contribute to overlooked findings [26]. When descriptions of pathology or trauma involve multiple organ systems, they may be segmented by the report structure items, such as being sorted by anatomical region or organ system, rather than presented as a unified disease process. This segmentation can make the reports harder to write and more difficult to understand [25]. Last but not least, a possibly extensive or exhaustive structured report may, while surely reducing risk of diagnostic error, incur substantial extra cost in time, so ultimately a trade-off between effort and cost in terms of time and diagnostic error may have to be found [5, 25].

35.2.5 Visual Report

A visual report requires mastering of basic visualization techniques. First, the correct reconstruction kernel is to be used. A hard kernel is beneficial for lung or bone slice viewing. But for all 3D rendering including the 3D visualization of fractures, as well as soft tissue or organ findings, soft kernel reconstructions are highly recommended. Second, axial or otherwise angled slice images (sagittal, coronal, MPR) will have to be properly windowed so as to provide the correct contrast to show the finding in question. Third, the finding has to be properly zoomed and panned, so it takes up the center of the presented image and so that it is not just a few pixels in size. So, for example, to demonstrate coronary artery calcification, it may be considered to provide an image that zooms in on the actual finding rather than using a full axial slice image of the whole thorax. Last but not least, maximum intensity projection, minimal intensity projection and volume rendering have to be used to their advantage.

35.2.6 Summary

As another option, one may consider 3D printing to convey forensic findings [27].

ACKNOWLEDGMENTS

The authors express their gratitude to Emma Louise Kessler, MD, for her generous donation to the Zürich Institute of Forensic Medicine, University of Zürich, Switzerland. This work released under CC BY 4.0. Figure 35.3 reused from [5] with permission via Elsevier license 5752990204395. While the authors have made efforts to ensure clarity and structure, the development of this manuscript occurred under the constraints of available time and timelines. The retrospective use of anonymized data was submitted to and evaluated by the Ethics Committee of the Canton of Zürich (reference number 15–0686).

Notes

1 A historically early Virtopsy promise was that "CT scanning to document any bone injuries and gross pathologies takes only minutes and can be used as a screening tool" [8].
2 What is not in the files is not in the world.

References

[1] Blender Online Community. Blender—a 3D modelling and rendering package. Blender Institute, Amsterdam; 2025. Available from: http://www.blender.org.
[2] Ebert LC, Franckenberg S, Sieberth T, Schweitzer W, Thali M, Ford J, et al. A review of visualization techniques of postmortem computed tomography data for forensic death investigations. International Journal of Legal Medicine. 2021;135(5):1855–1867.
[3] Suave Hat Man—Download free 3D model by oillut—sketchfab.com. [Accessed 04-12-2025]. Available from: https://sketchfab.com/3d-models/ suave-hat-man-76f79d333393414daff5cd541c699f14.
[4] Schweitzer W. swisswuff.ch—Image reading (3D) by Wolf Schweitzer [CC BY 4.0]; 2025. Available from: https://www.swisswuff.ch/wordpress/?p=2214.
[5] Schweitzer W, Bartsch C, Ruder TD, Thali MJ. Virtopsy approach: Structured reporting versus free reporting for PMCT findings. Journal of Forensic Radiology and Imaging. 2014;2(1):28–33.
[6] Blau S, Robertson S, Johnstone M. Disaster victim identification: New applications for postmortem computed tomography. Journal of Forensic Sciences. 2008;53(4):956–961.
[7] Xiong J, Ren J, Luo L, Horowitz M. Mapping histological slice sequences to the allen mouse brain atlas without 3D reconstruction. Frontiers in Neuroinformatics. 2018;12:93.
[8] Dirnhofer R, Thali M. Virtopsy after more than some 100 cases: Statement and perspectives of forensic imaging by using 3D optical and combined CT/MRI whole-body scanning. In: The virtopsy approach—3D optical and radiological scanning and reconstruction in forensic medicine. CRC Press, Boca Raton, FL; 2009.
[9] Sokolovskaya E, Shinde T, Ruchman RB, Kwak AJ, Lu S, Shariff YK, et al. The effect of faster reporting speed for imaging studies on the number of misses and interpretation errors: A pilot study. Journal of the American College of Radiology. 2015;12(7):683–688.
[10] Ranallo FN, Szczykutowicz T. The correct selection of pitch for optimal CT scanning: Avoiding common misconceptions. Journal of the American College of Radiology. 2015;12(4):423–424.
[11] Eidgenössisches Departement des Innern EDI AS Bundesamt für Gesundheit BAG. Anforderungen an medizinische Befundmonitore und die technische Qualitätssicherung. EDI. Available from: https://www.bag.admin.ch/dam/bag/de/dokumente/ str/str-wegleitungen/technische-qs/wegleitung-qs-bwg.pdf. download.pdf/wegleitung-med-befundmonitore-d.pdf.
[12] Quaghebeur G, Bhattacharya J, Murfitt J. Radiologists and visual acuity. European Radiology. 1997;7:41–43.
[13] Williams LH, Drew T. Distraction in diagnostic radiology: How is search through volumetric medical images affected by interruptions? Cognitive Research: Principles and Implications. 2017;2:1–11.
[14] Balint BJ, Steenburg SD, Lin H, Shen C, Steele JL, Gunderman RB. Do telephone call interruptions have an impact on radiology resident diagnostic accuracy? Academic Radiology. 2014;21(12):1623–1628.
[15] Schweitzer W, Thali M. Virtopsy of isolated bones: Better fracture morphology by reducing bone surface CT artifacts through using a water bed; 2023. Annual congress of the International Society of Forensic Radiology and Imaging, Toulouse, France, 2023. May 25–27.
[16] Mani MR, Christensen AM, Rafiq A. Effects of water and other filters on fracture resolution in industrial micro-CT scanning. Forensic Imaging. 2024;37:200588.
[17] Ampanozi G, Zimmermann D, Hatch GM, Ruder TD, Ross S, Flach PM, et al. Format preferences of district attorneys for postmortem medical imaging reports: Understandability, cost effectiveness, and suitability for the courtroom: A questionnaire based study. Legal Medicine. 2012;14(3):116–120.
[18] Suresh M. Accountability, authority and documentary fragility: Shadow files' in a trial in India. In: Sterett SM, Walker LD, editors. Research handbook on law and courts. Edward Elgar Publishing; 2019. p. 145–157.

[19] Treitl KM, Aigner LI, Gazov E, Fischer F, Schinner R, Schmid-Tannwald C, et al. Injuries of the isolated larynx-hyoid complex in postmortem computed tomography (PMCT) and postmortem fine preparation (PMFP)-a comparison of 54 forensic cases. European Radiology. 2020;30:4564–4572.
[20] Schwartz LH, Panicek DM, Berk AR, Li Y, Hricak H. Improving communication of diagnostic radiology findings through structured reporting. Radiology. 2011;260(1):174–181.
[21] Brough AL, Morgan B, Rutty GN. The basics of disaster victim identification. Journal of Forensic Radiology and Imaging. 2015;3(1):29–37.
[22] Rutty GN, Biggs MJ, Brough A, Morgan B, Webster P, Heathcote A, et al. Remote postmortem radiology reporting in disaster victim identification: Experience gained in the 2017 Grenfell Tower disaster. International Journal of Legal Medicine. 2020;134:637–643.
[23] Lee CS, Nagy PG, Weaver SJ, Newman-Toker DE. Cognitive and system factors contributing to diagnostic errors in radiology. American Journal of Roentgenology. 2013;201(3):611–617.
[24] Ganeshan D, Duong PAT, Probyn L, Lenchik L, McArthur TA, Retrouvey M, et al. Structured reporting in radiology. Academic Radiology. 2018;25(1):66–73.
[25] Srinivasa Babu A, Brooks ML. The malpractice liability of radiology reports: Minimizing the risk. Radiographics. 2015;35(2):547–554.
[26] Hall FM. The radiology report of the future. Radiology. 2009;251(2):313–316.
[27] Schweitzer W, Thali M, Aldomar E, Ebert L. Overview of the use of 3D printing in forensic medicine. Rechtsmedizin. 2020;30(5):292–299.

PART 8

STRATEGIES AND CASES

CHAPTER 36

THE FORENSIC EVALUATION OF EVIDENCE IN LEGAL SETTINGS

Dilan Seckiner

The purpose of forensic science[1] is to study crime and its trace(s),[2] whereby these 'silent witnesses' need to be detected, seen and understood in order to make reasonable inferences regarding criminal phenomena, investigation, intelligence gathering and court purposes (Roux *et al.*, 2012). Crispino *et al.* (2021) highlighted that Daubert only considered the Galilean dimension in terms of the development of general predictive models; in turn, Daubert emphasised the contribution of forensic science. It has been suggested that forensic science needs to be classified more accurately in the clinical approach and his needs to be integrated more to reconstruct a past event or presence of activity and the associated traces (Crispino *et al.*, 2021). This places great importance from the very start of any reconstruction of past events, which starts at documentation of the scene.

To facilitate this, it has been proposed that the forensic science focus needs to shift from means to purpose, where research and development should be refocused on shared principles and purposes; for instance, this includes the reconstructing, monitoring and preventing crime and security issues (Roux *et al.*, 2021). Both means and processes are dependant on the political and legal structures which differ between countries, jurisdiction and organisations, which in turn, makes it difficult agree on universal measures that can be implemented and effective long-term (Roux *et al.*, 2021). Therefore, it is recommended that the problem is tackled through focus on the discipline and less through organisational factors (Roux *et al.*, 2021).

The current 'organisation' model appears (discussed in the literature) as a laboratory with the purpose of producing 'evidence', where some of these laboratories work as a public or private business to provide the service asked of them (Roux *et al.*, 2021). The standardisation of the laboratories and police facilities across the world is necessary for forensic science (Figure 36.1). As a result of the forensic science refocus, this will allow the trace to be better understood and foster development of a forensic science culture that is unified by purpose rather than means (Roux *et al.*, 2021).

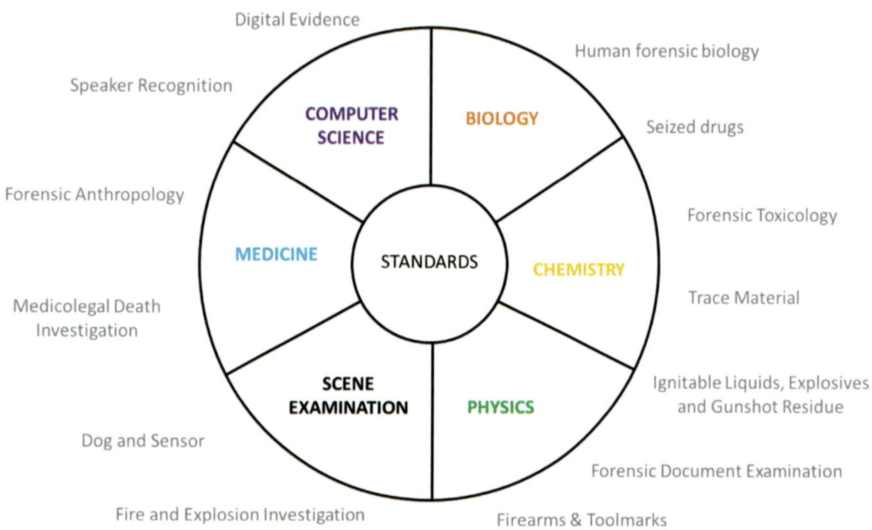

Figure 36.1 Multidisciplinary organisation (i.e. laboratories) model example of forensic science on enabling disciplines and standards (Roux *et al.*, 2021).

DOI: 10.4324/9781003413998-44

A critical part that forms the organisational model is Locard's principle, which highlights that criminal activities always leave multiple traces behind, in which the offender may leave traces at the scene by their actions or may alternatively pick up traces on their clothes/body from their location (Roux *et al.*, 2015). Ribaux *et al.* (2016) explains this definition further by dividing it into three main aspects. The first, is reliant on the nature of the criminal activity itself and how it impacts the material types that are exchanged and in turn distributed within the environment or taken by offender (Ribaux *et al.*, 2016; Roux *et al.*, 2015). The second, these remnants of activity, are traces that become signs when detected, recognised, collected and measured (Ribaux *et al.*, 2016; Roux *et al.*, 2015). Third, and finally, an interpretation process aims at transforming them into clues to reconstruct what occurred and also to potentially understand what activity caused the trace (Figure 36.2) (Ribaux *et al.*, 2016; Roux *et al.*, 2015).

In addition to this, Margot (2011) highlighted that traces are the most elementary information that result from crime, which is further reinforced by Kirk's individuality principle. If we assume that every object in the universe is unique, Kirk's principle relied on the focus that forensic science is the source of two items: questioned and known (trace and reference). In addition, to approach the comparison as closely as the present state of science allows one to determine if they originate from the same source (Roux *et al.*, 2012; Kirk, 1963). Furthermore, three types of investigations have been highlighted by Brodeur (2010), which include the identification problem (who?), the localisation problem (where?) and the problem of structuring evidence (what?) (Roux *et al.*, 2012). From here, the identity of the source can be made from the expert from a methodological approach known as the 'logical framework' or 'likelihood ratio', which provides unity within forensic science (Roux *et al.*, 2012).

Determining the identity of the deceased individual is necessary in any investigation, so the family and relatives of can be notified, as well as facilitate police investigations and fulfil the legal processes (Knight, 2004). The two main identification methods involve (1) primary, which is considered accurate identification of human remains (DNA, fingerprints, odontology, distinctive medical conditions, and (2) secondary identifiers, which includes personal effects, personal appearance, tattoos, and some medical findings which involve previous injuries, scars) (Burgess, 2014; INTERPOL, 2018). Forensic evidence, more commonly, referred to as a 'trace', and its subsequent examination, has faced substantial obstacles and criticism within legal settings, where the problem lies not within the discipline of the trace, but rather the lack of scientific validity and evaluation by the forensic examiner and poor understanding of the fundamental principles and of the forensic inference process. New and emerging technologies and validation studies have in turn improved the validity of the examination of the trace, but its use in legal settings and locations vary.

Internationally, legal systems do not all rely on admissibility criteria to accept evidence in court; for instance, in the inquisitorial system it remains the province of the trier-of-fact. The two main legal systems are adversarial and inquisitorial (Powell and Mitchell, 2007). In short, the adversarial system is dependent on the courts role of an impartial and unbiased referee between the prosecution and defence, whereas an inquisitorial system, depend on the court investigating the proof of facts of a case (Ambos, 2003). The adversarial law system is predominantly found in Australia, United Kingdom, America and Canada (Willke, 1986), while the inquisitorial law system is found in most of Europe, all of Latin America and parts of Africa and Asia (Merryman and Perez-Perdomo, 2007). However, instead of focusing on

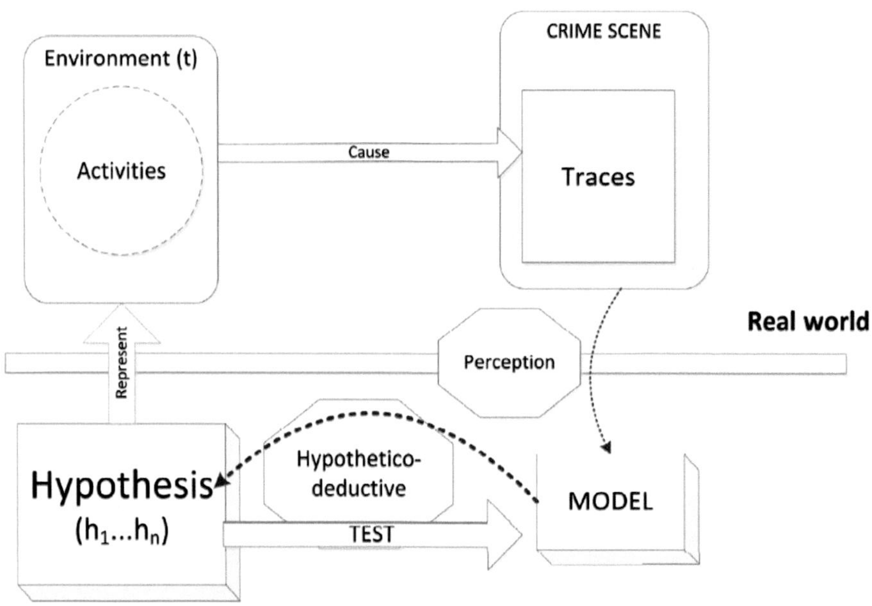

Figure 36.2 Model of reasoning (Roux *et al.*, 2012).

whether a rule is whichever adversarial or inquisitorial, importance of a fair trial, assisting tribunals to achieve tasks whilst abiding by fundamental fair trial standards (Ambos, 2003) forensic evaluation of the evidence whilst upholding scientific validity and reliability as the result of forensic evaluation is not a decision of identification but a description of the strength of evidence. Admissibility is not concerned with the formulation of the forensic evaluation conclusion but rather the legal rules in order to accept evidence. The result of the examination provides support to one of the alternative hypotheses; (H1) the trace and reference material originate from the same person, or (H2) the trace and reference material originate from different persons.

36.1 FORENSIC PROCESSES

Forensic examination can be divided into four steps overall, ACE-V, which are the analysis, comparison, evaluation and the verification process, which has become the general widespread stepwise approach to all forensic pattern evidence types that guides examiners throughout the examination process (Montani *et al.*, 2019). This needs to be applied to all criminal investigations, where the documentation, collection, examination and interpretation of physical evidence may serve to suggest or support possible scenarios of what may have happened at the scene of the crime and allow a reconstruction of events (Broeders, 2014). The following describes the forensic processes involved within this reconstruction process:

Forensic investigation—Biometric technology can contribute by comparing traces (such as fingermarks and prints) to the database—where the persons whose data are most similar to the mark are shortlisted (Meuwly and Veldhuis, 2012). Results are then refined where reference specimens are excluded based on the data, leaving only those individuals whose data most closely related to the mark (Meuwly and Veldhuis, 2012).

Forensic intelligence—Biometric technology can be applied in forensic intelligence scenarios to detect patterns of interest, especially when they are linked to abnormal and potentially security-related situations and behaviours. When forensic science is applied within the forensic intelligence context, the aim is to develop efficient strategies and concrete operations to disrupt or prevent criminal activity, as well as to link traces related to the same source (Julian *et al.*, 2011), which complements the conventional use of forensic science to be used in court outcomes (Julian *et al.*, 2011).

Forensic evaluation—Forensic evidence at source level comprises the application of biometric technology to exclude the POI or assess the strength of the evidence resulting from the comparison of the trace and reference specimens for reporting in court proceedings (Meuwly and Veldhuis, 2012). Development and application of a probabilistic approach within all disciplines under the forensic science umbrella, will allow the examiner to provide a description of the strength of evidence to the trier of facts in court (Meuwly and Veldhuis, 2012).

36.1.1 Likelihood Ratio

Upon determining the level of performance evaluation (and consequently the method validation), the developed protocol, system, model or method, must be tested. For instance, the testing of a biometric system, includes the calculation of scores for verification or 'identification' purposes (ISO/IEC, 2006). One important part of the process includes validation, where the results of an evaluation (determined based on the defined protocol) is used and subsequently compared to a threshold to conclude whether the model/method meets the threshold or not. Three types of evaluation are existent upon the testing of a biometric system: (1) technology evaluation; (2) scenario evaluation; and (3) operational evaluation (ISO/IEC, 2006).

Technology evaluation comprises the standardised testing of all algorithms and models to view performance (generally involves both environment and collected population) (ISO/IEC, 2006). It is important to note that this stage involves the development, in which the data is fine-tuned and testing is undertaken on data that is not previously tested; results should be repeatable (ISO/IEC, 2006). An example in relation to forensic gait analysis for instance, would be the examination for the morphological and anthropometric assessments.

Scenario evaluation involves the testing on a system that has been completed, where the environment imitates that of a real-world target application of interest (ISO/IEC, 2006). Each system that is tested, will require a combination of various evaluations and comparisons that is applied to the same population. Within these controlled conditions, the test results will be repeatable to the modelled scenario (ISO/IEC, 2006). An example for instance for forensic gait analysis, would include the development of an analytical method for forensic scenarios (i.e. morphometric techniques for forensic evaluation). This interpretation method consists of assigning likelihood ratios as measures of the probative value.

Operational evaluation is unable to be repeated since the undetermined and undocumented differences between operational environments, the ground truth is difficult to determine especially within the unverified and uncontrolled environments (ISO/IEC, 2006).To provide an example, using forensic gait analysis, this will be the implementation for cases.

Subsequently, the groundwork of the forensic evaluation logical approach as established by the ISO standard in preparation, pose the following question types: (1) question of source, (2) classification and identification, (3) quantification, and (4) reconstruction (ISO/IEC, 2006). When it comes to the question of source, the existing approach is to convey a probative value expressed in terms of likelihood ratios (LR). For evaluative reporting, probabilities are applied to express the degree of certainty and uncertainty within each proposition, which may be based on empirically derived data (NIFS, 2017).

It was highlighted by Williams and Maskell (2021) that all forensic practitioners and experts should follow three forensic interpretation principles of LR from Bayes Theorem, combined with the expert given case context. The three principles by Williams and Maskell (2021) are thought to minimise the chances of miscarriages of justice include:

1. Considering one alternative hypothesis at all times
2. Rather than considering the probability of the proposition given the evidence, the consideration needs to be placed on the probability of the evidence given the proposition
3. The framework of circumstance always needs to be considered

The logical framework as explained by Evett (2015), includes two main phases: (1) the investigative phase, where the scientist closely works with the investigator in search for the offender; and (2) the evaluative phase, in which the scientist(s) assists the court (Evett, 2015). The probability theory, or Bayes Theorem, contains the likelihood ratio, which is the strength of the evidence for a trace (i.e. evidence from CCTV footage, such as an image) and a reference (a suspected person where their data is collected as part of forensic procedures) with regard to a pair of mutually exclusive propositions (e.g., Aitken and Taroni, 2004; Robertson *et al.*, 2016). This is the quantity, or score reported by the forensic scientist within court proceedings, to highlight the weight of evidence to the judge and/or jury in the context of a case (e.g., Evett, 2015). Using the example of forensic gait analysis and the evaluation of CCTV images, the two mutually exclusive propositions are: the trace from the CCTV footage is of the suspected person and the trace from the CCTV footage is of a different person in the population.

The LR for the trace evidence from the CCTV footage with regard to these two propositions is the probability of the empirically derived (or other) examinations on the image from the CCTV footage (i.e., the trace) given that this image is of the suspected person. This is divided by the probability of these measurements given that this image is of a different person in the population. Signifying the examinations or measurements on the image from the CCTV footage is denoted by *s*, the first proposition with H_p, the second proposition H_d, the relevant background information for the case by *I*, and probabilities by *Pr()*. Together, the LR is given by the following mathematical expression:

$$LR = \frac{Pr(s \mid H_p, I)}{Pr(s \mid H_d, I)}$$

Following the forensic evaluation provided by the forensic practitioner/expert, the responsibility and duty then lies within the judge or jury (dependant on the adversarial and/or inquisitorial legal systems in place) to consider the proceedings of the case alongside other information sources and evidence put forth (e.g., Robertson *et al.*, 2016). This is where the Logical framework is calculated and applied: The forensic practitioner is tasked with the completion of (generally) empirical scientific analyses of the footage, followed by the quantification of the LR, whereas the judge or jury is responsible for the quantification of prior odds and posterior odds (e.g., Lindley, 1977). Figure 36.3, as adapted by Williams and Maskell (2021), highlights the roles of the court within each component of Bayes Theorem.

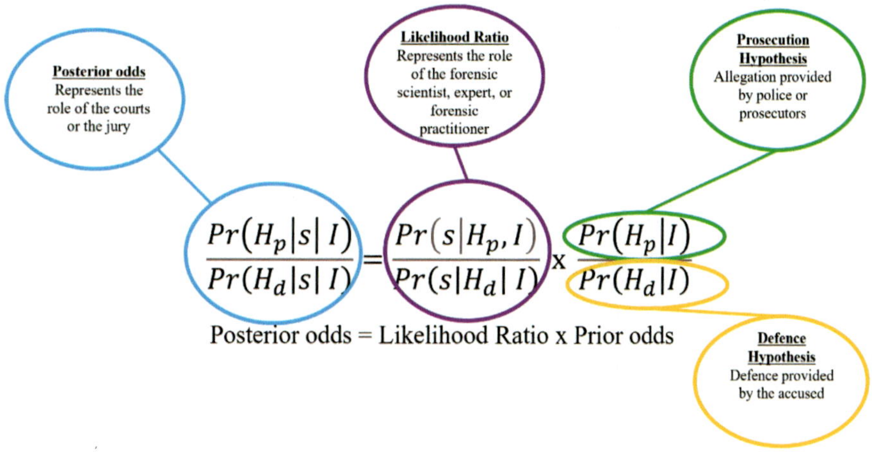

Figure 36.3 Bayes theorem and the roles of each component adapted by Williams and Maskell (2021).

Standards of evidence admissibility for emerging technologies, methods and/or new disciplines (of study in particular), should remain complex as stated by Macoveciuc *et al.* (2019). It is imperative to note that this complexity, should relate explicitly to the rigor of method testing, to allow high-quality results, rather than unnecessary complexities in the admissibility procedures (Macoveciuc *et al.*, 2019). The LR method development within each transdisciplinary discipline requires a combination of roles, in which the first role focuses on the development of forensic methodology, where forensic practitioners/experts devise solutions (Meuwly *et al.*, 2016). From here, the second role concentrates on further advancement, and consecutive validation of the new methods/solutions developed (Meuwly *et al.*, 2016). In this stage, a team of specialists/experts including forensic practitioners, statisticians and engineers establish new (or adapt existing) technologies for forensic evidence evaluation (Meuwly *et al.*, 2016). The third, and final role is the evaluation of forensic evidence in practice (Meuwly *et al.*, 2016). For forensic evaluation, the use of logical networks has been underlined repeatedly (Seckiner *et al.*, 2019; Meuwly *et al.*, 2016), but even with logic, rationality and transparency applied to the process, its complexity is the primary obstacle to its implementation (Meuwly and Veldhuis, 2012). It requires collaboration in many attributes, including (but not limited to) the relevant: (1) data, (2) inference models, (3) technology, and (4) evaluation framework (Meuwly and Veldhuis, 2012). Therefore, the development and implementation of a likelihood ratio model for all disciplines in forensic science, and the subsequent application of the methods is the first step towards a standardised probabilistic evaluation procedure for forensic examination.

If new evidence is presented to the juror, this is recognised as the 'prior odds' (Evett, 2015). The scientist provides scientific observations which is representative by *E*, where the juror will be given new information, this is referred to as the 'posterior odds' (Evett, 2015). The likelihood ratio, can be explained as the probability of the score (*P*) given the prosecution hypothesis is true, divided by the probability of the score given the defence hypothesis is true (Ali *et al.*, 2014). Moreover, the prosecution hypothesis states the source of the trace material is the suspected person, or that the trace material originates from the reference material of the suspected person, whereas the defence hypothesis states that the reference material does not originate from the trace (Ali *et al.*, 2014). These are viewed as Hp and Hd, respectively, and are 'two mutually exclusive and exhaustive source-level hypotheses' (Ali *et al.*, 2014, pg. 335). *I* subsequently describes background information, whereas *E* is the evidence by the scientist (Ali *et al.*, 2014; Evett, 2015).

Evett (2015), also developed the following principles in which the scientist: (1) should address a pair or propositions (prosecution and defence), (2) consider the probability of the scientific observations given that each in turn is true, (3) two probabilities that are conditioned by the two propositions are also conditioned by case circumstances (*I*) and (4) the ratio of the two probabilities determines weight of evidence. Typically, *I* is overlooked since the case circumstances at source level are not affected by whether Hp or Hd is true. However, this should also be implemented within all forensic disciplines, for instance, when an examiner compares the trace from CCTV footage to reference materials.

Referring back to surveillance footage, the European Network of Forensic Science Institutes (ENFSI) guidelines reinforce the implementation of LR and highlight that subsequent to the examination from CCTV footage, the findings are generally evaluated against two propositions assumed to be true: (1) the first is by the authorities; (2) and the alternative by the defendant (ENFSI, 2010). The LR of the two probabilities are then calculated and forms the value of the evidence through a graded conclusion—strengthening or weakening the prior opinions on the two propositions. The ENFSI guidelines also recommend the implementation of the *Guideline for Evaluative Reporting* as a twofold step, which involves overseeing the change and the training involved. It is emphasised that the framework of circumstance, propositions and likelihood ratio per forensic discipline should be supplied with the appropriate training and proficiency/competency testing.

The quality of the inference is dependent on the quantity and properties of the data that is used to examine the 'within' and 'between-source' variability (Meuwly and Veldhuis, 2012). As indicated by Meuwly and Veldhuis (2012), the classic 'forensic identification' disciplines depend principally on personal probabilities for assessment of the evidence this in turn, is becoming progressively challenged. Likelihood ratio methods are seen to be promising within forensic science and was originally applied within DNA modalities (Meuwly and Veldhuis, 2012). Other disciplines have since implemented likelihood ratio to numerous types of trace materials. For example, a study by Champod and Meuwly (2000), developed an interpretation framework (based on likelihood ratio) for speaker recognition, fingerprint recognition (Meuwly and Veldhuis, 2012). A PhD study by Seckiner (2021) developed an interpretation framework (based on likelihood ratio) for forensic gait analysis. Therefore, it is imperative for the implementation of likelihood ratios within the forensic disciplines to provide standardised probabilistic evaluation procedure for forensic examination for court processes.

Notes

1 Roux *et al.* (2018), define forensic science as 'a linear application of scientific disciplines in a legal context, primarily for court purposes'.

2 Trace evidence is defined as 'the analysis of a pattern, signal, or materials that, is transferred during an event (often unknowingly)' (Roux *et al.*, 2015, 2018). As a silent witness, it has the ability to provide information regarding what happened and who was involved, as it is 'the remnant (the memory) of the source (identity—who, with what?) and of the activity (what, how, when, why?) that produced it' (Roux *et al.*, 2015, 2018).

References

Ali, T., Spreeuwers, L. J., & Veldhuis, R. N. J. (2014). Biometric evidence evaluation: an empirical assessment of the sampling variability. IET Biometrics, 3(4), 238–245.

Ambos, K. 2003. International criminal procedure: "adversarial", "inquisitorial" or mixed? International Criminal Law Review 3: 1–37.

Bradley, D., N. Walker, R. Wilkie (1986), Managing the Police: Law, Organisation and Democracy, Brighton: Wheatsheaf Books.

Brodeur, J.P. The Policing Web, Oxford University Press, New York, 2010.

Broeders, T. Philosophy of Forensic Identification: Encyclopaedia of Criminology and Criminal Justice, Springer Science, New York, 2014, pp. 3513–3526.

Burgess, M.J.C. Identification: The coroner's perspective, in: X. Mallett, T. Blythe R. Berry (Eds.), Advances in Forensic Human Identification, CRC Press, Boca Raton, 2014, pp. 21–36.

Champod, C., & Meuwly, D. (2000). The inference of identity in forensic speaker recognition. Speech Communication, 31(2–3), 193–203.

Colin G.G. Aitken, Franco Taroni. 2004. Statistics and the Evaluation of Evidence for Forensic Scientists Print ISBN: 9780470843673 |Online ISBN:9780470011232 |DOI:10.1002/0470011238 Copyright © 2004 John Wiley & Sons, Ltd

Crispino, F., Weyermann, C., Delemont, O., Roux, C., Ribaux, O. Towards another paradigm for forensic science? WIREs Forensic Science, 2021: pp. 1–15.

Dennis V. Lindley.1977. "A problem in forensic science" Journal: Biometrika, Volume 64, Issue 2, August 1977, Pages 207–213

(ENFSI) European Network of Forensic Science Institutes. Strengthening the evaluation of forensic results across Europe (STEOFRAE), 2010. Version 3.0. p. 98.

A.V., Wiarda, W., Hordijk, M., Alberink, I., Vergeer, P. Implementation and assessment of a likelihood ratio approach for the evaluation of LA-ICP-MS evidence in forensic glass analysis. Journal of Science and Justice, 57: pp. 181–192.

Evett, I. The logical foundations of forensic science: Towards reliable knowledge. The Royal Society Publishing, 2015. 370: pp. 1–10.

INTERPOL. Disaster Victim Identification Guide, Lyon, 2018. https://share.google/qpSRC2THlb0ZqpYQN

I. Macoveciuc, C.J. Rando, and H. Borrion, 2019. Forensic Gait Analysis and Recognition: Standards of Evidence Admissibility. Journal of Forensic Sciences. Volume 64, Issue 5. 1294–1303. DOI: 10.1111/1556-4029.14036

ISO/IEC. Information technology—Biometric performance testing and reporting. Part 1: Principles and framework, 2006. Reference number: ISO/IEC 19795–1:2006(E).

Julian, R.D., Kelty, S.F., Roux, C., Woodman, P., Robertson, J., Davey, A., Hayes, R., Margot, P., Ross, A., Sibly, H., White, R. What is the value of forensic science? An overview of the effectiveness of forensic science in the Australian criminal justice system project. Australian Journal of Forensic Sciences, 2011. 43(4): pp. 217–229.

Kirk, P.L. The ontogeny of criminalistics. Journal of Criminal Law and Criminology, 1963. Article 17, 54(2): pp. 235–238.

Knight, B. Knight's Forensic Pathology, 3rd ed., Hodder Arnold, London, 2004.

Margot, P. 2011. Forensic science on trial - What is the law of the land? Australian Journal of Forensic Sciences 43(2-3):89–103

Merryman J. H., Pérez-Perdomo R., (2007). The civil law tradition. Stanford, Calif: Stanford University Press.

Meuwly, D., Ramos, D., & Haraksim, R. (2016). "A guideline for the validation of likelihood ratio methods used for forensic evidence evaluation." Forensic Science International, 276, 142–153

Meuwly, D., Veldhuis, R. Forensic biometrics: From two communities to one discipline, in 2012 BIOSIG—Proceedings of the International Conference of Biometrics Special Interest Group (BIOSIG). 2012: Darmstadt, Germany. Journal of Forensic Science, 2011. 43(2): pp. 89–103.

Montani, I., Marquis, R., Anthonioz, N.E., Champod, C. Resolving differing expert opinions. Journal of Science and Justice, 2017. 59: pp. 1–8.

NIFS National Institute of Forensic Science Australia New Zealand. An introductory guide to Evaluative Reporting, 2017.

Powell, E.J., and Mitchell, S.M. 2007. The International Court of Justice and the World's Three Legal Systems. Journal of Politics. Volume 69, Issue 2, Pages 397–415,

Ribaux, O., Crispino, F., Delémont, O., & Roux, C. (2016). The progressive opening of forensic science toward criminological concerns. Security Journal, 29(4), 543-560. doi:10.1057/sj.2015.29

Robertson, B., Vignaux, T., Berger, C.E.H. 2016. Interpreting Evidence - Evaluating Forensic Science in the Courtroom. DOI:10.1002/9781118492475.fmatter Edition: 2nd EditionPublisher: WileyISBN: 978-1-118-49243-7

Roux, C., Crispino, F., Ribaux, O. From forensics to forensic science. Current Issues in Criminal Justice, 2012. 24(1): pp. 7–24.

Roux, C., Ribaux O., Crispino, F. Forensic science 2020—the end of the crossroads? Australian Journal of Forensic Sciences, 2018. 50(6): pp. 607–618.

Roux, C., Talbot-Wright, B., Robertson, J., Crispino, F., Ribaux, O. The end of the (forensic science) world as we know it? The example of trace evidence. Philosophical Transactions of the Royal Society, 2015. 370.

Roux, C., Willis, S., Weyermann, C. Shifting forensic science focus from means to purpose: A path forward for the discipline? Journal of Science and Justice, 2021. 61: pp. 678–686.

Seckiner, D., Mallett, X., Maynard, P., Meuwly, D., Roux, C. 2019. Forensic gait analysis — Morphometric assessment from surveillance footage. Forensic Science International Volume 296, March 2019, Pages 57–66

Williams, G.A., Maskell, P.D. Embracing likelihood ratios and highlighting the principles of forensic interpretation. Forensic Science International: Reports, 2021. 3: pp. 1–3.

CHAPTER 37

POSTMORTEM IMAGING WORKFLOWS

Lars Ebert and Wolf Schweitzer

Postmortem forensic imaging encompasses a large variety of different image modalities such as postmortem computed tomography (PMCT), PMCT angiography, PMCT ventilation, postmortem magnetic resonance imaging (PMMR) and a variety of different surface documentation techniques such as photogrammetry or structured light scanning. Applying all scanning techniques to each body admitted to an Institute of forensic medicine is not productive, as many of the techniques are specifically valuable for certain findings and some of them can be time consuming. Streamlining the decision making process as well as the documentation process and integrating it with the conventional autopsy procedures is therefore valuable to increase efficiency and yield better imaging results. In this chapter, we describe the imaging workflow for postmortem cases as it is implemented at the Zürich Institute of Forensic Medicine of the University of Zürich (IRM-UZH)—from the time the forensic pathologists arrive at the incident scene to the end of the autopsy. All procedures are regulated by the Institute's quality management system and various documents defining indications, procedures, responsibilities and communication pathways. These documents are updated and certified annually. The postmortem imaging workflow employed at the IRM-UZH is a multi-disciplinary effort, including forensic pathologists, radiologists, engineers, autopsy technicians and radiographers (Figure 37.1).

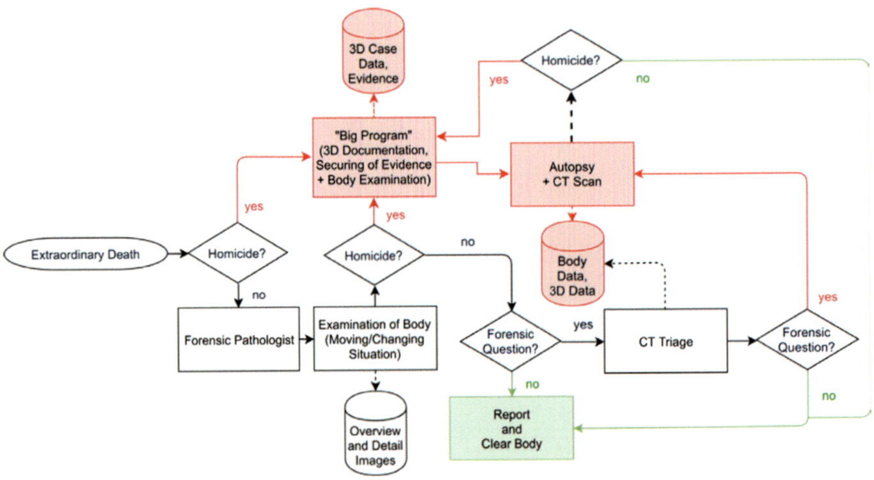

Figure 37.1 Workflow at the Institute of Forensic Medicine in Zürich.

DOI: 10.4324/9781003413998-45

37.1 FACILITY SETUP

The setup of the facility also plays an important role in minimizing transport distances and therefore times. At the IRM-UZH, fridges for the bodies, autopsy halls, postmortem imaging devices and the offices for radiological analysis of the images are all positioned on the same floor, allowing for a better, more direct communication between forensic pathologists in charge of the case and forensic radiologists. This type of close collaboration is relevant for two reasons. First, having quick access to radiological findings can help in autopsy planning. Second, the demonstration of relevant autopsy findings to the radiologists can help in quality control and improvement of the radiological readouts. At the IRM-UZH, the two autopsy halls with a total of three autopsy tables are positioned in a room adjacent to the cooling and entry room with the fridges. In each autopsy room, a PACS workstation and large ceiling mounted screens for each autopsy table are available. These give a forensic pathologist direct access to the imaging data during autopsy. Opposing the hallway, opposite to the autopsy halls, the imaging facilities are located. A dedicated room for a CT scanner including the surface scanning equipment, a 3 Tesla MRI scanner and a reading room in between the two, with multiple PACS workstations for reading.

37.2 GENERAL WORKFLOWS

37.2.1 At the Scene

While the in-depth documentation of a body is performed after admittance to the IRM-UZH, main findings are being documented at the scene on forensic photographs. The body position and location is documented as well as injuries, blood flow marks and other findings that can be photographically documented. Any damages or contamination of the clothes is also documented prior to undressing the body. In cases, where a laser scan of the scene is performed by the forensic science department, it is usually conducted with the body in place, prior to the postmortem legal investigation at the scene by a forensic pathologist. For these tasks, all relevant equipment including camera, light, rulers and other required materials such as thermometers are compiled in a case.

37.3 ARRIVAL AT THE INSTITUTE

Once a body arrives at the Institute, the forensic pathologist in charge conducts an external entry examination. If not yet done, evidence and traces are being secured. The body is placed on a stretcher, and weighed using scales embedded into the floor. A ceiling mounted digital camera is used to create a standardized, top down, full body, high resolution photo of the body. This overview photograph is meant to document the state of the body at entry and before cleaning, including items such as medical installations. If the external examination reveals injuries that might require surface documentation, in collaboration with experts from the 3D Center Zürich, the decision might be made to perform a surface documentation [1]. Such a surface documentation might be indicated in all cases, where superficial injuries exist that might be relevant for a 3D reconstruction. This includes patterned injuries that might be matched to a weapon or other structure, but also gunshot injuries or stabbing wounds, that might be required for trajectory estimations.

37.4 CT

Without exception, each body that is admitted to the Institute will get full PMCT documentation, amounting to approximately 500 cases per year in Zürich. Increasingly, PMCT is used as a triage tool, where under certain conditions the postmortem investigation ends after a clear cause of death has been determined. The PMCT scans are performed timely, so the forensic radiologists have sufficient time to analyze the images and inform the forensic pathologist in charge about the findings. This can help in deciding whether additional scans such as PMMRI or PMCTA have to be performed. In clothes an autopsy is conducted, the information gathered can help in autopsy planning and warn of potential safety hazards such as tuberculosis or sharp foreign bodies.

PMCT is conducted with the body in the body bag using a standardized protocol in supine position. The protocol consists of a full body CT that is reconstructed using an extended field of view, to document the body in its entirety [2]. Because on this scan detail can be lost due to a larger pixel spacing, additional high-resolution scans of thorax/abdomen as well as the head are performed. For identification purposes, an additional dental scan is sometimes conducted. To avoid photon starvation artifacts caused by the humeri, an additional thorax/abdomen scan might be required with elevated arms. All scans are reconstructed using a hard and a soft reconstruction kernel. Reconstructed datasets are automatically stored to the PACS and are immediately available for image reading and visualization. The forensic radiologists analyzes the images and communicates their findings to the forensic pathologists managing the case. In addition, an image report is created visualizing the forensically most relevant findings for the state attorney in charge of the case. Depending on the case circumstances and in agreement with the state attorney, PMCT can be used as a triage tool [3]. In such cases, if the cause of death can clearly be determined using imaging alone, the case can be closed and no additional autopsy is required. In all other cases, it might be decided to conduct additional cross sectional imaging procedures such as PMCTA [4], ventilated PMCT [5] or PMMRI [6].

37.5 SURFACE DOCUMENTATION

Approximately 30 to 40 cases a year require a postmortem surface documentation at the IRM-UZH. While in principle, CT data can be uses to generate surface reconstructions, CT scans do not contain color weighed and the resolution is much lower than what can be achieved with structured light scanners (approx. 0.6mm vs 0.03mm for a full body CT scan with extended CT scale). The surface documentation procedure is conducted after the standard PMCT scans, but before other procedures that might alter the surface of the body such as PMCTA or image-guided biopsies. This is because the PMCT documentation documents the body "as is", whereas the surface documentation procedure requires some preparation. For this preparation, the body is placed on the autopsy table. If there are no visible relevant patterns in the dirt on the body surface, the body is cleaned and freed of dirt and blood. Medical installations such as tubes, vascular access, ventilation or defibrillation devices and other foreign bodies that protrude externally are being removed, as their position has already been documented by the overview photograph as well as the PMCT scans. Because optical scanning devices such as the structured light scanner used at the IRM-UZH have problems scanning through hair, relevant injuries are being shaved using an electrical shaver. Special care is taken to avoid skin artifacts during this procedure. To minimize bleeding during the scanning procedure, medically induced injuries such as drainage access can be sewn. Larger injuries such as larger cuts are being adapted, if necessary using glue. Because structured light scanners, like all optical surface scanning methods, can produce artifacts with reflective surfaces, the body surface is sufficiently dried.

Once the body is prepared, it is moved back to the CT room for the actual surface documentation. The body is placed on a vacuum mattress to minimize moving during the scanning process. A thick plastic sheet protects the mattress from body fluids. The body is placed feet first, because the ceiling mounted surface scanner cannot scan through the CT gantry, which leaves the underside of the feet undocumented rather than the head. The body is further prepared by applying uncoded reference markers. These markers are randomly placed on the body surface with a distance of approximately 30 cm to each other. These uncoded reference markers play a role in automatically fusing multiple surface scans to one whole surface scan. Postmortem surface documentation is always done in conjunction with a dedicated full body PMCT scan with extended field-of-view in the same position, with a slice thickness of 0.6 mm and an increment of 0.4 mm. This allows for a registration of CT and surface scan, which can either be used to fill missing parts of the surface scan such as the back side of the body with low resolution surface models extracted from CT, or to create reconstructions using information from inside and outside the body [7]. (Figure 37.2) Examples would be gunshot injuries or blunt trauma from traffic accidents. After the CT scan, marker boards are attached to the mounting rails of the CT table [8]. These marker boards allow for an automatic scaling of the scans. The surface scan itself consists of two parts—photogrammetry and surface scanning, which are described in detail in other chapters. In addition to these two steps, which are automated using the Virtobot system, overview photos as well as high resolution close up photos using a digital camera are taken of all relevant wounds. These can be used for texturing, in case the standardized photogrammetry photos do not offer sufficient resolution. If there are injuries present on the back side as well, the body is turned and the procedure is repeated. In rare cases, larger superficial injuries cross from the front to the back, which might require a scan with the body lying on the side, in order to get a complete documentation of the injury. Recent advances may include extreme extent of flash light and rapid sequence manual photography with images as fast as 10 or 20 per second, or, redundant low light RAW photographs. This allows for higher image qualities than with previously published methods [Schweitzer W, Fukuda H, Thali M, Bolliger S, Ebert L. Mobile forensic photogrammetry in the field: Conservative approach to non-collaborative surfaces. Forensic Imaging. 2024 Sep 1;38:200597., Lux C, Ebert L, Schweitzer W, Ampanozi G, Chatzaraki V, Bösch M, Kreitmeier FF, Kettner M. Photography techniques for 3D surface reconstruction during autopsy. Forensic Science International. 2026 Feb 1;378:112702.].

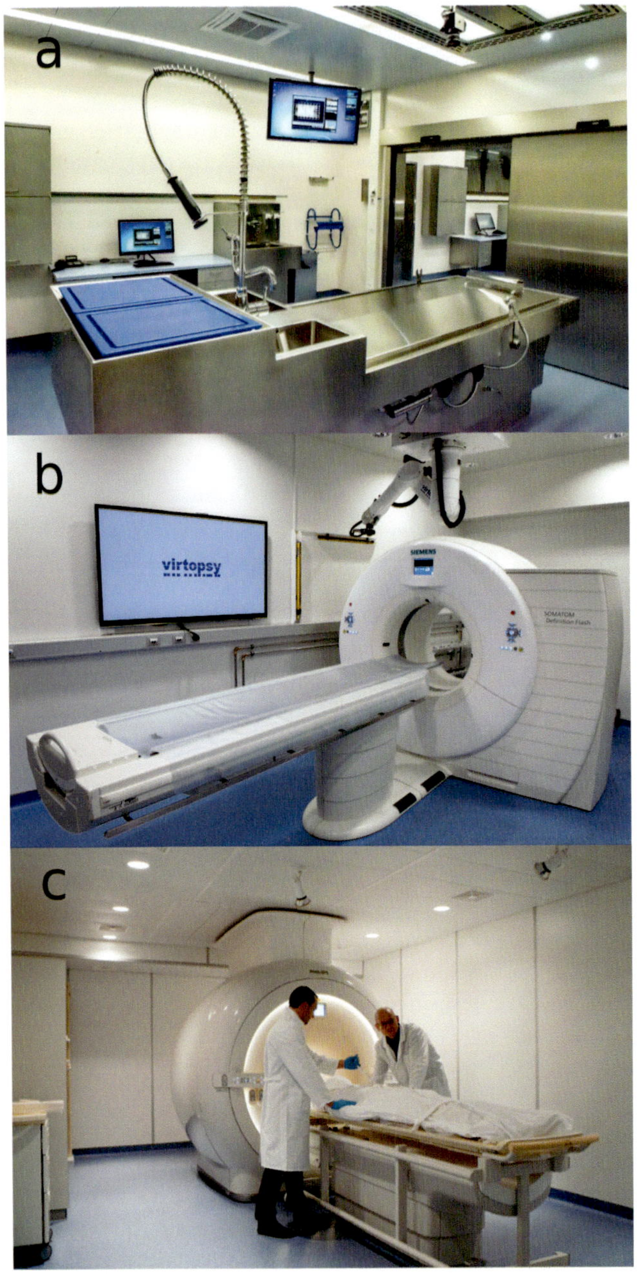

Figure 37.2 Investigation facilities. a: Autopsy hall. b: Dual energy CT scanner with Virtobot system. c: 3 Tesla MRI scanner.

37.6 CT ANGIOGRAPHY

PMCT angiography (PMCTA) can be conducted to increase vessel contrast in cases with suspected vascular lesions i.e. to localize a source of bleeding or identify an aortic dissection. At the IRM-UZH, we use a mixture of a water soluble iodine contrast mixed with polyethylen glycol (PEG) 200 as a carrier medium. The contrast/carrier mixture is calibrated to 350 Hounsfield units. Access to the vessel system is gained through a cannulation of the femoral vessels. The contrast mixture is pumped through the vessel system using an immersion pump and several custom made 3D printed parts [9]. The entire setup is affordable, easy to maintain and gives good control over pressure and flow rates. After filling the arterial system, thorax/abdomen as well as a head and full body scan are conducted, following the standard protocols. The procedure is then repeated for the venous system.

37.7 VENTILATION

Normal postmortem changes of the lung such as inner livores can often obfuscate pulmonary pathologies. The differentiation between pathological pulmonary changes and normal postmortem changes such as inner livores can be partially overcome by inflating the lungs using ventilation methods. In cases where this seems necessary on PMCT images, endotracheal tubes, larynx masks or continuous positive airway pressure (CPAP) masks can be used to inflate the lungs while performing a CT scan. While endotracheal tubes and larynx masks are often in place after medical interventions prior to death, a postmortem placement of these might be difficult due to rigor mortis. In such cases, tightly fixed CPAP masks can be an alternative method. The ventilation uses a continuous positive airway pressure to maintain the pressure during the scan [10].

37.8 MRI

Postmortem MRI (PMMR) is conducted in cases of suspected strangulation.

37.9 AUTOPSY PHOTOS

During autopsy, every finding that is deemed relevant to the case is documented using digital photography. This includes overview photos as well as portrait photos. For all documented external injuries, it must be possible to localize them on the body, so overview photos are created. For close-ups, a non-reflective ruler containing name and case number is used. Photos are made, if possible, perpendicular to the surface. For photos of organs, the organs are placed in front of a neutral background.

37.10 CONSIDERATIONS FOR NEW FACILITIES

Establishing new facilities for forensic imaging can present challenges due to the diverse array of factors that must be considered, which can vary depending on the legal context. Some of the key considerations include available space and the imaging modalities that will be employed. Additionally, factors such as whether CT triage or autopsy will be used for each case, the existing workflows, and the potential for incorporating imaging workflows in an efficient manner should also be taken into account. It is also important to consider which image modality is most appropriate for a given case and the optimal parameters for the scans in light of scanning time and available resources. The order in which scanning procedures are conducted, as well as the regulations and documentation involved in the process, should also be addressed. Given the involvement of multiple disciplines, including forensic pathologists, radiologists, engineers, autopsy technicians, and radiographers, it is important to clearly delineate each party's responsibilities and to establish effective communication protocols. This includes determining who has the final decision-making authority and how image data will be accessed and used. The amount of data generated per case must also be considered, along with issues related to data storage, backup, deletion scheduling, and post-recording analysis. Appropriate expertise and software are also essential for effective data management and analysis.

Setting up new forensic imaging facilities can be a complex and multifaceted process, requiring careful consideration of factors such as available space, imaging modalities, workflows, communication protocols, data management, and expertise.

References

[1] Ebert LC, Flach P, Schweitzer W, Leipner A, Kottner S, Gascho D, et al. Forensic 3D surface documentation at the Institute of Forensic Medicine in Zürich—Workflow and communication pipeline. J Forensic Radiol Imaging 2016;5:1–7. https://doi.org/10.1016/j.jofri.2015.11.007.

[2] Flach PM, Gascho D, Schweitzer W, Ruder TD, Berger N, Ross SG, et al. Imaging in forensic radiology: An illustrated guide for postmortem computed tomography technique and protocols. Forensic Sci Med Pathol 2014;10:583–606. https://doi.org/10.1007/s12024-014-9555-6.

[3] Baumeister R, Thali MJ, Ampanozi G. Postmortem radiological imaging of natural causes of death in adults: A review. Forensic Imaging 2021:200473. https://doi.org/10.1016/j.fri.2021.200473.

[4] Schweitzer W, Flach PM, Thali M, Laberke P, Gascho D. Very economical immersion pump feasibility for postmortem CT angiography. J Forensic Radiol Imaging 2016;5:8–14. https://doi.org/10.1016/j.jofri.2015.11.009.

[5] Germerott T, Preiss US, Ebert LC, Ruder TD, Ross S, Flach PM, et al. A new approach in virtopsy: Postmortem ventilation in multislice computed tomography. Leg Med Tokyo Jpn 2010;12:276–9. https://doi.org/10.1016/j.legalmed.2010.07.001.

[6] Thali MJ, Yen K, Schweitzer W, Vock P, Boesch C, Ozdoba C, et al. Virtopsy, a new imaging horizon in forensic pathology: Virtual autopsy by postmortem multislice computed tomography (MSCT) and magnetic resonance imaging (MRI): A feasibility study. J Forensic Sci 2003;48:386–403.

[7] Ebert LC, Ptacek W, Breitbeck R, Fürst M, Kronreif G, Martinez RM, et al. Virtobot 2.0: The future of automated surface documentation and CT-guided needle placement in forensic medicine. Forensic Sci Med Pathol 2014. https://doi.org/10.1007/s12024-013-9520-9.

[8] Sekula AD, Thali MJ, Breitbeck R. Improving postmortem surface documentation with a CT mounted marker board. J Forensic Radiol Imaging 2014;2:213–6. https://doi.org/10.1016/j.jofri.2014.06.005.

[9] Schweitzer W, Enders M, Thali M. Very affordable post mortem CT angiography kit: Feasibility study using immersion pump and 3D printed parts. J Forensic Radiol Imaging 2019;16:11–8. https://doi.org/10.1016/j.jofri.2018.12.002.

[10] Germerott T, Flach PM, Preiss US, Ross SG, Thali MJ. Postmortem ventilation: A new method for improved detection of pulmonary pathologies in forensic imaging. Leg Med Tokyo Jpn 2012;14:223–8. https://doi.org/10.1016/j.legalmed.2012.03.003.

CHAPTER 38

PMCT FOR TRIAGE

Wolf Schweitzer, Vasiliki Chatzaraki, and Garyfalia Ampanozi

DOI: 10.4324/9781003413998-46

38.1 MEDICOLEGAL AND FORENSIC BACKGROUND

38.1.1 Sequence of Examinations in Medicolegal Cases Appears to Depend on Jurisdiction and Local Practices

The sequence of examinations in medicolegal cases appears to depend on jurisdiction and local practices [1]. More specifically, depending on jurisdiction, the examination of a death first requires an external examination of the body by a doctor. This may be performed on location, at the place where the body was found, or close to it—which in itself has been identified to also be mentally and physically demanding [2, 3]. Depending on the case, the place where the body is examined in detail may be the death scene and even the accident, crime or suicide scene. Conversely, these locations may not coincide—particularly if death occurred in a different place than the initial events that ended up causing it, such as after a body was disposed of or after hospital treatment.

38.1.2 Specific Decisions in Specific Cases May Require a Specific Focus Using Specific Methods

It may be impossible to issue general recommendations, but specific cases require a specific approach—which, if one has a mortuary CT scanner installed, may always be to perform postmortem computed tomography (PMCT) of the whole body and to immediately read and report [4, 5] the images. After an external examination, the investigating authorities may decide to release the body or order a further examination, such as a second external examination, PMCT [6–9] or autopsy. While some basis for such a decision may result from police investigation, another method to see if an autopsy may be able to, or possibly have to, identify relevant results, is PMCT. It is obvious that a PMCT investigation, due to inherent limitations, will overlook and miss some findings [10], while a conventional autopsy may be more correct in various ways (by majority vote, by opinion by acknowledged doyen or by a scientific approach to findings using an evidence base) while clearly having to face sampling error—but the question of PMCT adequacy is with regard to its specific relevance in a given case. So, one may formulate a reverse question by asking whether a PMCT scan can positively identify fractures, relevant hemorrhages or other indicators of particular forensic relevance in a given specific case. Often, the absence of findings that indicate or even prove a violent and particularly possibly homicidal death is far more relevant than the identification of a single finding indicating acute ischemic damage to the heart [11–13], the subspecification of metal artifacts [14] or conclusively identifying a particular natural cause of death [15, 16].

Forms of external influence that may be viewed under a violent death attribution also include hypothermia [17] and obstructive asphyxia [18], which may be particularly difficult to identify in any given instance.

Interpretation of results is strongly suggested to be performed with various experts working on the case, not in isolation. There, the full medicolegal external examination has to be taken into account, and its findings have to be considered when reading and discussing the PMCT because there, it may not be as important to identify any vessel injuries, small as they may be, in a medicolegal triage PMCT that aims to decide whether to perform an autopsy, as long as there are no signs or suspicion of a possibly fatal hemorrhage.

38.2 RADIOLOGICAL AND TECHNICAL BACKGROUND

38.2.1 Choice of Triage Examination Methods Means to Weigh Cost and Effort Required against Benefits for a Specific Case

While many practically and pragmatically employed examination techniques offer a great many particular insights and observations, no single examination appears to provide all present findings in a single sweep, particularly not in a cost-effective way [10, 19]. The question then is how to proceed. PMCT has now gained worldwide recognition and relatively widespread acceptance.

38.2.2 Clinical Radiology Uses CT for Triage

One point of discussion seems to be that specific body region scans are an alternative to whole-body CT scans—which is interesting when evaluated for emergency blunt trauma patients [20, 21]. Technically, radiological diagnosis does not seem to be of a lower quality if an observation-based decision is made to restrict the scope of a CT examination, such as by restricting the scan to head, neck and/or thorax/abdomen, for example. But a whole-body CT may be more convenient in that logistics (transferring a body only once to the scan facility) and overall data acquisition speed seem to be better for a whole-body PMCT [21]. Second, overall, consideration may play a role into what CT dose to use, but from a purely medicolegal view, triage cases may require an focus on best quality [22–24] before anything else—particularly if small injuries should also be excluded. Since even small lesions and soft tissue contrast–related changes may play a significant role, the examiner is strongly recommended to maximize CT data quality, particularly for postmortem CT triage cases. Using a maximum CT dose—alongside other improvements such as smallest applicable field-of-view—appears to actually improve delineation, particularly of small organ and soft tissue findings (Figure 38.1). Interestingly, not a single study mentioning such soft tissue limitations of PMCT so far ever mentioned to simply and fully *maximize* (not just moderately increase or optimize) CT-dose and to provide other data quality improving measures [25, 26]. Obviously, CT scanning has clear limitations and clear advantages—but for a technically good comparison, a technical focus is required. We found that studies critically discussing an allegedly low quality of computed tomography scans typically seemed to lack specification of comprehensive detailed technical parameters with regard to actual quality [27, 28]—including actual resolution, slice thicknesses and intervals, as well as dose-related parameters and attempts to quantify signal-to-noise ratio in relation to reconstruction kernel; this, then, may not so much constitute a problem of CT scanning as such but a problem of slippage.

38.2.3 Technical Approach for Better PMCT Triage from a Morphological View

The ideal setting for both bone and soft tissue spatial resolution is to minimize the width of reconstructed field of view to a value of F [mm], then get the value for the reconstructed image width W [pixels] (typically 512 pixels in the year 2024), and set the z-resolution or slice distance Z to $Z \approx \frac{F}{W}$. To get better contrast resolution, image noise needs to be reduced at the level of data acquisition (not during any of the "garbage to gospel" postprocessing software filtering attempts). This is achieved either by increasing CT-dose (mAs value) (Figure 38.1) and/or by performing repeated scans. Settings or devices with high noise are not desirable. Viewing the scan, and considering its diagnoses, is a key step in forensic PMCT scanning: once a lower-quality scan has resulted in a relevant finding, be it pathology or trauma, focusing in on the region of interest by first reducing reconstructed field of view and then possibly scanning again with increased mAs is recommended. From a purely technical forensic medicine view, no trauma case should leave the CT table without maximal spatial resolution, and no trauma or pathology case should be scanned with submaximal or lower CT dose. This means that the person performing decision relevant PMCT may have to "think on their feet". Conversely, less X-ray energy may be less hard on the tube, and thicker slices (lower Z-resolution) mean less storage used.

38.2.4 Triage in Context of Using a PMCT May Not Mean Perfect Reconstruction or Perfect Diagnosis

Some texts suggest a role of PMCT in being used to "reconstruct" bullet trajectories or being able to exhibit depth and direction of a stab wound [33].

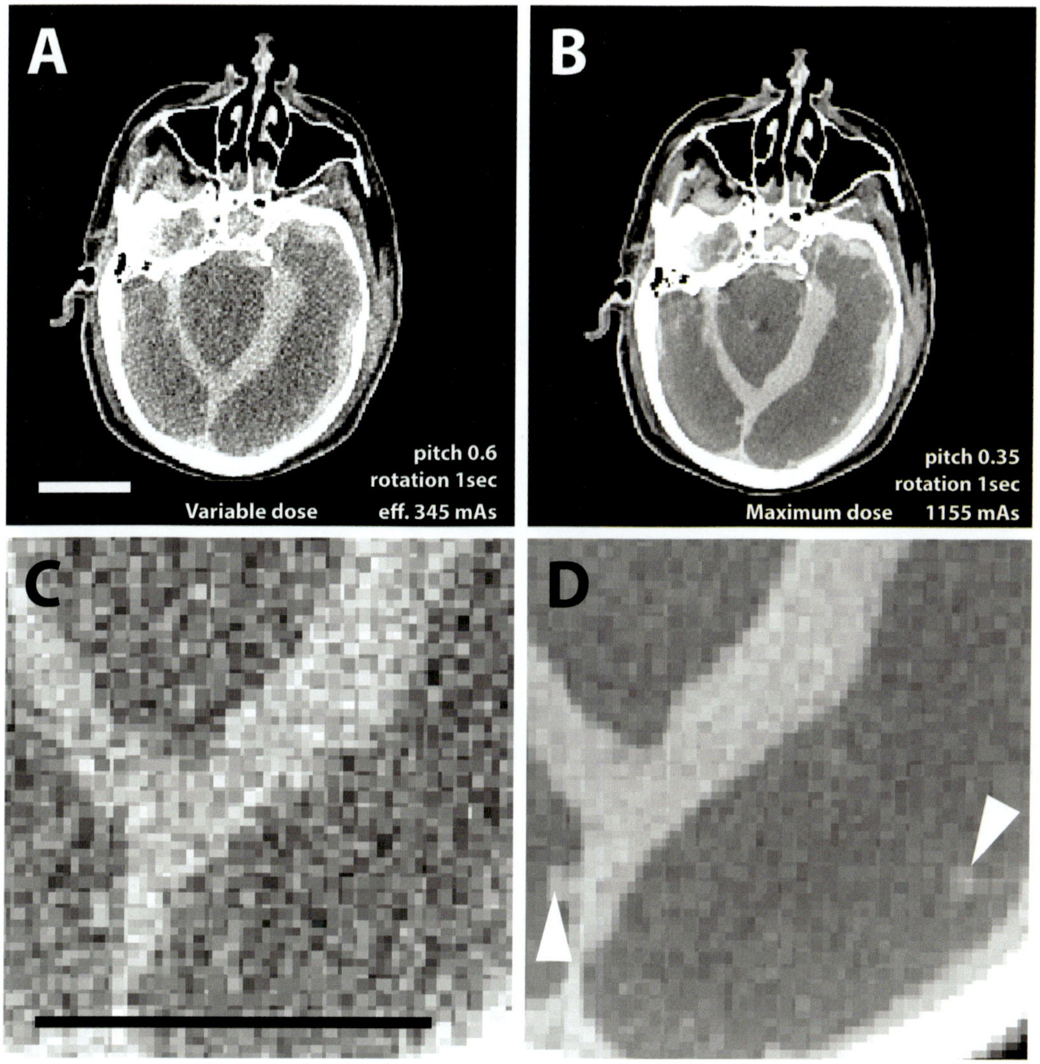

Figure 38.1 Relevance of CT data quality for triage CT cases. CT of the brain at variable dose (overview: A, detail: C, variable dose setting (CareDose, Siemens, Erlangen, Germany) with an effective dose of 345 mAs, spiral pitch factor (ratio of the table increment over the detector collimation) of 0.6, rotation time 1 sec, tube voltage of 120 kVp) allows one to easily delineate a subdural hemorrhage. The same scanner was used at higher dose (overview: B, detail: D, fixed CT dose of 1155 mAs, spiral pitch factor (ratio of the table increment over the detector collimation) of 0.35, rotation time 1 sec, tube voltage 120 kVp) and a number of additional regions exhibiting blood density become apparent: not only does it appear to be a lot easier to ascertain left temporal subdural and subarachnoidal hemorrhage, but also small areas of intracerebral hemorrhage (see D: arrows) can now be discerned better. Conversely, why complain about image noise or low quality before exhausting a device's options for best image quality? Bar 4cm.

However, consequences of such forms of injury—small fractures [34, 35] as well as fatal internal hemorrhage, such as cerebral hemorrhage, hemothorax, abdominal or retroperitoneal hemorrhage or pneumothorax or pneumopericardium, may be more relevant for first decision making, while a purely imaging-based reconstruction of natural disease morphology, or trauma findings such as actual sharp force or firearm injuries, may appear exceedingly difficult at an adequate level.

38.2.5 The Benefit of a Whole-Body CT for Triage Purposes Lies in Its Value for Wider Team Discussion

The benefit of a whole-body CT for triage purposes lies in its value for a wider team discussion [20], whereas in both clinical routine and in our experience in postmortem examination of forensic cases, determination of next steps should not be done by a radiologist in isolation. Instead, discussion of results and further steps should be performed by all specialists involved in the case. How to read and document the PMCT is described in another chapter of this book.

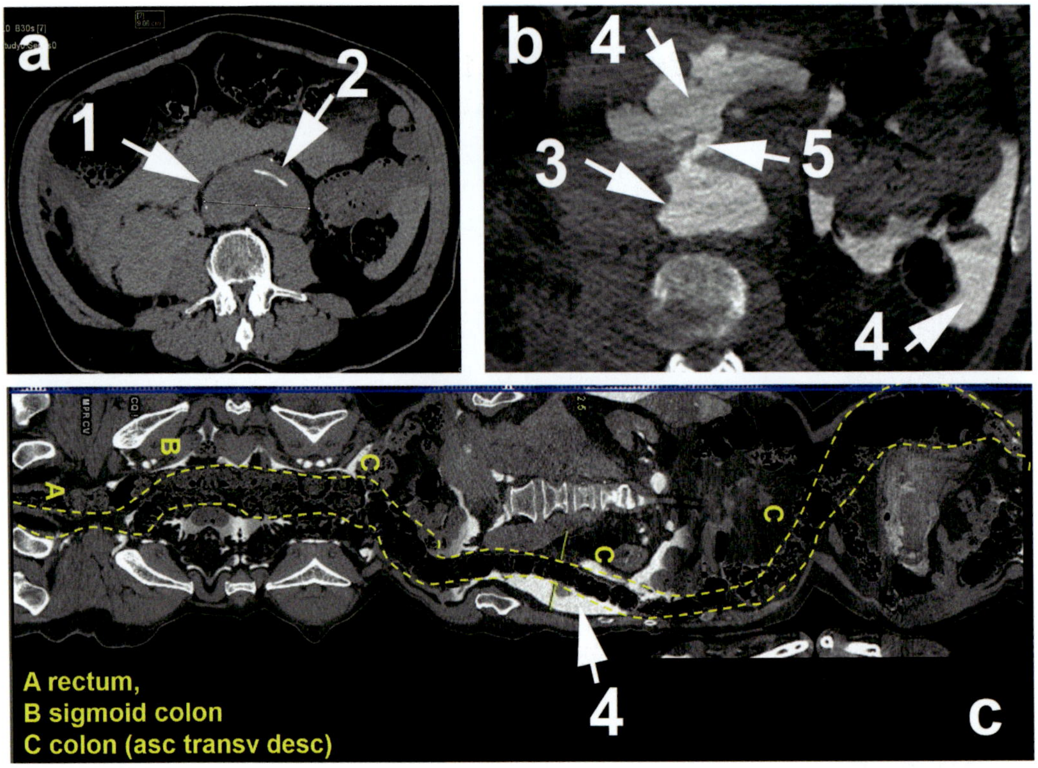

Figure 38.2 Case 1: The principal finding was a large aortic aneurysm. On native PMCT, the outer contour of the aneurysm (1) appeared displaced outward relative to the calcified original aortic wall (2). Suspected blood components within and around the aneurysm were already visible on the non-contrast scan (a). After administration of contrast for PMCTA, the true aortic lumen (3) became clearly distinguishable from areas of contrast leakage (4), allowing visualization of a direct connection (5) between the lumen and the extravasation site. For assessment of the colon (c), a curved multiplanar reconstruction was generated along a manually placed centerline, extending from the rectum (A) through the sigmoid colon (B) to more proximal segments (C). Throughout this course, the colonic wall showed no focal thickening that would indicate abscess formation, nor were there signs of intraluminal bleeding or contrast leakage. a: axial native PMCT; b: PMCTA; c: curved multiplanar reconstruction along the colonic lumen. Figure reused from [29] with Elsevier license 5723680322978.

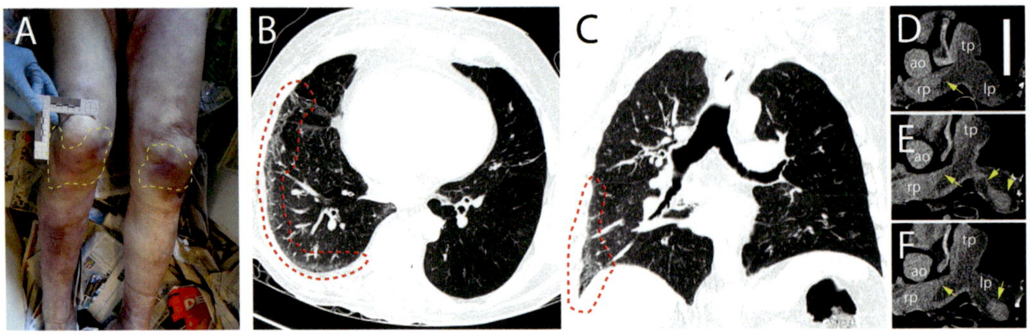

Figure 38.3 Case 2: A: Purple discoloration of the skin over both knees (yellow dashed lines), a finding compatible with (but not specific for) fatal hypothermia. B/C: Axial (B) and coronal (C) multiplanar reconstructions of the lungs showing markedly low parenchymal attenuation, a pattern typical of (though not diagnostic for) death due to hypothermia. Areas of sparse posterior hypostasis (red dashed lines) correspond to the documented position of the body, with the upper torso leaning to the right. D–F: Multiplanar reconstructions of the truncus pulmonalis (tp) and the right (rp) and left (lp) pulmonary arteries demonstrate inhomogeneous intra-luminal contents. This appearance is considered suggestive of pulmonary thromboembolism in the appropriate medicolegal context. Figure reused from [29] with Elsevier license 5723680322978.

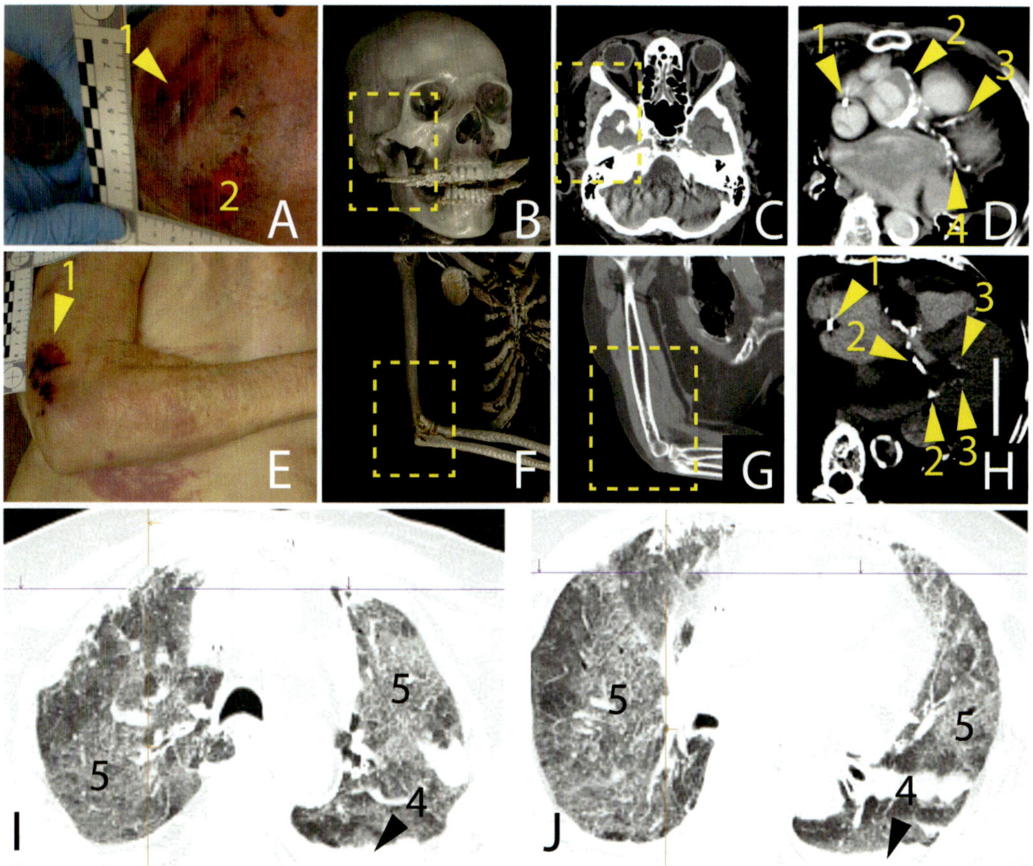

Figure 38.4 Case 3: External examination at the death scene showed a monocle hematoma over the right eye (A-1) and an abrasion with associated hemorrhage on the right cheek (A-2). A further superficial abrasion was present on the lateral aspect of the distal right upper arm (E-1). Three-dimensional VRT reconstruction of the skull (B; facial injury region marked by a yellow dashed rectangle) and the axial PMCT image of the head (C; same region marked) did not reveal any significant underlying injury, and no intracranial trauma was identified. The abrasion at the right elbow (E-1) was also confirmed to be superficial on PMCT, as shown in the corresponding 3-D VRT view (F) and the multiplanar reconstruction (G) (injury regions indicated by yellow dashed rectangles). Cardiac findings included the presence of a pacemaker lead (D-1, H-1), calcified coronary arteries (D-3, D-4), and calcifications of the aortic valve (D-2), mitral valve (H-2), and papillary muscles (H-3). The lungs did not show features typically associated with hypothermia or obstructive asphyxia (I, J). Visible findings included pleural effusion (4) and scattered ground-glass opacities (5), both of which were considered nonspecific. Bar in H 5 cm, bar in I letter I. Figure reused from [29] with Elsevier license 5723680322978.

38.3 FORENSIC IMAGING EXPERIENCE

From our case-load, a number of illustrative cases may be selected to illustrate the role that PMCT obtained in examining these cases [29].

38.3.1 Case 1—Examining for Possible Medical Complications

A 67-year-old man died approximately 2 hours after reportedly experiencing acute pain radiating from the back to the abdomen. External examination revealed no relevant injuries; specifically, the body showed no visible trauma, not even superficial abrasions. Based on the symptomatology and the perimortem clinical course, the forensic pathologist on scene considered an acute rupture of an abdominal aortic aneurysm the most plausible working diagnosis. The decedent's medical history included type II diabetes mellitus, colonic diverticulosis and endoscopic removal of 14 colonic polyps 1 week prior to death. He had not complained of fever, abdominal pain or hematochezia in the interval. Nonetheless, the possibility of delayed complications related to the recent polypectomy became a key investigative question. Both PMCT and postmortem CT angiography demonstrated a ruptured infrarenal aortic aneurysm with signs of dissection and rupture, reaching a maximal diameter of approximately 9 cm. The abdominal cavity contained fresh blood, with CT characteristics (i.e., density layering, sedimentation phenomena) consistent with at least 1.1 L of hemoperitoneum. No PMCT evidence of colonic wall thickening suggestive of abscess formation was present, nor was intraluminal colonic blood identified; after PMCTA, the colon remained free of contrast agent (Figure 38.2). The subcutaneous tissue and the skin, both radiologically and on external inspection, lacked any indications of bruising or subcutaneous hemorrhage that would speak to traumatic rupture. The rupture location itself was morphologically typical for a spontaneous event in the context of an atherosclerotic aneurysm rather than for high-energy blunt trauma. Toxicological screening detected only metformin in urine, in accordance with his prescribed antidiabetic regimen. Given the absence of findings suggestive of trauma and the coherent radiologic evidence for a spontaneous aneurysm rupture, the body was released following imaging and toxicological analysis without autopsy. The manner of death was classified as natural, and the cause of death as fatal hemorrhage due to acute rupture of an atherosclerotic infrarenal abdominal aortic aneurysm.

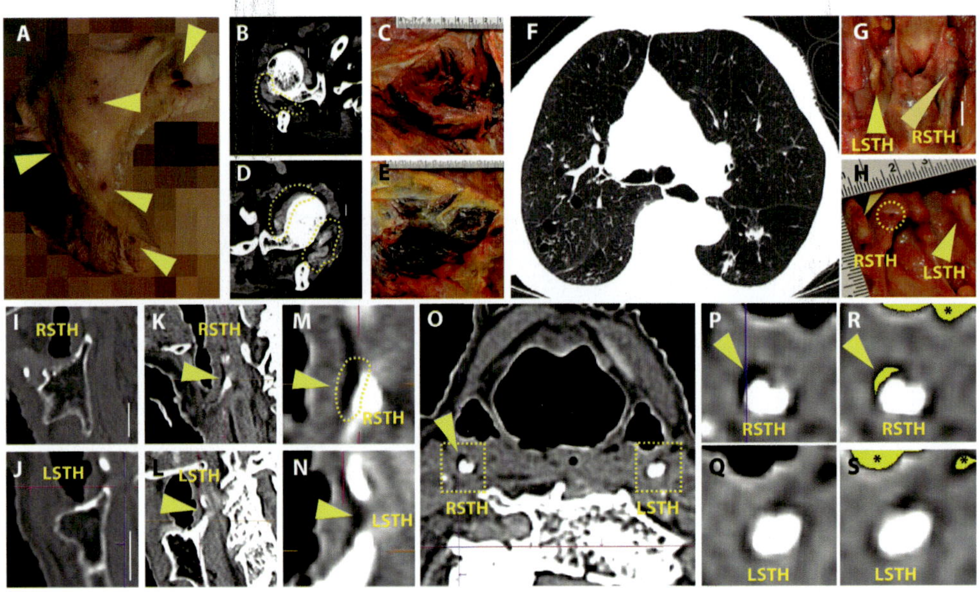

Figure 38.5 Case 4: A: Facial bruises and a small abrasion at the nostril (yellow arrows). B to E: Hemorrhagic soft-tissue and muscle contusions of the shoulders. On PMCT (B, D; yellow dashed outlines), hyperdense areas were visible, corresponding to what appeared to be symmetric contusion-related hemorrhages documented at autopsy (C, E) whereas the surface the woman had been found on had been a hard surface. These injuries were located in positions typical for compression of the shoulder girdle, where soft tissues may be trapped between hard structures such as the humerus and scapula. F: PMCT of the lungs showed very low attenuation of the pulmonary parenchyma, with no internal lividity. The lung lobes were in contact behind the sternum, and bullous emphysema was present. GS (laryngeal structures): The left superior horn of the thyroid cartilage (LSTH in G, H, J, L, N, O, Q, S) showed no skeletal injury on PMCT or autopsy. Adjacent hypodense soft tissue areas (Q, S) did not reveal any pathological changes. The right superior horn (RSTH; G, H, I, K, M, O, P, R) appeared intact on PMCT; however, adjacent hypodense areas (P, R; arrows) contained what was interpreted as gas. This finding is highlighted in R and S using a CT threshold of −95 HU (yellow regions). Intraluminal laryngeal gas (star in R and S) was clearly demarcated, and the presence of soft-tissue gas near the cartilage was considered indirect correlate of a fracture [30]. Autopsy confirmed fresh hemorrhagic fractures of the thyroid cartilage (G, H), with an adjacent soft-tissue hemorrhage (yellow circle in H). Figure reused from [29] with Elsevier license 5723680322978.

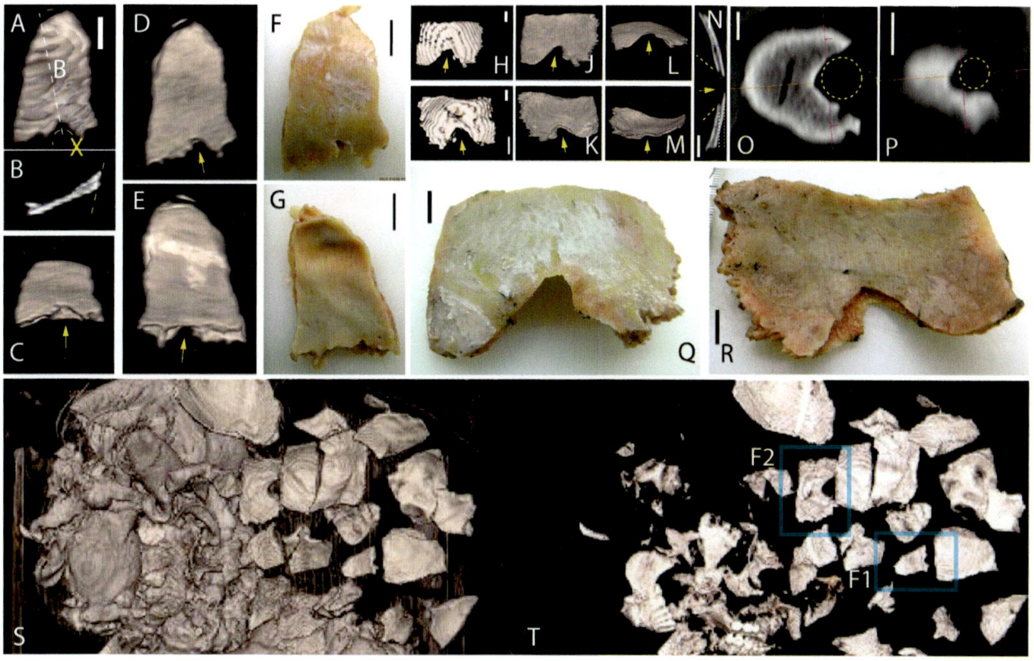

Figure 38.6 Case 5: complex train suicide with pistol headshot preceding train collision [31]. S: initial PMCT; T: PMCT after sorting fragments, marked fragment 1 (F1) and fragment 2 (F2). Fragment 1: On initial whole-body PMCT (A), this bone fragment appeared to exhibit a semicircular defect (yellow dashed circle labeled x) that appeared consistent with the assumption of a perforating pistol shot injury with outward beveling (B, yellow dashed line, reference to fragment of image B see white dashed line in A). More refined PMCTA reconstructions from same CT raw data (C–E) correlated better with the physical bone fragment margins whereas curvature on outer (D, F) (but not inner, E and G) compact table was that of a ruptured bone suture line. Bar 10 mm. Fragment 2: This fragment exhibited a semicircular defect visible on initial whole-body PMCT (H, I, yellow arrow) that appeared to be consistent with a gunshot injury. Refined PMCT reconstructions (VRT: J–M; MPR: O–P) correlated well with the macroscopic (photos: Q, R) appearance whereas a slightly irregularly shaped round semi-circular defect of approximately 8 to 12 mm diameter of the outer cortical bone table (O and P, yellow dashed circle) contained a cone shaped (yellow dashed lines, N–P) defect shape towards a larger defect of the inner cortical bone table (N through P, yellow dashed lines/circles) with several actual dark gray discolorations, also on the assumed gunshot injury circumference, consistent with gunshot residue (no further testing was requested). Bar 10 mm. Figure from CC BY 4.0 licensed publication [31].

38.3.2 Case 2—Indication of Hypothermia and Pulmonary Artery Thromboembolism

An 87-year-old man was found deceased after police forcibly entered his locked apartment when he failed to respond to repeated telephone calls. The body was located seated on the toilet, leaning to the right, wearing only a ribbed tank top. The ambient temperature at the scene was 22.7°C. Approximately 20 days prior to death, the decedent had been hospitalized for pulmonary edema. The body was transferred to the Institute of Forensic Medicine for further evaluation. External examination revealed bruises of varying coloration on the extensor surfaces of both arms, the lateral aspect of the left hip and over the left knee. The skin of the hands and knees showed a purple discoloration compatible with (though not diagnostic of) fatal hypothermia (Figure 38.3A). PMCT demonstrated severe coronary artery calcifications, cardiomegaly and markedly decreased pulmonary parenchymal attenuation (mean approximately −880 HU) with minimal hypostasis, findings that support the possibility of hypothermia-related changes [17] (Figure 38.3B, C). The pulmonary arterial contents appeared heterogeneous, with an oblique interface of CT-dense blood components in the right pulmonary artery and a thin intraluminal hyper-dense strand in the left pulmonary artery (Figure 38.3D, F). Additionally, hyperdense subcutaneous tissue in the right calf suggested venous insufficiency, if not thrombosis. In combination, these findings were interpreted as indicative of, or at least consistent with, central pulmonary thromboembolism. No traumatic injuries were detected, including no abnormalities of the laryngeal structures. Qualitative toxicological analysis of peripheral blood detected antihypertensive and analgesic agents consistent with prescribed medication. As no evidence suggestive of third-party involvement was identified, a forensic autopsy was not mandated by the district attorney. The cause of death was certified as unascertained, with the possibilities of pulmonary artery thromboembolism and hypothermia considered. While hypothermia is classically associated with outdoor temperatures below 15°C, mortality risks increase once ambient temperatures fall outside the 20–25°C range; in frail elderly

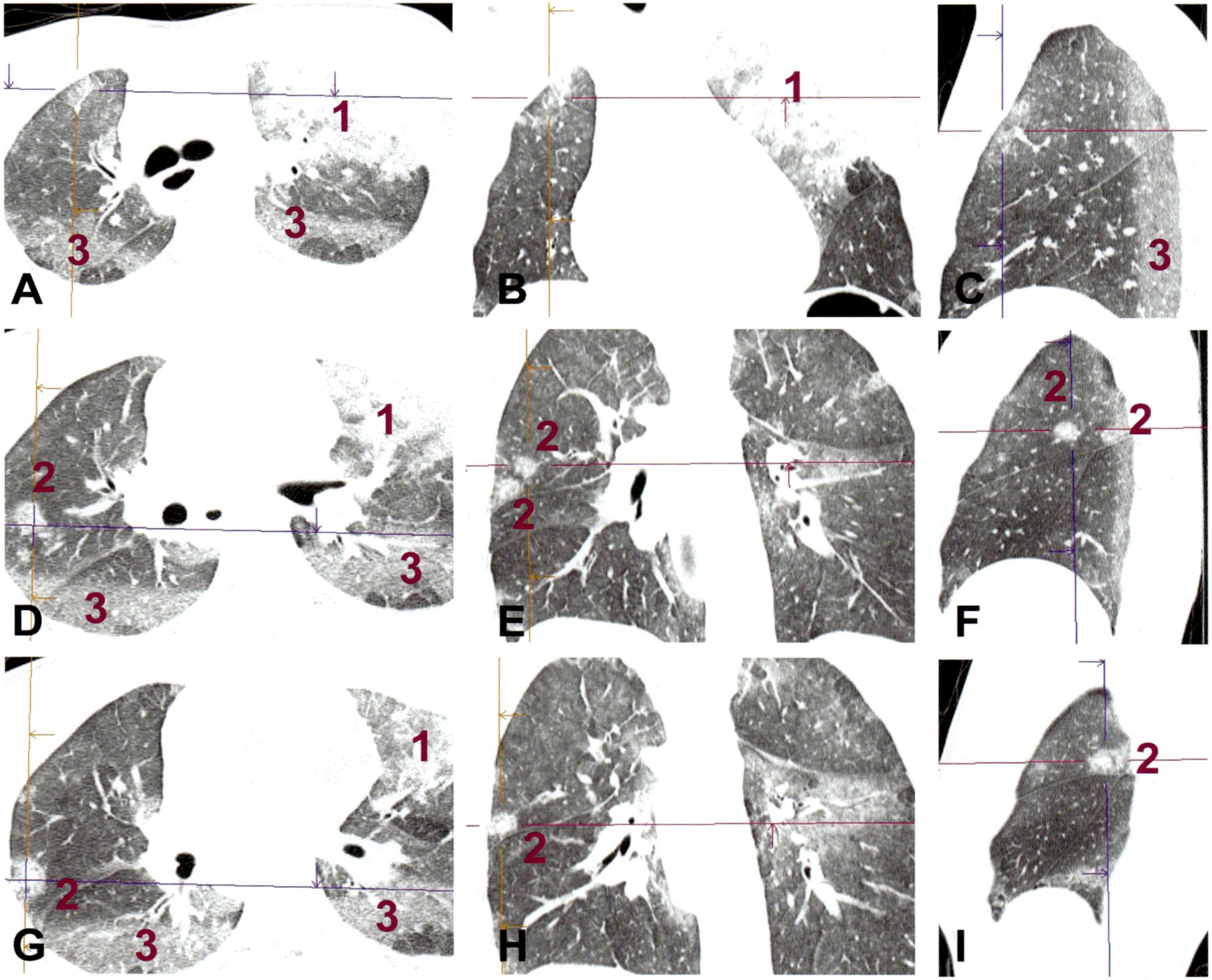

Figure 38.7 Case 6: The lungs in what here was issued as a death by drug intoxication (mixed oral/intravenous) showed coalescent and partly consolidated ground glass opacity (1) and patchy, in part subpleural ground glass opacities (2) as well as horizontally and linearly demarcated more subtle ground glass opacities (3) interpreted as postmortem hypostasis. Seven readers achieved a median score of CO-RADS 5 [32].

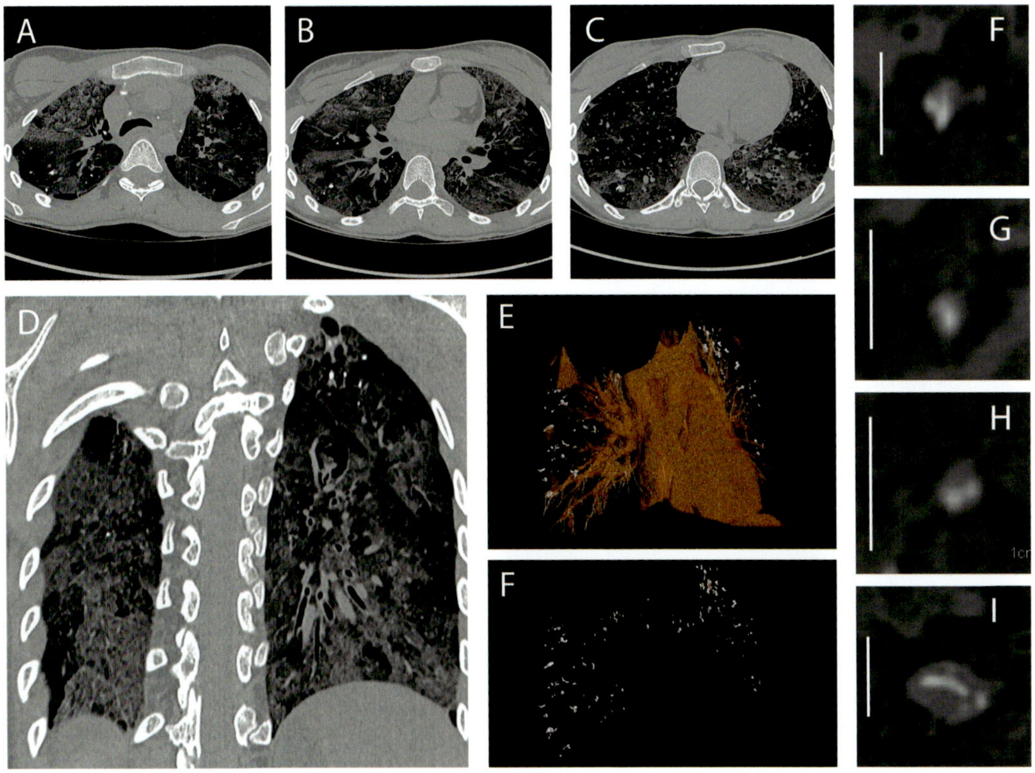

Figure 38.8 Case 7: PMCT images of both lungs are shown in axial (A–C) and coronal (D) planes, supplemented by 3D volume-rendered reconstructions (VRT) both with (E) and without (F) overlying organs and soft tissues. The latter view facilitates visualization of the spatial distribution of numerous calcified pulmonary nodules. Enlarged views of selected nodules (F–I) demonstrate dense central calcification surrounded by parenchymal tissue of lower attenuation, a pattern compatible with chronically evolved granulomatous disease. At autopsy, both lungs contained multiple caseous granulomas, and microbiological analysis confirmed infection with Mycobacterium tuberculosis complex. Figure reused from [29] with Elsevier license 5723680322978.

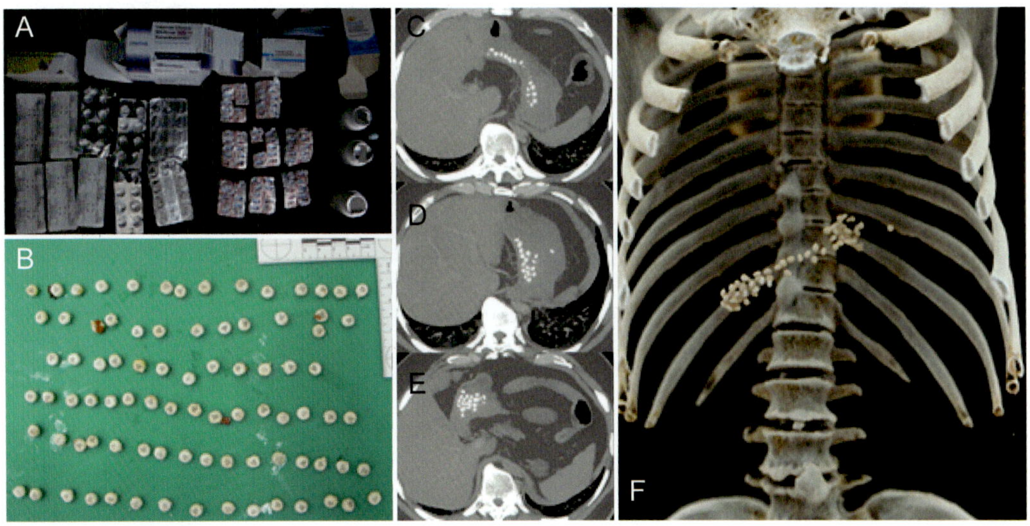

Figure 38.9 Case 8: A: A: Pill packets and opened blister packs were found at the death scene. B: Autopsy revealed a total of 85 pills located in the stomach and proximal duodenum. C, D, E, and F: Corresponding abdominal PMCT cross-sections and a three-dimensional reconstruction of the torso show numerous radiopaque pills within the stomach and proximal duodenum, confirming the imaging–autopsy correlation. Figure reused from [29] with Elsevier license 5723680322978.

individuals with limited mobility, even temperatures below approximately 24°C may contribute to hypothermia. Based on the documented findings and the absence of indicators of homicide or suicide the manner of death was classified as accidental. Although underlying natural disease was significant, the pattern of pulmonary and cutaneous findings would not be expected had an acute, fulminant pulmonary thromboembolism caused immediate death.

38.3.3 Case 3—Unwitnessed Death at Home: PMCT Assessment of Minor Trauma and Natural Cardiac Etiology

A 93-year-old woman was found deceased by home care providers in her bathroom. The entrance door of her house was unlocked. Her medical history included mitral valve re-gurgitation, arterial hypertension, chronic atrial flutter with oral anticoagulation, implantation of a cardiac pacemaker and chronic renal failure (grade III). External examination revealed a large monocle hematoma on the right (Figure 38.4A). The sleeve of the right arm of her pajamas showed red discolorations consistent with blood traces, and the underlying skin exhibited superficial abrasions (Figure 38.4B). According to the home care assistant, the woman had fallen to the ground several days prior. Given the constellation of findings and the possibility (though not the demonstration) of third-party involvement, the district attorney ordered further examination. PMCT demonstrated mild soft-tissue contusion in the right temporal and periorbital regions corresponding to the external monocle hematoma (Figure 38.4C). The right arm showed no significant soft-tissue changes and no fractures (Figure 38.4F,G). No skeletal fractures were identified elsewhere, and no hemorrhages or other injuries relevant to the cause of death were observed. The cardiac pacemaker was in correct position. Extensive arterial calcifications were present, including coronary artery calcification. Calcifications of the aortic and mitral valves as well as both papillary muscles were also evident (Figure 38.4D, H). Pulmonary edema was present, with no PMCT features indicative of hypothermia or obstructive asphyxia (Figure 38.4I, J). An old, pre-existing cerebral infarction was also noted. The cervical spine showed osteoporotic vertebral bodies without displacement, malalignment or signs of spinal cord injury or associated hemorrhage. The thyroid cartilage, hyoid bone and soft tissues of the neck were unremarkable and showed no gas inclusions [30]. No injuries of the lips were detected, and petechial hemorrhages were absent. In the absence of findings that would substantiate assault or other violent mechanisms, and with the documented injuries compatible with a fall occurring days earlier, the body was released without autopsy. The manner of death was certified as consistent with a possible natural death, with no detectable indicators of violent death based on external inspection and PMCT. The cause of death was recorded as consistent with possible cardiac failure in the context of the decedent's pre-existing cardiovascular disease, without demonstrable evidence for a violent etiology.

38.3.4 Case 4—Indication of Relatively Subtle Facial Injury #2

A 93-year-old woman (height: 144 cm, weight: 33.8 kg, BMI: 17) was found deceased in her bed, fully dressed and covered with a duvet. Moist hand towels with blood were present beside the bed. External examination at the scene revealed diffuse bruising and abrasions of the nose, left lower jaw angle and left side of the neck (Figure 38.5A). These findings prompted concern for possible violent death, and a medicolegal autopsy was ordered. Pre-autopsy PMCT demonstrated only subtle abnormalities of the head and neck, including the laryngopharyngeal skeleton (Figure 38.5). Notably, the gas-bubble sign [30] was present (Figure 38.5P, arrow; Figure 38.5R, arrow), with thresholding highlighting CT regions below −95 HU. Bilateral hyperdense dependent components were observed in the shoulder soft tissues (Figure 38.5B,D). Additional findings included hyperinflated lungs with chronic emphysematous parenchymal changes (Figure 38.5F), as well as heart enlargement and coronary artery calcifications. Autopsy revealed soft-tissue hemorrhages and muscle contusions of both shoulders (Figure 38.5C,E) and a fresh fracture of the right superior horn of the thyroid cartilage (Figure 38.5G,H), corresponding to the PMCT gas-bubble sign adjacent to the right thyroid cartilage horn. Chemical analysis of cardiac blood demonstrated elevated β-hydroxybutyric acid, and muscle tissue showed markedly increased acetone concentrations, findings compatible with a starvation-related metabolic state. Swabs obtained from the neck and chin yielded a complete male DNA profile. In the aggregate, i.e., fresh laryngopharyngeal skeletal injury, associated soft-tissue trauma, metabolic context and foreign DNA, the findings were interpreted as indicative of obstructive asphyxiation due to strangulation. The manner of death was certified as homicide.

38.3.5 Case 5—Complex Suicide with Pistol Head Shot and Subsequent Train Collision

In a rural alpine canton (previously published under CC BY 4.0 license [31]), a 52-year old man placed a wooden chair onto railway tracks, sat on it, then presumably shot himself through his head using a pistol, when the train hit and ran over him. The time of day was night time, with the place poorly lit. According to the first information about this case, the train driver had seen the man sitting on a chair right on the train track, the body slumped over, consistent with a position right after a head shot. Police investigation also yielded information according to which it was reasonable to assume that the man had previously examined that particular site on the train tracks by himself a week before.

The examination of the body was performed by a board certified forensic pathologist, the death scene was also documented and examined by forensic scientists. Rhodizonate-positive filter paper stain test of both hands returned a negative result while the examiners observed also severe covering of the hands with both blood and dirt. The examination of the collision scene identified a pistol (SIG P210 pistol, caliber 7,65, by SIG, Neuhausen, SH, Switzerland), and a single case/shell, matching the pistol's caliber, was identified nearby. The weapon appeared to be severely damaged and exhibited a hammer in a cocked position while the sleigh appeared jammed, magazine missing and the whole weapon contaminated with blood. The ownership of the weapon reportedly could be attributed to the man's belongings. The chair was broken. Close to where the pistol was found, one of the two rails exhibited blood drops with slightly jagged edges and a blood drop diameter range from ~0.5 to 20 mm, in keeping with the assumption of a head shot with a victim's head wound elevated from the ground.

There was a highly fragmented head and skull, amounting to decapitation (see Figure 38.6 S and T). Also, there was amputation of the left arm at the shoulder and both legs at the thighs, with numerous additional fractures. No apparent and definitive soft tissue defect consistent with a perforating gunshot injury was found on the remaining portions of the scalp (or remaining portion of the skin) as well as elsewhere on the body. Forensic test of the hands for gunshot residue was negative (Rhodizonate-stained filter paper).

Time of death estimation based on body temperature and ambient temperature was consistent with the indication of the train driver that had reported seeing the man sitting on that chair, and the associated collision time.

No written suicide note was identified at the scene or at the victim's home. For further examination including identification and CT scanning, the body was admitted to our institute.

The case-specific forensic question was that of identifying consequences of gunshot-specific injuries to the head. The whole-body PMCT with separate reconstructions as well as both death scene and within-institute second examination did not yield any indication for a firearm injury to the neck, thorax, abdomen or the extremities and no projectile or foreign bodies suspicious of projectiles or their fragments.

Screening of the whole-body PMCT identified two bone fragments that came into closer consideration for possible perforating pistol shot injuries. They were labeled fragment 1 and 2 (Figure 38.6).

Fragment 1 (Figure 38.6 A–G and T) measured 40 × 26 × 3, 5 mm. It contained a semi-circular indent on one edge that was apparent already on whole-body PMCT. There, it appeared to be a plausible candidate for a perforating skull defect due to a 7,65 mm caliber pistol. Subsequently higher resolution data analysis and macroscopic validation showed this finding to be a ruptured bone suture line.

Fragment 2 (Figure 38.6 H–R and T) measured 70 × 43, 5 × 4, 7 mm. It exhibited a semi-circular defect whose characteristics also on higher resolution PMCTA and macroscopic examination matched specifics of a pistol shot perforation of the skull; most notably, the defects of the outer (~ 8–12 mm) and inner (~ 15–20 mm)) cortical bone table of this skull fragment provided a constellation one might expect typical for a gunshot entry wound in a case such as this (oblique bone edges: see yellow dashed lines in Figure 38.6 N). In the context of the extensive blunt trauma due to the body being run over by the train, the lungs exhibited multiple pneuomatoceles. The lower lobe of the right lung appeared to contain a nodular patchy ground-glass pattern, consistent with aspiration of gastric content an subsequent bronchiolitis [36]. Trachea and bronchi exhibited no fluid on PMCT, and there was no fluid (such as blood) on macroscopic inspection.

Due to the information brought about by the police investigation and medicolegal examination in conjunction, no evidence for homicidal involvement was found. This pre-disclosed further detailed examination of a range of technical aspects. Tests or examinations that would have been possible but were not found necessary and thus no requested included full autopsy, including histology and toxicology, anthropological study and assembly of bone fragments particularly of the skull, evaluation of all DNA evidence, and a more refined examination for gunshot residue on the samples taken. The information and all findings were found to be consistent with (albeit not proof of) a complex suicide with pistol headshot and train collision.

38.3.6 Case 6—Unexpected Infectious Disease #1

During the early days of the COVID-19 pandemic in Switzerland [37–40], when cases went from a ~1000 cases per day to ~1400 cases per day a week later, which was when a 32-year old man was found dead in a locked public toilet next to what appeared to be materials commonly used for intravenous drug consumption, including a syringe. He allegedly had been a known drug addict having consumed cocaine, heroine, ecstasy, various benzodiazepines and amphetamines over the years prior to his death. In the preceding days and weeks he reportedly was said to have consumed excessively, but no one had mentioned that he had been ill with an infection. He had been with close friends moments before he left them saying he needed to use that toilet. External examination and autopsy identified what appeared to be a fresh needle mark in the right cubital fossa and included findings of a severe brain edema, pulmonary edema with whitish foam in the trachea, a distended full urinary bladder, endocardial and subpleural pinpoint hemorrhages, as well as macroscopic suspicion of a pneumonia. Toxicology results of peripheral blood and urine were interpreted towards an effective to fatal blood levels of morphine and codeine in combination with a therapeutic blood level of alprazolam and diazepam. The initial PMCT however had shown bilateral, multifocal, partly consolidated findings (Figure 38.7), that seven readers classified as CO-RADS 5 (median result) [32]. While the investigation of this case concluded that findings were most conclusive of an accidental or suicidal manner of death and while the lung diagnosis was not pursued further, Swiss

Penal Code does not require further specification of findings once a reasonable effort to examine the case was performed and no evidence of homicide was identified. While a COVID-19 PCR test based on a nasal swab was considered, the autopsy team in this in-stance decided to forgo this test also because of cost, time delay (in the early days and weeks of the COVID-10 pandemic the tests were not returned within hours) and a possibly false-negative result [37], when clearly, the actual role of the pulmonary PMCT findings here were to inform the autopsy team of a possible COVID-19 infection. So, increased levels of protections were chosen based on PMCT findings.

38.3.7 Case 7—Unexpected Infectious Disease #2

A 26-year-old woman was found deceased on her bed by family members. She had a known history of epilepsy, and prescription medication packages were present at the scene. The body was fully clothed; after undressing, external forensic inspection revealed no traumatic injuries. Because both the cause and manner of death were initially unascertained, the body was transported to our institute for further examination. PMCT demonstrated calcified lymph nodes in the axillary, hilar and mediastinal regions. Numerous diffuse, round calcified nodules were present bilaterally within the pulmonary parenchyma and mediastinum (Figure 38.8). Additional findings included pulmonary edema and a markedly distended urinary bladder. Autopsy revealed diffusely distributed caseous granulomas throughout both lungs. Fibrous adhesions between lung and pleura were present, together with pronounced pulmonary edema and enlarged cervical, axillary and mediastinal lymph nodes. Polymerase chain reaction (PCR) analysis of peripheral blood detected DNA of the *Mycobacterium tuberculosis* complex. In light of these findings, the manner of death was classified as natural. The cause of death was recorded as cardiac failure in the setting of post-primary pulmonary tuberculosis. As tuberculosis is a notifiable disease, the death was reported to the Department of Health.

38.3.8 Case 8—Would It Make Sense to Perform an Autopsy with Toxicology?

A 53-year-old man was found deceased in a public garden plot, adjacent to a handwritten note interpreted as a possible suicide note and several empty medication containers (Figure 38.9A). PMCT demonstrated cerebral and pulmonary edema, a markedly distended urinary bladder, and numerous radiopaque tablets within the distal stomach and duodenum (Figure 38.9C–F). These imaging findings suggested, though did not conclusively prove, recent ingestion of a substantial quantity of tablets with suicidal intent, prompting a full medicolegal autopsy. Autopsy identified more than 80 tablets corresponding to the PMCT-visible material (Figure 38.9B). Macroscopic appearance and toxicological results were consistent with bupropion retard, the medication listed on the recovered prescription packaging. Despite the large number of tablets present in the gastrointestinal tract, peripheral blood analysis revealed only low, subtherapeutic concentrations of bupropion and its metabolites. In contrast, toxicological examination of peripheral blood demonstrated markedly elevated concentrations of mirtazapine and trimipramine, both within a range associated with potentially fatal intoxication. Notably, tablets corresponding to these substances exhibited no detectable PMCT correlate—no focal hyperdensities or gravity-dependent radiodense layering—highlighting the modality's limited sensitivity for certain pharmaceutical formulations. Based on the integration of scene findings, autopsy results, and toxicological evidence, the manner of death was classified as suicide, and the cause of death as fatal intoxication due to mirtazapine and trimipramine. The ingestion of a large quantity of bupropion tablets, although radiologically conspicuous and strongly suggestive of suicidal behavior, did not represent the physiologically lethal agent in this case. Retrospectively, the PMCT findings are best interpreted as indicative of suicidal intent rather than as predictors of the pharmacological mechanism of death.

38.4 DISCUSSION

38.4.1 Triage in Cases of Assumed Medical Complications and Liability Cases

Usually, a full autopsy may have to be considered depending on specifics—as it may be relevant to rule out diagnoses as hard or impossible to find, adequately characterize or exclude on PMCT, such as peripheral pulmonary artery thromboembolism or acute myocardial infarction, infections or tumors. With a few exceptions with regard to the concise issue being discussed in any particular case (e.g., case 1 above), a comprehensive postmortem diagnosis including (but not restricted to) full autopsy, histology, microbiology and toxicology may be needed for case analysis.

38.4.2 Triage for Manner of Death—or for Identifying a Possible Violent Death

Some forms of violence, particularly those that do not leave readily identifiable external injuries, may remain difficult to detect or even to suspect. These include mechanisms such as obstructive asphyxia or poisoning, in which overt morphological markers may be minimal or entirely absent. A structured approach involving meticulous death scene investigation, contributions from forensic scientists, comprehensive postmortem imaging, and full autopsy with ancillary analyses is therefore essential.

Cases 3 and 4 were selected to illustrate how challenging the delineation of subtle or equivocal findings on PMCT can be. Case 8 was included to demonstrate how the presence or lack of PMCT correlates for tablets may shape the initial interpretive framework of a case, potentially in a misleading direction. The detection or apparent absence of PMCT-visible tablets must, therefore, always be interpreted within the broader context of the case rather than taken as an isolated indicator.

More generally, it is a fallacy to assume that the manner of death can be logically inferred from what appears to be a complete evidentiary dataset such as an apparently full case description and what may come across as full PMCT dataset: forensic datasets are, by their nature, logically incomplete, and there commonly remains the possibility of an undetectable or unmeasurable component.

For example, a homicidal application of a mobile carbon dioxide source could hypothetically result in fatal asphyxia without leaving any discernible external or internal traces. In such a scenario, conclusions would necessarily rely solely on the positive findings avail-able, while accepting that crucial negative evidence might be inherently unattainable. As another example, a projectile striking an intermediate target may exploit the Vinogradov phenomenon [41], producing wound characteristics mimicking an intermediate-range shot despite the actual firing distance being much greater.[1]

With these considerations in mind, conclusions regarding the manner of death must be formulated as interpretations of the available findings, representing the scenario judged most likely under a given set of assumptions. At the same time, it must be acknowledged that certain cases legitimately warrant more extensive forensic investigation, even in the absence of explicit criminalistic findings, due to the inherent limitations of the data and the potential for otherwise undetectable mechanisms. The central question, therefore, becomes how to identify such cases.

38.4.3 Triage for Cause of Death

Identifying an acute cause of death—or even establishing a plausible candidate mechanism—frequently requires the demonstration of findings indicative of vitality, such as fat embolism, infarction, inflammation or hemorrhage [42]. Coronary artery calcification is a commonly encountered PMCT observation and may hold contextual relevance; however, it cannot, in isolation, substantiate an acute cardiac cause of death. Decisions to forgo autopsy or toxicological analysis based solely on the presence of calcified coronary arteries therefore warrant careful consideration.

In a study comparing fatal opioid intoxications with age- and sex-matched controls, we [43] found that coronary artery calcification was more frequent in the poisoning group (39%) than among controls (29%). This underscores the limited discriminative value of coronary calcification when assessing acute lethality. A commonly voiced argument that "the individual nonetheless had coronary artery atherosclerosis" does not meaningfully resolve the issue. Individuals who undergo assisted suicide, for example, frequently have advanced stages of various chronic diseases; interpreting their PMCT scans without knowledge of natrium-pentobarbital ingestion could similarly lead to over-attribution of death to underlying morphology rather than to the actual toxicological mechanism. Consequently, postmortem diagnosis of what may appear to be an easy to solve, e.g. cardiac, cause of death is generally most reliable when integrated with comprehensive toxicological and possibly other tests, for a better evaluation, particularly in cases lacking unequivocal morphological markers [44].

38.4.4 Triage for Possibly Infectious Disease

Relatively advanced putrefaction, whether generalized or localized [45], as well as characteristic pulmonary alterations (as illustrated in Cases 6 and 7) may provide indications suggestive of underlying infectious processes. However, such findings alone do not define the role of infection in the causation of death. Depending on the broader case context and the degree of diagnostic uncertainty, a full autopsy with appropriate ancillary examinations may therefore be required to clarify the contribution of infection to the fatal outcome.

38.4.5 Discriminating Causes of Injury

The delineation of one form of injury against another—such as locating a firearm injury in the remains of what appears to be a train suicide (Case 5) can be easy, daunting or impossible, depending on specifics. However there are situations where this is a relevant question. There, we do recommend to consider the highest data quality for the morphological findings that is available—if that is manual examination, possibly before and after removal of soft tissues from bone fragments, with high resolution photography, or whether a PMCT delivers an adequate data quality, may have to be decided based on case specifics.

ACKNOWLEDGMENTS

The authors express their gratitude to Emma Louise Kessler, MD, for her generous donation to the Zürich Institute of Forensic Medicine, University of Zürich, Switzerland. This work released under CC BY 4.0. While the authors have made efforts to ensure clarity and structure, the development of this manuscript occurred under the constraints of available time and timelines. Case data was used and published after submitting the process to the local ethics committee (Ethics Committee of Canton of Zürich, waiver 15–0686). Figures and text reused from [29] with Elsevier license 5723680322978 and 5723680201952.

Note

1 Doctor Kaufman: "My name is Doctor Kaufman. I am an outstanding pistol marksman. Take my word for it, ya?" Bond: "It won't look like suicide if you shoot me from over there." Doctor Kaufman: "I am a professor of forensic medicine. Believe me, Mr Bond, I could shoot you from Stuttgart und still create ze proper effect."—*James Bond: Tomorrow Never Dies* (1997).

References

[1] Bedford PJ, Oesterhelweg L. Different conditions and strategies to utilize forensic radiology in the cities of Melbourne, Australia and Berlin, Germany. Forensic Science, Medicine, and Pathology. 2013;9(3):321–326.
[2] Sollie H, Kop N, Euwema MC. Mental resilience of crime scene investigators: How police officers perceive and cope with the impact of demanding work situations. Criminal Justice and Behavior. 2017;44(12):1580–1603.
[3] Schweitzer W, Thali MJ, Egger D. Case-study of a user-driven prosthetic arm design: Bionic hand versus customized body-powered technology in a highly demanding work environment. Journal of Neuroengineering and Rehabilitation. 2018;15(1):1–27.
[4] Ampanozi G, Zimmermann D, Hatch GM, Ruder TD, Ross S, Flach PM, et al. Format preferences of district attorneys for postmortem medical imaging reports: Understandability, cost effectiveness, and suitability for the courtroom: A questionnaire based study. Legal Medicine. 2012;14(3):116–120.
[5] Schweitzer W, Bartsch C, Ruder TD, Thali MJ. Virtopsy approach: Structured reporting versus free reporting for PMCT findings. Journal of Forensic Radiology and Imaging. 2014;2(1):28–33.
[6] Thali MJ, Yen K, Schweitzer W, Vock P, Boesch C, Ozdoba C, et al. Virtopsy, a new imaging horizon in forensic pathology: Virtual autopsy by postmortem multislice computed tomography (MSCT) and magnetic resonance imaging (MRI)-a feasibility study. Journal of Forensic Sciences. 2003;48(2):386–403.
[7] O'Donnell C, Rotman A, Collett S, Woodford N. Current status of routine postmortem CT in Melbourne, Australia. Forensic Science, Medicine, and Pathology. 2007;3(3):226–232.
[8] Decker SJ, Braileanu M, Dey C, Lenchik L, Pickup M, Powell J, et al. Forensic radiology: A primer. Academic Radiology. 2019;26(6):820–830.
[9] Rutty GN. What has postmortem computed tomography even done for forensic pathology? Diagnostic Histopathology. 2020 June;26(8). https://doi.org/10.1016/j.mpdhp.2020.05.005
[10] Chatzaraki V, Thali MJ, Ampanozi G, Schweitzer W. Fatal road traffic vehicle collisions with pedestrian victims: Forensic postmortem computed tomography and autopsy correlation. The American Journal of Forensic Medicine and Pathology. 2018;39(2):130–140.
[11] Schweitzer W, Dangel S, Schaepman M, Wagen M, Maass D, Cordner S, et al. Comparison of maximal left ventricular intramyocardial artery lumen diameter in 2 aneurysmectomy specimen after TMR and 44 control patients. Laser-Medizin. 1998;13(3–4):83–101.
[12] Schweitzer W, Yen K, Thali M, Scheurer E, Vock P, Boesch C, et al. Virtopsy: Dokumentation von Herzbefunden bei 35 Autopsien im Vergleich mit CT und MRI. Rechtsmedizin. 2001;11(4):160.
[13] Steffen C, Schaepman M, Hardmeier T, Schweitzer W. Digital 3D reconstruction of two parahissian accessory bundles in a case of Wolff-Parkinson-White syndrome. International Journal of Legal Mmedicine. 2004;118(2):101–105.
[14] Winklhofer S, Stolzmann P, Meier A, Schweitzer W, Morsbach F, Flach P, et al. Added value of dual-energy computed tomography versus single-energy computed tomography in assessing ferromagnetic properties of ballistic projectiles: Implications for magnetic resonance imaging of gunshot victims. Investigative Radiology. 2014;49(6):431–437.
[15] Ampanozi G, Thali YA, Schweitzer W, Hatch GM, Ebert LC, Thali MJ, et al. Accuracy of non-contrast PMCT for determining cause of death. Forensic Science, Medicine and Pathology. 2017;13(3):284–292.
[16] Ebert LC, Heimer J, Schweitzer W, Sieberth T, Leipner A, Thali M, et al. Automatic detection of hemorrhagic pericardial effusion on PMCT using deep learning: A feasibility study. Forensic Science, Medicine and Pathology. 2017;13(4):426–431.
[17] Schweitzer W, Thali M, Giugni G, Winklhofer S. Postmortem pulmonary CT in hypothermia. Forensic Science, Medicine, and Pathology. 2014;10(4):557–569.
[18] Schweitzer W, Thali M. Fatal obstructive asphyxia: Transpulmonary density gradient characteristic as relevant identifier in postmortem CT. Journal of Forensic Radiology and Imaging. 2019;19:100337.

[19] Poletti PA, Mirvis SE, Shanmuganathan K, Takada T, Killeen KL, Perlmutter D, et al. Blunt abdominal trauma patients: Can organ injury be excluded without performing computed tomography? Journal of Trauma and Acute Care Surgery. 2004;57(5):1072–1081.
[20] Leidner B, Beckman M. Standardized whole-body computed tomography as a screening tool in blunt multitrauma patients. Emergency Radiology. 2001;8(1):20–28.
[21] Healy DA, Hegarty A, Feeley I, Clarke-Moloney M, Grace PA, Walsh SR. Systematic review and meta-analysis of routine total body CT compared with selective CT in trauma patients. Emergency Medicine Journal. 2014;31(2):101–108.
[22] Flach PM, Gascho D, Schweitzer W, Ruder TD, Berger N, Ross SG, et al. Imaging in forensic radiology: An illustrated guide for postmortem computed tomography technique and protocols. Forensic Science, Medicine, and Pathology. 2014;10(4):583–606.
[23] Ford JM, Decker SJ. Computed tomography slice thickness and its effects on three-dimensional reconstruction of anatomical structures. Journal of Forensic Radiology and Imaging. 2016;4:43–46.
[24] Gascho D, Thali MJ, Niemann T. Postmortem computed tomography: Technical principles and recommended parameter settings for high-resolution imaging. Medicine, Science and the Law. 2018;58(1):70–82.
[25] Ampanozi G, Held U, Ruder TD, Ross SG, Schweitzer W, Fornaro J, et al. Pulmonary thromboembolism on unenhanced postmortem computed tomography: Feasibility and findings. Legal Medicine. 2016;20:68–74.
[26] Shenton A, Kralt P, Suvarna SK. Post Mortem CT for Non-Suspicious Adult Deaths: An Introduction. Springer Nature; 2021.
[27] Kiesel E, Hanzlick R, of the College of American Pathologists AC, et al. Case of the month: The autopsy and new technology: All that glitters is not a gold standard. Archives of Internal Medicine. 2000;160(13):1901–1902.
[28] Molina DK, Nichols JJ, DiMaio VJ. The sensitivity of computed tomography (CT) scans in detecting trauma: Are CT scans reliable enough for courtroom testimony? Journal of Trauma and Acute Care Surgery. 2007;63(3):625–629.
[29] Chatzaraki V, Heimer J, Thali M, Dally A, Schweitzer W. Role of PMCT as a triage tool between external inspection and full autopsy–case series and review. Journal of Forensic Radiology and Imaging. 2018;15:26–38.
[30] Schulze K, Ebert LC, Ruder TD, Fliss B, Poschmann SA, Gascho D, et al. The gas bubble sign: A reliable indicator of laryngeal fractures in hanging on postmortem CT. The British Journal of Rradiology. 2018;91(1084):20170479.
[31] Schweitzer W, Thali MJ, Ampanozi G. Planned complex suicide combining pistol head shot and train suicide and Virtopsy examination. Forensic Imaging. 2022 Mar;28:200485.
[32] Prokop M, Van Everdingen W, van Rees Vellinga T, Quarles van Ufford H, Stöger L, Beenen L, et al. CO-RADS: A categorical CT assessment scheme for patients suspected of having COVID-19—definition and evaluation. Radiology. 2020;296(2):E97–E104.
[33] Chandy PE, Murray N, Khasanova E, Nasir MU, Nicolaou S, Macri F. Postmortem CT in trauma: An overview. Canadian Association of Radiologists Journal. 2020;71(3):403–414.
[34] Schweitzer W, Ruder T, Thali M, Ringl H. Skull fractures in postmortem CT: VRT, flat and skin surface projections in comparison. Journal of Forensic Radiology and Imaging. 2015;3(4):214–220.
[35] Gunz S, Erne S, Rawdon EJ, Ampanozi G, Sieberth T, Affolter R, et al. Automated rib fracture detection of postmortem computed tomography images using machine learning techniques. arXiv preprint. 2019;arXiv:190805467. https://doi.org/10.48550/arXiv.1908.05467
[36] Prather AD, Smith TR, Poletto DM, Tavora F, Chung JH, Nallamshetty L, et al. Aspiration-related lung diseases. Journal of Thoracic Imaging. 2014;29(5):304–309.
[37] Schweitzer W, Ruder T, Baumeister R, Bolliger S, Thali M, Meixner E, et al. Implications for forensic death investigations from first Swiss postmortem CT in a case of non-hospital treatment with COVID-19. Forensic Imaging. 2020;21:200378.
[38] Esposito M, Salerno M, Scoto E, Di Nunno N, Sessa F. The impact of the COVID-19 pandemic on the practice of forensic medicine: An overview. In: Healthcare. vol. 10. MDPI; 2022. p. 319.
[39] Bogdanović M, Atanasijević T, Popović V, Mihailović Z, Radnić B, Durmić T. Is the role of forensic medicine in the covid-19 pandemic underestimated? Forensic Science, Medicine and Pathology. 2021;17:136–138.
[40] Verhoff MA, Püschel K, Ondruschka B. Infectiousness of SARS-CoV-2 positive corpses during external post mortem examination. MMW Fortschritte der Medizin. 2021;163(12):56–58.
[41] Sellier K, editor. Shot Range Determination. Springer Science & Business Media; 2012.
[42] Chatzaraki V, Heimer J, Thali MJ, Ampanozi G, Schweitzer W. Approaching pulmonary fat embolism on postmortem computed tomography. International Journal of Legal Medicine. 2019;133(6):1879–1887.
[43] Winklhofer S, Surer E, Ampanozi G, Ruder T, Stolzmann P, Elliott M, et al. Postmortem whole body computed tomography of opioid (heroin and methadone) fatalities: Frequent findings and comparison to autopsy. European Radiology. 2014;24(6):1276–1282.
[44] Oliva A, Brugada R, D'Aloja E, Boschi I, Partemi S, Brugada J, et al. State of the art in forensic investigation of sudden cardiac death. The American Journal of Forensic Medicine and Pathology. 2011;32(1):1–16.
[45] Baumeister R, Gauthier S, Schweitzer W, Thali MJ, Mauf S. Small–but fatal: Postmortem computed tomography indicated acute tonsillitis. Journal of Forensic Radiology and Imaging. 2016;6:52–56.

INDEX

Note: Page numbers in *italics* indicate a figure and page numbers in **bold** indicate a table on the corresponding page.

A

B

C

D

E

F

G

H

I

J

K

L

M

N

O

P

Q

R

S

T

U

V

For Product Safety Concerns and Information please contact our EU representative GPSR@taylorandfrancis.com Taylor & Francis Verlag GmbH, Kaufingerstraße 24, 80331 München, Germany

Batch number: 10756625

Printed by Printforce, the Netherlands